Evolutionary Cognitive Neuroscience

Cognitive Neuroscience
Michael S. Gazzaniga, editor

Gary Lynch, *Synapses, Circuits, and the Beginning of Memory*
Barry E. Stein and M. Alex Meredith, *The Merging of the Senses*
Richard B. Ivry and Lynn C. Robertson, *The Two Sides of Perception*
Steven J. Luck, *An Introduction to the Event-Related Potential Technique*
Roberto Cabeza and Alan Kingstone, eds., *Handbook of Functional Neuroimaging of Cognition*
Carl Senior, Tamara Russell, and Michael S. Gazzaniga, eds., *Methods in Mind*
Steven M. Platek, Julian Paul Keenan, and Todd K. Shackelford, eds., *Evolutionary Cognitive Neuroscience*

Evolutionary Cognitive Neuroscience

Edited by Steven M. Platek, Julian Paul Keenan, and Todd K. Shackelford

The MIT Press
Cambridge, Massachusetts
London, England

MIT Press books may be purchased at special quantity discounts for business or sales promotional use. For information, please email special_sales@mitpress.mit.edu or write to Special Sales Department, The MIT Press, 55 Hayward Street, Cambridge, MA 02142.

This book printed and bound in the United States of America.

Library of Congress Cataloging-in-Publication Data
Evolutionary cognitive neuroscience / edited by Steven M. Platek, Julian Paul Keenan, and Todd K. Shackelford.
p. cm.—(Cognitive neuroscience)
Includes bibliographical references and index.
ISBN 13: 978-0-262-16241-8
ISBN 10: 0-262-16241-5
1. Cognitive neuroscience. 2. Brain–Evolution. 3. Evolutionary psychology. I. Platek, Steven M. II. Keenan, Julian Paul. III. Shackelford, Todd K. IV. Series.

QP360.5.E97 2006
612.8′233–dc22

2006048171

10 9 8 7 6 5 4 3 2 1

Contents

Contributors

Editors:
Steven M. Platek, PhD, Department of Psychology, Drexel University, Philadelphia, Pennsylvania

Julian Paul Keenan, PhD, Associate Professor, Director, Cognitive Neuroimaging Laboratory, Department of Psychology, Montclair State University, Upper Montclair, New Jersey

Todd K. Shackelford, PhD, Director, Evolutionary Psychology Lab, Department of Psychology, Florida Atlantic University, Davie, Florida

Chapter Contributors:
C. Davison Ankney, PhD, Professor Emeritus of Biology, University of Western Ontario, London, Ontario, Canada

Simon Baron-Cohen, PhD, Director, Autism Research Centre (ARC), Cambridge University, Cambridge, United Kingdom

S. Marc Breedlove, PhD, Neuroscience Program, Michigan State University, East Lansing, Michigan

William Christiana, BA, Cognitive Neuroimaging Laboratory, Montclair State University, Upper Montclair, New Jersey

Michael C. Corballis, PhD, Professor, Department of Psychology, University of Auckland, Auckland, New Zealand

Robin I. M. Dunbar, PhD, School of Biological Sciences, University of Liverpool, Liverpool, United Kingdom

Russell D. Fernald, PhD, Professor of Biological Sciences, Psychology Department, Program in Human Biology, Neuroscience Program, Stanford University, Stanford, California

Helen E. Fisher, PhD, Visiting Research Professor, Department of Anthropology, Rutgers University, New Brunswick, New Jersey

Jonathan Flombaum, Department of Psychology, Yale University, New Haven, Connecticut

Farah Focquaert, Department of Philosophy, Universiteit Gent, Belgium

Steven J. C. Gaulin, PhD, Professor, Department of Anthropology, Center for Evolutionary Psychology, University of California–Santa Barbara, Santa Barbara, California

Aaron T. Goetz, MA, Department of Psychology, Florida Atlantic University, Davie, Florida

Kevin Guise, Department of Psychology, Montclair State University, Upper Montclair, New Jersey

Ruben C. Gur, PhD, Professor, Departments of Psychiatry and Neurology, School of Medicine, University of Pennsylvania, Philadelphia, Pennsylvania

William D. Hopkins, PhD, Regular Faculty, Living Links Center, Research Associate, Yerkes National Primate Research Center; Associate Professor, Department of Psychology, Berry College, Mount Berry, Georgia

Farzin Irani, MA, Department of Psychology, Drexel University, Philadelphia, Pennsylvania

Michael B. Kimberly, Department of Bioethics, University of Pennsylvania, Philadelphia, Pennsylvania

Stephen M. Kosslyn, PhD, Professor, Department of Psychology, Harvard University, Cambridge, Massachusetts

Austen L. Krill, Department of Psychology, Drexel University, Philadelphia, Pennsylvania

Monisha Kumar, BA, Cognitive Neuroimaging Laboratory, Montclair State University, Upper Montclair, New Jersey

Sarah L. Levin, Department of Psychology, Drexel University, Philadelphia, Pennsylvania

Euginia Mamikonyan, Department of Psychology, Drexel University, Philadelphia, Pennsylvania

Lori Marino, PhD, Faculty Affiliate, Living Links Center, Yerkes National Primate Research Center; Senior Lecturer in Neuroscience and Behavioral Biology, Department of Psychology, Emory University, Atlanta, Georgia

David B. Newlin, PhD, National Institutes of Health–National Institute on Drug Abuse, Baltimore, Maryland

Ivan S. Panyavin, MA, Department of Psychology, Drexel University, Philadelphia, Pennsylvania

Shilpa Patel, Department of Psychology, Drexel University, Philadelphia, Pennsylvania

Webb Phillips, Department of Psychology, Yale University, New Haven, Connecticut

David Andrew Puts, PhD, Neuroscience Program, Michigan State University, East Lansing, Michigan

Katie Rodak, MS, Department of Psychology, Drexel University, Philadelphia, Pennsylvania

J. Philippe Rushton, PhD, Professor, Department of Psychology, University of Western Ontario, London, Ontario, Canada

Laurie Santos, PhD, Assistant Professor, Department of Psychology, Yale University, New Haven, Connecticut

Kyra Singh, Department of Psychology, Drexel University, Philadelphia, Pennsylvania

Sean T. Stevens, Department of Psychology, Montclair State University, Upper Montclair, New Jersey

Valerie E. Stone, PhD, Adjunct Professor, Department of Psychology, University of Denver, Denver, Colorado

J. Anderson Thomson, Jr., MD, Staff Psychiatrist, Counseling and Psychological Services, University of Virginia Student Health Services; Staff Psychiatrist, Institute of Law, Psychiatry, and Public Policy, University of Virginia, Charlottesville, Virginia

Jaime W. Thomson, MS, Adjunct Research Professor, Department of Psychology, Drexel University, Philadelphia, Pennsylvania

Paul Root Wolpe, PhD, Assistant Professor of Sociology, Assistant Professor of Psychiatry, Faculty Associate, Center for Bioethics, Chief, Bioethics and Human Subjects Protections, NASA, Senior Fellow, The Leonard Davis Institute for Health Economics, University of Pennsylvania, Philadelphia, Pennsylvania

Preface

Steven M. Platek, Julian Paul Keenan, and Todd K. Shackelford

Cognitive neuroscience, the study of brain-behavior relationships, is historically old in its attempt to map the brain. As a discipline it is flourishing, with an increasing number of functional neuroimaging studies appearing in the scientific literature daily. Unlike biology and even psychology, however, the cognitive neurosciences have only recently begun to apply evolutionary theory and methods. Approaching cognitive neuroscience from an evolutionary perspective allows scientists to apply a solid theoretical guidance to their investigations, one that can be carried out in both human and nonhuman animals. This book represents the first formal attempt to document the burgeoning field of evolutionary cognitive neuroscience.

Introduction to *Evolutionary Cognitive Neuroscience*

All organisms were and continue to be subject to the pressures of natural and sexual selection. These pressures are what formed all biological organs and hence also carefully crafted animal nervous systems—the seat of animal and human behavior, and the means by which organisms employ information-processing programs to adaptively deal with their environment. This theory was first formalized by Darwin (1859) in his seminal book, *On the Origin of Species by Natural Selection*. Unlike the theoretical work of early psychologists and behavioral scientists such as Skinner and Watson, which envisioned organisms as "blank slates" capable of making an infinite number of associations, evolutionary metatheory is beginning to shed light on this flawed theoretical approach to behavior analysis (see Barkow, Cosmides, & Tooby, 1992; Buss, 2005; Cosmides & Tooby, 2005). In fact, many of the emerging studies are contending directly with the standard social science model of psychology, namely, that organisms possess general-purpose learning mechanisms and that biology plays little if any role in the manifestation of behavior. Some of the first psychological studies to demonstrate that learning is not mediated by general-purpose learning mechanisms were conducted several decades ago and mark what might be considered the beginning of evolutionary thinking in psychology; they also contributed greately to what has become known as the *cognitive revolution.*

In his landmark study, Garcia discovered that animals learned to avoid novel food products that made them ill in as little as one learning conditioning trial, something that had not been demonstrated with any other stimulus class previously. Labeled *conditioned taste aversion*, this effect describes an adaptive problem that has since been demonstrated in almost every species tested (the exception to this rule appears to be crocodilians; see Gallup & Suarez, 1988). This adaptation serves an important function: don't eat food that makes you ill, or you might not survive to reproduce and pass on your genes. In other words, being ill could result in a number of fitness disadvantages such as death, inability to avoid predation, inability to search and secure mates, and loss of mate value.

In a similar discovery, Seligman demonstrated what he referred to as prepared learning. Prepared learning is a phenomenon in which it is easier to make associations between stimuli that possess a biological predisposition to be conditioned because of a role these stimuli played in an organism's evolutionary history. Seligman and his colleagues demonstrated that it was much easier for humans (and animals) to form conditioned emotional responses and associative fear responses to evolutionarily relevant threats such as snakes, insects, and heights than it took to condition fear to present-day threatening stimuli that subjects were much more likely to be have encountered and be harmed by, such as cars, knives, and guns. In other words, it was easier to condition humans to fear snakes, spiders, and heights than it was to condition them to fear guns, cars, and knives.

These two series of studies demonstrated that psychological traits, like the design of bodily organs, were crafted by evolutionary forces into adaptations that allowed our ancestors to flourish. That is, the information-processing mechanisms designed to deal with situations such as poisonous food or potential threats to survival evolved as part of our ancestors' recurrent experience with such situations. These studies refute a key premise of the standard social science model, emphasizing that there is no general-purpose learning mechanism. Rather, all learning is a consequence of carefully crafted modules dedicated to solving specific evolutionary problems (see Barkow, Cosmides, & Tooby, 1992; Pinker,

2002). Our brains have evolved to be efficient problem solvers, and the problems they are designed to solve are those that our ancestors recurrently faced over human evolutionary history. Hence, those among our ancestors who were psychologically adaptated to solve these problems survived and passed the genes for those traits on to offspring.

Recently, evolutionary metatheory has been applied directly to investigations of the cognitive neuroscience kind. For example, O'Doherty, Perrett, and colleagues (2003) have begun to investigate neural correlates of facial attraction. O'Doherty and colleagues discovered that the orbitofrontal cortex appears to be activated when a person finds a face attractive, which suggests that facial attractiveness activates a reward system in the brain. Further, Baron-Cohen and colleagues have demonstrated that there is a neural module dedicated to processing socially relevant information. Baron-Cohen and colleagues demonstrated that the ability to conceive of others' mental states appears to be (1) a highly modularized neurocognitive process and (2) affected by certain neuropsychiatric pathologies (e.g., autism). Platek and colleagues have extended initial behavioral findings of sex differences in reaction to children's faces to the cognitive neuroscience arena, demonstrating sex differences in functional neural activation associated with reactions to children's faces. They found that males but not females showed activation in left frontal regions of the brain when viewing self-resembling child faces, suggesting that males inhibit negative responses to children's faces as a function of facial (phenotypic) resemblance.

Perhaps the most convincing set of studies demonstrating evolved structures or modules dedicated to social interaction and exchange has come from Cosmides, Tooby, and colleagues at the Center for Evolutionary Psychology in Santa Barbara, California. By modifying a logic problem known as the Wason Selection Task to reflect evolutionarily important social interactions (e.g., cheater detection), Cosmides, Tooby, and colleagues have demonstrated that the human brain appears to have evolved a cheater detection mechanism that is extremely efficient. They have furthered the evidence for a cheater detection module by showing that one can incur impairment (i.e., brain trauma) of performance on cheater detection problems but remain relatively unimpaired on other types of problem solving. Their data suggest that parts of the limbic system are implicated in the ability to detect cheaters in social interactions.

The investigation of an evolutionary cognitive neuroscience extends beyond humans, however. Hauser, Hare, and a number of other researchers have been studying social behavior and social exchange in nonhuman primates and have demonstrated an apparent cognitive continuity among primate phyla in the ability to understand the mental states of others. Daniel Povinelli's ongoing research program has been particularly powerful at demonstrating phylogenetic and ontogenetic trajectories for the capacity for theory of mind and self-awareness among nonhuman primates.

These new investigations, by applying cognitive neuroscientific methods to answer questions posed from an evolutionary theoretical perspective, are crafting a new understanding of how the mind and brain evolved. In fact, they call into question much of the psychological investigation that was conducted throughout the twentieth century. This book is the first to present, in an organized overview, the way in which researchers are beginning to wed the disciplines of evolutionary psychology and cognitive neuroscience in order to provide new data on and insights into the evolution and functional modularity of the brain.

Each of the six sections in this book addresses a different adaptive problem. Part I consists of three chapters that outline the basic tenets of an evolutionarily informed cognitive neuroscience. These chapters discuss evolutionary theory as it can be applied to behavior and cognition, as well as modern technological advances and methods that are available to the cognitive neuroscientist for the investigation of the adapted mind.

In Chapter 1, Aaron Goetz and Todd Shackelford present an overview of the basic principles of evolution—natural and sexual selection, fitness, and adaptation—as they apply to behavior and cognition. In Chapter 2, Robin Dunbar expands on this presentation by describing a theory known as the social brain hypothesis and discusses the major social evolutionary forces that gave rise to big-brained humans and adaptive brains. Chapter 3, by Shilpa Patel and colleagues, outlines the current methodological approaches used in evolutionary cognitive neuroscience.

Part II broaches the topic of neuroanatomy from an ontogenetic and phylogenetic perspective. In Chapter 4, Valerie Stone considers why big-brained organisms have extended ontogenetic and brain developmental periods. In Chapter 5 William Hopkins considers hemispheric specialization in our closest living relative, the chimpanzee. In Chapter 6, J. Philippe Rushton and C. Davison Ankney review their studies on the relationship between brain size and intelligence. To close Part II, Lori Marino in Chapter 7 discusses the current state of the science in cetacean brain evolution.

Part III tackles the topic of reproduction and kin recognition. Chapter 8, by Russell Fernald, discusses the degree to which social environments can exert effects on reproductive behaviors. He draws on studies in his own laboratory on fishes and other nonhuman organisms, as well as on classic studies of this effect. In Chapter 9, Steven Platek and Jaime Thomson describe their recent findings supporting a sex difference in neural substrates involved in the detection of facial resemblance, and discuss what these findings might mean for kin selection or detection and paternal uncertainty. In Chapter 10, Helen Fisher and J. Anderson Thomson, Jr., summarize their recent studies with fMRI to identify the neural correlates of romantic attraction and lust. In Chapter 11, David Newlin outlines his SPFit model for drug addiction, which posits that drugs of addiction capitalize on evolutionary predispositions for reward- and reproductive-based behavioral and neural mechanisms.

Part IV addresses two well-known and well-researched areas: spatial cognition and language. David Puts, Steven Gaulin, and Marc Breedlove in Chapter 12 discuss sex differences in spatial abilities, paying particular attention to the endocrinological aspects associated with sex differences. In Chapter 13, Ruben Gur and colleagues extend the discussion of the evolution of sex differences in spatial cognition by summarizing current literature showing sex differences in neural substrates involved in solving spatial tasks. To conclude Part IV, Michael Corballis in Chapter 14 describes a theory of language evolution that draws on recent findings in animal and human neuroscience, especially the discovery of mirror neurons.

Part V takes up the topic of self-awareness and social cognition. In Chapter 15, Laurie Santos and her colleagues summarize their recent research showing that nonhuman primates possess the capability for social cognition, such as rudimentary theory of mind. In Chapter 16, Farah Focquaert and Steven Platek discuss their theory about the evolution of self-processing, introducing evidence from the nonhuman primate literature as well as from their own functional neuroimaging studies. Simon Baron-Cohen in Chapter 17 then presents his systemizing-empathizing theory for the development of theory of mind and describes how the model can be used to help classify individuals along this spectrum, with particular reference to autism and autism spectrum conditions. In Chapter 18, the discussion of self-awareness and social cognition takes a different direction describing the evolution of deception. Sean Stevens and colleages outline the "dark side of consciousness" theory, which links the capacity for deception to an intact self-awareness. Finally, Stephen Kosslyn in Chapter 19 presents a new theory for human motivation in which he describes social prostheses and reconsiders the self in light of this social network.

The volume concludes with Part VI, which considers the ethical implications for evolutionary cognitive neuroscience.

References

Barkow, J., Cosmides, L. & Tooby, J. (1992). *The adapted mind: Evolutionary psychology and the generation of culture*. New York: Oxford University Press.

Baron-Cohen, S. (2004). The cognitive neuroscience of autism. *Journal of Neurology, Neurosurgery and Psychiatry, 75*, 945–948.

Buss, D. M. (2005). *The handbook of evolutionary psychology*. New York: Wiley.

Cosmides, L. & Tooby, J. (2005). Neurocognitive adaptations designed for social exchange. In D. M. Buss (Ed.), *The handbook of evolutionary psychology* (pp. 584–627). New York: Wiley.

Garcia, J., & Koelling, R. A. (1966). Relation of cue to consequence in avoidance learning. *Psychonomic Science*, *4*, 123–124.

Hare, B., & Tomasello, M. (2005). Human-like social skills in dogs? *Trends in Cognitive Science, 9*, 439–444.

Hauser, M. (2005). Our chimpanzee mind. *Nature, 437*, 60–63.

Pinker, S. (2002). *The blank slate: The modern denial of human nature*. New York: Viking.

Platek, S. M., Keenan, J. P., & Mohamed, F. B. (2005). Sex differences in the neural correlates of child facial resemblance: An event-related fMRI study. *Neuroimage*, *25*, 1336–1344.

Povinelli, D. J., & Preuss, T. M. (1995). Theory of mind: Evolutionary history of a cognitive specialization. *Trends in Neurosciences, 18*, 418–424.

Suarez, S. & Gallup, G. G., Jr. (1981). Self-recognition in chimpanzees and orangutans, but not gorillas. *Journal of Human Evolution*, *10*, 157–188.

Winston, J. S., O'Doherty, J., Kilner, J. M., Perrett, D. I., & Dolan, R. J. (2006). Brain systems for assessing facial attractiveness. *Neuropsychologia*, (in press).

Acknowledgments

The editors would like to thank the authors who contributed chapters to this volume. These scientists are leaders in the growing field of evolutionary cognitive neuroscience and we appreciate them sharing their work in this volume.

Special thanks to Jaime Thomson for assistance with editorial organization. Additionally, special thanks goes out to our families—

Jonathan Platek, Viviana Weekes-Shackelford, and Ilene Kalish for patience and support during the development of this project.

Special thanks also to Barbara Murphy and Kate Blakinger for their continued support and patience during the completion of this volume. Without their guidance *Evolutionary Cognitive Neuroscience* may very well have never seen the light of day.

I Introduction and Overview

There are few absolutes in evolution, neuroscience, or the cognitive sciences. Models are created that are generally predictive, with an understanding that there are numerous exceptions to most "rules." Evolution may create more complex organisms, for example, but gradual change also may prune or reduce complexity. Behavior may become more advanced as brain size increases, but this claim has many exceptions, even when relative scaling is employed (Striedter, 2004). The cognitive sciences present many challenges as well: language, memory, and consciousness, for example, are complex and difficult concepts to define and research. Examining and detailing higher-order cognitive abilities such as these and others (e.g., deception, abstract reasoning, planning) presents challenges at all levels of research. Finally, the brain and its related systems remain largely mysterious, as progress in discovering and elucidating brain functions has been painfully slow. With these challenges in mind, we present an initial examination of the field of evolutionary cognitive neuroscience.

Genetics, the environment, the brain, and cognition (and behavior) interact with each other in complex ways. It is a combination of the environment and genetics that molds the central and peripheral nervous system across all organisms. Yet the brain also interacts (as a physical and functional structure) with its genes and environments, above and beyond behavior and cognition. Further, what an organism does and thinks will influence its genes, its environment, and its brain. Note also that these systems interact with themselves. The brain is not a singular, static entity any more than the environment is. Within the brain, there are countless interactions across and within all levels of analysis. What happens at the neurotransmitter level in region X will likely have an influence at the molar level in region Y. The brain is a system that acts not unlike a mobile (Keenan et al., 2003), with countless interactions

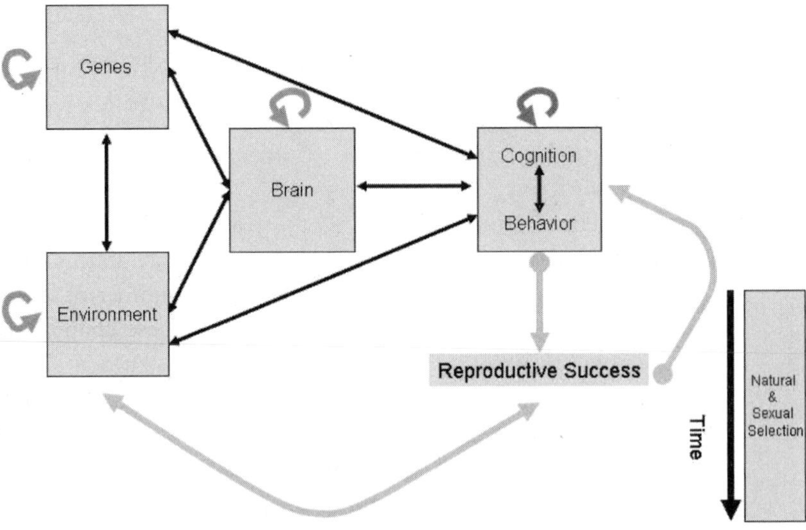

Figure I.1
Evolutionary cognitive neuroscience involves complex systems that not only interact with each other but in which each entity also interacts with itself.

that are inter- and intradependent. Events occurring in a given region (or neuron) will likely have influences at other regions, both proximal and distal. The same interactive systems can be used to describe the environment, the cognitive process, and genetic transmission.

At an ultimate level, in species that possess a nervous system it is this series of interactions that determines the replication of genes and provides the grist for the evolutionary mill. Some simplification must occur, such as describing the brain as if it were a singular unit (Figure I.1). In light of this, it is impressive that research has identified any statistically significant relationships among these systems.

Evolutionary psychology, which does not necessarily focus on cognitive issues directly, is fraught with challenges. Applying evolutionary principles, even to noncognitive behaviors, is difficult. From the fact that "behavior does not fossilize" to the notion that our behaviors are influenced by ultimate causes, evolutionary psychology faces a number of experimental challenges. It is not surprising that evolutionary psychology has faced criticism in terms of its applied methods. However, if we assume that nothing makes sense in biology that is not cast in terms of evolution, the same must apply to psychology. Behavior can be no different from biology because behavior is the manifestation of biology. Psychology, which is the scientific study of behavior and cognition, has largely neglected ultimate questions; neuroscience has not.

The three chapters that make up Part I of this volume address the intricacies of understanding cognitive neuroscience from an evolutionary perspective. Two related questions are presented: How did the human brain get to be the way it is? And how did our cognitive schema come to be? The relationships between the brain and cognition remain not well understood, but more is being revealed every day. In this section we learn of the findings and the progress made on a number of levels, as well as an overview of methodologies and theory. These chapters lay the basic groundwork for the remaining book.

The neocortex is of particular interest to cognitive neuroscientists and evolutionary neuroanatomists because it is extensive in the primates and appears to be related to many higher-order cognitive processes. However, the ultimate reason for its expansion has been debated. Did our metabolically expensive neocortex evolve to facilitate social behavior or foraging behavior, or both? These questions—simplified here for lack of space rather than lack of interest—are the ultimate questions that cognitive neuroscientists might do well to address; the answers will provide us with a greater understanding of the brain and behavior. By looking at the ultimate origins of behavior and the brain, we are in a better position to predict and explain modern human behavior. Part I introduces some of the methods and basic theories that are typically considered in evolutionary cognitive neuroscience, as well as some of the controversies and questions that are being addressed.

References

Keenan, J. P., Gallup, G. G., Jr., & Falk, D. (2003). *The face in the mirror. The search for the origins of consciousness*. New York: HarperCollins/Ecco.

Striedter, G. (2004). *Principlesof brain evolution*. Sunderland, Mass.: Sinauer.

1 Introduction to Evolutionary Theory and Its Modern Application to Human Behavior and Cognition

Aaron T. Goetz and Todd K. Shackelford

Darwin (1859) was not the first to suggest that species evolve. In fact, one of the first discussions of evolution predates Darwin by two and a half millennia. Anaximander, a Greek philosopher, suggested that "in water the first animal arose covered with spiny skin, and with the lapse of time some crawled onto dry land and breaking off their skins in a short time they survived." What Darwin (1859) provided, however, was a viable working mechanism of evolution: natural selection. Darwinian selection has become the centerpiece of biology, and in the last few decades, many psychologists and anthropologists have recognized the value of employing an evolutionary perspective in their work (e.g., Barkow, Cosmides, & Tooby, 1992; Chagnon & Irons, 1979; Daly & Wilson, 1983; Symons, 1979). With a focus on evolved psychological mechanisms and their information processing, evolutionary psychology has risen as a compelling and fruitful approach to psychological science. This chapter provides an introduction to evolution by natural selection and its modern application to the study of human behavior and cognition.

Natural Selection and Sexual Selection

Evolution by natural selection is the process that results when (1) individuals of a population vary in their characteristics, (2) much of the variation is heritable, and (3) resources are limited, so that individuals reproduce differentially (Darwin, 1859; Mayr, 1982). Individuals can vary morphologically, physiologically, psychologically, and behaviorally—no two individuals are exactly the same. Because of these variations, some individuals may be better able to survive and reproduce in their current environment than other individuals. If the variations are heritable (i.e., if they have a genetic component), the characteristics can be passed down from parents to offspring. Limited resources (e.g., food, available mates) result in a competition between individuals, and those individuals who have inherited characteristics that allow them to compete more effectively will produce more offspring. Thus, all organisms are subject to evolution by natural selection. As long as the ingredients of natural selection are present—variation, heredity, and competition resulting in differential reproduction—organisms will evolve. An example of natural selection follows.

The peppered moth (*Biston betularia*) is typically white with black spots. This coloration provides an effective camouflage for the moths as they rest on certain birch trees. There exists variation in the coloration of moths so that some are very white and some very black. In a series of studies, Kettlewell (1955, 1956) documented that when the white trees on which the moths rested became dark from industrial pollution, birds ate more of the white moths because they were now conspicuous on the soot-covered trees. In polluted areas, the population of darker, or melanic, moths replaced the lighter form, but in unpolluted areas, more of the light-colored moths survived. Kettlewell showed that the environment in which the moths were better camouflaged contributed to better survival and reproduction. Kettlewell's work is a classic demonstration of natural selection in action.

Herbert Spencer's summary of natural selection, "survival of the fittest," has, unfortunately, caused more confusion than clarification (Gaulin & McBurney, 2004). Reproduction is a much larger component of natural selection than is survival. If an individual had characteristics that enabled it to survive for hundreds of years yet it never reproduced, those characteristics could not be favored by selection because without transmission to offspring, characteristics cannot become more common in a population. Survival, therefore, functions only to enable individuals to reproduce (directly or indirectly). Second, Spencer's adage suggests that an individual may evolve to be the "fittest." What determines whether an individual is fit is its design in relation to competing designs in the current environment. What is fit in one generation may be unfit in another generation. Also, fit is often taken to imply physically fit. Fitness, in an evolutionary context, is an organism's success in producing offspring that survive to reproductive age (Williams, 1966).

Sexual selection is the process that favors an increase in the frequency of alleles associated with reproduction (Darwin, 1871). Darwin distinguished sexual selection from natural selection, but today most evolutionary scientists combine the two concepts under the label natural selection. Sexual selection is composed of intrasexual competition (competition between members of the same sex for sexual access to members of the opposite sex) and intersexual selection (differential mate choice of members of the opposite sex). Under sexual selection, even a trait that is a liability to survival can evolve. When the sexual attractiveness, for example, of a trait outweighs the survival costs to maintain it, the trait may be sexually selected. The epitome of a sexually selected trait is the peacock's tail. Maintaining and maneuvering an unwieldy tail is metabolically costly for peacocks, and it is often the target of predators. The cumbersome tail evolved, however, because it was attractive to peahens. The mass and brightness of the plumage are attractive to peahens because this appearance signals a modicum of parasites (Hamilton & Zuk, 1982). Peacocks with smaller, lackluster tails are more susceptible to parasites and have a higher parasite load. Thus, the large, bright tail feathers are an honest signal of health, and peahens would be reproductively wise to select as mates males with such tails (who sire offspring that share their high-quality genes).

In many species, particularly polygynous species where male reproductive variance is high and female reproductive variance is low, sexual selection is responsible for prominent sexual dimorphism. In such species, intrasexual competition between males for sexual access to females is fierce, and a size advantage is adaptive. It is often difficult to establish whether a trait evolved via natural selection or sexual selection, but, as mentioned previously, this distinction is not often necessary.

In summary, the core premise of natural selection as a mechanism for evolution is that individual variation exists among traits in a population as a result of random mutations. Those individuals who have traits that better enable them to survive and reproduce will propagate the genes associated with those traits throughout the population.

The Modern Synthesis and Inclusive Fitness Theory

The details of modern evolutionary theory, or neo-Darwinian theory, are the result of the *modern synthesis*. From the early 1930s to the 1950s, advances in genetics, systematics, and paleontology aligned Darwin's theory with the facts of genetics (Mayr & Provine, 1980). The modern synthesis is so called because it was the integration or synthesizing of Darwinian selection with Mendelian genetics. R. A. Fisher, J. B. S. Haldane, Sewall Wright, Ernst Mayr, and Theodosius Dobzhansky are considered the primary authors of the modern synthesis (Mayr &

Provine, 1980). With a more precise understanding of inheritance, Darwin's theory of evolution by natural selection took flight as a powerful explanatory model.

Following the modern synthesis, evolution by natural selection was extended once more to include *inclusive fitness theory* (Hamilton, 1964). Hamilton reasoned that selection could operate through classic fitness (i.e., the sum of an individual's own reproductive success) and inclusive fitness, which includes the effects of an individual's actions on the reproductive success of genetic relatives. That is, a trait will be naturally selected if it causes an individual's genes to be passed on, regardless of whether the individual directly produces offspring. This addendum to natural selection produced a "gene's eye" view of selection and could now explain the evolution of altruistic behavior (i.e., behavior that is beneficial to others but costly for the actor). Genes associated with producing an alarm call when sighting a predator, for example, may spread throughout a population even when is detrimental to the caller if the alarm call is emitted in the presence of genetic relatives and has an overall benefit to those relatives (e.g., Sherman, 1977). Hamilton's inclusive fitness theory is considered the most important advance in our understanding of natural selection, so much so that the term inclusive fitness theory is synonymous with the term evolution by natural selection.

The Products of Evolution: Adaptations, Byproducts, and Noise

Although natural selection is not the only mechanism of evolution—others include mutation, migration, gene flow, genetic drift—it is the primary means of modification and the only creative evolutionary force capable of producing functional organization (Fisher, 1954; Mayr, 1963; Williams, 1966). The creative force of natural selection, acting on random genetic variation, designs three products: adaptations, byproducts of adaptations, and noise.

Adaptations are central to the study of evolution. Through the process of natural selection, small, incremental phenotypic changes that enhance an organism's ability to survive and reproduce (relative to competing designs) accumulate to form an adaptation. Adaptations are inherited, they develop reliably, they are usually species-typical, and they were selected for because they were economic, efficient, and reliable solutions to adaptive problems (Buss, Haselton, Shackelford, Bleske, & Wakefield, 1998; Thornhill, 1997; Tooby & Cosmides, 1990; Williams, 1966). An adaptive problem is an obstacle or impediment that was recurrent during a species's evolutionary history and whose solution affected the survival and reproduction (i.e., genetic propagation) of an organism. Furthermore, adaptive problems are not necessarily "problems," they are the "regularities of the physical, chemical, developmental, ecological, demographic, social, and informational environments encountered by ancestral populations during the course of a species's or population's evolution" (Tooby & Cosmides, 1992, p. 62). In sum, natural selection designs adaptations that solve adaptive problems associated with survival and reproduction. The function of the heart, the production of sweat, and sexual arousal are all adaptations designed by natural selection. The heart is an anatomical adaptation designed to circulate blood throughout an organism's body. The production of sweat is a physiological adaptation designed to thermoregulate an organism. Sexual arousal is a psychological adaptation designed to motivate sexual behavior.

Not all products of natural selection are adaptations. *Byproducts* of adaptations are characteristics of a phenotype that are functionless and do not solve adaptive problems. They are called byproducts because they are incidentally tied to adaptations and are therefore "carried along" with them. Identifying byproducts is as rigorous a process as identifying adaptations because the allegation that a trait is a byproduct requires the alleger to state the adaptations of which it is a byproduct. The human navel and the whiteness of bone are byproducts of adaptations—they do not contribute in any way to an individual's survival or reproduction. In keeping with our mandate: the human navel is a byproduct of an umbilical cord and the whiteness of bone is a byproduct of the calcium in bones.

The third product of evolution is *noise*, or random effects. Noise is also functionless and cannot solve adaptive problems. Noise can be produced by random changes or perturbations in the genetic or developmental environment or by chance mutations. Noise, unlike a byproduct, is not linked to the adaptive aspect of a characteristic. The random shape of an individual's navel is an example of noise.

In summary, the evolutionary process produces three products: adaptations, byproducts, and noise. Adaptations are the product of natural selection and are functionally organized features that contribute to a species's reproductive success, however indirectly. Byproducts and noise do not solve adaptive problems and are not subject to natural selection themselves. In the following section, we discuss how the study of psychological adaptations has changed the study of human behavior and cognition.

Evolutionary Psychology

Evolutionary psychology attempts to make sense of current human thought, emotion, and behavior by careful consideration of human evolutionary history. Over the course of our evolutionary history, humans have faced many adaptive problems that needed to be solved to survive and reproduce. Generation after generation, over millions of years, natural selection slowly shaped the human brain, favoring circuitry that was good at solving these adaptive problems of our ancestors. The study of psychological adaptations (or evolved psychological mechanisms) is central to evolutionary psychology.

Because the focus of evolutionary psychology is on describing adaptations, some have charged its practitioners as being hyperadaptationists. Assuming a priori that a trait may be an adaptation is an experimental heuristic that guides research questions and methodology. Biologists have been conducting their research this way for over 70 years. Moreover, byproducts and noise typically are identifiable only after the adaptations of which they are a byproduct or noise have been discovered and described (Tooby & Cosmides, 1990).

Although modern evolutionary psychological theories are relatively new, all psychological theories are evolutionary in nature: "All psychological theories—be they cognitive, social, developmental, personality, or clinical—imply the existence of internal psychological mechanisms" (Buss, 1995, p. 2). If the internal psychological mechanisms implied in any psychological theory were not the product of the evolutionary process, then they would be, by default, unscientific theories.

Psychological Mechanisms as Information-Processing Modules

An evolved psychological mechanism is an information-processing module that was selected throughout a species's evolutionary history because it reliably produced behavior that solved a particular adaptive problem (Tooby & Cosmides, 1992). Evolved psychological mechanisms are understood in terms of their specific input, decision rules, and output (Buss, 1995). Each psychological mechanism evolved to take in a narrow range of information—information specific to a specific adaptive problem. The information (or input) that the organism receives signals the adaptive problem that is being confronted. The input (either internal or external) is then transformed into output (i.e., behavior, physiological activity, or input relayed to another psychological mechanism) via a decision rule, an if-then procedure. An example illustrating the input, decision rules, and output of a psychological mechanism is appropriate here.

Fruit can be either ripe or unripe. Because ripe fruit is more nutritious (calorically dense) than immature fruit, humans have developed a preference for ripe fruit. The decision rule regarding the selection of fruit might go something like, "If the fruit tastes sweet, then eat it." If all fruit were maximally saturated with sugar all of the time, then that particular decision rule would not exist. The output associated with this mechanism might be to eat the ripe fruit or disregard the unripe fruit. This example illustrates the fact that psychological mechanisms develop and operate without any conscious awareness or formal learning, and we are blind to their underlying logic. (Do you enjoy calorically dense fruit because it provides nutrition needed to carry out activities related to survival and reproduction, or do you simply enjoy sweet fruit?)

Tooby and Cosmides (1992) have written that the causal link between evolution and behavior is made through psychological mechanisms. That is, the filter of natural selection operates on psychological mechanisms that produce behavior. Natural selection cannot operate on behavior directly but instead operates on the genes associated with the psychological mechanisms that produce the behavior. Williams (1966) spoke similarly: "The selection of genes is mediated by the phenotype [psychological mechanism], and in order to be favorably selected, a gene must produce phenotypic reproductive success [adaptive behavior]" (p. 25).

Psychological Mechanisms and Domain Specificity

The vast majority of psychological mechanisms are presumed to be domain-specific. That is, the mind is composed of content-dependent machinery (i.e., physiological and psychological mechanisms) that is presumed to have evolved to solve a specific adaptive problem. Psychological mechanisms can also be expressed as cognitive biases that cause people to more readily attend to or make sense of some pieces of information relative to others. This presumption of domain specificity or modularity contrasts with the traditional position that humans are endowed with a general set of learning or reasoning mechanisms that are applied to any problem regardless of specific content (e.g., Atkinson & Wheeler, 2004). A system that is domain-general or content-independent, however, is a system that lacks a priori knowledge about specific situations or problem domains (Tooby & Cosmides, 1992). Such a system, when faced with a choice in a chain of decisions, must select from all behavioral possibilities (e.g., wink, jump, remember father, smile, point finger, scream). This problem of choosing among an infinite range of possibilities when only a small subset are appropriate has been described by researchers in artificial intelligence, linguistics, and other disciplines (see Tooby & Cosmides, 1992, for a review).

Not only are there theoretical arguments against a content-independent system, myriad evidence for domain specificity comes from, among other areas, evolutionary psychological theory and research (e.g., Cosmides, 1989; Cosmides & Tooby, 1994; Flaxman & Sherman, 2000; Pinker & Bloom, 1990), cognitive research (e.g., Hirschfeld & Gelman, 1994), studies of animal learning (e.g., Carey & Gelman, 1991; Garcia, Ervin, & Koelling, 1966), and the clinical neurological literature (e.g., Gazzaniga & Smylie, 1983; Ramachandran, 1995; Sergent, Ohta, & MacDonald, 1992). Practitioners of evolutionary psychology concede that relatively domain-general mechanisms that function, for example, to integrate and relay information between domain-specific mechanisms may exist, but the vast majority of mechanisms are presumed to be domain-specific.

Some of the controversy surrounding the modularity of the mind seems to be rooted in the use of the term *domain*. Psychologists have often used the term to refer to particular domains of life, such as the mating domain, kinship domain, and parenting domain. Many have assumed subsequently that labeling a mechanism as domain-specific restricts the proposed mechanism to a particular domain, and if evidence can be garnered to show that the mechanism functions in more than one domain (e.g., the mating domain and the kinship domain), then it is taken as evidence for the domain generality of the proposed mechanism. This, however, is incorrect. A domain, when referring to a psychological mechanism, is a selection pressure, an adaptive problem (Cosmides & Tooby, 1987). Domain, then, is synonymous with *problem*. That is, a domain-specific mechanism refers to a problem-specific mechanism—a mechanism that evolved to solve a specific adaptive problem. So, although evolutionary and cognitive psychologists use the term *domain-specific*, perhaps some confusion could be avoided if the more accurate term *problem-specific* were employed instead. Although some psychological mechanisms cut across different domains of life (e.g., face recognition, working memory, processing speed), they still solve specific problems. Working memory, for example, solves the specific problem of holding information in the mind for a brief period of time. It has been suggested that evolutionary and cognitive psychologists might be better off avoiding these contentious labels and simply describing the proposed mechanism and its function (D. M. Buss, personal communication, January 2005).

Evolutionary Time Lags and the Environment of Evolutionary Adaptedness

Because evolution is an excruciatingly slow process, extant humans and their minds are designed for the earlier environments of which they are a product. Our minds were not designed to solve the day-to-day problems of our modern society but instead were designed to solve the day-to-day problems of our evolutionary past. Examples of evolutionary time lags abound: our difficulty in learning to fear modern threats, such as guns and cars, and our near effortless learning to fear more ancient threats, such as snakes and spiders (Öhman & Mineka, 2001); children's ease in learning biologically primary mathematic abilities, such as counting, and their difficulty in learning biologically secondary mathematic abilities, such as arithmetic (Geary, 1995); women not conceding to intercourse indiscriminately, even though modern contraception can eliminate the reproductive costs associated with intercourse; and our preference for sugar and fat, which was once adaptive, owing to their scarcity, but has now become maladaptive. These few examples illustrate that our modern behavior is best understood when placed in the context of our environment of evolutionary adaptedness.

The environment of evolutionary adaptedness is not a place or a time in history but a statistical composite of the selection pressures (i.e., the enduring properties, components, and elements) of a species's ancestral past, and more specifically the *adaptations* that characterize that past (Tooby & Cosmides, 1990). That is, each adaptation evolved as a result of a specific set of selection pressures. Each adaptation, in principle, has a unique environment of evolutionary adaptedness, but there likely would have been significant overlap in the environments of related adaptations. Tooby and Cosmides (1990) and other practitioners of evolutionary psychology, however, use "Pleistocene" to refer to the human environment of evolutionary adaptedness because this time period, approximately 1.81–0.01 million years ago, was appropriate for virtually all adaptations of *Homo sapiens sapiens*.

Although our evolutionary past is not available for direct observation, the discovery and description of adaptations allows us to make inferences about our evolutionary past, and the characterization of adaptations is arguably the single most reliable way of learning about the past (Tooby & Cosmides, 1990). Some adaptations provide unequivocal information about our ancestral past. Our cache of psychological mechanisms associated with navigating the social world tells us that our ancestors were a social species (e.g., Cosmides, 1989; Cummins, 1998; Kurzban, Tooby, & Cosmides, 2001; Pinker & Bloom, 1990; Trivers, 1971). A multitude of psychological mechanisms associated with cuckoldry avoidance tell us that female infidelity was a recurrent feature of our evolutionary past (Buss, Larsen, Westen, & Semmelroth, 1992; Buss & Shackelford, 1997; Goetz et al., 2005; Platek, 2003; Shackelford et al., 2002).

Some adaptations, however, do not make clear (at least on first inspection) their link with our ancestral past. There exists, for example, a mechanism present in the middle ear of all humans that is able to reduce sound intensity by as much as 30 db in 50 ms. The attenuation reflex, as it is known, acts by contracting muscles that pull the stirrup away from the oval window of the cochlea, preventing strong vibrations from damaging the inner ear. The attenuation reflex meets the characteristics of an adaptation (e.g., economic, efficient, reliable), yet it is not obvious what selection pressures drove the evolution of this adaptation. That is, what specific noises did our ancestors recurrently hear that would create this noise-reducing mechanism? That the muscles appear to contract as we are about to speak suggests that our own loud voices might have been the impetus for this adaptation. Moreover, sound attenuation is greater at low frequencies than at high ones (and humans speak at low frequencies), also suggesting that ululating was a recurrent (enough) feature of our evolutionary past. Thus, from discovering and describing adaptations, we can tentatively characterize aspects of our evolutionary environment.

Evolutionary Psychology Is Not Sociobiology

Those less familiar with evolutionary psychology often construe the approach as "sociobiology reborn." Although sociobiology, ethology, behavioral ecology, and evolutionary psychology share with each other evolution as a guiding framework, the programs are conceptually distinct, for at least two reasons (Buss, 1995; Crawford, 2000). First, the focus on evolved psychological mechanisms and their information processing is a unique and defining feature of evolutionary psychology. The input, decision rules, and output of psychological mechanisms are central to the analysis. Second, evolutionary psychology does not measure individuals' reproductive success or fitness and views this endeavor as fruitless. Because many sociobiologists have advocated measuring an individual's reproductive success to understand the adaptive value of behavior, the pejorative label "baby counting" has been applied to sociobiology. Evolutionary psychology rejects the premise that measuring fitness in a recent or current environment provides any information about a particular behavior. The information needed to measure fitness correctly becomes known only generations later because there is no guarantee that selection pressures remain stable over time. Practitioners of evolutionary psychology hold that "humans are adaptation executers, not fitness maximizers" (Tooby & Cosmides, 1990, p. 420). While many agree that evolutionary psychology is a separate field from other adaptationist programs, others hold that it is sociobiology in camouflage (e.g., Silverman, 2003).

Evolutionary Psychology's Future

Although this modern approach to human behavior and cognition is relatively young, only about 25 years old, evolutionary psychology's impact is already permeating all areas of psychology and opening up lines of research missed entirely by previous psychologists.

Evolutionary psychology's merit and future are also demonstrated in the fact that the number of publications using an evolutionary psychological approach is growing exponentially (Durrant & Ellis, 2003). Some have even suggested that in the foreseeable future, the psychological equivalent to *Gray's Anatomy* will be possible, describing a number of evolved psychological mechanisms, their information processing, and their neural substrates (Tooby & Cosmides, 1992).

As new psychologists are impartially introduced to evolutionary psychology, as "traditional" (i.e., antievolutionary) psychologists retire, as evolutionary psychology's empirical harvest grows, as findings from genetics corroborate findings from evolutionary psychology (e.g., Cherkas et al., 2004), as the neural substrates underlying hypothesized psychological mechanisms are discovered (e.g., Platek, Keenan, & Mohamed, 2006), and as cross-disciplinary frameworks of evidence are utilized (Schmitt & Pilcher, 2004), evolutionary psychology is expected to emerge as *the* metatheory for psychological science.

Conclusion

In this chapter we introduced evolutionary theory and its modern impact on psychological science. We discussed how, with a focus on evolved psychological mechanisms and their information processing, evolutionary psychology has arisen as a compelling and fruitful approach to the study of human behavior and cognition.

Because the design of the mind owes its functional organization to a natural, evolutionary process, an evolutionarily psychological approach is a logical framework on which to base all psychological theories. Evolutionary psychological theories specify what problems our cognitive mechanisms were designed to solve, thereby providing important information about what their design features are likely to be. In other words, "Is it not reasonable to anticipate that our understanding of the human mind would be aided greatly by knowing the purpose for which it was designed?" (Williams, 1966, p. 16).

It is possible to do research in psychology with little or no knowledge of evolution. Most psychologists do. But without an evolutionary perspective, psychology becomes a disparate set of fields. Evolutionary explanations pervade all fields in psychology and provide a unifying metatheoretical framework within which all of psychological science can be organized.

References

Atkinson A. P., & Wheeler, M. (2004). The grain of domains: The evolutionary-psychological case against domain-general cognition. *Mind and Language*, *19*, 147–176.

Barkow, J. H., Cosmides, L., & Tooby J. (1992). *The adapted mind: Evolutionary psychology and the generation of culture.* New York: Oxford University Press.

Buss, D. M. (1995). Evolutionary psychology: A new paradigm for psychological science. *Psychological Inquiry*, 6, 1–20.

Buss, D. M., Haselton, M. G., Shackelford, T. K., Bleske, A. L., & Wakefield, J. C. (1998). Adaptations, exaptations, and spandrels. *American Psychologist*, *53*, 533–548.

Buss, D. M., Larsen, R., Westen, D., & Semmelroth, J. (1992). Sex differences in jealousy: Evolution, physiology, and psychology. *Psychological Science*, *3*, 251–255.

Buss, D. M., & Shackelford, T. K. (1997). From vigilance to violence: Mate retention tactics in married couples. *Journal of Personality and Social Psychology*, *72*, 346–361.

Carey, S., & Gelman, R. (1991). *The epigenesis of mind: Essays on biology and cognition*. Hillsdale, NJ: Erlbaum.

Chagnon, N. A., & Irons, W. (1979). *Evolutionary biology and human social behavior: An anthropological perspective*. North Scituate, MA: Duxbury Press.

Cherkas, L. F., Oelsner, E. C., Mak, Y. T., Valdes, A., & Spector, T. (2004). Genetic influences on female infidelity and number of sexual partners in humans: A linkage and association study of the role of the vasopressin receptor gene (*AVPR1A*). *Twin Research*, *7*, 649–658.

Cosmides, L. (1989). The logic of social exchange: Has natural selection shaped how humans reason? Studies with the Wason selection task. *Cognition*, *31*, 187–276.

Cosmides, L., & Tooby. J. (1987). From evolution to behavior: Evolutionary psychology as the missing link. In J. Dupre (Ed.), *The latest on the best: Essays on evolution and optimality* (pp. 277–306). Cambridge, MA: Bradford Books/MIT Press.

Cosmides, L., & Tooby, J. (1994). Origins of domain specificity: The evolution of functional organization. In L. A. Hirschfeld and S. A. Gelman (Eds.), *Mapping the mind: Domain specificity in cognition and culture* (pp. 85–116). New York: Cambridge University Press.

Crawford, C. (2000). Evolutionary psychology: Counting babies or studying information processing mechanisms. *Annals of the New York Academy of Sciences*, *907*, 21–38.

Cummins, D. D. (1998). Social norms and other minds: The evolutionary roots of higher cognition. In D. D. Cummins & C. Allen (Eds.), *The evolution of mind* (pp. 30–50). New York: Oxford University Press.

Daly, M., & Wilson, M. (1983). *Sex, evolution, and behavior* (2nd ed.). Boston: Willard Grant.

Darwin, C. (1859). *On the origin of species*. London: Murray.

Darwin, C. (1871). *The descent of man and Selection in relation to sex*. London: Murray.

Durrant, R., & Ellis, B. J. (2003). Evolutionary psychology: Core assumptions and methodology. In M. Gallagher & R. J. Nelson (Eds.), *Comprehensive handbook of psychology, Vol. 3: Biological psychology* (pp. 1–33). New York: John Wiley & Sons.

Flaxman, S. M., & Sherman, P. (2000). Morning sickness: A mechanism for protecting mother and embryo. *The Quarterly Review of Biology*, *75*, 113–148.

Fisher, R. A. (1954). Retrospect of the criticisms of the theory of natural selection. In J. S. Huxley, A. C. Hardy, & E. B. Ford (Eds.), *Evolution as a process* (pp. 84–98). London: Allen & Unwin.

Futuyma, D. J. (1986). *Evolutionary biology* (2nd ed.). Sunderland, MA: Sinauer Associates.

Garcia, J., Ervin, F. R., & Koelling, R. A. (1966). Learning with prolonged delay of reinforcement. *Psychonomic Science*, *5*, 121–122.

Gaulin S. J. C., & McBurney, D. H. (2004). *Evolutionary psychology* (2nd ed.). Upper Saddle River, NJ: Pearson Education.

Gazzaniga, M. S., & Smylie, C. S. (1983). Facial recognition and brain asymmetries: Clues to underlying mechanisms. *Annals of Neurology*, *13*, 536–540.

Geary, D. C. (1995). Reflections on evolution and culture in children's cognition: Implications for mathematical development and instruction. *American Psychologist*, *50*, 24–37.

Goetz, A. T., Shackelford, T. K., Weekes-Shackelford, V. A., Euler, H. A., Hoier, S., Schmitt, D. P., & LaMunyon, C. W. (2005). Mate retention, semen displacement, and human sperm competition: Tactics to prevent and correct female infidelity. *Personality and Individual Differences*, *38*, 749–763.

Hamilton, W. D. (1964). The genetical evolution of social behaviour. *Journal of Theoretical Biology*, *7*, 1–52.

Hamilton, W. D., & Zuk, M. (1982). Heritable true fitness and bright birds: A role for parasites? *Science*, *218*, 384–387.

Hirschfeld, L. A., & Gelman, S. A. (1994). *Mapping the mind: Domain specificity in cognition and culture*. New York: Cambridge University Press.

Kettlewell, H. B. D. (1955). Selection experiments on industrial melanism in the Lepidoptera. *Heredity*, *9*, 323–342.

Kettlewell, H. B. D. (1956). Further selection experiments on industrial melanism in the Lepidoptera. *Heredity*, *10*, 287–301.

Kurzban, R., Tooby, J., & Cosmides, L. (2001). Can race be erased? Coalitional computation and social categorization. *Proceedings of the National Academy of Sciences, U.S.A.*, *98*, 15387–15392.

Mayr, E. (1963). *Animal species and their evolution*. Cambridge, MA: Harvard University Press.

Mayr, E. (1982). *The growth of biological thought: Diversity, evolution, and inheritance*. Cambridge, MA: Harvard University Press.

Mayr, E., & Provine, W. B. (1980). *The evolutionary synthesis: Perspectives on the unification of biology*. Cambridge, MA: Harvard University Press.

Öhman, A., & Mineka, S. (2001). Fears, phobias, and preparedness: Toward an evolved module of fear and fear learning. *Psychological Review*, *108*, 483–522.

Pinker, S., & Bloom, P. (1990). Natural language and natural selection. *Behavioral and Brain Sciences*, *13*, 707–727.

Platek, S. M., Keenan, J. P., & Mohamed, F. B. (2005). Sex differences in the neural correlates of child facial resemblance: An event-related fMRI study. *Neuroimage*, *25*, 1336–1344.

Platek, S. M. (2003). An evolutionary model of the effects of human paternal resemblance on paternal investment. *Evolution and Cognition*, *9*, 189–197.

Ramachandran, V. S. (1995). Anosognosia in parietal lobe syndrome. *Consciousness and Cognition*, *4*, 22–51.

Schmitt, D. P., & Pilcher, J. J. (2004). Evaluating evidence of psychological adaptation. *Psychological Science*, *15*, 643–649.

Sergent, J., Ohta, S., & MacDonald, B. (1992). Functional neuroanatomy of face and object processing. *Brain*, *115*, 15–36.

Shackelford, T. K., LeBlanc, G. J., Weekes-Shackelford, V. A., Bleske-Rechek, A. L., Euler, H. A., & Hoier, S. (2002). Psychological adaptation to human sperm competition. *Evolution and Human Behavior*, *23*, 123–138.

Sherman, P. W. (1977). Nepotism and the evolution of alarm calls. *Science*, *197*, 1246–1253.

Silverman, I. (2003). Confessions of a closet sociobiologist: Personal perspectives on the Darwinian movement in psychology. *Evolutionary Psychology*, *1*, 1–9.

Symons, D. (1979). *The evolution of human sexuality*. New York: Oxford University Press.

Thornhill, R. (1997). The concept of an evolved adaptation. In G. R. Bock & G. Cardew (Eds.), *Characterizing human psychological adaptations* (pp. 4–22). West Sussex, U.K.: John Wiley & Sons.

Tooby, J., & Cosmides, L. (1990). The past explains the present: Emotional adaptations and the structure of ancestral environments. *Ethology and Sociobiology*, *11*, 375–424.

Tooby, J., & Cosmides, L. (1992). The psychological foundations of culture. In J. H. Barkow, L. Cosmides, & J. Tooby (Eds.), *The adapted mind: Evolutionary psychology and the generation of culture* (pp. 19–136). Oxford: Oxford University Press.

Trivers, R. L. (1971). The evolution of reciprocal altruism. *Quarterly Review of Biology*, *76*, 35–57.

Williams, G. C. (1966). *Adaptation and natural selection*. Princeton, NJ: Princeton University Press.

2 Brain and Cognition in Evolutionary Perspective

Robin I. M. Dunbar

Humans share with their primate cousins an intense sociality that is unique in the animal kingdom. That sociality is premised on forms of cognition that emphasize both the capacity to read behavior in a more sophisticated way than is typical of most other animals and forms of cognition that appear to be explicitly social in focus. These, in turn, appear to be underpinned by brains that are significantly larger for body weight than those of any other group of animals. Although each of these aspects of primate biology has been studied in considerable detail over the past several decades, we still have very little real understanding of how they relate to each other. Moreover, although evolution is given tacit acknowledgment in the neurosciences, in practice, neuroscience is virtually an evolution-free zone in that few neuroscientists have more than a nodding acquaintance with evolutionary theory and its implications.

In this chapter, I sketch out an integrated theory of primate social cognition, drawing together the elements of the story from all four principal components: evolution, neuroanatomy, cognition, and behavior. This story will necessarily be incomplete, but I hope to show that by drawing these components together into a single coherent account, we might gain measurably in our understanding of the processes involved. I will begin with neuroanatomy, then try to relate this to cognition, and finally draw out some of the implications for understanding behavior.

Evolution and the Brain

It is now 30 years since Jerison (1973) pointed out that primates have unusually large brains for body size. Understanding why this should be so is more than an esoteric exercise, because brain tissue is among the most expensive to grow and maintain of any tissues in the body (Aiello & Wheeler, 1995). This is mainly a consequence of the fact that neurotransmitters are expensive to create, and the constant need to replenish them as neurons go about their business means that brain tissue consumes roughly 8–10 times more energy than we would expect for its mass. Evolving large brains thus is not a trivial exercise and is unlikely to be undertaken without good cause by any lineage. The significance of this is that some very pressing advantages must have existed if a species has an unusually large brain. It is not enough to note that all animals require brains to be able to survive. That may explain adequately why vertebrates have brains at all, but not why some have larger brains than others. The explanation for the unusually large brains of primates has to be sought in functions that go beyond those that apply to all species.

By the same token, it is not enough to suggest that large brains are a functionally uninteresting byproduct (or, in Stephen Jay Gould's term, a spandrel) of having a large body. There may well be developmental constraints on growing or servicing a large brain that impose a requirement for a large body: the allometric scaling of the energetic costs of tissue function against body mass is less than 1 (known as Kleiber's law), thereby yielding energy savings in larger species that can be channeled into extra brain tissue (see Martin, 1981). However, this is not a functional explanation and does not—and cannot—explain why some taxa should be prepared to pay the additional costs of having larger brains for body size than others. The energetic costs of servicing the extra tissue still have to be met by additional feeding, thereby adding significantly to the animal's evolutionary burden by exposing it to an increased risk of predation. Given the costs of large brains, the steepness of the evolutionary gradient up which selection has to drive brain growth means that there will always be intense selection against evolving brains of larger size than that minimally necessary to maintain life.

The point at issue here, then, is the fact that developmental constraints and evolutionary (or functional) explanations are two quite different types of explanation, and both are needed for a full and complete understanding of any given phenomenon. In organismic biology, these conventionally constitute two of Tinbergen's Four Why's,[1] the other two being mechanistic explanations and historical (or phylogenetic) accounts

---

1 In fact, the point was originally made by Aristotle in the fourth century B.C., and was well understood by the founding fathers of evolutionary biology in the first half of the twentieth century. However, they recognized only three levels of explanation; Tinbergen added phylogenetic history to complete the modern quartet.

(see Tinbergen, 1963). These different kinds of explanations are complementary and not alternatives, and all four are ultimately needed to provide a complete and comprehensive explanation for any given biological phenomenon. Showing that there is a developmental mechanism that produces large brains (as Finlay, Darlington, & Nicastro [2001] do, for example) is not an alternative to offering an evolutionary explanation in terms of selection factors selecting for large brains. Both kinds of explanations are necessary, precisely because the high costs of brain tissue mean that some pressure must exist for paying the costs involved. It is important to appreciate that these four levels of explanation are logically independent of each other: our conclusions about any one of them do not constrain or limit the range of options on any of the others (a point that conveniently allows us to deal with each in isolation).

One of the key points of Jerison's (1973) seminal analysis was the distinction between what is minimally necessary to run the cellular activities of the body and what is left over as spare computing capacity that can be devoted to other, less physiologically functional activities. His point was that while larger-bodied animals would surely need larger brains to manage somatic activity, this should be a constant factor (i.e., one that is proportional to body mass) across all organisms; hence, by scaling brains to body size, we can ask whether individual species have relatively larger or smaller brains than might be expected for an animal of that particular size. Jerison captured this with his encephalization quotient (EQ), defined as a simple ratio of total brain volume to that which would be expected for an average mammal of that body size. In the absence of any better way of estimating the true minimum requirements for sustaining life, Jerison used the average for all species as his baseline, a strategy that can be justified fairly easily on the grounds that the average is likely to scale isometrically with the minimum requirements, and that it is enough to be able to distinguish between those species that have more than the average and those that have less. In the absence of any understanding of the real minimum brain size needed to sustain bodily life, this at least provides us with a first approximation to the question of whether all species's brains are proportional to body size.

In most analyses of this kind, brains are viewed as essentially homogeneous. But this, of course, is not true, for two reasons. First, not all parts of the brain are necessary to support life. Notoriously in clinical neuropsychology, frontal lobe functions can be lost almost in their entirety without threatening the patient's physical existence (Stuss, Eskes, & Foster, 1994). The patient's ability to function effectively in a human *social* environment may be severely compromised, but in a strictly non-social environment such individuals can (and do) cope just fine (although often at very heavy cost to their careers). Second, as Finlay and Darlington (1995), de Winter and Oxnard (2001), and many others have pointed out, different parts of the brain do not share a common allometric relationship to each other. Certain parts (in particular, the neocortex itself) scale much more steeply to total brain volume than others. Although there may be sound mechanistic reasons why this has to be so (e.g., the number of processing units is necessarily some kind of power function of the number of sensory input units because processing requires a constant ratio of neural units for every unit of input; see Stevens, 2001), this does not entirely explain the fact that, in primates at least, it is frontal lobe volume that increases disproportionately. In part, this reflects the fact that brains evolve (and, indeed, grow: Gogtay et al., 2004) from back (the visual areas) to front (the so-called executive brain), so that increases in brain volume across the primates are driven largely by a disproportionate increase in frontal lobe volume (see also Semendeferi, Damasio, Frank, & van Hoesen, 1997[2]).

Jerison's original argument—that baseline brain volume should be proportional to body mass because larger bodies need more brain tissue to manage them—raises the question of whether total brain volume is the appropriate variable to scale against body size. If Jerison's argument is valid (and there is no reason to assume that it is not), then it may in fact be more appropriate to scale only those components of the brain that are principally involved in somatic management (mainly but not exclusively the subcortical components). The balance (indexed as EQ in just the same way as Jerison previously did, but against a different baseline) would then reflect more accurately what Jerison had in mind, namely, how much free neural capacity (in effect, extra computing

2 This paper has been interpreted as implying that frontal lobes are not disproportionately larger in humans than in other species of primates, in part because this is how Semendeferi et al. phrase their conclusion. What they intended to mean (and what their analysis shows) is that human frontal lobes scale to total brain volume on the same allometric relationship as that of for anthropoid primates as a whole. However, the scaling factor on a log-log plot is $b \approx 1.115$, meaning that frontal lobels become disproportionately larger (as a proportion of total brain volume) as brain volume increases. Human frontal lobes *are* disproportionately larger than those of other primates, though not more so than we would expect for the allometric relationship for all anthropoid primates.

power) was available for other functions over and above pure somatic management.

Even without introducing this added sophistication, it is perhaps already obvious that the steepness of the scaling ratio for neocortex implies that some additional kind of processing must be going on. It is now widely accepted that the brain relies heavily on parallel processing on a massive scale. One reflection of this seems to be the bolting on of added frontal lobe mass in large-brained primates (especially hominoids), since this appears to provide added layers of processing not available to species with smaller brains. In humans, for example, the processing of visual inputs seems to involve at least three levels of analysis located in different parts of the brain (in this case, forming a sequential stream from back to front): pattern recognition (in the primary visual areas of the occipital lobe), recognition of the relationships between patterns (the association areas in the parietal lobes), and understanding the *meaning* of those relationships (in the frontal lobes) (Frith, 1996).

Before concluding this section, it is essential to emphasize one final point, not least because it has already become the source of some confusion in the literature. Finlay et al. (2001) argued that increasing brain volume is largely controlled by the timing of neurogenesis. The longer precursor cells are produced during brain development, the more cells can be grown and the larger will be the final volume of the brain. This finding is important for three reasons. First, it provides us with a simple account of how larger brains are produced. Second, it reinforces the view from comparative studies that brain tissue can only be laid down at a constant rate, and hence that evolving large brains necessitates proportionately extending the period of development in order to allow sufficient time for neurogenesis. Third, and perhaps most important, it demonstrates that the process involved in evolving large brains may depend on a single "switch" gene that controls the timing of neurogenesis. This is important because it means that the business of evolving larger brains may be very simple: rather than requiring many mutations in different chromosomal locations, only a single mutation may be needed, and single mutations are much easier to engineer than coordinating the simultaneous occurrence of several (or even many) mutations. This does not, of course, obviate the fact that evolving large brains is an energetically expensive business and therefore requires a proportionately advantageous fitness benefit, but it does remove one of the main *genetic* constraints that would otherwise make the evolution of larger brains difficult (if not impractical).

What Selects for Large Brains?

Conventional wisdom has always assumed that brains exist to allow organisms to find their way about in the world, and so survive successfully. This is undoubtedly part of the story, and it surely lies at the root of the circumstances that initiated brain evolution in the vertebrates. However, in the long evolutionary period since then, brains have come to serve other functions. Even though many of these may have ecological survival as their ultimate function, they achieve this end by other means (e.g., solving a social as opposed to an ecological problem). In addition, however, much of what the higher vertebrates (birds and mammals) do is explicitly social, with rather little direct ecological or survival relevance: mate choice, parental care, or even servicing friendly relationships with allies. Nowhere is this more true than among the higher primates (monkeys and apes), where social solutions to ecological problems may be regarded as the norm rather than the exception.

Indeed, the argument that primate brain evolution may have been driven by the computational demands of living in a complex social system (put forward originally by Jolly [1969] and Humphrey [1976], and, ultimately in fully fledged form as the "Machiavellian intelligence hypothesis" by Byrne & Whiten [1988]) builds on just this observation. Subsequently renamed the "social brain hypothesis" in recognition of the fact that the term Machiavellian often gave rise to an unintended emphasis on political scheming at the expense of social affiliation, this hypothesis argued that the unusually complex nature of primate sociality, involving both the formation of intense social relationships and the use of coalitions in cooperative defense, imposed unusually heavy demands on animals' capacities to make inferences about the future behavior of other group members.

While there was circumstantial evidence to support both ecological and social hypotheses, once concerted attempts were made to test formally *between* them, the social hypothesis was confirmed at the expense of the various alternative ecological hypotheses (Barton & Dunbar, 1997; Dunbar, 1992, 1998; Joffe & Dunbar, submitted). All these studies used social group size as their proxy for social complexity. However, more direct tests of the social hypothesis have subsequently used other measures of social behavior, including grooming clique size (Kudo & Dunbar, 2001), the proportion of play that is social (Lewis, 2001), the capacity to exploit subtle mating strategies (Pawłowski, Lowen, & Dunbar, 1998), and the frequency of tactical deception (Byrne & Corp, 2004). In addition, these initial findings for primates have been extended to other mammalian taxa, including advanced insectivores and carnivores (Dunbar & Bever, 1998), cetaceans (Marino, 1996), and African bovids (unpublished).

Significantly, these analyses have focused on neocortex volume rather than on total brain volume. Analyses using total brain volume on its own often do not produce significant relationships with social group size. This makes sense, for two reasons. First, as noted earlier, if Jerison's argument about encephalization is correct, then the subcortical parts of the brain may simply add random error variance to the analysis, thereby distorting any relationship that might exist with the cortical components. Second, as was made clear by Finlay and Darlington's (1995) analysis, primate brain evolution is principally about the neocortex: the steepness of the scaling relationship between neocortex volume and total brain volume means that the greater part of the increase in brain size that has occurred during the course of primate evolution is due to a disproportionate increase in neocortex volume rather than to increases in the volume of subcortical components. In effect, when we ask what factors selected for large brains in primates, we are really asking what factors selected for a large neocortex. As we shall see in the following section, this view is in accord with the neurocognitive evidence, which shows that those aspects of cognition that are most intimately involved with sociality are located in the neocortex (principally the frontal lobe).

Deaner, Nunn, and van Schaik (2000) have argued that the case against the ecological hypotheses remains unproven, since they were able to demonstrate that, on multiple regression analysis, which of two putative independent variables (social group size or home range size) was the better predictor of relative neocortex volume depended on how the data were analyzed. Although their argument appears plausible, it is undermined by three key facts.

First, it seems odd, on face value, that evolutionary increases in a large component of the primate brain (neocortex or total brain volume) should be driven by a rather specific (and computationally quite modest) aspect of behavior (mental mapping) that in both birds and humans is associated with a very specific (and volumetrically rather small) component of the brain (the hippocampus; Krebs, Sherry, Healy, Perry, & Vaccarino, 1989). One might, perhaps, argue, with Finlay and Darlington (1995), that the scaling relationships between brain components are so tight that selection for one component (say, the hippocampus) would inevitably result in a proportionate increase in all brain components (including the neocortex). However, we can discount any such claim because, as we have already noted, such an explanation flies in the face of basic biological and evolutionary principles (the cost of brain tissue and the relative efficiency of natural selection, respectively).

Second, even if it were the case that a relationship with range size could not be excluded, the balance of probabilities must lie with the social hypothesis. All other versions of the ecological hypothesis that have been tested (e.g., frugivory and extractive foraging; Gibson 1990) produce unequivocally negative results (Dunbar, 1992, 1995), whereas tests with many other social variables consistently yield positive results.

Third, and perhaps most important, Deaner et al. (2000) used log-transformations of all variables (as is conventional in comparative analyses) in two of their three scaling methods (residuals against body weight, residuals against brain volume, but not ratios of neocortex volume to subcortical brain volume). Failure to do so in the last case is particularly unfortunate, for two reasons. First, Dunbar's (1992) ratio method is mathematically identical to conventional residuals methods when the ratios are logged (see Dunbar, 1998); they differ only in the base against which the residual is taken (a predicted species value based on the average relationship against all taxa against the actual species value). Failure to log-transform the ratio values may explain why previous analyses (e.g., Barton & Dunbar, 1997) have obtained essentially identical results when using alternative methods. Second, and more important, analyses using raw values have repeatedly demonstrated that the relationship between social group size and any number of alternative indices of neocortex volume is strongly exponential in shape. Using linear regression analysis will thus inevitably yield a better fit with an ecological variable, such as home range size, that typically has a linear relationship with body size (and hence brain size across the range of primates). In any case, since many would argue that the only correct method of analysis is to use residuals against brain volume (see Barton & Dunbar, 1997), one might reasonably conclude from Deaner et al. (2000) that the social brain hypothesis is vindicated, since this method yielded consistent results in its favor at the expense of the mental mapping alternative.

An alternative version of the ecological hypothesis has been offered by Reader and Laland (2002), who argued that the significance of sociality might lie in social learning rather than in social knowledge about other individuals. They were able to show that neocortex size correlated with the relative frequencies of innovative behavior and tool use and an index of social learning in primates (as recorded in the primary literature). Since most of these innovations refer to feeding behavior, they reasonably conclude that social intelligence (meaning the ability to copy the successful innovations of others) is primarily ecological in focus. However, all the examples in their database were derived from foraging contexts, the easiest contexts in which to observe and recognize innovations. Hence, to conclude from this, as they do, that ecological factors must be at least as important as, if not more so than, social contexts in the evolution of large brains is, at best, premature. It may be safe to conclude that social learning per se is not the function that has been selected for by sociality in primates, but this is a rather different claim from that being made on behalf of the social brain hypothesis.

A Role for Cognition

If brain size is driven by the demands of sociality, then we need to ask what kinds of cognitive mechanisms bridge the gap between brain and behavior. The simple conclusion on this question is that we have absolutely no idea. However, there are two possible positions on this that we can usefully explore here which might point the way for future research. One is that the evolution of larger brains is associated with qualitative differences in cognitive mechanisms: species that have bigger brains do so because novel cognitive modules (instantiated in specific neural circuits) have been bolted on to the ancestral brain. The other is that the differences between species with smaller and larger brains is not qualitative at the cognitive level, but strictly quantitative: in simple terms, they just have a bigger computer. In this view, what appears to us as distinctive psychological modules are really just the emergent properties of being able to execute basic universal cognitive processes on a larger scale. I am philosophically neutral with respect to these two alternatives; the issue will (and must) be decided empirically. However, it seems to me that there is at least a prima facie case for the second, and I will sketch this out below.

First, however, I will begin by saying something about the higher-order cognitive mechanisms that are likely to be involved, irrespective of which of these hypotheses is correct. There is a growing consensus, at least among those who work on primate cognition and human developmental psychology, that whatever processes might be involved at the cognitive level in the primates, these are likely to be distinctively social in nature; hence the distinction that is often made between cognition in general and *social* cognition in particular. In practice, we know almost nothing about social cognition in any of these species, except for one phenomenon that appears to be unique to humans and is usually referred to as *theory of mind*. Since we know very little about what is actually involved in social cognition other than theory of mind, it is not possible to give anything remotely resembling a complete account here. However, theory of mind does at least provide us with a sufficiently concrete example, and I will use it as a case study for the kinds of processes that might be involved.

Theory of mind is the capacity to understand the mental states of other individuals. Sometimes also known as *mindreading* or *mentalizing*, it depends critically on an appreciation that others have minds similar to one's own, and in particular that they can hold views that differ significantly from those one believes to be true. Children lack this capacity at birth, but over a period of time, beginning in late infancy, they gradually build their way toward fully fledged theory of mind at around age 4–5 years (Astington, 1993; Tomasello, 2001). In this sense, the development of theory of mind can be seen as the progressive scaffolding of levels of understanding through experience and practice (as well, perhaps, as some level of natural neurogenic development) that build toward a qualitative shift when theory of mind is finally achieved. Theory of mind thus appears as the end point of a developmental process that may involve a complex interaction between biological hard-wiring and culturally entrained experience (Tomasello, 2001).

Philosophers of mind introduced an alternative way of casting mindreading in this sense as a reflexive hierarchy of intentional states. Intentionality, in the technical philosophical sense, is the term used to cover those mind states that we usually describe by words such as *intend*, *think*, *believe*, *suppose*, *know*, and so on; the intentional stance represents the capacity to reflect on one's own mind states. This way of viewing theory of mind provides us with a natural hierarchy in which theory of mind itself is equivalent to second-order intentionality. An organism that has first-order intentionality is capable of reflecting on its own mind (it knows something to be the case), but a second-order-intentional organism is capable of reflecting on another individual's mind state (it knows that another individual believes something to be the case). This leads to a natural hierarchy that is, in principle, infinitely reflexive: I *believe* that you *suppose* that I *think* that you *intend* that I *understand* that you *know*. . . .

That children acquire theory of mind (second-order intentionality) at age 4–5 years is not of itself especially worthy of comment. It undoubtedly has lots of implications for their capacity to operate in the human social world: the acquisition of theory of mind, for example, seems to be intimately associated with the development of the ability to engage in fictional (or pretend) play and to lie in such a way as to influence another individual's beliefs about the world. But aside from this, theory of mind is perhaps more interesting for two separate reasons. One is whether other species share this capacity with us; the other is how children's competences at age 4 compare with those of adults. Both questions remain surprisingly understudied.

Very young children can be said to be first-order-intentional agents: they understand the contents of their own minds (at least, at some point fairly early on, even if this is unlikely to be true at or in the months immediately following birth). The transition to second-order intentionality at around age 4 years or so marks a significant rubicon in the social development of the child, because the child then begins to engage interactively in the adult social world in a way that he or she has not really been capable of doing hitherto. This inevitably raises the question of whether humans are alone in this respect or whether other species share with us these important social cognitive capacities. It has not been easy to design appropriate tests for theory of mind for use with animals: developmental psychologists have argued that the only certain assay for theory of mind is the false belief task (understanding that another individual holds a belief that you believe to be false) because such a task can only be successfully solved by an organism that can aspire to second-order intentionality.

There have been two published attempts to use false belief tasks with chimpanzees and one as yet unpublished study with dolphins; all three used a design that was first benchmarked against children. Of the two chimpanzee studies, one (Tomasello & Call, 1997) yielded unequivocally negative results, but the other (using a slightly different design: O'Connell & Dunbar, 2003) obtained modestly positive results (chimpanzees did about as well as 4-year-old children on the threshold of acquiring theory of mind and significantly better than autistic[3] adults, although much worse than 5- and 6-year-old children, whose theory-of-mind capacities are beyond doubt).

---

3 Lack of theory of mind is now considered to be one of the defining clinical features of autism.

There is, however, a growing consensus that the kinds of tasks used with the chimpanzees might be too opaque for them to solve even if they did have fully fledged theory of mind. In other words, chimpanzees might not understand the point of the task because it invariably involves cooperative interactions with humans, something that chimpanzees find hard to do even with their own kind, never mind with another species. A more chimpocentric task might therefore yield more promising results. This has led to a shift in emphasis in more recent studies to tasks that involve interactions with other chimpanzees and that focus on competition rather than cooperation (an emphasis more in line with everyday chimpanzee behavior and experience). These studies have focused on knowledge states rather than beliefs, using paradigms such as the *guesser-knower* distinction and the *seeing implies knowing* paradigm. Studies using this kind of design have been relatively successful: chimpanzees do appreciate that another individual can have a different perspective on a situation (e.g., because they see a food source from a different side) and readily adjust their behavior accordingly (e.g., by withdrawing if they know a more dominant animal can also see where a food item is hidden, or acting rapidly to take it if they think that it probably has not: Hare, Call, Agnetta, & Tomasello, 2000; Hare, Call, & Tomasello, 2001). Developmental psychologists would probably insist that these experiments do not test explicitly for false belief (the only criterion that they are really prepared to accept, perhaps not entirely without justification, as unequivocal evidence for theory of mind). Nonetheless, even if these studies do not confirm that great apes have theory of mind, they do at least point to a level of social skills based on understanding that is quite sophisticated.

This position contrasts markedly with the consensus on all other species (including monkeys). In this case, the view would be fairly uncompromising: no nonape species, primate or otherwise, can aspire to anything more than first-order intentionality. This conclusion has, in some sense, been borne out by our as yet unpublished study of dolphins. After initially promising results, a tighter experimental design failed to show any evidence for theory-of-mind capacity in dolphins on a false belief task based on a design that 4-year-old children pass with ease. Of course, we might express the same concern as that noted above for the great apes, namely, that the point of the task (based on the same design as used by Call and Tomasello [1999] with chimpanzees) was not sufficiently clear to the animals (and indeed, this view was expressed by the dolphins' handlers). At present, all we can say is that chimpanzees seem to do rather better than dolphins on mentalizing tasks, a conclusion that is at least in line with observed differences between them in neocortex volume and organization.

In concluding this discussion of the social cognitive capacities of animals, one caveat is perhaps in order. Even though both monkeys and autistic humans lack theory of mind, it is clear that monkeys and autistic humans differ significantly in their respective social competences. This would seem to imply that there is something more to primate sociality than merely the ability to apply theory of mind to false belief tasks. I will return to this point later. First, however, we have to deal with the second question raised above, namely, how do children's competences on theory of mind compare with those of adults?

The hierarchical structure of intentionality provides us with a natural metric that we can use to explore the developmental sequence that takes place after age 4 years. A number of studies have examined adult performance on multilevel (or "advanced") theory-of-mind tasks, and these all indicate that (1) adults can work at higher levels of intentional reflexivity than children and (2) the limit for most people seems to lie at fifth-order intentionality (I *believe* that you *suppose* that I *imagine* that you *expect* me to *believe* that something is the case) (Kinderman et al., 1998;[4] Stiller & Dunbar, in press). Henzi et al. (submitted) have shown that there is a natural developmental progression in achievable level of intentionality (i.e., the level at which individuals typically start to get the answer wrong) across the first decade or so of life, such that by the early teenage years, children are performing at adult levels. A number of other studies have demonstrated that clinical conditions such as schizophrenia and bipolar disorder are associated with loss of higher levels of theory of mind in adults (Kerr, Dunbar & Bentall, 2003; Swarbrick, 2000). Kinderman et al. (1998) provided some evidence to suggest that mentalizing tasks were more demanding (in a computational sense) than comparable problems involving causal sequences about the observable physical world: the same subjects show no tendency for success rates to fall on causal chains that were at least one order longer than those at which they failed on mentalizing tasks.

The fact that human adults are so much more competent in these terms than either young children or great apes raises the question of why

---

4 Kinderman et al. (1998) state that limit at fourth-order intentionality; however, they did not include the subject's own perspective or mind state in the calculation. Including the subject adds one extra level.

humans should need such computationally demanding capabilities. This is particularly pertinent because there is some evidence to suggest that the achievable level of intentionality might be linearly related to relative frontal lobe volume (but no other measure of brain volume) across the sequence monkey-ape-human (Dunbar, 2003). While this result must be regarded as preliminary until confirmed by more extensive data, it does imply that social cognitive abilities such as mentalizing may be genuinely expensive in neural wetware terms. Once again, we are forced to ask what could be so advantageous about these capacities as to promote their evolution. The fact that they appear to be explicitly related to the social world suggests that this selective advantage lies in some aspect of sociality (in effect, group size and maintaining the coherence through time of large numbers of relationships). We have very little idea what this might be—indeed, we have very little idea how even theory of mind is used in everyday social interactions (Roth & Leslie, 1998). However, Stiller and Dunbar (submitted) have shown that individuals' performance on multilevel intentionality tasks does correlate with the size of the primary social network (their sympathy group with whom they have most of their regular social interaction, typically around 12–15 individuals in size).

Let me now return to the question I raised at the start of this section, namely, whether these mentalizing capacities represent qualitative differences (in effect, are modular in the conventional sense) or quantitative (in other words, an emergent property of the scale at which certain fundamental cognitive processes can be brought to bear on a task). Either option is perfectly possible, since both in principle imply a requirement for additional wetware. Consequently, nothing about the relationships between brain volumes and social group size (or any other aspect of social behavior) allows us to discriminate between these two hypotheses. Nonetheless, the distinction does represent two camps in the child development literature, one of which argues that theory of mind is a module that more or less springs into action at a certain point in child development (see, e.g., Baron-Cohen, Leslie, & Frith, 1985; Leslie, 1987; Perner, 1991), while the other argues that theory of mind is in reality just a product of executive (or frontal lobe) function whose emergence is governed by natural brain growth and the child's developing competences in this respect (see, e.g., Mitchell, 1997; Ozonoff, 1995).

Barrett, Henzi, and Dunbar (2003) have suggested that social cognitive abilities in primates in general (including humans) can be understood as being the outcome of the scale at which core basic (and hence universal) cognitive abilities can be recruited in problem-solving tasks, particularly in the social domain. These cognitive abilities are likely to include things like causal reasoning, analogical reasoning, the temporal scale across which events can be forecast, the capacity to understand perspective, and the extent to which two or more scenarios can be compared directly in the mind, but others may also be involved. What may be socially specialized about these abilities may thus not be the fact that they depend on some purpose-specific module responsive only (or mainly) to social contexts but that there is a kind of social affordance that triggers the engagement of these more generalized cognitive processes. Barrett et al. argued that theory of mind (as a recognizable phenomenon) pops out when all of these can be brought to bear at the same time on a large enough scale. The extent to which this can be done, they suggest, is a simple consequence of the size of the computer available. Computer size (in this crude sense) might be synonymous with total brain volume, neocortex volume, or frontal lobe volume (or the size of any other brain component or combination of components).

Our capacity to decide between alternative explanations is severely limited by the lack of comparative experimental data from key species. However, we can perhaps ask whether the socially cognitive skills of primates in general, and humans in particular, depend on any particular neural circuitry within the brain. The executive function model of mentalizing, for example, would imply that frontal lobe volume may be critical, and this would receive some support from the tentative correlation between achievable level of intentionality and frontal lobe volume in primates (see Dunbar, 2003). Some additional support for this position would be the finding that, across the hominoids (great apes and humans), there is a linear relationship between frontal lobe volume (and only frontal lobe volume) and at least one core cognitive capacity (the speed with which tasks can be solved; Dunbar, McAdam, & O'Connell, submitted).

The role of the frontal lobes has, of course, been the subject of intense interest in clinical neuropsychology, although the advent of brain imaging has by no means helped to clarify what remains a very opaque if important brain region (Stuss et al., 1994). There is now fairly general agreement in the neuropsychology literature that damage to the frontal lobes does not have a significant effect on most conventional cognitive functions (sensorimotor integration, attention, learning, memory, even aspects of IQ) but does influence processes such as impulsivity, behavior sequencing on complex tasks, hypothesis formation, and, in particular, social skills (Anderson, Bechara, Damasio, Tranel, & Damasio, 1999; Stuss et al., 1994; see also Kolb & Wishaw, 1996). However, clinical interest in social skills has so far been rather unsophisticated (some might say absent altogether), and a great deal more work needs to be done in this area in particular.

Aside from the perhaps rather obvious case of the frontal lobes, there has in recent years been some interest in the role that other brain regions might play in the advanced social capacities exhibited by humans and, to a lesser but still interesting extent, the great apes. Three regions of the brain have received particular attention: the anterior cingulate cortex (ACC), the amygdala (the latter because of the significance of emotional responses in managing social exchanges), and the cerebellum. In addition, the recent discovery of a novel cell type (spindle cells) that appears to be unique to the hominoids (great apes and humans) (Nimchinsky et al., 1999) has attracted considerable interest, not least because they occur at unusually high densities in the ACC.

The ACC has received particular attention because it seems to be involved in a number of cognitive processes that involve behavioral and neural conflict monitoring. Neuroanatomical evidence from monkeys suggests that the ACC has a strong dopaminergic projection and receives input from the amygdala (though this may be specific to negative, fear-related stimuli), and is particularly active in a wide variety of contexts involving learning, error recognition, and tasks requiring focused concentration (Allman, Hakeem, Erwin, Nimchinsky, & Hop, 2001). More interestingly, Posner and Rothbart (1998) have suggested that the ACC may play a particularly important role in the process of self-control. They point out that in individuals that lack self-control (those with ADHD syndrome), the ACC does not exhibit its normal response in tasks that require a subject to separate out two different kinds of stimuli (e.g., the counting Stroop task, in which the subject has to discount the meaning of the word three in order to say how many times the word actually appears on a screen). This finding is reinforced by evidence that the postnatal development of spindle cells mirrors the acquisition of self-control in young children. Similarly, chimpanzees have been shown to be poor at self-control (they are invariably distracted by the presence of a large reward such that they will persistently disadvantage themselves on choice tasks: Boysen & Berntson, 1995) and this might be seen as a functional consequence of the difference between humans and great apes in ACC volume. However, there is as yet no evidence for an equivalent difference in behavior between monkeys and apes.

The discovery of spindle cells in the ACC of great apes and humans but not monkeys (Allman et al., 2001) has aroused considerable interest in the role that these cells may play in the uniquely complex cognition and behavior of the hominoids. Of particular relevance to our concerns here is the fact that Allman et al. (2001) reported that spindle cell volume is linearly related to relative brain size (indexed as Jerison's EQ) across the hominoids (all four species of great apes, plus humans). Allman et al. suggest that the high volume of spindle cells is related to high levels of axonal arborization, suggesting widespread connections throughout the brain. However, important as spindle cells must be, their small absolute volume means that cannot alone explain why hominoids have larger neocortices than monkeys, nor can it explain why there are grade shifts in the relationship between neocortex volume and all social indices so far examined.

The amygdala would seem to be an obvious candidate for a subcortical unit that might be heavily involved in the social brain. Indeed, this suggestion receives added weight from the fact that lack of affect and inappropriate emotional response are key diagnostic features of autism and seem to correlate with amygdala neurological abnormality in autistic individuals (Bauman & Kemper, 1988). Joffe and Dunbar (1997) plotted data on amygdala volume in primates against social group size but found no relationship; they argued that this suggests that it is not sensitivity to emotional cues per se that is responsible for the social brain effect (the correlation between neocortex volume and social group size) but rather the fact that the latter is underpinned by some kind of genuinely high-level cognitive processing. Hence, although part of the input to these higher social cognition processes may involve emotional cues via the amygdala, this cannot on its own explain the correlation between neocortex volume and social group size.

Subsequently, Emery and Perrett (2000) questioned this suggestion, arguing for a more central role for the amygdala (and hence emotional sensitivity) on the ground that there was a correlation between the amygdala (in particular the basolateral lobe) and social group size. Some support for this suggestion is offered by the fact that there is a rich array of connections between the amygdala (and the basolateral lobe, in particular) and the frontal lobe of the the neocortex, suggesting close functional integration between these areas of the brain. However, their results in fact offer little support for their claim: in the main, absolute and relative volume of the amygdala and its components either did not correlate with group size or did so negatively, and what correlations there were disappeared when they controlled for the volume of the neocortex. The only significant positive correlation was between group size and the basolateral nuclei of the amygdala when this was residualized against the rest of the amygdala. Thus, there is a relationship between group size and the amygdala, but it concerns the proportional distribution of tissue within the amygdala, not the amygdala as a whole. Their results are thus in line with more detailed analyses showing no correlation between social group size and the volume of either the amygdala (Joffe & Dunbar, submitted) or the "emotional brain" (hypothalamus + septum: Keverne, Martel, & Nevison, 1996) in primates. What Emery and Perrett (2000) appear to have shown is not that the amygdala is a major player in the social brain process but that, within the amygdala taken in isolation, one lobe makes a particularly important contribution to the social brain process—which is rather a different claim.

The cerebellum has also attracted attention in the context of the social brain, again because there are important neurological pathways connecting at least some of its components (principally its lateral lobes) to the frontal lobes of the neocortex (McLeod, 2004; McLeod, Zilles, Schleicher, Rilling, & Gibson, 2003; Whiting & Barton, 2003). While the cerebellum has been found to play a role in a wider range of cognitive functions than had previously been supposed, most of these functions in fact seem to be concerned with motor coordination in various guises (see McLeod, 2004), and it is by no means clear what role it plays (if any) in complex social interaction. Moreover, more detailed analysis has failed to reveal any statistical relationship between social group size and relative cerebellum size. Similarly, although the cerebellum is relatively large in humans compared both to great apes and monkeys (reflecting the demands of bipedalism?), cerebellar volume is just not large enough to account for the larger (relative or absolute) size of some species's brains.

This is not to deny that any of these subcortical units play a role in social cognition; they almost certainly do. Rather, it is to say that the explanation for the remarkably sophisticated social cognition of primates (and their unusually large brains) is unlikely to lie in any of these specific neural units. Attempts to tie the social brain phenomenon (and particularly the differences between hominoids and other primates) down to particular neural units (in particular, small neural complexes) seems at best naive. For one thing, it hardly makes sense to explain the massive increases in the size of the brain as a whole during the course of primate evolution by reference to what are, in fact, extremely small neural components, especially given the fact that primate brain expansion is driven mainly, as Finlay and Darlington (1995) have noted, by a disproportionate increase in the size of the neocortex. In addition, the fact that the correlations with group size seem to involve the whole neocortex suggests that the real issue is one of the integration of sensory information across a wide range of inputs (including memory of past events) and the ability to manipulate these in real time in a mental virtual world, a phenomenon that is likely to involve massive parallel processing that is dependent more on the scale of the computational machinery than on any single component.

A neurologically more plausible explanation may thus be that these social cognitive abilities are a function of total computer size (i.e., the amount of spare neural capacity that can be brought to bear on a task). Two recent comparative studies provide evidence (albeit indirect) suggesting that it may be gross neocortical volume that is the critical issue. One (Joffe & Dunbar, submitted) found evidence for increasing integration across brain units in the evolutionary sequence from insectivores through prosimians to anthropoid primates. A factor analysis of the coevolution of brain unit volumes across the insectivores yielded five core dimensions, which were reduced to four in the prosimians and two in the anthropoids. Since convergence on the neocortex seemed to be the key factor in this progressive reduction in the number of dimensions required to account for the observed variance in brain component volumes, this could be interpreted as evidence that mentalizing capacities reflect the ability to focus a number of cognitive processes on the same task. The second study (Bush & Allman, 2004) showed that, in primates, social group size does not correlate with the proportion of gray matter in the frontal lobe (that this, the ratio of gray matter in the frontal lobe to that in the rest of the neocortex). Although they did not test whether frontal or nonfrontal gray matter volume correlates with social group size (it does: Dunbar, submitted), their analysis does support the conclusion that it is total neocortex volume that is critical, not the volume of the neocortex in any one region. The only region that can definitely be excluded is the primary visual cortex (area V1 in the occipital lobe): Joffe and Dunbar (1997) demonstrated that social group size in primates does not correlate with V1 volume but does correlate with the volume of the rest of the neocortex (i.e., total neocortex minus V1). This makes some evolutionary sense, because the brain has evolved from back (occiput) to front.

Closing the Evolutionary Loop

In this final section, I return to the evolutionary question and ask why it is that there are such core differences in cognitive and social abilities even within the primates. In part, this question is prompted by the fact that the relationship between indices of social complexity (for example, group size) and indices of brain volume is not a simple one. Although all species seem to lie on the same slope, different taxonomic groups within the primates (prosimians, monkeys, apes) lie on slopes that are displaced from one another. In other words, there are very striking grade shifts (regression lines whose intercepts differ significantly) within these patterns. The most conspicuous of these are the separation of the primates into three separate and distinct grades on the group size/neocortex size graph: prosimians lie to the left of the simians and the hominoids (including both gibbons and humans) lie to the right (Dunbar, 1993, 1998, 2003). This pattern is evident in all the different analyses that have so far been carried out, irrespective of what behavioral index or what index of brain volume is used.

The implication of this finding is that some species have to work harder to support social groups of a given size than others do. There appears to be something about the computational demands of hominoid social life that requires apes (and humans) to need a significantly larger neocortex to support groups of a size that monkeys (both of the New and Old World varieties) can support with significantly less. Since we know that this cannot simply be a memory problem, it implies that it must have something to do with either the nature of the relationships involved or how the cognition involved handles these tasks.

It is clear from the comparative analyses that the grade shift must have occurred after the time at which the Old World monkeys split off from the hominoid lineage (around 23 mya) but before the gibbons branched off from the great ape lineage (around 16 mya). This much is clear from the fact that the gibbons consistently lie with the great apes rather than with the monkeys in all these analyses. We now know that at least one relevant genetic event did occur during this interval: a retrotransposition onto the X chromosome of a widely expressed housekeeping gene, *GLUD1*, gave rise to a new mutation (Burki &

Kaessmann, 2004). The gene's function is to mop up the neurotransmitter glutamate after neurons have fired, so preparing the way for the neuron to be able to fire again (it cannot do so while swamped with glutamate from a previous event). The new mutation, *GLUD2*, has additional amino acids that facilitate glutamate removal. This mutation is present only in hominoids, and Burki and Kaessmann (2004) argue that it may play an important role in facilitating the greater cognitive abilities of hominoids compared to those of simians by allowing shorter neuron recovery times after firing.

But once again, we are drawn back to the functional question of why this should have been selected for in the hominoid lineage. Barrett et al. (2003) have suggested that the only defining feature of hominoids as a group relative to simians is that great ape (and human) social groups tend to be dispersed over much larger geographical areas than simian groups (which tend to be spatially more compact). They suggest that the computational demands of having to factor virtual individuals (those not physically present) into their calculations about relationships with individuals that are physically present may have been the trigger that precipitated the hominoid trajectory. This needs to be seen in the context of hominoid and simian ecology. While both groups rely extensively on fruits as a dietary source, apes and monkeys differ in one key respect: apes lack a key gene that, in monkeys, produces an enzyme that allows monkeys to deal with the toxins (mainly trypsin and other phenolic compounds) in unripe fruits (Andrews, 1981). (Ripening is associated with the leaching out of these toxins so as to make the fruit palatable to seed dispersers: see Cowlishaw & Dunbar, 2000.) Because of this, apes (including humans) find unripe fruits bitter and are unable to digest them. Monkeys are thus able to exploit fruit crops long before apes can, and this competitive edge may in part explain why apes have been in terminal decline since the rise of the Old World monkeys during the Miocene (Cowlishaw & Dunbar, 2000; Fleagle, 1999). The combination of larger body size (and hence larger absolute energy demands) and the inability to exploit fruits crops until they are fully ripe means that apes are forced to disperse over wider areas than monkeys.

Other hypotheses can, of course, be envisaged. One of these is that the larger brains/neocortices of hominoids are associated with solving the ecological problems involved in dispersed foraging regimes (principally, the location of temporally ephemeral fruit patches round a large spatial area, and how animals should disperse themselves around these so as to minimize ecological competition and social conflict resulting from crowding in small foraging patches). This view has been advocated by Milton (1988, 2000), who has resolutely argued that brain size evolution within the primates generally (and the hominoids in particular) is a consequence of the need to solve ecological problems associated with dispersed food supplies. Note the contrast between this view and that proposed by Barrett et al. (2003), who have argued that the critical selection pressure is the need to maintain social cohesion (principally to ensure cooperative defence against predators) when animals are forced to live in dispersed social groups. Milton's view emphasizes the primacy of ecological considerations, that of Barrett et al. the primacy of social considerations. Clearly, the fact that there is no relationship between neocortical volume and any of the ecological variables tested (notably degree of dietary frugivory) tends to militate rather strongly against this particular version of the ecological hypothesis.

Acknowledgments This work was supported by a British Academy Research Professorship.

References

Aiello, L. C., & Wheeler, P. (1995). The expensive tissue hypothesis: The brain and the digestive system in human evolution. *Current Anthropology*, *36*, 199–221.

Allman, J. M., Hakeem, A., Erwin, J. M., Nimchinsky, E., & Hop, P. (2001). The anterior cingulate cortex: The evolution of an interface between emotion and cognition. *Annals of the New York Academy of Sciences*, *935*, 107–117.

Anderson, S. W., Bechara, A., Damasio, H., Tranel, D., & Damasio, A. R. (1999). Impairment of social and moral behavior related to early damage in human prefrontal cortex. *Nature Neuroscience*, *2*, 1032–1037.

Andrews, P. (1981). Species diversity and diet in monkeys and apes during the Miocene. In C. Stringer (Ed.), *Aspects of human evolution* (pp. 25–62). London: Taylor & Francis.

Astington, J. W. (1993). *The child's discovery of the mind*. Cambridge: Cambridge University Press.

Baron-Cohen, S., Leslie, A. M., & Frith, U. (1985). Does the autistic child have a "theory of mind"? *Cognition*, *21*, 37–46.

Barrett, L., Henzi, S. P., & Dunbar, R. I. M. (2003). Primate cognition: From "what now?" to "what if?" *Trends in Cognitive Science*, *7*, 494–497.

Barton, R. A., & Dunbar, R. I. M. (1997). Evolution of the social brain. In A. Whiten & R. W. Byrne (Eds.), *Machiavellian intelligence II* (pp. 240–263). Cambridge: Cambridge University Press.

Bauman, M., & Kemper, T. L. (1988). Limbic and cerebellar abnormalities: Consistent findings in infantile autism. *Journal of Neuropathology and Experimental Neurology*, *47*, 369.

Boysen, S. T., & Berntson, G. G. (1995). Responses to quantity: Perceptual versus cognitive mechanisms in chimpanzees (*Pan troglodytes*). *Journal of Experimental Psychology: Animal Behavior Processes*, *21*, 82–86.

Burki, F., & Kaessmann, H. (2004). Birth and adaptive evolution of a hominoid gene that supports high neurotransmitter flux. *Nature Neuroscience*, *36*, 1061–1063.

Bush, E. C., & Allman, J. M. (2004). The scaling of frontal cortex in primates and carnivores. *Proceedings of the National Academy of Sciences, U.S.A.*, *101*, 3962–3966.

Byrne, R. W., & Corp, N. (2004). Neocortex size predicts deception rate in primates. *Proceedings, Biological Sciences/The Royal Society*, *271*, 1693–1699.

Byrne, R. W., & Whiten, A. (1988). *Machiavellian intelligence*. Oxford: Oxford University Press.

Call, J., & Tomasello, M. (1999). A nonverbal theory of mind test: The performance of children and apes. *Child Development*, *70*, 381–395.

Cowlishaw, G., & Dunbar, R. I. M. (2000). *Primate conservation biology*. Chicago: University of Chicago Press.

de Winter, W., & Oxnard, C. E. (2001). Evolutionary radiations and convergences in the structural organisation of mammalian brains. *Nature*, *409*, 710–714.

Deaner, R. O., Nunn, C. L., & van Schaik, C. P. (2000). Comparative tests of primate cognition: Different scaling methods produce different results. *Brain, Behavior and Evolution*, *55*, 44–52.

Dunbar, R. I. M. (1992). Neocortex size as a constraint on group size in primates. *Journal of Human Evolution*, *22*, 469–493.

Dunbar, R. I. M. (1993). Coevolution of neocortex size, group size and language in humans. *Behavioral and Brain Sciences*, *16*, 681–735.

Dunbar, R. I. M. (1995). Neocortex size and group size in primates: A test of the hypothesis. *Journal of Human Evolution*, *28*, 287–296.

Dunbar, R. I. M. (1998). The social brain hypothesis. *Evolutionary Anthropology*, *6*, 178–190.

Dunbar, R. I. M. (2003). Why are apes so smart? In P. Kappeler & M. Pereira (Eds.), *Primate life histories and socioecology* (pp. 285–298). Chicago: University of Chicago Press.

Dunbar, R. I. M., & Bever, J. (1998). Neocortex size predicts group size in carnivores and some insectivores. *Ethology*, *104*, 695–708.

Dunbar, R. I. M., McAdam, M. R., & O'Connell, S. (submitted 2005). Mental rehearsal in great apes and humans. *Behavioral Processes*, *69*, 323–330.

Emery, N. J., & Perrett, D. I. (2000). How can studies of the monkey brain help us understand "theory of mind" in autism in humans? In S. Baron-Cohen, H. Tager-Flusberg, & D. J. Cohen (Eds.), *Understanding other minds: Perspectives from developmental neuroscience* (pp. 274–305). Oxford: Oxford University Press.

Finlay, B. L., & Darlington, R. B. (1995). Linked regularities in the development and evolution of mammalian brains. *Science*, *268*, 1578–1584.

Finlay, B. L., Darlington, R. B., & Nicastro, N. (2001). Developmental structure in brain evolution. *Behavioral and Brain Sciences*, *24*, 263–308.

Fleagle, J. (1999). *Primate adaptation and evolution* (2nd ed.). New York: Academic Press.

Frith, C. (1996). Brain mechanisms for having a "theory of mind." *Journal of Psychopharmacology*, *10*, 9–16.

Gibson, K. R. (1990). Cognition, brain size and the extraction of embedded food resources. In J. G. Else & P. C. Lee (Eds.), *Primate ontogeny, cognition and social behaviour* (pp. 93–103). Cambridge: Cambridge University Press.

Gogtay, N., Giedd, J. N., Lusk, L., Hayashi, K. M., Greenstein, D., Vaituzis, A. C., Nugent, T. F., Herman, D. H., Clasen, L. S., Toga, A. W., Rapoport, J. L., & Thompson, P. M. (2004). Dynamic mapping of human cortical development during childhood through early adulthood. *Proceedings of the National Academy of Sciences, U.S.A.*, *101*, 8174–8179.

Hare, B., Call, J., Agnetta, B., & Tomasello, M. (2000). Chimpanzees know what conspecifics do and do not see. *Animal Behavior*, *59*, 771–785.

Hare, B., Call, J., & Tomasello, M. (2001). Do chimpanzees know what conspecifics know? *Animal Behavior*, *61*, 139–151.

Henzi, S. P., de Sousa Pereira, L. F., Barrett, L., Hawker-Bond, D., Stiller, J., & Dunbar, R. I. M. (submitted). Social cognition and developmental trends in the size of conversational cliques.

Humphrey, N. L. (1976). The social function of intellect. In P. P. G. Bateson & R. A. Hinde (Eds.), *Growing points in ethology* (pp. 303–317). Cambridge: Cambridge University Press.

Jerison, H. J. (1973). *Evolution of the brain and intelligence*. New York: Academic Press.

Joffe, T. H., & Dunbar, R. I. M. (1997). Visual and socio-cognitive information processing in primate brain evolution. *Proceedings of the Royal Society of London, Series B, Biological Sciences*, *264*, 1303–1307.

Joffe, T. H., & Dunbar, R. I. M. (submitted). Allometry and the organisation of insectivore and primate brains: Implications for encephalisation and brain evolution. *Journal of Human Evolution*.

Jolly, A. (1966). Lemur social behaviour and primate intelligence. *Science*, *153*, 501–506.

Kerr, N., Dunbar, R. I. M., & Bentall, R. (2003). Theory of mind deficits in bipolar affective disorder. *Psychology and Medicine*, *73*, 253–259.

Keverne, E. B., Martel, F. L., & Nevison, C. M. (1996). Primate brain evolution: Genetic and functional considerations. *Proceedings of the Royal Society of London, Series B, Biological Sciences*, *262*, 689–696.

Kinderman, P., Dunbar, R. I. M., & Bentall, R. P. (1998). Theory-of-mind deficits and causal attributions. *British Journal of Psychology*, *89*, 191–204.

Kolb, B., & Wishaw, I. Q. (1996). *Fundamentals of human neuropsychology*. San Francisco: Freeman.

Krebs, J. R., Sherry, D. F., Healy, S. D., Perry, V. H., & Vaccarino, A. L. (1989). Hippocampal specialization of food-storing birds. *Proceedings of the National Academy of Sciences, U.S.A.*, *86*, 1388–1392.

Kudo, H., & Dunbar, R. I. M. (2001). Neocortex size and social network size in primates. *Animal Behavior*, *62*, 711–722.

Leslie, A. M. (1987). Pretence and representation in infancy: The origins of theory of mind. *Psychological Review*, *94*, 84–106.

Lewis, K. (2001). A comparative study of primate play behaviour: Implications for the study of cognition. *Folia Primatologica*, *71*, 417–421.

Marino, L. (1996). What can dolphins tell us about primate evolution? *Evolutionary Anthropology*, *5*, 81–85.

Martin, R. D. (1981). Relative brain size and metabolic rate in terrestrial vertebrates. *Nature*, *293*, 57–60.

McLeod, C. E. (2004). What's in a brain? The question of a distinctive brain anatomy in great apes. In A. E. Russun & D. R. Begon (Eds.), *The evolution of thought: Evolutionary origins of great ape intelligence* (pp. 105–121). Cambridge: Cambridge University Press.

McLeod, C. E., Zilles, K., Schleicher, A., Rilling, J. K., & Gibson, K. R. (2003). Expansion of the neocerebellum in Hominoidea. *Journal of Human Evolution*, *44*, 401–429.

Milton, K. (1988). Foraging behavior and the evolution of primate intelligence. In R. Byrne & A. Whiten (Eds.), *Machiavellian intelligence* (pp. 285–305). Oxford: Oxford University Press.

Milton, K. (2000). Quo vadis? Tactics of food search and group movement in primate and other animals. In S. Boinski & P. Garbe (Eds.), *On the move: How and why animals travel in groups*. Chicago: University of Chicago Press.

Mitchell, P. (1997). *Introduction to theory of mind*. London: Arnold.

Nimchinsky, E. A., Gilissen, E., Allman, J. M., Perl, D. P., Erwin, J. M., & Hof, P. R. (1999). A neuronal morphologic type unique to humans and great apes. *Proceedings of the National Academy Sciences, U.S.A.*, *96*, 5268–5273.

O'Connell, S., & Dunbar, R. I. M. (2003). A test for comprehension of false belief in chimpanzees. *Evolution and Cognition*, *9*, 131–139.

Ozonoff, S. (1995). Executive functions in autism. In E. Schopler & G. B. Mesibov (Eds.), *Learning and cognition in austism* (pp. 199–218). New York: Plenum Press.

Pawłowski, B. P., Lowen, C. B., & Dunbar, R. I. M. (1998). Neocortex size, social skills and mating success in primates. *Behaviour*, *135*, 357–368.

Perner, J. (1991). *Understanding the representational mind*. Cambridge, MA: MIT Press.

Posner, M. I., & Rothbart, M. K. (1998). Attention, self-regulation, and consciousness. *Philosophical transactions of the Royal Society of London B*, *353*, 1915–1927.

Reader, S. M., & Laland, K. (2002). Social intelligence, innovation and advanced brain size in primates. *Proceedings of the National Academy of Sciences, U.S.A.*, *99*, 4436–4441.

Roth, & Leslie, A. M. (1998). Solving belief problems: Toward a task analysis. *Cognition*, *66*, 1–31.

Semendeferi, K., Damasio, H., Frank, R., & van Hoesen, G. W. (1997). The evolution of the frontal lobes: A volumetric analysis based on three-dimensional reconstructions of magnetic resonance scans of human and ape brains. *Journal of Human Evolution*, *32*, 375–388.

Stevens, C. F. (2001). An evolutionary scaling law for the primate visual system and its basis in cortical function. *Nature*, *411*, 193–195.

Stiller, J., & Dunbar, R. I. M. (in press). Perspective-taking and social network size in humans. *Social Networks*.

Stuss, D. T., Eskes, G. A., & Foster, J. K. (1994). Experimental neuropsychological studies of rontal lobe functions. In F. Boller & J. Grafman (Eds.), *Handbook of neuropsychology, Vol. 9* (pp. 149–185). Amsterdam: Elsevier.

Swarbrick, R. (2000). *A social cognitive model of paranoid delusions*. PhD thesis, University of Manchester.

Tinbergen (1963). On aims and methods of ethology. *Zeitschrift für Tierpsychologie*, *20*, 410–433.

Tomasello, M. (2001). *The cultural origins of human cognition*. Cambridge, MA: Harvard University Press.

Tomasello, M., & Call J. (1997). *Primate cognition*. Oxford, UK: Oxford University Press.

Whiting, B. A., & Barton, R. A. (2003). The evolution of the cortico-cerebellar complex in primates: Anatomical connections predict patterns of correlated evolution. *Journal of Human Evoution*, *44*, 3–10.

3 Introduction to Evolutionary Cognitive Neuroscience Methods

Shilpa Patel, Katie L. Rodak, Euginia Mamikonyan, Kyra Singh, and Steven M. Platek

Evolutionary cognitive neuroscience is the empirical integration of cognitive neuroscience and evolutionary psychology. The aim of the discipline is to provide an evolutionary framework for the investigation of brain-behavior relationships—in other words, to explicitly apply Darwinian theoretical understandings to the methodology used by cognitive neuroscientists (e.g., functional magnetic resonance imaging, electroencephalography, transcranial magnetic stimulation). To understand evolutionary cognitive neuroscience, it is imperative to become familiar with the history of antecedent methods and approaches. In this chapter we briefly review (and refer readers to more expansive sources) the key methodological approaches that have given rise to evolutionary cognitive neuroscience.

Introduction to Domain Specificity

The evolution of the brain has been a long process marked by modulation to accommodate new, specialized cognitive tasks, meaning that the mind has been broken into units that have the specific purpose of carrying out a cognitive process. In *Modularity of Mind*, Fodor expresses what it means for this evolved brain to function in modular systems for lower-level cognitive processes; this idea has been extrapolated by evolutionary psychologists to understanding the domain specificity of neural processing and architecture (e.g., Pinker, 2002; Tooby & Cosmides, 1992), which is hypothesized to be a set of hierarchically organized specialized information-processing mechanisms.

Memory, particularly declarative memory, is one such modulated process that is controlled primarily by the hippocampus and surrounding neural tissues of the medial temporal lobe. Declarative memory consists of facts and information gathered through learning. The hippocampus is critical in forming episodic memories and is involved in forming semantic memories. The CA3 area of the hippocampus is where declarative memories are thought to be formed. Convergent afferents carry inputs about stimuli from cortical regions to CA3. The principal neurons in this region then send projections to other CA3 principal cells. These principal cells produce weak activating inputs that, in conjunction with a strong, separate, but convergent activating input, surpass the threshold for activation of NMDA receptors. This causes long-term potentiation (LTP), or the formation of a memory. Since only one pathway is active at a time, the LTP is input-specific and mostly unidirectional (Bliss & Collingridge, 1993; Eichenbaum, 2004).

Concrete evidence for the role of the hippocampus in the encoding and retrieval of memories comes from case studies. Perhaps the most famous is that of H.M. H.M. underwent bilateral medial temporal lobe resection to relieve his epilepsy. Even though the posterior portion of the hippocampal formation remained, a 1997 study showed much of the tissue to have atrophied. Following the resection, H.M. had normal memory for semantic knowledge acquired prior to 1953, the year of onset of his amnesia, but a deficit in forming new memories requiring semantic knowledge. More recent studies of H.M. show that although his ability to form new semantic memories is far below that of controls, he still has some ability. He can sketch the floor plan of the house he lived in 5 years after he became amnesic, as well as pick fictitious names from among those who became famous after the onset of his amnesia. With another patient, E.P., Bayley and Squire (2002) demonstrated that the remaining ability for semantic memory of amnesic patients differs from that of normal subjects. E.P.'s newly acquired semantic information was rigidly categorized and not consciously available; Bayley and Squire showed it to be more like perceptual memory than declarative memory (Bayley & Squire, 2002; Eichenbaum, 2004; Gabrieli, Cohen, & Corkin, 1988; Schmolck, Kensinger, Corkin, & Squire, 2002).

Although the degree to which semantic memory is contingent on a working hippocampus is debatable, episodic memory almost certainly requires it. Episodic memory works through associative representations, sequential organization, and rational networks. Associative representations are constructed when two discrete events, places, or stimuli are linked together. The case of K.C. suggests just how important the hippocampus is to episodic memory. K.C. sustained severe injury to his medial temporal lobes in a motorcycle accident and almost complete bilateral hippocampal loss, as confirmed by magnetic resonance imaging

(MRI). K.C. had no past episodic memory, although he was able to learn some past personal knowledge through what might be thought as a semantic learning mechanism. He also could not form new long-term episodic memories. However, his new knowledge was usually inflexible and often could not be accessed in novel situations. K.C. did retain the ability to learn new information or skills, such as computer knowledge, through a method using "vanishing cues." This knowledge was gained either incidentally or in a highly controlled environment and was usually fragmented, or not associated with other knowledge (Eichenbaum, 2004; Rosenbaum et al., 2005).

It is the ability to create an association that sets the human brain apart from the animals in the evolutionary line. Each separate, distinct memory can be linked in a relational network with overlapping yet distinct features. Within the network are cells with firing patterns associated with one or an overlapping sequence; these cells fire only within the network and have no external inputs. Each time a memory with overlapping features is made it becomes part of the relational network. Many studies have found higher levels of hippocampal activation with the associations generated through relational networks than with independent events or entities, showing the critical need for the hippocampus in forming associations in humans. A few studies have even demonstrated that even if recognition of single items is not impaired, the recognition of associations is (Eichenbaum, 2004).

Another example of the modularity of the mind is the detection of cheaters in social contracts. For humans to have evolved in a group, they had to have engaged in social exchanges, defined by Cosmides as cooperation between two or more individuals for mutual benefit. An individual pays a cost to or meets a requirement benefiting another individual or group and in return receives a benefit. This would help significantly with hunting and protecting the young during the days as Pleistocene hunter-gatherers. Therefore those individual who were able to detect cheaters, those who did not pay but still benefited, were likely to have had higher fitness, and so cheater detection has evolved (Cosmides, 1989; Gigerenzer & Hug, 1992).

There is research to support the theory that the processing of social information is distinct from the processing of other types of information and that the human mind exhibits processes specialized for detecting cheaters. Wason's four-card selection task is most often used to test for this theory. Participants are given a rule, if *P* then *Q*, and must turn over the smallest number of cards to find out if the rule is being violated. By far, participants give more correct answers when the test deals with a social contract than when it describes something else. Stone et al. place the percentages at 65%–80% correct for social contracts and 5%–30% correct for some state of the world (Gigerenzer & Hug, 1992; Stone, Cosmides, Tooby, Kroll, & Knight, 2002).

The area of the brain for processing social contracts was discovered in part from the case of R.M. R.M. sustained extensive bilateral limbic system damage to the medial orbitofrontal cortex and the anterior temporal cortex after a bicycle accident. Damage to the anterior temporal cortex was severe, such that the right amygdala was disconnected from the left. R.M. underwent a series of tests, one of which was a set of 65 reasoning problems using the Wason selection test. He made significantly more errors on problems of social contract than on problems involving precautions or descriptive rules. Similar results have been obtained in other patients with bilateral amygdala lesions. Two patients with lesions similar though different in volume to that of R.M. performed similarly on the same tasks. A study by Adolphs, Tranel, and Damasio (1998) on three subjects with complete bilateral amygdala damage tested for accurate social judgment of others based on facial appearance for approachability and trustworthiness. The patient subjects were more apt to judge strangers to be approachable and trustworthy than were the control subjects, even for the faces the control subjects rated most negative on approachability and trustworthiness. These findings clearly indicate that the amygdala and orbitofrontal cortex are the components used for decifering social contracts (Adolphs et al., 1998; Stone et al., 2002).

Although there is no question that many functions of the brain show modularity of the mind, in some instances many modules working together incorrectly can create a neurological disorder such as autism. Autism (Baron-Cohen, 2005; see also Chapter 17, this volume) is a neurodevelopmental disorder that usually manifests at the toddler age. Although there is no cut-and-dried list of symptoms, social, behavioral, cognitive, sensorimotor, and communicative defects are usually present in some form.

Many brain studies have been done on autistic individuals in an attempt to better understand the disorder. Some of the anatomical abnormalities have been found in the cerebellum, brainstem, hippocampus, and frontal and parietal lobes, as well as the amygdala. The abnormalities in these regions may help to explain many of the individual symptoms that together constitute autism. The abnormalities may in part be caused by growth rates. Although they are born with normal to small brain size, infants who later develop autism experience excessive brain growth for a short time, followed by years of reduced brain growth.

The frontal lobe is particularly affected by the dysregulation of brain growth. This region typically undergoes growth and selection much longer than other brain areas. The microcolumns and surrounding neuropil space are significantly smaller in volume and contain far greater amounts of gray matter than usual, reflecting poor regulation of neurogenesis and apoptosis. Asymmetry between the frontal lobes and temporal lobes has been found in subjects with language impairment. Empathising tasks have shown reduced activity in the left medial frontal cortex and orbitofrontal cortex. Functional studies usually find abnormal function in areas pertinent to theory of mind, memory, attention, and language tasks.

Not only does growth dysregulation affect the frontal cortex, the excessive growth is also detrimental to the Purkinje cells and granular cells found in the cerebellum. On postmortem examinations, loss of cerebellar Purkinje cells was reported in 95% of cases of autism. The low numbers of Purkinje cells may lead to disinhibition of deep cerebellar nuclei and overexcitement of the thalamus, which has a smaller volume in relation to total brain volume than is found in unaffected people, and cerebral cortex. Many MRI and postmortem studies have shown a reduction in the size of the vermis, specifically lobules VI to VII. A smaller ratio of gray to white cerebellar matter has also been reported. Cerebellar lesioning studies in both humans and animals have shown cognitive and behavioral impairment. The lateral posterior cerebellum is connected to the prefrontal cortex region. This connection, involved in language and other executive functions, is now impaired. Cerebellar dysfunction can explain, in part, deficits in language, cognition, and emotion.

Dysregulation of brain growth also affects the limbic system. The hippocampus, amygdala, and other areas of the limbic system have shown reduced neuronal cell size and cell density. The amygdala was reported to have the densest packing of neurons, particularly in the medial section. The arbors of pyramidal cells also are decreased. These regions of the limbic system are involved in memory, social, and affective functions.

Autism cannot be solely attributed to deficits in any single region. Rather, it reflects the compilation of deficits in all regions. Because the dysregulation of brain growth occurs at such a young age, it affects not only the particular brain region but also the connections to other regions of the brain. Therefore, deficits are not limited to the cerebral, cerebellar, thalamic, and limbic areas but affect the connections between these regions and the rest of the brain. For this reason it is impossible to show that autism represents a modularity of mind; rather, each module can be construed as a building block, and all of the blocks together produce autism.

Ethology: Naturalistic Observation of Behavior

The field of ethology, a branch of zoology, developed under the influence of strict behaviorism. Behaviorism focuses on observable behavior and discounts the importance of internal events. Much like behaviorists, ethologists focus on observing an animal's outward behaviors. However, in contrast to strict behaviorists, they are also concerned to document species-specific behaviors that result from the animal's genetic programming (e.g., instinct). In other words, ethologists do not discount the application of evolutionary metatheory to the understanding of behavior. Ethologists primarily observe behaviors under ecologically valid conditions (in the animal's natural environment) and attempt to understand the complexities of the behaviors from an evolutionary perspective. Behaviors that have been classically studied by ethologists include aggression, communication, migration, parent-offspring interaction, mating, and territoriality. This approach allows scientists to comprehend how species-specific behaviors allow animals to survive in their environments without the manipulative aspects of a laboratory setting.

Ethological observations yield a great deal of information about the evolution of behaviors and their neuroscientific implications. The monumental research of Karol von Frisch, Konrad Lorenz, and Nikolaas Tinbergen, widely regarded as the founders of the discipline, relied heavily on naturalistic observations. For these contributions, they shared the 1973 Nobel Prize in Medicine or Physiology. Their work led to the development of a subspecialty of ethology known as neuroethology, an innovative branch of science that seeks to identify the individual neurons responsible for animals' behaviors, or the neural correlates of naturalistic behavior.

Karol von Frisch began his ethological research career by observing fish engaging in species-specific behaviors. He discovered that goldfish (*Carassius auratus*) had the unique ability to hear (Smith et al., 2003). Von Frisch confirmed that their auditory acuity is far superior to that of humans. Following his initial success, he turned to observing honeybees (*Apis mellifera)* and ultimately discovered the dance of the honeybee—a landmark discovery (von Frisch, 1974). His research confirmed that *A. mellifera* communicate the location of food to their colonymates by the waggle dance, or the tremble dance, depending on the distance. Notably, *A. mellifera* engage in dancing behaviors only when they find a food supply (Dewsbury, 2003) ample enough to warrant the adjustment of the colony's labor efforts (Thom, 2003).

Konrad Lorenz's behavioral observations similarly contributed greatly to the development of the field of ethology. Lorenz's observations focused on the animal's instinctual behavior in its natural environment, primarily the graylag goose's. Among the processes discovered by Lorenz and his colleagues were such masterful concepts as the critical period, imprinting, and the fixed action pattern. Lorenz determined that the first few hours after hatching influence the behavioral development of graylag geese and other birds. He dubbed this time span "the critical period," the time during which imprinting, the rapid development of a genetically programmed response to a specific stimulus, occurs. In the critical period, the bird becomes irrevocably attached to the first moving entity it encounters and resorts to following behaviors that last a lifetime (Baker, 2001). Furthermore, Lorenz and Tinbergen, through careful observation, learned that graylag geese exhibit fixed action patterns, a series of movements triggered by a biological stimulus. If a graylag goose egg rolls out of the nest, the goose will invariably waddle to the egg and roll it back to the nest with her bill. However, even if the egg moves ever so slightly, the goose will still carry out all the movements as if the egg had rolled out of the nest. The egg's movement triggers an irreversible reaction in the goose that must be completed regardless of its necessity (Baker, 2004).

Nikolaas Tinbergen, influenced by von Frisch and a pupil of Konrad Lorenz, also made seminal contributions to the field of ethology by employing behavioral observation techniques. Tinbergen focused his efforts on observing the herring gull (*Larus argentatus*). His careful observations revealed that as soon as the young hatch, the presiding gull removes the remaining shell from the nest so that it does not attract predators to the nest. As soon as the gull's bill comes in contact with the jagged remains of the shell, it is stimulated to remove it immediately. Further observations revealed that newly hatched gulls instinctively, upon seeing their parent's bill, engage in begging behaviors, pecking at the adult's bill. In response, the parent, whether experienced or not, regurgitates food for the chick (Dewsbury, 2003).

The unique observations made by the brilliant leaders of the ethological revolution prompted questions regarding which brain structures, or more specifically which neurons, might be responsible for the recorded behaviors. Neuroethology is concerned with uncovering the neural mechanisms responsible for fixed action patterns, imprinting, instincts, and so on. Single-cell recording is a widely used method for studying the animal brain in action. This technique entails attaching recording devices to the living brain and noting action potentials from single nerve cells or a group of neurons in response to a behavior or stimulus. Other methods include electroencephalography (EEG) and brain lesions. In contrast to single-cell recording and EEG, lesions entail the physical removal, or ablation, of brain tissue or the severing of connections between two brain regions. These techniques are extremely useful in determining the exact neurological connections of innate release mechanisms. This allows for the isolation of single neurons or distinct brain areas that are responsible for fully integrated pieces of behavior (Pflüger & Menzel, 1999). As a result, scientists are able to comprehend the evolutionary practicality of brain function in animals.

The behavioral and neurological study of animals leads naturally to comparisons between humans and other species, which is the prime concern of comparative psychology. In animals and humans alike, topical behavior results from certain kinds of motivational states and its intensity and specific external stimulus (Klein, 2000). Comparisons between various species are made from an evolutionary perspective. For example, a comparative psychologist might compare social behaviors of humans with those of other species to determine how each species's unique behavioral program contributes to its survival and subsequent reproductive efforts. Furthermore, from a neurological perspective, the human brain may be compared with the brains of various species (with respect to size, organization, and so on). Scientists isolate the differences and similarities between the organs and hypothesize why differences exist and how they contribute to the survival of the particular species. For example, the human brain has a significantly more developed neocortex than the brain of other species. This may be a result of selection for cortical processing in humans that did not take place in other animals (Hawkins & Blakeslee, 2004). Selection of this sort might have led to the evolution of consciousness and intelligence (see later chapters in this volume).

Neuroethology has focused primarily on nonhuman species. This is probably the result of an inability to investigate neural mechanisms of humans under ecologically valid conditions. Recent discoveries in biomedical engineering, however, have made it possible to study the living, thinking human brain fairly easily.

Functional Neuroimaging: Peering into Living Brains

Neuroimaging is widely used throughout the comparative and cognitive neuroscience fields. Neuroimaging involves the use of several techniques to measure brain form and function. This approach allows investigation of the relationship between integrity, morphology, and activity in certain areas of the brain and the cognitive or behavioral functions associated with them. These techniques are beginning to be used as research methods in evolutionary cognitive neuroscience to compare brain function across species, sex, and other variables.

Neuroimaging includes several different kinds of studies. Positron emission tomography (PET) and functional magnetic resonance imaging (fMRI) are two methods commonly used to measure changes in blood oxygen utilization as a function of brain activity. Some studies have also used EEG and magnetoencephalography (MEG) or transcranial magnetic stimulation (TMS) to look at electrical and magnetic potentials associated with cognitive activity. The most recent imaging technique involves the use of diffuse optical technology, for example, functional near-infrared spectroscopy (fNIRS). In an attempt to shed light on the various neuroimaging procedures in the cognitive neurosciences, we provide a brief look at these popular methods.

PET is an important tool used especially in the clinical applications of evolutionary cognitive neuroscience. PET measures the decay of radioactive atoms that release positrons. Each positron removes a nearby electron, which results in the release of a pair of oppositely directed high-energy gamma rays. The joint detection of the pair of gamma rays on each side of the head provides the data for localization of the corresponding brain activation (Herholz & Heiss, 2004). PET is not limited to the study of neural activation but can also be used to study the physiology of brain function, including such processes as glucose metabolism, Krebs cycle function, and protein synthesis. PET can also be used to study blood flow, blood volume, and oxygen utilization as a function of hemodynamic changes associated with activity in the brain (Cherry & Phelps, 1996).

EEG and MEG are superlative tools to ensure that the brain is functioning correctly. EEG and MEG measure the electrical activity of neuronal cells. The source of these electrical signals is thought to be in the apical dendrites of pyramidal cells in the cerebral cortex. Simultaneous activation of a large number of pyramidal cells in small areas of cortex is known as an equivalent current dipole (ECD), which is found normally to its surface. The current dipole is therefore used as the basic element representing neural activation in EEG and MEG techniques. The activated pyramidal cells are found to lie within the gray matter in the cortex. EEG measures the potential differences on the scalp resulting from ohm currents induced by electrical activity in the brain. In this technique, a set of scalp electrodes is connected to amplifiers and a data acquisition system. Since the signals are produced by the ohm current flow, they are sensitive to the conductivity of the brain and surrounding areas. MEG, on the other hand, measures the magnetic field outside the head that is induced by the flow within the brain. This signal is dependent on the neural current generators or primary currents. The primary currents are localized to regions where the brain becomes active during evoked responses (Darvas, Pantazis, Kucukaltun-Yildirim, & Leahy, 2004). EEG and MEG are complementary modalities and are used to make simultaneous data analysis. In evolutionary cognitive neuroscience studies, the EEG/MEG method is ecologically valid in studies investigating brain function as a function of event-related potentials (Michel et al., 2004).

Our next topic of review is transcranial magnetic stimulation (TMS). The TMS technique is based on Faraday's principles of electromagnetic induction. A pulse of current flowing through a wire generates a magnetic field. This wire is held over the subject's head. A brief pulse of current is passed through the wire, and a magnetic field is formed that passes through the subject's head and skull. When the magnitude of the magnetic field changes, a nearby conductor induces a secondary current. The rate of change of the field determines the size of the current induced. Generally, single pulses of stimulation are applied in experiments. Each of them lasts about 100 μs. This pulse technique has minimal risk factors when used in healthy subjects. Other experiments call for a series of pulses at rates of up to 50 Hz to be applied. This pulse technique is considered a bit risky (Pascual-Leone, Walsh, & Rothwell, 2000).

Optical techniques have been used to monitor changes in cerebral oxygenation and metabolism for several decades. fMRI has been the gold standard in neuroimaging. fMRI measures hemodynamic changes that are associated with brain activity. MRI allows the creation of images of soft tissue without using the radiation in PET or x-ray images. Since MRI is widely available in the clinical settings of hospitals and laboratories, it is an attractive modality for brain mapping.

MRI consists of nuclear magnetic resonance (NMR). Magnetic fields are used to generate an NMR signal from the hydrogen nuclei of water molecules in the body. Specific magnetic fields are used to receive NMR signals from various areas in the three-dimensional parts of the body. This creates the MR image. MRI uses a pulse sequence, or a pattern of time and intensity weighting of the electromagnetic sources and sensors. Specific pulse sequences are used to create images exhibiting hemodynamic changes associated with neural activation (Savoy, 2000). This is known as fMRI.

In the last decade, fNIR has been introduced as a new neuroimaging modality for conducting functional brain imaging studies. Brain activity induces increases in local cerebral blood volume (CBV), blood flow (CBF), and blood/hemoglobin oxygenation. Thus, measurement of light absorption by means of fNIRS can record the changes in oxyhemoglobin and deoxyhemoglobin concentrations that occur during functional brain activation (Villinger & Chance, 1997). Oxyhemoglobin and deoxyhemoglobin have different absorption patterns of light in the near-infrared range. Changes in the amount of light reaching the detectors correspond to changes in light absorption and scattering. The probe is made up of the interface between the control system and the subject. It holds the light source and the detector. The light source and detectors are operated by the control circuit, which consists of a transmitter and a receiver. Computer software for detecting the two wavelengths controls the transmitter and receiver. The computer also stores and displays received light information after signal-processing schemes are applied. The spacing between the sources and the detectors is set at 3 cm to ensure penetration of about 2 cm from the scalp surface. Absolute changes in the concentrations of oxyhemoglobin, deoxyhemoglobin, and total hemoglobin are calculated and averaged over 1-second time periods (Izzetoglu et al., 2004).

Imaging plays an important role in evolutionary cognitive neuroscience approaches. Each method has unique characteristics, and together, these imaging studies can be applied to almost every question of interest in evolutionary cognitive neuroscience. Moreover, neuroimaging is beneficial to comparative psychology, and specifically to evolutionary cognitive neuroscience, because it allows neural function and form to be compared between sexes, across the menstrual cycle, across species, and so on. Evolutionary cognitive neuroscience, and human neuroethology, may be able to capitalize on optical techniques because they allow wireless, portable measurement of brain activity in humans under ecologically valid conditions.

Neurogenetics

For decades, investigators have studied the differences between humans and other species in an attempt to find the molecular changes responsible for these differences. Genetics of the human brain, or neurogenetics, is one approach used to try to clarify our understanding of the many genetic and evolutionary changes that distinguish human brains from that of our ancestors.

One type of neurogenetic method used to study the evolution of the nervous system entails calculating the K_a/K_s ratio, or the ratio of nonsynonymous to synonymous gene substitution. This ratio is calculated over a defined evolutionary period. Nonsynonymous substitutions basically involve changes in nucleotides of the coding regions in genes; these changes eventually alter encoded amino acids. Synonymous substitutions, on the other hand, do not alter encoded amino acids because they occur in degenerate positions of codons. Nonsynonymous changes are generally subject to selection because the changes modify the biochemical properties of the protein product, while synonymous substitutions do not, and thus are functionally neutral. Therefore, the K_a/K_s ratio is a calculation of the rate of the protein product of a gene that has evolved in relation to the expected rate with selective neutrality. If the ratio is relatively high, then theoretically the protein encoded by the gene evolved quickly, and if the ratio is low, then the protein encoded by the gene evolved slowly (Gilbert, Dobyns, & Lahn, 2005).

Gilbert et al. (2005) used this method to study differences between humans and other species. They calculated K_a/K_s ratios separately for primates (human and macaque) and rodents (rat and mouse). The K_a/K_s ratio was higher by about 30% in the primates compared with the rodents, indicating that the proteins encoded evolved faster in primates by nearly 30%. Gilbert et al. suggest that many genes are involved, rather than a few exceptional outliers. Additionally, genes that function primarily in the routine physiology and maintenance of the nervous system showed much less discrepancy between primates and rodents. This finding might indicate the evolution of the human nervous system, which can be correlated with accelerated evolution of the genes.

Of the 214 genes studied, the ones that showed higher K_a/K_s ratios were referred to as the primate fast outliers. These outliers seem to regulate brain size or behavior. Three of them are particularly important because the loss of their function can lead to mutations in humans and decreased brain size, known as microcephaly. These outliers are known as abnormal-like spindles (*asp*), microcephaly-associated (*ASPM*); microcephaly primary autorecessive 1, known as *microcephalin* (*MCPH1*); and *sonic hedgehog* (*SHH*) (Gilbert et al., 2005). Microcephaly is a developmental defect present at birth that is found in less than 2% of newborns, but it occurs in more than 400 genetic syndromes. The range of head sizes found with microcephaly is similar to the range of head sizes observed in our hominid ancestors, *Australopithecus*, and the present great apes (Gilbert et al., 2005).

The evolutionary geneticist Bruce Lahn and his colleagues (Evans et al., 2004) are studying the ongoing adaptive evolution of *ASPM*, which is a specific regulator of brain size. They report that its evolution in the lineage to *Homo sapiens* is based on a strong positive selection and that it might still be under selective pressure. The team found one genetic variant of *ASPM* that arose fairly recently (about 5,800 years ago). This suggests that the human brain is still undergoing adaptive evolution. Another gene they are investigating, microcephalin (*MCPH1*), regulates brain size. Microcephalin continues to evolve adaptively in humans, suggesting the ongoing plasticity of the human brain. Because of the very recent appearance of the genetic variant (less than 37,000 years ago), microcephalin is a good candidate for studying the genetics of human variation in brain-related phenotypes (Gilbert et al., 2005).

Some investigations of genetic selection have uncovered subsets of human genes that illustrate positive evolutionary selection; for example, mutant alleles of *FOXP2* can cause a severe disorder of articulation and speech. Changes in *FOXP2* can also be important in the evolution of language. It has been noted that *FOXP2* shares a common ancestral sequence in the human population, which suggests that evolutionary selection may be very recent as well (Hill & Walsh, 2005).

Christopher A. Walsh and his colleague, Anjen Chenn, suggest that the cerebral cortex of genetically engineered mice had a large brain surface area and a wrinkled appearance, similar to what is seen in humans. These researchers believe that the wrinkled appearance did not need any special genetic evolution and was a passive response when the brain was bigger than the head. To test their theory, they modified a gene, encoding a protein called beta-catenin. This protein regulates cell division in many tissues and in the brain progenitor cells, in which the nerve cells of the cerebral cortex originate. The resulting beta-catenin accumulation in the mutant mice increased the size of the cerebral cortex, resulting in brain sizes two to three times larger than normal (Walsh & Chenn, 2003). Because the mice did not survive past birth, more studies are needed to determine whether beta-catenin actually plays a role in the size of the brain (Travis, 2002).

Santiago Ramon y Cajal, widely regarded as the founder of modern neuroscience, recognized early on that the human brain is not only large but also different in circuitry from our ancestors' brains. The understanding of gene function with regard to evolutionary changes is just beginning, such that for each gene, and functional polymorphis in the genome, there is a human carrying a mutated allele gene. Many neurological diseases are processes with mutated genes. With further study of these mutated genes, researchers may have a new tool for investigating the recent evolutionary history of humans by means of different neurogenetic methods (Hill & Walsh, 2005).

Conclusion

Taken together, the approaches adopted by evolutionary cognitive neuroscience—the application of *human* neuroethological approaches, functional neuroimaging, and neurogenetics—allow researchers to investigate hypotheses generated from an evolutionary biological framework using modern methods. Without the theoretical guidance of evolutionary metatheory, however, cognitive neuroscience, like psychology, runs the grave risk of going awry, that is, of building an incorrect understanding of the processes of the brain, behavior, and their interactions.

References

Adolphs, R., Tranel, D., & Damasio, A. R. (1998). The human amygdala in social judgement. *Nature*, *393*, 470–474.

Baker, L. (2001). Learning and behavior: Biological, psychological, and sociological perspectives.

Baron-Cohen, S. (2005). The cognitive neuroscience of autism. *Journal of Neurology, Neurosurgery, and Psychiatry*, *75*, 945–948.

Bayley, P. J., & Squire, L. R. (2002). Medial temporal lobe amnesia: Gradual acquisition of factual information by nondeclarative memory. *Journal of Neuroscience*, *22*, 5741–5748.

Bliss, T. V. P., & Collingridge, G. L. (1993). A synaptic model of memory: Long-term potentiation in the hippocampus. *Nature*, *361*, 31–39.

Cherry, S., & Phelps, M. (1996). Imaging brain function with positron emission tomography. In *Brain mapping: The methods* (pp. 191–221). San Diego: Academic Press.

Cosmides, L. (1989). The logic of social exchange: Has natural selection shaped how humans reason? Studies with the Wason selection task. *Cognition*, *31*, 187–276.

Darvas, F., Pantazis, D., Kucukaltun-Yildirim, E., & Leahy, R. M. (2004). *NeuroImage*, *23*, S289–S299.

Dewsbury, D. A. (2003). The 1973 Nobel Prize for Physiology or Medicine. *American Psychologist*, *58*, 747–752.

Eichenbaum, H. (2004). Hippocampus: Cognitive processes and neural representations that underlie declarative memory. *Neuron*, *44*, 109–120.

Evans, P. D., Anderson, J. R., Vallender, E. J., Gilbert, S. L., Malcolm, C. M., Dorus, S., & Lahn, B. T. (2004). Adaptive evolution of ASPM, a major determinant of cerebral cortex size in humans. *Human Molecular Genetics*, *13*, 489–494.

Fodor, J. (1983). *The modularity of mind*. Cambridge, MA: MIT Press.

Gabrieli, J. D. E., Cohen, N. J., & Corkin, S. (1988). The impaired learning of semantic knowledge following bilateral medial temporal-lobe resection. *Brain and Cognition*, *7*, 157–177.

Gigerenzer, G., & Hug, K. (1992). Domain-specific reasoning: Social contracts, cheating, and perspective change. *Cognition*, *43*, 127–171.

Gilbert, S. L., Dobyns, W. B., & Lahn, B. T. (2005). Genetic links between brain development and brain evolution. *Nature Reviews: Genetics*, *6*(7), 581–590.

Hawkins, J., & Blakeslee, S. (2004). *On intelligence*. New York: Times Books.

Hill, R. S., & Walsh, C. A. (2005). Molecular insights into human brain evolution. *Nature*, *437*, 64–67.

Herholz, K., & Heiss, W. D. (2004). Positron emission tomography in clinical neurology. *Molecular Imaging and Biology*, *6*(4), 239–269.

Izzetoglu, M., Izzetoglu, K., Bunce, S., Ayaz, H., Devaraj, A., Onaral, B., & Pourrezaei, K. (2004). Functional near-infrared neuroimaging. *Engineering in Medicine and Biology Society*, *7*, 5333–5336.

Klein, Z. (2000). The ethological approach to the study of human behavior. *Neuroendocrinology Letters*, *12*, 477–481.

Lorenz, K. (1981). *The foundations of ethology*. New York: Springer.

Michel, C. M., Murray, M. M., Lantz, G., Gonzalez, S., Spinelli, L., & Grave dePeralta, R. (2004). EEG source imaging. *Clinical Neurophysiology*, *11*, 2195–2222.

Pascual-Leone, A., Walsh, V., & Rothwell, J. (2000). Transcranial magnetic stimulation in cognitive neuroscience: Virtual lesion, chronometry, and functional connectivity. *Current Opinion in Neurobiology*, *10*, 232–237.

Pflüger, H., & Menzel, R. (1999). Neuroethology, its roots and future. *Journal of Comparative Physiology, A, Sensory, Neural, and Behavioral Physiology*, *185*, 389–392.

Pinker, S. (2002). *The blank slate: The modern denial of human nature.* New York: Penguin.

Rosenbaum, R., Shayna et al. (2005). The case of K.C.: Contributions of a memory-impaired person to memory theory. *Neuropsychologia, 43,* 989–1021.

Savoy, R. L. (2000). History and future directions of human brain mapping and functional neuroimaging. *Acta Psychologica, 107,* 9–42.

Schmolck, H., Kensinger, E. A., Corkin, S., & Squire, L. (2002). Semantic knowledge in patient H.M. and other patients with bilateral medial and lateral temporal lobe lesions. *Hippocampus, 12,* 520–533.

Smith, M., et al. (2003). Noise-induced stress response and hearing loss in goldfish (*Carassius auratus*). *Journal of Experimental Biology, 207,* 1–9.

Stone, V. E., Cosmides, L., Tooby, J., Kroll, N., & Knight, R. T. (2002). Selective impairment of reasoning about social exchange in patient with bilateral limbic system damage. *Proceedings of the National Academy of Sciences, U.S.A., 99,* 11531–11536.

Thom, C. (2003). The tremble dance of honeybees can be caused by hive-external foraging experience. *Journal of Experimental Biology, 206,* 2111–2116.

Tooby, J., & Cosmides, L. (1992). The psychological foundations of culture. In J. Barkav, L. Cosmides, J. Tooby (Eds.), *The adapted mind: Evolutionary Psychology 3, The generation of culture.* Oxford: Oxford University Press.

Travis, J. (2002). Sizing up the brain. *Science News Online, 162*(20), November 16.

Villinger, A., & Chance, B. (1997). Non-invasive optical spectroscopy and imaging of human brain function. *Trends in Neuroscience, 20,* 435–442.

von Frisch, K. (1974). Decoding the language of the bee. *Science, 185,* 663–668.

Walsh, C. A., & Chenn A. (2003). Increased neuronal production, enlarged forebrains, and cytoarchitectural distotions in β-Catein overexpressing transgenic mice. *Cerebral Cortex, 13,* 599–606.

II Neuroanatomy: Ontogeny and Phylogeny

In the last century, "human" cognition has become more difficult to identify. Traces and even full-blown repertoires of "unique" human cognition have been discovered in nonhuman primates and in nonprimates as well. Tool making, tool use, abstract reasoning, language, deception, self-awareness, number manipulation, and a host of other cognitive abilities can be traced throughout the animal kingdom. Describing how human brains and cognitive abilities evolved has advanced significantly in the past century, almost explosively so in the past two to three decades.

There has been a particular emphasis on certain lines of research, for obvious reasons. The human brain is lateralized, convoluted, large (in terms of both absolute and relative scaling), and dominated by executive regions. Human cognition is flexible and adaptive, and has a number of hallmarks such as language and abstract reasoning. However, these attributes are not uniquely human, nor are they generally uniquely primate. Yet there are notable differences between humans and all other species. We are the only species to draft manuscripts, create a sustained written language, or drive on the expressway while talking on cell phones, listening to satellite radio while our children watch videos on their handheld devices in the back seat.

What are the general principles that have created such a brain? Most researchers have started by investigating anatomical differences, comparing human neuroanatomy to that of other primates, mammals and nonmammals. These differences are then scaled against cognitive abilities. The first three chapters in Part II examine these relations in a general manner, with specific reference to lateralization in chimpanzees and cetaceans. The basics and general rules of brain-cognition relationships are presented and reconsidered. For example, Lori Marino points out in Chapter 7 that "the underlying cytoarchitectural and organizational scheme of the dolphin neocortex is unique and highly different from that in primates. These differences further support the notion that the same cognitive capacities in primates and dolphins are underwritten by different neurobiological 'themes,' resulting in convergent cognition." Her claim that similar cognitive abilities can arrive from divergent brain anatomy adds a level of complexity to evolutionary cognitive neuroscience, one that deserves emphasis.

The final chapter in this part examines human cranial capacity. Although one of the editors of this volume (J.P.K.) has argued against the race/IQ relationship on scientific grounds (e.g., the heterogeneity of race on a genetic level), Philippe Rushton and Davison Ankney provide a significant and important review of their own and others' research on the topic of brain and cognition as measured (mainly) via IQ and cranial capacity.

Does brain size matter in terms of cognition? This simple question is parsed throughout this section, as the question becomes increasingly complicated in terms of how and what we measure in the brain, brain size, and cognition.

4 The Evolution of Ontogeny and Human Cognitive Uniqueness: Selection for Extended Brain Development in the Hominid Line

Valerie E. Stone

Humans are remarkable among primates for both our large brains relative to body size and our complex cognitive skills (Darwin, 1871; for a review, see Oxnard, 2004). Thus, any evolutionary approach to cognitive neuroscience should give an account of how humans came to have these unique features. Several different fields can contribute to such an account. Comparative neuroscience can give us information about systematic variations in the size, development, and connectivity of different brain regions across primate species. Archaeology can give us clues about body size, brain size, and development for extinct hominid species. Evolutionary psychology can provide insight into which cognitive mechanisms we might share with other mammals or primates and which are likely to be unique to our species. Cognitive neuroscience can investigate the brain systems that underlie uniquely human cognitive abilities, through patient studies and neuroimaging. The challenge for evolutionary cognitive neuroscience is to weave together these approaches in a way that illuminates human cognition.

There are several accounts of which cognitive abilities are unique to humans, with authors tending to put forward their favorite candidate ability as *the* defining feature of humanity. However, it is likely that there are several overlapping abilities that uniquely define human cognition. There is a growing recognition that this is the case in discussions of human uniqueness, with lists of these abilities including language, executive function, long-term memory and future planning, recursion, complex categorization and problem solving, abstraction, and theory of mind (Byrne, 2001; Corballis, 2003; Dunbar, 1998; Hoffecker, 2005; Pinker & Bloom, 1990; Stone, 2005; Stone & Gerrans, 2006; Suddendorf, 1999, 2004; Tooby & DeVore, 1987). Depending on the writer, "uniquely human" can mean that humans are the only current species that possesses a certain cognitive ability, or it can mean that

Homo sapiens sapiens is unique compared to our extinct hominid ancestors in possessing a certain cognitive ability. Claims about our abilities relative to those of extant primates can be tested empirically in the laboratory; claims about our abilities relative to those of other hominids are tested using inferences from artifacts such as tools and hunted animal bones associated with fossilized hominids. With each of these abilities, of course, certain aspects may be shared with other species, and other aspects may be unique (Stone, 2005; Suddendorf, 2004). Each unique ability builds on other cognitive abilities that we share with other primates. Some basic ability to associate a symbol with a meaning may well be an ability that we share with other primates (Snowdon, 2002). However, complex syntax and recursion are aspects of language that seem to be uniquely human (Corballis, 2003; Pinker & Bloom, 1990). Our ability to monitor others' eye gaze is a building block of theory of mind that we share with other primates; however, inferences about others' belief and knowledge, "theory of mind proper," appears to be unique to our species (for a review, see Stone, 2005). Nevertheless, there is a basic set of uniquely human cognitive abilities that seem to be uncontroversial: recursion, episodic memory and future planning, theory of mind, complex problem solving requiring high levels of executive function, and language that involves complex syntax (Byrne & Whiten, 1988; Corballis, 2003; Pinker & Bloom, 1990; Suddendorf, 1999; Tooby & DeVore, 1987). This list is certainly not exhaustive; it is merely a minimal set for which there is evidence.

Cognitive neuroscience has already given us information on the brain areas involved in these abilities in humans. The most complex levels of executive function seem to be mediated by lateral prefrontal cortex (Cummings, 1993; Knight & Grabowecky, 1995). We know that simpler aspects of executive function, such as basic working memory, are also mediated by prefrontal cortex in primates (Goldman-Rakic, Bourgeois, & Rakic, 1997). We know that the storage and retrieval of episodic memory and future planning depend on the frontal and temporal lobes, though memories may be stored throughout the cortex (Knight & Grabowecky, 1995; Rowe, Owen, Johnsrude, & Passingham, 2001; Shimamura, 2000; Shimamura, Janowsky & Squire, 1990; Tulving, 1995; Wood & Grafman, 2003). We know that language, syntax, and recursion also depend on the frontal and temporal lobes (Caplan et al., 2002; Cooke et al., 2001). Furthermore, humans are not merely designed to process information, but to act on that information. Cognitive abilities such as those in the list above are of no use without the ability to execute sequences of action as the output of cognition. We also know from cognitive neuroscience that executing sequences of actions depends on frontal regions and striatum, possibly also parietal lobes, for their involvement in body and action representations (Cummings, 1993; Krams et al., 1998; Reed, Stone, & McGoldrick, 2005; Rowe et al., 2001). In very rough terms, then, we can identify brain structures that subserve the uniquely human aspects of cognition and action: prefrontal cortex, temporal cortex, parietal cortex and striatum, perhaps frontal and temporal cortex particularly. By inference, these structures would have been under selection in the evolution of the primate and hominid line.

To create brains with more complex abilities, natural selection can act on two factors: the number of neurons and the connectivity of those neurons. Connectivity includes not only the "wiring diagram" but also which neurotransmitters are used where. Connectivity is probably the most important factor; however, it is also the most difficult to study, because we do not have a complete map of neural connections in the human brain, nor in most other primate brains. Neuroscience often discusses research on "the monkey brain" or "the primate brain," but such phrases almost always refer to rhesus macaques, and sometimes vervets or squirrel monkeys. Thus, our comparative knowledge of connectivity patterns within and between brain structures in primate and human brains is incomplete, limited to a tiny number of species. Number of neurons is a little easier to study. Having more neurons in a brain structure generally means either a larger structure or a more convoluted structure. Size or convolution of a particular structure is a much rougher measure of a structure's function than is connectivity, but it has the advantage of being information that is available for a greater number of primate species (Rilling & Insel, 1999; Semendeferi, Armstrong, Schleicher, Zilles, & Van Hoesen, 2001; Semendeferi & Damasio, 2000; Stephan, Frahm & Baron, 1981). Furthermore, we can also make inferences about size of brain structures from analyzing fossilized skulls of extinct hominid species (Falk, 1987). While acknowledging that size of brain structures is one of the roughest possible measures of function, I would nevertheless like to review comparative research on the size of frontal and temporal lobes, cortex, and striatum.

One study has looked at a link between function and size of brain structures comparatively. The executive brain is defined as the neocortex plus striatum (i.e., basal ganglia), to denote those parts of the brain involved in executing complex actions (Keverne, Martel, & Nevison,

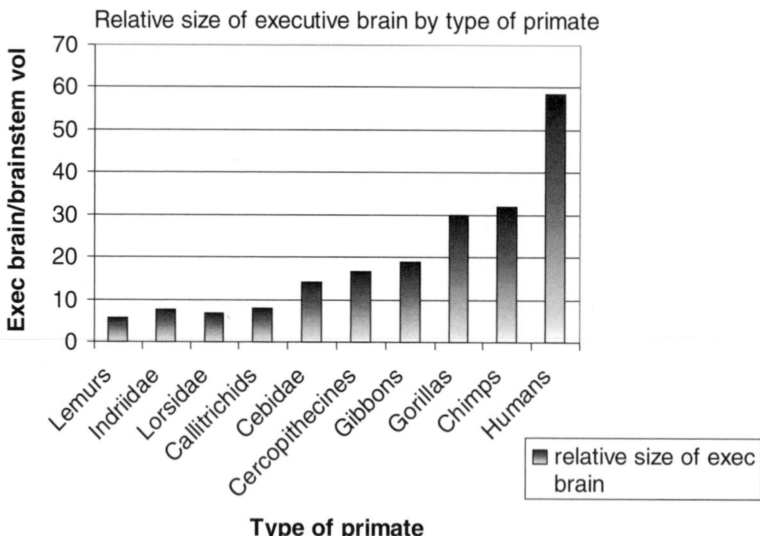

Figure 4.1
The executive brain is defined as the volume of neocortex plus striatum (Keverne et al., 1996; Reader & Laland, 2002). Taking a ratio of this volume to brainstem volume produces a measure of executive brain size that is corrected for the influence of body size and overall brain size (Barton, 1990; Reader & Laland, 1996). The relative size of the executive brain appears larger in apes than in other primates, and especially large in humans. (Data from Stephan, Frahm, & Baron, 1981.)

1996; Reader & Laland, 2002). Neocortex and striatum are closely linked genomically and neuroanatomically (Reader & Laland, 2002). (Note that the term executive has nothing to do with executive function or the frontal lobes in this context, but rather refers to the execution of action.) When corrected for body size, the executive brain is much larger in humans than in other primates (figure 4.1). In great apes and humans, compared to monkeys, one can see a greater capacity for perceiving and using innovative sequences of actions to solve complex problems (Byrne, 2001; Tooby & DeVore, 1987). This capacity includes both social problem solving, such as political maneuvering, social learning, and theory of mind, and physical problem solving, such as tool use and innovative strategies for foraging. Our own species has these abilities in the extreme, as demonstrated by the variety and flexibility of human cultures, tool manufacture and use, and the number of ecological niches in which we can forage successfully. Reader and Laland (2002) attempted to index complex and flexible problem-solving skills in primates by counting reported instances of innovation, social learning, and tool use in 116 primate species. Executive brain size data were available for 32 species. Even with such an approximate measure, they demonstrated a significant relationship between the size of the executive brain (from Stephan et al., 1981), corrected for body size, and instances of problem solving. Thus, increases in the size of the executive brain over evolution seem to be associated with functional increases in intelligence.

Many analyses of the size of primate brain regions are based on a published data set of postmortem analyses of the brains of primates that either died naturally in captivity or were recovered from poachers (Stephan et al., 1981). An advantage of this data set is that it contains information on over 40 primate species. A disadvantage is that often a "species" is represented by one individual. Data on the size of brain regions in living primates, based on multiple individuals, would obviously be an improvement. New technologies have made this ethically possible. Recently, other researchers have begun using volumetric analysis of structural MRI scans to determine the size of various brain regions in living, sedated primates. Such data currently exist for only a few species, but it is to be hoped that in the future, they will become available on a larger number of species. Semendeferi and colleagues have scanned macaques, gibbons, orangutans, gorillas, chimps, bonobos, and humans to determine overall brain volume and the size of frontal, temporal, and parieto-occipital regions (Semendeferi & Damasio, 2000; Semendeferi et al., 2001) (figure 4.2). Insel and colleagues have analyzed the volume of brain regions and the degree of cortical convolution in 11 primate species (Rilling & Insel, 1999). These analyses show clear increases in the size of some cortical regions across apes and humans, and little increase in others. Although the executive brain overall clearly seems larger in humans, particular subdivisions of the cortex—temporal, frontal, and parietal lobes—show little evidence of disproportionate expansion specific to those regions over species that have diverged at various points over the past 18 million years. However, subdivisions of these areas may be important. Within the frontal lobes, the subdivisions of dorsolateral frontal, medial frontal, and orbitofrontal also do not show strong evidence of disproportionate size in humans (Semendeferi et al., 1997). However, the frontal pole, Brodmann's area 10, shows a dramatic increase in size in humans compared to our closest relatives (figure 4.3). The frontal pole is known to be involved in executive function, problem solving, future planning, and episodic memory, all things

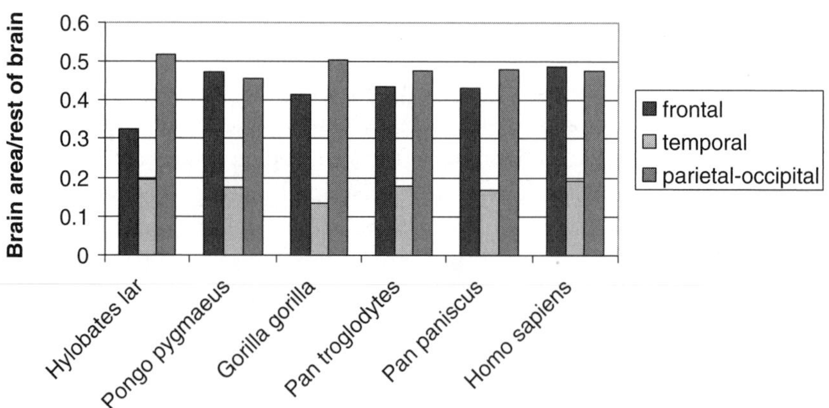

Figure 4.2
Size of three major divisions of cortex relative to whole brain size in primates most closely related to humans. With such gross divisions between cortical areas, no clear increase in size for any one area is evident for humans relative to non-human primate species. (Data from Semendeferi et al., 2000.)

that contribute to the cognitive uniqueness of humans (Braver & Bongiolatti, 2002; Lepage, Ghaffer, Nyberg, & Tulving, 2000; Tulving, 1995; Wood & Grafman, 2003).

If, as Reader and Laland (2002) propose, our complex cognitive abilities require larger cortical areas, then how has evolution produced these large brains? Selection would have had to occur in the genes regulating brain ontogeny. Prenatal neural development is the first aspect of ontogeny under genetic control. Finlay and Darlington (1995) have pointed out that there are strong developmental constraints on the size of different brain structures, such that the size of one structure is tightly correlated with the size of other structures, with approximately 96% of the variance in structure size accounted for by the size of other brain structures. This linkage between structure sizes appears to be strongly related to the length of time spent generating neurons prenatally, known as neurogenesis ($r = 0.94$; Finlay & Darlington, 1995). To the extent that a particular part of neocortex, such as the frontal pole, is larger in humans, it may also be the case that the rest of the neocortex and subcortical structures such as the thalamus are larger as well. Thus, one place that natural selection can act on the genome to produce bigger brain structures is on genes that regulate the extent of neurogenesis.

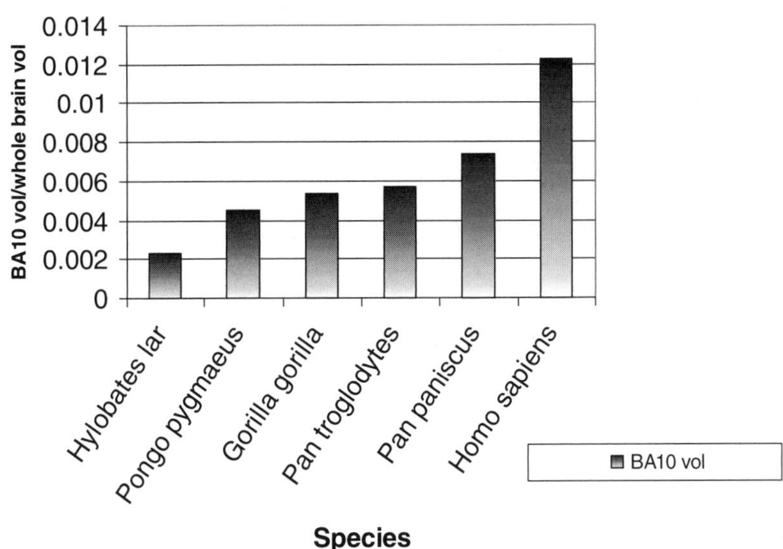

Figure 4.3
In contrast to the cortical areas in figure 4.2, the volume of the frontal pole relative to the whole brain appears to have undergone expansion in the primate line, with the frontal pole in humans disproportionately larger than in the primates most closely related to humans. (Data from Semendeferi et al., 2001.)

However, there would be no advantage conferred by genes for the extension of neurogenesis to create bigger brains unless those brains could develop and wire themselves up correctly.[1] Genes for the extension of neurogenesis would have had to be closely linked, in primate and hominid evolution, with genes for the extension of postnatal brain ontogeny. Neural development from the fetal stage to the adult stage is quite an extended process in humans. Prenatally, there is a period of neurogenesis, during which precursor cells divide and then later differentiate into neurons. Postnatally, neurons grow, dendritic branches extend, and synaptic density increases. After a plateau period of peak synaptic

1 Saying that a brain is wired "correctly" is shorthand for "in a way that would make the resulting organism inclusively fit"—that is, in a way that enables the developing organism to solve the adaptive problems it is confronted with at each developmental stage.

density, unused synapses are pruned over the course of late childhood and adolescence. Myelination of axons also continues throughout childhood and adolescence.

Large brains require longer maturation times than small ones (Passingham, 1985). In particular, the relatively large brains of great apes and humans require a longer period of "postnatal development" (also known as "experience") in order to become fully functioning adult brains (Allman & Hasenstaub, 1999; Smith & Tompkins, 1995). Early in infancy and childhood, neurons grow and form synapses. As childhood and adolescence progress, synapses decrease in density in a process known as pruning. Thus, experience is necessary for the correct wiring up of the human brain. Natural selection could have acted on genes regulating the overall extent of postnatal brain development in primates and hominids as part of the evolution of hominids' large brains.

Neural Development in Humans and Primates

As noted by Finlay and Darlington (1995), neurogenesis is one key way to affect the size of brain structures. The longer the period of prenatal neurogenesis for a structure, the more neurons are in that structure. One of their key points was that the extent of neurogenesis is very tightly linked across structures (Finlay, Darlington, & Nicastro, 2001). Extending this period of neurogenesis or increasing the rate of precursor cell division during neurogenesis is one of the primary developmental changes that natural selection could have acted on to produce large hominid brains (Finlay & Darlington, 1995; Finlay et al., 2001). Finlay and Darlington (1995) argue that if selection acted on genes regulating the extent of neurogenesis to enlarge any particular brain structure, then, because this process is so tightly linked across neurons in all areas, all brain structures would also have become larger. Neurogenesis also occurs in the adult brain, though as of now, the functional consequences of adult neurogenesis are unknown (Djavadian, 2004; Duman, 2004).

Certain stages of postnatal neural development are quite extended in humans. During the 1–3 years after birth, there is an initial period of cortical neuron growth and promiscuous synapse formation, in which it appears that any neurons that come into contact and share the same neurotransmitters will form a synapse (Huttenlocher & Dabholkar, 1997a). (This same developmental period extends only from 2–3 months in rhesus macaques; Goldman-Rakic et al., 1997.) Synaptic density peaks latest in prefrontal cortex and temporal cortex (Giedd et al., 1999; Huttenlocher & Dabholkar, 1997a). Dendritic branches continue to grow and lengthen from birth up to 2 years in occipital cortex, and beyond that period in frontal cortex (Huttenlocher & Dabholkar, 1997a). There is a plateau period of peak synaptic density in the frontal lobes from 1–3 years to 7–8 years, the period of greatest plasticity in cognition (Huttenlocher & Dabholkar, 1997a). (The plateau of peak synaptic density lasts from 2–3 months to 3–4 years in macaques, and may not show such pronounced differences between frontal and occipital regions; Goldman-Rakic et al., 1997; Huttenlocher & Dabholkar, 1997a.)

Pruning is the next stage of synaptic development. From ages 7–8 years through adolescence and into young adulthood, to about age 20, there is a period of decrease in synaptic density in both frontal and posterior regions, down to a lower plateau that is maintained until after age 60 (Huttenlocher, 1979). Adult synaptic density is about 60% of that during the peak plateau period (Giedd et al., 1999; Goldman-Rakic et al., 1997; Huttenlocher & Dabholkar, 1997a). During late childhood and adolescence, while synaptic density is decreasing, it is thought that unused synaptic connections disappear, or are "pruned" (Changeux, 1993; Huttenlocher & Dabholkar, 1997a; Moody, 1998). Although skeletal growth in humans undergoes an adolescent growth spurt (Bogin, 1999), developmental changes in synaptic density appear to proceed gradually through this period (Giedd et al., 1999).

The cognitive consequences of synaptic pruning are speculative at this point. Synapses that are used more often remain, while those that are rarely used appear to be pruned (Johnston, 1995; Moody, 1998). It has been proposed that pruning increases the efficiency of information processing in the cortex (Goldman-Rakic et al., 1997; Moody, 1998; Pribram, 1997). Goldman-Rakic et al. (1997) and Huttenlocher and Dabholkar (1997b) both stress that it is important to separate learning from changes in synaptic density, and state that learning appears to result from changes in the strength of *existing* synaptic connections, not from the *formation or loss* of synaptic connections. Since learning continues throughout adulthood and changes in synaptic density appear to plateau at adult levels by about age 20, this seems to be an important possible distinction. However, refinement of a cognitive process or skill during development, which they both say is what synaptic pruning allows, would seem to be procedural learning. Furthermore, the discovery of adult neurogenesis and pruning of neurons in the hippocampus, with some proposed role in learning and memory (Djavadian, 2004), means we should be cautious in asserting that learning can only occur through one kind of synaptic change.

Myelination, which allows neural signals to travel more rapidly, also proceeds in stages. It begins near birth in central subcortical white matter, spreads posteriorly, and only later anteriorly, not reaching prefrontal regions until after 6 months (Benes, 1997; Huttenlocher & Dabholkar, 1997b; Yakovlev & LeCours, 1967). Myelination of frontal cortex continues through childhood, adolescence, and into young adulthood (Benes, 1997; Yakovlev & LeCours, 1967). Myelination has traditionally been thought to be an excellent index of brain maturity. However, because some areas never fully myelinate even in the adult (e.g., callosal connections between frontal areas), some claim that myelination is less useful as a marker of brain maturation than changes in synaptic density (Goldman-Rakic et al., 1997).

This general pattern of brain ontogeny—neurogenesis, early, dense synapse formation followed by synaptic pruning, and sequential myelination of different regions—seems to be a general pattern of primate brain development, although the extent and timing of different phases of development differ between species. Primates have capitalized on the extension of development. For a species to have an adult brain of a particular size, that species must go through the necessary ontogenetic stages to wire up that brain appropriately, and for larger brains, these stages of brain development will take longer.

Late Development of Complex Skills

Although many cognitive abilities in humans first emerge during the period of peak synaptic density, many of our more complex cognitive functions do not reach adult levels of efficiency and competence until late adolescence or early adulthood, when synaptic pruning and myelination are complete (Huttenlocher & Dabholkar, 1997b; Pribram, 1997). For example, while by age 4 children may be able to produce and understand the syntax required by embedded sentences, the more subtle skills required for more complex syntax and conversation take longer to master (Bosacki, 2003; De Villiers & Pyers, 2002; Smith, Apperly, & White, 2003). Children at age 4 show some evidence of understanding the future (Suddendorf & Busby, 2003), and many adolescents can master frontal executive function tasks (Welsh, Pennington, & Groisser, 1991). However, adolescents in general are not known for their ability to plan ahead and anticipate future consequences, a major function of planning and executive abilities. Development in these cognitive abilities proceeds throughout adolescence (Pribram, 1997). It takes about 19–21 years to develop a functioning adult *Homo sapiens sapiens* brain. Thus, extended childhood seems to allow for an extended period of plasticity in neural development, while extended adolescence allows for a longer period of synaptic pruning—fine-tuning of cortical skills and processes through experience—and an increase in processing speed through myelination.

Extension of Development During Hominid Evolution

Recent data on dental development indicate that this extension of childhood and adolescence is a recent evolutionary phenomenon in the hominid line (Dean et al., 2001). The cyclical deposition of enamel on teeth results in periodic markings in dental tissue, called striae of Retzius, and the surface ridges on the tooth formed by these striations are called perikymata. Counting perikymata allows an estimate of crown formation times and thus an estimate of the rate of development (Dean et al., 2001). Dental development rates in australopithecine species and *Paranthropus* overlap with those of great apes, whereas Neanderthals demonstrate crown formation times more closely resembling those of modern humans. One might expect that early species of *Homo*, *Homo habilis* and *Homo erectus*, would show an intermediate pattern of development, but the surprising fact is that crown formation times in these species overlap completely with those of australopithecines (Dean et al., 2001). Thus, the pattern of extended ontogeny appears to be a recent evolutionary change, occurring since *Homo erectus*, and coincides in evolutionary time with the expansion of brain size since *Homo erectus*. Archaic *Homo sapiens* species had cranial capacities of 1,000 cc or more, compared to ~800 for *Homo erectus*, and Neanderthals and early modern humans had cranial capacities of 1,200–1,500 cc (Falik, 1987; Foley, 1997; Smith, Gannon, & Smith, 1995). It is these larger-brained species that seem to have longer childhoods, based on dental data (Dean et al., 2001).

Complex Cognition and the Executive Brain

One would expect to see a close relationship between the time it takes the brain to develop and the size of the brain. For which structures should this relationship be strongest? Humans' most complex, cortically mediated skills take the longest to develop; thus the size of brain structures subserving these abilities should be most closely related to maturation time. Long-lived organisms may need to be particularly adept at complex problem solving. The longer-lived a species is, the more the environment will change during its lifetime and the more the organism will have to be able to adapt flexibly to change (Allman & Hasenstaub, 1999). Thus, if it is our flexible problem-solving abilities (whether social or physical) and the motor behavior based on those abilities that need the most time to develop, one would expect the size of the executive brain to be most closely linked to the length of time it takes the brain to mature. Other researchers have demonstrated relationships between brain size and measures of maturation in primates (Allman & Hasenstaub, 1999; Allman, McLaughlin, & Hakeem, 1993; Sacher & Staffeldt, 1974; Smith, 1989), but none has investigated the maturation of the executive brain per se, or focused on linking maturational variables to key points in neural development. We can ask whether there is a relationship between the time it takes the brain to develop through childhood and the adolescence and the size of the adult executive brain across primate species. This is a methodologically difficult question to answer because of a lack of available data on brain development in most primate species, but we can get a rough answer by using existing data to approximate the necessary variables.

Measuring Time to Reach Adulthood

For humans and macaques, developmental milestones in life history correspond approximately to milestones in neural development. Peak synaptic density begins to decline at 7.5 years for humans and at 3 years for macaques (Goldman-Rakic et al., 1997; Huttenlocher & Dabholkar, 1997a), whereas the emergence of second incisors occurs at 7.1 years for humans and 2.6 years for macaques (Smith, Crummett & Brandt, 1994; Smith et al., 1995). The time to reach adulthood in neural development for humans and macaques—that is, to reach an adult plateau for synaptic density—is 19–21 years for humans and 4 years for rhesus macaques (Goldman-Rakic et al., 1997; Huttenlocher & Dabholkar, 1997a). Humans first reproduce at an average age of 18 in industrial cultures and 19 in hunter-gatherer cultures (Trinkaus & Tompkins, 1990). Rhesus macaques first reproduce at an average age of 3.7 years (Goldman-Rakic et al., 1997; Harvey & Clutton-Brock, 1985; Huttenlocher & Dabholkar, 1997a). Data on these milestones in neural development are available only for these two primate species. Thus, to essay whether the time to reach adulthood and the size of the executive brain are related, it is necessary to find a proxy variable that corresponds well to the age at which adult levels of synaptic density are reached. Given the correspondences already noted for humans and macaques in relation to average age at first reproduction, or generation time, this seems a reasonable proxy variable to use for time to reach adulthood at the neural level.[2] For data on average age at first breeding, I used published data sets on life history variables in primates and humans (Godfrey, Samonds, Junger, & Sutherland, 2001; Harvey & Clutton-Brock, 1985).

Measuring Executive Brain Size

Data were compiled from published sources on brain volume and volume of different brain structures for 47 species of primates, including *Homo sapiens sapiens* (Godfrey et al., 2001; Harvey & Clutton-Brock, 1985; Rilling & Insel, 1999; Semendeferi & Damasio, 2000; Semendeferi et al., 2001; Stephan et al., 1981). Not all data points were available for all species. Size of neocortex, striatum (basal ganglia), and brainstem were available for 30 species for which life history data were also available. Following Reader and Laland (2002), executive brain size was calculated as a measure of size of neocortex plus striatum.

Controlling for Body Size

Since brain size covaries strongly with body size, some measure of body size must be used as a reference variable to control for allometric effects

2 There are several possible measures of time to adulthood. Time to reach one's full growth in stature is one possibility but is problematic. Adolescents may reach their full adult stature well before they have reached adult levels of cognitive and social maturity. Also, full growth in stature can precede the attainment of full adult weight (Dainton & Macho, 1999). Time to sexual maturity is another possible index of time to adulthood but will underestimate the correct figure. Human females reach menarche at age 13 (industrial cultures) to age 16 (hunter-gatherer cultures) (Trinkaus & Tompkins, 1990), but may not be fully grown at that point. Furthermore, it is often several years after menarche that a first child would be born (5 years in industrial cultures, 3 in hunter-gatherer cultures) (Trinkaus & Tompkins, 1990). For an evolutionary analysis such as this one, only data for hunter-gatherers are relevant. In great apes and many primate species as well, there is a significant delay (1–4 years) between onset of menarche and birth of first offspring (Harvey & Clutton-Brock, 1985). Of these measures of adulthood, only generation time seems to correspond to the data on neural development.

on overall brain size. Brainstem volume is considered the most conservative way to control for this (Barton, 1999; Keverne et al., 1996; Reader & Laland, 2002). Stephan et al. (1981) present data on the size of the medulla and midbrain, but unfortunately not on the size of the pons; thus, medulla + midbrain was used as an estimate of brainstem size, as in Reader and Laland (2002). The ratio of executive brain volume to brainstem volume was calculated as:

$$\frac{(\text{Volume of neocortex}) + (\text{Volume of striatum})}{(\text{Volume of medulla}) + (\text{Volume of midbrain})}$$

Predicting Executive Brain Size from Time to Reach Adulthood

I ran a linear regression of this ratio of executive brain/brainstem volume on generation time for 30 primate species, including humans. There was a strong linear relationship (r = 0.95, r^2_{adj} = 0.90, $F_{1,28}$ = 248.4, $P < 0.0001$) (figure 4.4A). For primates, including humans, time to reach adulthood explains 90% of the variance in size of the executive brain. The linkage between time to reach adulthood and size of cortex is tight.

One possibility is that our species's data represent such an extreme outlier in executive brain size and time to reach adulthood that they inflate the strength of this linear relationship for primates. Figure 4.4A shows that the human data point lies far from those of other primates on both variables. Are humans typical primates in the relationship between development and brain size? To reduce the effect of the outlier status, I ran a log transformation on both variables, executive brain volume and generation time, and repeated the regression. The results of the analysis do not change substantively with this transformation (r = 0.90, r^2_{adj} = 0.80, $F_{1,28}$ = 118.0, $P < 0.0001$).

To test whether we are typical primates for this developmental pattern, I ran the same regression analysis relating the ratio of executive brain/brainstem volume to time to reach adulthood for the remaining 29 primate species (excluding humans). There was still a strong relationship ($r = 0.88$, $r^2_{adj} = 0.77$, $F_{1,27} = 95.3$, $P < 0.0001$). The data point for humans lies almost exactly on the regression line for nonhuman primates (figure 4.4B). Studentized deleted residuals for the regression in figure 4.4A also show that humans are not significantly different from the value that would be predicted from the regression equation for nonhuman primates (for humans, studentized deleted residual t_{27} = 1.62, $P > 0.10$, N.S.; Stevens, 1984). We are indeed like other primates in this respect.

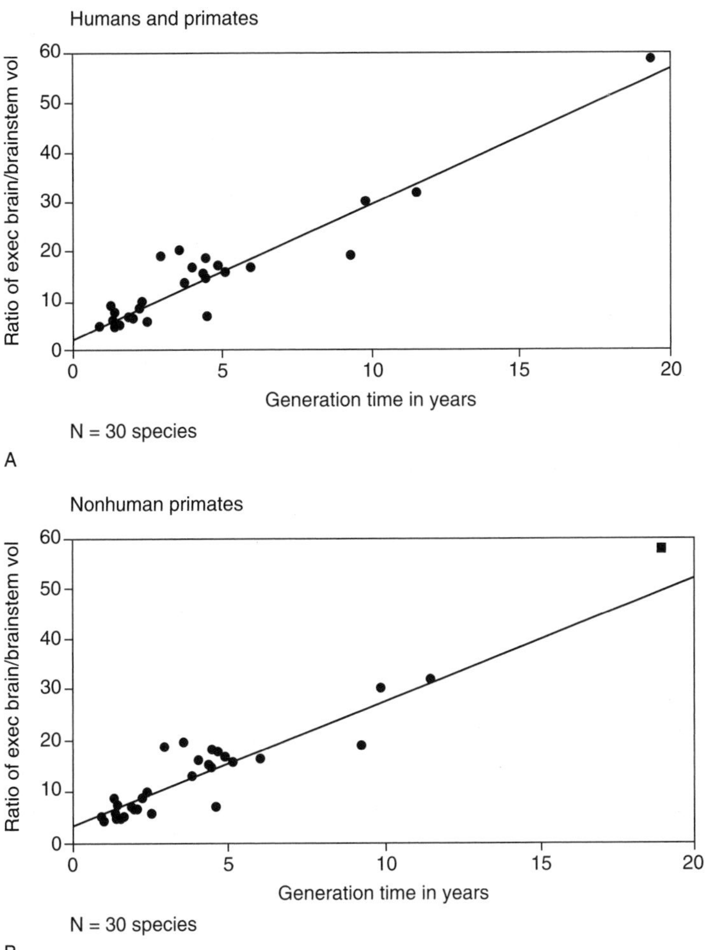

Figure 4.4
Regression lines predicting relative size of executive brain in primates and humans (A) and in nonhuman primates alone (B). The regression line looks much the same in the two cases. When the regression line for nonhuman primates is extrapolated, human data are seen to lie very close to the line. Thus, in this developmental relationship, humans appear to show a typical primate pattern, just a more extended one.

How Specific Is This Developmental Relationship to the Executive Brain?

Does the linear relationship between executive brain size and generation time exist simply because the executive brain is such a large part of total brain volume? This does not seem to be the case, since this linear relationship also holds true for smaller structures that are involved in executing action. To test the specificity of the relationship between specific brain structure sizes and time to reach adulthood, one can use the volume of several distinct brain structures to predict generation time in a multiple regression. The contribution of each structure to the relationship with generation time can then be determined. The major subdivisions of the brain reported in the data set of Stephan et al. (1981) are neocortex, striatum, brainstem, limbic system, cerebellum, hypothalamus, and thalamus. Accordingly, I calculated the ratio of the volume of each of the following structures to brainstem volume: executive brain, limbic system, cerebellum, hypothalamus, and thalamus. These values were entered into a simultaneous multiple regression predicting time to adulthood. Only relative size of executive brain and thalamus were significantly related to time to reach adulthood (table 4.1). Relative size of the limbic system appears particularly unrelated to time to adulthood. Just as olfactory and limbic structures show some differences in how they scale with other brain structures (Finlay et al., 2001), the size of the limbic system is unrelated to generation time (see table 4.1). One key fact about the executive brain is that the ultimate result is behavior, action. The basal ganglia

Table 4.1
Relationship of Brain Structures (Relative to Brainstem Size) to Generation Time

Variable Tested for Relationship to Generation time	Semipartial r	β	t	P
Executive brain vol./brainstem vol.*	0.21	0.862	4.24	0.001
Limbic system vol./brainstem vol.	0.05	0.056	1.01	0.33
Cerebellum vol./brainstem vol.	−0.04	−0.215	−0.87	0.40
Hypothalamus vol./brainstem vol.	0.04	0.061	0.84	0.42
Thalamus vol./brainstem vol.*	0.13	0.311	2.52	0.025

*$P < 0.05$.

Note: Structures involved in complex problem solving and execution of skilled action show a stronger relationship. Olfactory and limbic structures show no relationship. Limbic system size was calculated as size of hippocampus + ento- and perirhinal cortex + pre- and parasubicular cortices + amygdala + palaeocortex (Stephan et al., 1981).

are included because of their key role in generating action. However, the thalamus could be seen as important for executing skilled actions. Three different frontal-subcortical circuits connect different regions of frontal cortex (dorsolateral, orbitofrontal, and anterior cingulate) to particular parts of the basal ganglia and the thalamus (Cummings, 1993). Thus, the thalamus, in addition to being a sensory relay station for the cortex, is an important structure for behavior generated from frontal lobe computations. One could make an argument for extending the definition of the executive brain to include the thalamus. Maturation time appears to be related primarily to the size of the executive brain and its associated action-related structures.

How Specific Is the Relationship of Brain Size to Time to Reach Adulthood?

Generation time as a measure of time to adulthood was chosen because it is a rough match to time for synaptic density to reach adult levels in humans and macaques. Thus, it seemed the most appropriate choice for testing the relationship between the time for the brain to mature and the size of brain structures. However, many developmental and life history variables are closely related to each other (table 4.2), and thus it may be an extension of development per se that shows a close relationship to executive brain size, rather than only time to reach adulthood. Other developmental variables might also show strong relationships. The age of emergence of second incisors is roughly matched to the age at which synaptic density begins to decrease for humans and macaques (7.5 vs. 7.1, and 4.0 vs. 3.6, respectively). Although data on both age at second incisor emergence and executive brain size are available for only 10 species, the relationship of this developmental variable to the relative size of the executive brain still appears quite strong ($r = 0.92$, $r^2_{adj} = 0.82$, $F_{1,8} = 41.1$, $P < 0.0001$).

To determine if there is a general developmental factor, I chose a set of developmental variables, each coded in years, for which any pair of variables in the set had at least 15 observations in common. This set was gestation time, age at sexual maturity, generation time, and longevity, available for only 20 species. I ran a principal components factor analysis for a two-factor solution on these four variables to determine a primary factor for development. The first factor accounted for 88% of the variance. Generation time loaded most heavily on this factor, gestation time the least (factor loadings: gestation time, 0.87; age at sexual maturity, 0.97; generation time, 0.98; longevity, 0.93).

Table 4.2
Relationship between Variables Measuring Key Points in Development for Primates and Humans

Time Point	GST	LD	M1	M2	I2	M3	SM	GT
GST (gestation time)	1.0							
Tooth emergence: LD (last deciduous [“baby”] teeth)	0.81* (N = 21)	1.0						
M1 (first molars)	0.92* (N = 22)	0.96* (N = 21)	1.0					
M2 (second molars)	0.88* (N = 19)	0.96* (N = 19)	0.98* (N = 20)	1.0				
I2 (second incisors)	0.97* (N = 17)	0.84* (N = 17)	0.96* (N = 18)	0.92* (N = 18)	1.0			
M3 (third molars [“wisdom teeth”])	0.89* (N = 16)	0.96* (N = 14)	0.98* (N = 16)	0.99* (N = 15)	0.91* (N = 13)	1.0		
SM (age at sexual maturity)	0.80* (N = 37)	0.91* (N = 13)	0.92* (N = 14)	0.98* (N = 11)	0.84* (N = 9)	0.96* (N = 10)	1.0	
GT (generation time)	0.81* (N = 49)	0.94* (N = 19)	0.95* (N = 22)	0.96* (N = 17)	0.90* (N = 17)	0.95* (N = 16)	0.97* (N = 36)	1.0
LG (longevity)	0.65* (N = 41)	0.88* (N = 15)	0.90* (N = 15)	0.87* (N = 13)	0.84* (N = 12)	0.91* (N = 11)	0.87* (N = 32)	0.87* (N = 38)

*$P < 0.001$.

Note: Data from Godfrey et al. (2001); Harvey and Clutton-Brock (1985); Smith, Crummett, and Brandt (1994); and Smith, Gannon, and Smith (1995).

I then ran a regression of relative executive brain size (executive brain volume/brainstem volume) on this developmental factor to determine how well development in general explained executive brain size. As with generation time, the relationship was strongly linear ($r = 0.96$, $r^2_{adj} = 0.93$, $F_{1,18} = 227.1$, $P < 0.0001$). A general developmental factor that included more life history and developmental variables would undoubtedly be an even stronger predictor, but estimates of a developmental factor based on fewer than 20 observations would not likely be stable. Although time to reach adulthood is clearly an important developmental variable, strongly related to executive brain size, the relationship may be to length of development overall. Given the strong correlations between developmental variables, genes extending one phase of development may extend all phases.

Are Particular Subdivisions of the Executive Brain Important in Determining This Developmental Relationship?

The executive brain includes the entire neocortex. Cognitive neuroscience, of course, focuses on much more fine-grained distinctions between the functions of different cortical regions. In a perfect world, data on the size of many specific cortical regions would be available for a large number of primate species, and we could ask focused questions about how the size of different cortical regions might be related to development. Multiple regression analysis could be used to determine if certain subdivisions of neocortex contribute more to the developmental relationship with time to reach adulthood than do others. However, currently this can be done with only a few data points, and so it must be considered provisional and speculative at best. Size data for subdivisions of the neocortex are not available for many species. Although some excellent imaging work has been done in recent years to measure gyrification and the size of the whole brain and of the frontal, temporal, and parieto-occipital regions, this has only been done for 6–11 species (Rilling & Insel, 1999; Semendeferi & Damasio, 2000; Semendeferi et al., 2001). Of these, data on time to reach adulthood are available for only six species, and data on the general developmental factor calculated above are available for only five. Furthermore, brainstem size from Stephan et al. (1981) as a control for body size is available on only four of the six species. To be able to analyze data for even six species, a different way of controlling for overall size must be used here. Though dividing by brainstem size is preferable, some authors argue for using brain region size as a proportion of total brain size (e.g., Barton & Dunbar, 1997; Clark, Mitra, & Wang, 2001). Using brainstem size is considered more conservative and avoids the problem of using the size of the structure itself as part of the control variable (Keverne et al., 1996; Oxnard, 2004). However, using proportion of total brain volume allows us to analyze six species instead of four. If one part of the cortex contributes more to the relationship with time to adulthood than do other parts of the cortex, a multiple regression using relative size of cortical areas to predict time to adulthood should show the relative contributions of these different areas.

From the data published by Semendeferi and Damasio (2000), I computed the ratio of each subdivision of the executive brain to the volume of the whole brain for six species of apes and humans. I ran a simultaneous multiple regression predicting time to adulthood (generation time) from the relative size of frontal, temporal, and parieto-occipital cortex. For these six species, this regression did not reach significance at the 0.05 level for predicting time to adulthood ($r = 0.97$, $r^2_{adj} = 0.86$, $F_{3,2} = 11.2$, $P = 0.083$). Within this regression, the beta weight for frontal volume showed a significant relationship to time to adulthood ($\beta = 1.86$, $P = 0.038$), with the beta weight for temporal lobe volume, approaching significance ($\beta = 0.74$, $P = 0.059$).

However, when a similar regression was run substituting relative frontal pole volume (BA10) for relative frontal lobe volume (from Semendeferi & Damasio, 2000; Semendeferi et al., 2001), the result was highly significant ($r = 0.99$, $r^2_{adj} = 0.99$, $F_{3,2} = 177.3$, $P = 0.006$; relative frontal pole volume, temporal lobe volume, and parieto-occipital volume predicting time to adulthood). Relative frontal pole and temporal lobe volume were significantly related to time to adulthood, parieto-occipital volume was not (table 4.3). These analyses are on a number of variables on a small number of species, and thus have few degrees of freedom and little variability to work with. They can say very little beyond being suggestive of fruitful lines of inquiry for the future if data on more species become available. Nevertheless, the suggestion is there in the data that the frontal and temporal lobes, and the frontal pole in particular, may be the outer limit for the relationship with time to reach adulthood. Because the frontal and temporal lobes are the latest to develop (Giedd et al., 1999; Huttenlocher & Dabholkar, 1997a), it makes sense that size of frontal lobes might have the strongest relationship to total length of time spent in development.

Table 4.3
Relationship of Cortical Areas (Relative to Total Brain Volume) to Generation Time

Variable Tested for Relationship to Generation Time	Semipartial r	β	t	P
BA10 vol./total brain vol.*	0.87	0.95	20.07	0.002
Temporal lobe vol./total brain vol.	0.25	0.25	5.74	0.03
Parieto-occipital vol./total brain vol.	0.021	0.023	0.48	0.68

*$P < 0.05$.
Note: Volume of frontal pole (Brodmann's area 10) shows a stronger relationship to time to reach adulthood than does volume of temporal lobes or parieto-occipital cortex. Both frontal pole and temporal lobe volume are significantly related to time to reach adulthood. Data from Semendeferi and Damasio (2000) and Semendeferi et al. (2001).

Size Isn't Everything

A structure can be complex and have more neurons without becoming much larger over the course of evolution if it becomes more convoluted. Rilling and Insel (1999) measured the degree of whole brain gyrification in 11 primate species. If size of cortex shows a significant relationship to time to reach adulthood, degree of gyrification should as well. For eight species for which data on both time to adulthood and degree of gyrification[3] were available, the regression predicting time to adulthood from degree of whole brain gyrification was highly significant ($r = 0.94$, $r^2_{adj} = 0.87$, $F_{1,6} = 49.1$, $P < 0.0001$). Even with so few species' data, the relationship was quite strong. Gyrification as an index of number of neurons might be as good as or better than structure size. Again, it is to be hoped that such data will become available on more species in the future.

Relationship Between Executive Brain Size and Development for Extinct Hominids

How do our ancestors fit into this developmental relationship? Time to reach adulthood and executive brain size can be estimated independently

3 The gyrification index in Rilling and Insel (1999) is already corrected for whole brain size, so it could be entered directly into the regression without having to take any kind of ratio.

for a handful of extinct hominids. We can then see whether these values fall close to the regression line derived from humans and primates. Since there is no way to get an estimate of brainstem size in hominids, there is no way to place hominids on a graph relating generation time and executive brain/brainstem ratio. However, we can use the regression relating the log transformation of executive brain volume and time to reach adulthood (generation time).

Although no fossilized brains are contained in the archaeological record, we do have several skulls that housed our ancestors' brains, and so we can measure cranial capacity for our ancestors. Cranial capacity is directly related to executive brain size, so that executive brain volume for our extinct ancestors can be estimated using regression techniques (figure 4.5). Furthermore, teeth fossilize and can give us an accurate estimate of development. Dental development rates for fossilized hominid teeth have been used to compare development in apes, australopithecines, *Homo erectus*, Neanderthals, and modern humans, and to derive estimates of the age at which first molars would have emerged for a few species of hominids (Dean et al., 2001; Dean, quoted in Pearson, 2001). As the age of emergence for first molars is strongly related to generation time (see table 4.2), this data can be used to estimate generation time for extinct hominids. When values for executive brain size and for time to reach adulthood are estimated for *Australopithecus afarensis*, Asian *Homo erectus*, and Neanderthals, the data are seen to lie very close to the regression line derived for primates and humans (see figure 4.5). These estimates are quite rough. However, although the extent of childhood and adolescence has changed over hominid evolution (Dean et al., 2001; Smith & Tompkins, 1995), the developmental pattern linking brain size and rate of development appears to have remained the same.

Discussion

Although humans have much larger executive brains and much longer development times than other primates, we are quite typical primates in the developmental pattern linking size of executive brain to length of childhood and adolescence. As shown in the preceding discussion, our time to reach adulthood and our executive brain size can be closely predicted by extrapolating the regression line for other primates. This appears to be true for extinct hominid species as well. If this developmental pattern is common to primates, then relatively small changes in genes regulating the extent of childhood and adolescence may have been

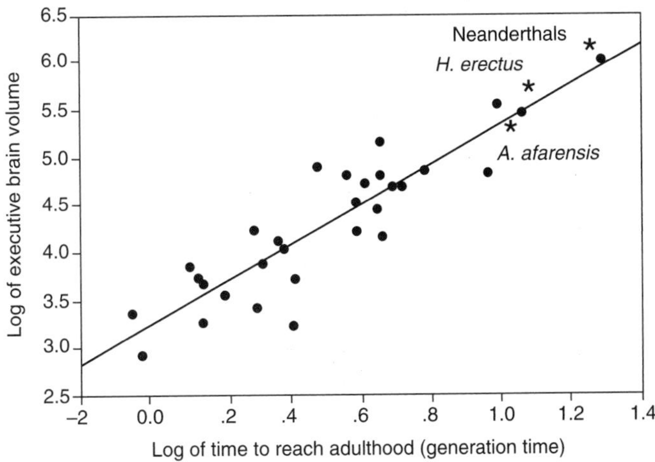

Figure 4.5
Placing extinct hominids on the regression line relating time to adulthood and executive brain size for primates and humans. *Australopithecus afarensis*, Asian *Homo erectus*, and Neanderthals all lie close to the line for primates and humans. Executive brain volume for extinct hominids was estimated using the regression equation: $\log_{10}$ executive brain volume = 2.707 + 1.063 × $\log_{10}$ cranial capacity, derived from primates and humans (N = 38 species). Time to reach adulthood was estimated using the regression equation: time to adulthood = 1.049 + 2.966 × age at first molar emergence, derived from primates and humans (N = 22 species). (Dental data and cranial capacity data from Dean et al., 2001; Falk, 1987; Foley, 1997; Harvey & Clutton-Brock, 1985; Smith, Crummett, & Brandt, 1994; Smith, Gannon, & Smith, 1995.)

key events in the speciation of *Homo sapiens sapiens*. Genes involved in regulating the timing of events in neural development may have been particularly important.

Finlay and Darlington (1995) demonstrated that the size of many brain structures is highly intercorrelated across mammalian species, and perhaps explained by some of the same underlying factors, such as extent of neurogenesis. The above analyses imply that in spite of the strong intercorrelations in size among structures, the size of only certain brain structures, the executive brain and thalamus, is related to time to mature to adulthood. Emphasizing the linkage between *sizes* of different brain structures may neglect key differences in developmental patterns between these structures.

Lieberman, McBratney, and Krovitz (2002) have shown that the size of frontal and temporal lobes may significantly differentiate *Homo sapiens sapiens* from our closest hominid relatives, archaic *Homo sapiens* and Neanderthals. The key differences in skull morphology that significantly differentiate *Homo sapiens sapiens* from archaic *Homo sapiens* and Neanderthals are in the bones surrounding the temporal lobes, allowing for greater temporal lobe volume, and the high, domed forehead of our species, allowing more room for the frontal pole (Lieberman et al., 2002).[4] The size of these cortical structures, the frontal pole and temporal lobes, may show this relationship to time to reach adulthood most strongly. As these may be relatively late-developing structures, they place an outer limit on when neural maturation is complete (Giedd et al., 1999; Huttenlocher & Dabholkar, 1997a).

The picture of brain evolution that emerges from a comparative analysis of brain development, linked with insights from hominid archaeology, can inform cognitive neuroscience. Evolutionary perspectives on the human brain suggest lines of research relevant to understanding human cognitive uniqueness. Neuroimaging can produce more detailed information about the functions of the frontal pole. Neurological and developmental disorders that disproportionately affect the frontal and temporal lobes, such as schizophrenia or frontotemporal dementia, may provide a window into understanding unique aspects of human cognition. Conversely, the fact that these brain areas have expanded so recently in evolution may explain why they are vulnerable to these particularly human diseases.

Finally, not only our purely cognitive abilities but also our social capacities have made our brains unique. Raising slow-developing offspring is a task requiring significant parental and kin investment. If a large brain requires a long childhood and adolescence, then, as the time to reach adulthood grew longer in the hominid line, the amount of investment in offspring had to increase as well. Parents must invest more time and resources to raise altritial offspring successfully. The high investment required by our extended altritiality would call for multiadult cooperation—the typical mammalian pattern of investment only from the mother would no longer be sufficient to ensure the success of the offspring. Such investment could come from monogamy and paternal investment, but it could also come from looser social arrangements involving multiple adults, such as kin members or polyamorous mating arrangements

4 Although the data of Semendeferi and Damasio (2000) do not show that the temporal lobes of humans are disproportionately larger than would be expected for other apes based on whole brain size, the analyses of Lieberman et al. (2002) point to the temporal lobes significantly differentiating *Homo sapiens sapiens* from our closest hominid relatives.

(Beckerman & Valentine, 2002). Whatever the mating system involved, changes in the attachment system—that is, in the capacity of adults other than the mother to become attached to an infant—must have been among the changes that took place in hominid evolution as well. High parental investment tends not to occur in primates with large size differences between the sexes and high levels of male-male competition (Plavcan & Van Schaik, 1997). *Homo erectus,* subsequent species of *Homo*, and possibly *Australopithecus afarensis* as well are characterized by relatively minor size differences between the sexes (Plavcan & Van Schaik, 1997; Reno, Meindl, McCollum, & Lovejoy, 2003), consistent with a system of multiadult cooperation and high parental investment (Larsen, 2003). Extended development means that selection pressures on brain systems mediating adult care for children would have been significant in hominid evolution. Thus, clues to human uniqueness can also be sought in social neuroscience, in brain systems involving oxytocin and vasopressin, regulating social bonding. Both social neuroscience and cognitive neuroscience can contribute to an understanding of hominid brain evolution.

Ontogeny does not recapitulate phylogeny. However, to understand the ontogeny of the human brain, it is essential to consider phylogeny, because the course of ontogeny has undergone evolutionary change in the hominid line. The analyses presented here point to an important set of methods that can be used in evolutionary cognitive neuroscience. Our closest relatives over the past 6 million years are all extinct. We cannot analyze the brains of our extinct ancestors and cousins directly, but by using comparative analyses such as the ones described in this chapter, we can do so indirectly. Many variables are known to covary in primates. If one variable known for primates is also known for hominids, such as cranial capacity, regression analyses using interrelated variables allows the interpolation of data for extinct hominids. Comparative neuroanatomy and comparative studies on development can provide us with a rich database from which to draw inferences about what was being selected for in the brain over the course of primate and hominid evolution. Brains themselves do not fossilize, but knowing that general developmental pattern for primates and humans, we can begin to understand brain development in our extinct hominid ancestors.

References

Allman, J. M., & Hasenstaub, A. (1999). Brains, maturation times, and parenting. *Neurobiology of Aging, 20*, 447–454.

Allman, J. M., McLaughlin, T., & Hakeem, A. (1993). Brain structures and lifespan in primate species. *Proceedings of the National Academy of Sciences, U.S.A., 90*, 3559–3563.

Barton, R. (1999). The evolutionary ecology of the primate brain. In P. C. Lee (Ed.), *Comparative primate socioecology* (pp. 167–194). Cambridge: Cambridge University Press.

Barton, R. A., & Dunbar, R. I. (1997). Evolution of the social brain. In *Machiavellian intelligence II: Social expertise and the evolution of intellect in monkeys, apes and humans* (pp. 240–263). Oxford: Oxford University Press.

Beckerman, S., & Valentine, P. (2002). *Cultures of multiple fathers: The theory and practice of partible paternity in lowland South America.* Gainesville: University Press of Florida.

Benes, F. M. (1997). Corticolimbic circuitry and the development of psychopathology during childhood and adolescence. In N. A. Krasnegor, G. R. Lyon, & P. S. Goldman-Rakic (Eds.), *Development of the prefrontal cortex: Evolution, neurobiology, and behavior* (pp. 211–239). Baltimore: Paul H. Brookes.

Bogin, B. (1999). Evolutionary perspective on human growth. *Annual Review of Anthropology, 28*, 109–153.

Bosacki, S. L. (2003). Psychological pragmatics in preadolescents: Sociomoral understanding, self-worth, and school behavior. *Journal of Youth and Adolescence, 32*(2), 141–155.

Braver, T. S., & Bongiolatti, S. R. (2002). The role of frontopolar cortex in subgoal processing during working memory. *NeuroImage, 15*(3), 523–536.

Byrne, R. (2001). Social and technical forms of primate intelligence. In F. B. M. de Waal (Ed.), *Tree of origin: What primate behavior can tell us about human social evolution* (pp. 145–172). Cambridge, MA: Harvard University Press.

Byrne, R., & Whiten, A. (Eds.). (1988). *Machiavellian intelligence.* Oxford: Oxford University Press.

Caplan, D., Vijayan, S., Kuperberg, G., West, C., Waters, G., Greve, D. D., et al. (2002). Vascular responses to syntactic processing: Event-related fMRI study of relative clauses. *Human Brain Mapping, 15*(1), 26–38.

Changeux, J. P. (1993). A critical view of neuronal models of learning and memory. In P. Anderson (Ed.), *Memory concepts* (pp. 413–433). Amsterdam: Elsevier.

Clark, D. A., Mitra, P. P., & Wang, S. (2001). Scalable architecture in mammalian brains. *Nature, 411*, 189–193.

Cooke, A., Zurif, E. B., DeVita, C., Alsop, D., Koenig, P., Detre, J., et al. (2001). Neural basis for sentence comprehension: Grammatical and short-term memory components. *Human Brain Mapping, 15*, 80–94.

Corballis, M. (2003). Recursion as the key to the human mind. In K. Sterelny & J. Fitness (Eds.), *From mating to mentality: Evaluating evolutionary psychology* (pp. 155–171). New York: Psychology Press.

Cummings, J. L. (1993). Frontal-subcortical circuits and human behavior. *Archives of Neurology*, *50*(8), 873–880.

Dainton, M., & Macho, G. A. (1999). Heterochrony: Somatic, skeletal and dental development in *Gorilla*, *Homo* and *Pan*. In R. D. Hoppa & C. M. Fitzgerald (Eds.), *Human growth in the past: Studies from bones and teeth* (pp. 32–64). Cambridge: Cambridge University Press.

Darwin, C. (1871). *The descent of man and Selection in relation to sex*. Reprinted in *The origin of species and The descent of man*. New York: Modern Library.

De Villiers, J., & Pyers, J. (2002). Complements to cognition: A longitudinal study of the relationship between complex syntax and false-belief understanding. *Cognitive Development*, *17*, 1037–1060.

Dean, C., Leakey, M. G., Reid, D., Schrenk, F., Schwartz, G. T., Stringer, C., & Walker, A. (2001). Growth processes in teeth distinguish modern humans from *Homo erectus* and earlier hominins. *Nature*, *414*, 628–631.

Djavadian, R. L. (2004). Serotonin and neurogenesis in the hippocampal dentate gyrus of adult mammals. *Acta Neurobiologiae Experimentalis*, *64*(2), 189–200.

Duman, R. S. (2004). Depression: A case of neuronal life and death? *Biological Psychiatry*, *56*(3), 140–145.

Dunbar, R. I. (1998). The social brain hypothesis. *Evolutionary Anthropology*, *6*(5), 178–190.

Falk, D. (1987). Hominid paleoneurology. *Annual Review of Anthropology*, *16*, 13–30.

Finlay, B. L., & Darlington, R. B. (1995). Linked regularities in the development and evolution of mammalian brains. *Science*, *268*(5217), 1578–1584.

Finlay, B. L., Darlington, R. B., & Nicastro, N. (2001). Developmental structure in brain evolution. *Behavioral and Brain Sciences*, *24*, 263–308.

Foley, R. (1997). *Humans before humanity*. Oxford, Blackwell.

Giedd, J. N., Blumenthal, J., Jeffries, N. O., Castellanos, F. X., Liu, H., Zijdenbos, A., et al. (1999). Brain development during childhood and adolescence: A longitudinal MRI study. *Nature Neuroscience*, *2*(10), 861–863.

Godfrey, L. R., Samonds, K. E., Jungers, W. L., & Sutherland, M. R. (2001). Teeth, brains and primate life histories. *American Journal of Physical Anthropology*, *114*, 192–214.

Goldman-Rakic, P. S., Bourgeois, J., & Rakic, P. (1997). Synaptic substrate of cognitive development: Life-span analysis of synaptogenesis in the prefrontal cortex of the nonhuman primate. In N. A. Krasnegor, G. R. Lyon, & P. S. Goldman-Rakic (Eds.), *Development of the prefrontal cortex: Evolution, neurobiology, and behavior* (pp. 27–47). Baltimore: Paul H. Brookes.

Harvey, P. H., & Clutton-Brock, T. H. (1985). Life history variation in primates. *Evolution*, *39*, 559–581.

Hoffecker, J. F. (2005). A prehistory of the north: Human settlement of the higher latitudes. New Brunswick, NJ: Rutgers University Press.

Huttenlocher, P. R. (1979). Synaptic density in human frontal cortex: Developmental changes and effects of aging. *Brain Research*, *163*(2), 195–205.

Huttenlocher, P. R., & Dabholkar, A. S. (1997a). Regional differences in synaptogenesis in human cerebral cortex. *Journal of Comparative Neurology, 387*, 167–178.

Huttenlocher, P. R., & Dabholkar, A. S. (1997b). Developmental anatomy of prefrontal cortex. In N. A. Krasnegor, G. R. Lyon, & P. S. Goldman-Rakic (Eds.), *Development of the prefrontal cortex: Evolution, neurobiology, and behavior* (pp. 69–83). Baltimore: Paul H. Brookes.

Johnston, M. V. (1995). Neurotransmitters and vulnerability of the developing brain. *Brain Development, 17*(5), 301–306.

Keverne, E. B., Martel, F. L., & Nevison, C. M. (1996). Primate brain evolution: Genetic and functional considerations. *Proceedings of the Royal Society of London, B, Biological Sciences, 263*(1371), 689–696.

Knight, R. T., & Grabowecky, M. (1995). Escape from linear time: Prefrontal cortex and conscious experience. In M. S. Gazzaniga (Ed.), *The cognitive neurosciences*. Cambridge, MA: MIT Press.

Krams, M., Rushworth, M. F., Deiber, M. P., Frackowiak, R. S., & Passingham, R. E. (1998). The preparation, execution and suppression of copied movements in the human brain. *Experimental Brain Research, 120*(3), 386–398.

Larsen, M. S. (2003). Equality for the sexes in human evolution? Early hominid sexual dimorphism and implications for mating systems and social behavior. *Proceedings of the National Academy of Sciences, U.S.A., 100*(16), 9103–9104.

Lepage, M., Ghaffar, O., Nyberg, L., & Tulving, E. (2000). Prefrontal cortex and episodic memory retrieval mode. *Proceedings of the National Academy of Sciences, U.S.A., 97*(1), 506–511.

Lieberman, D. E., McBratney, B. M., & Krovitz, G. (2002). The evolution and development of cranial form in *Homo sapiens*. *Proceedings of the National Academy of Sciences, U.S.A., 99*(3), 1134–1139.

Moody, W. J. (1998). Control of spontaneous activity during development. *Journal of Neurobiology, 37*(1), 97–109.

Oxnard, C. E. (2004). Brain evolution: Mammals, primates, chimpanzees and humans. *International Journal of Primatology, 25*(5), 1127–1158.

Passingham, R. E. (1985). Rates of brain development in mammals including man. *Brain, Behavior and Evolution, 26*, 167–175.

Pearson, H. (2001). Ancestors skip adolescence. *Nature Science Update*, December 6, 2001. Retrieved from http://www.nature.com/nsu/011206/011206-10.html on March 11, 2002.

Pinker, S., & Bloom, P. (1990). Natural language and natural selection. *Behavioral and Brain Sciences, 13*(4), 707–784.

Plavcan, J. M., & Van Schaik, C. P. (1997). Interpreting hominid behavior on the basis of sexual dimorphism. *Journal of Human Evolution, 32*(4), 345–374.

Pribram, K. H. (1997). The work in working memory: Implications for development. In N. A. Krasnegor, G. R. Lyon, & P. S. Goldman-Rakic (Eds.), *Development of the prefrontal cortex: Evolution, neurobiology, and behavior* (pp. 359–378). Baltimore: Paul H. Brookes.

Reader, S. M., & Laland, K. N. (2002). Social intelligence, innovation, and enhanced brain size in primates. *Proceedings of the National Academy of Sciences, U.S.A., 99*(7), 4436–4441.

Reed, C., Stone, V. E., & McGoldrick, J. (2005). Not just posturing: Configural processing of the human body. In M. Shiffrar et al. (Eds.), *Perception of the human body from the inside out*. Oxford: Oxford University Press.

Reno, P. L., Meindl, R. S., McCollum, M. A., & Lovejoy, C. O. (2003). Sexual dimorphism in *Australopithecus afarensis* was similar to that of modern humans. *Proceedings of the National Academy of Sciences, U.S.A., 100*(16), 9404–9409.

Rilling, J. K., & Insel, T. R. (1999). The primate neocortex in comparative perspective using magnetic resonance imaging. *Journal of Human Evolution, 37*, 191–223.

Rowe, J. B., Owen, A. M., Johnsrude, I. S., & Passingham, R. E. (2001). Imaging the mental components of a planning task. *Neuropsychologia, 39*(3), 315–327.

Sacher, G. A., & Staffeldt, E. F. (1974). Relation of gestation time to brain weight for placental mammals. *American Naturalist, 108*, 593–616.

Semendeferi, K., Armstrong, E., Schleicher, A., Zilles, K., & Van Hoesen, G. W. (2001). Prefrontal cortex in humans and apes: A comparative study of area 10. *American Journal of Physical Anthropology, 114*, 224–241.

Semendeferi, K., & Damasio, H. (2000). The brain and its main anatomical subdivisions in living hominoids using magnetic resonance imaging. *Journal of Human Evolution, 38*, 317–332.

Semendeferi, K., Damasio, H., & Frank, R. (1997). The evolution of frontal lobes: A volumetric analysis based on three-dimensional reconstructions of magnetic resonance scans of human and ape brains. *Journal of Human Evolution, 32*, 375–388.

Shimamura, A. P. (2000). Toward a cognitive neuroscience of metacognition. *Consciousness & Cognition, 9*, 313–323.

Shimamura, A. P., Janowsky, J. S., & Squire, L. R. (1990). Memory for the temporal order of events in patients with frontal lobe lesions and amnesic patients. *Neuropsychologia, 28*(8), 803–813.

Smith, B. H. (1989). Dental development as a measure of life history in primates. *Evolution, 43*, 683–688.

Smith, B. H., Crummett, T. L., & Brandt, K. L. (1994). Ages of eruption of primate teeth: A compendium for aging individuals and comparing life histories. *Yearbook of Physical Anthropology, 37*, 177–231.

Smith, B. H., & Tompkins, R. L. (1995). Toward a life history of the hominidae. *Annual Review of Anthropology, 24*, 257–279.

Smith, M., Apperly, I., & White, V. (2003). False belief reasoning and the acquisition of relative clause sentences. *Child Development, 74*(6), 1709–1719.

Smith, R. J., Gannon, P. J., & Smith, B. H. (1995). Ontogeny of australopithecines and early *Homo*: Evidence from cranial capacity and dental eruption. *Journal of Human Evolution, 29*, 155–168.

Snowdon, C. T. (2002). From primate communication to human language. In F. B. M. de Waal (Ed.), *Tree of origin: What primate behavior can tell us about human social evolution.* Cambridge, MA: Harvard University Press.

Stephan, H., Frahm, H., & Baron, G. (1981). New and revised data on volumes of brain structures in insectivores and primates. *Folia Primatologica, 35*, 1–29.

Stevens, J. P. (1984). Outliers and influential data points in regression analysis. *Psychological Bulletin, 95*(2), 334–344.

Stone, V. E. (2005). Theory of mind and the evolution of social intelligence. In J. Cacciopo (Ed.), *Social neuroscience: People thinking about people* (pp. 103–130). Cambridge, MA: MIT Press.

Stone, V. E., & Gerrans, P. (2006). Does the normal brain have a theory of mind? *Trends in Cognitive Sciences, 10*(1), 3–4.

Suddendorf, T. (1999). The rise of the metamind. In M. C. Corballis & S. Lea (Eds.), *The descent of mind: Psychological perspectives on hominid evolution* (pp. 218–260). London: Oxford University Press.

Suddendorf, T. (2004). How primatology can inform us about the evolution of the human mind. *Australian Psychologist, 39*(3), 180–187.

Suddendorf, T., & Busby, J. (2003). Mental time travel in animals? *Trends in Cognitive Sciences, 7*(9), 391–396.

Tooby, J., & DeVore, I. (1987). The reconstruction of hominid behavioral evolution through strategic modeling. In W. G. Kinzey (Ed.), *The Evolution of human behavior: Primate models* (pp. 183–237). Albany: State University of New York Press.

Trinkaus, E., & Tompkins, R. L. (1990). The Neandertal life cycle: The possibility, probability and perceptibility of contrasts with recent humans. In C. J. DeRousseau (Ed.), *Primate life history and evolution* (pp. 153–180). New York: Wiley-Liss.

Tulving, E. (1995). Organization of memory: Quo vadis? In M. S. Gazzaniga (Ed.), *The cognitive neurosciences* (pp. 839–845). Cambridge, MA: MIT Press.

Welsh, M. C., Pennington, B. F., & Groisser, D. B. (1991). A normative-developmental study of executive function: A window on prefrontal function in children. *Developmental Neuropsychology, 7*(2), 131–149.

Wood, J. N., & Grafman, J. (2003). Human prefrontal cortex: Processing and representational perspectives. *Nature Reviews Neuroscience, 4*(2), 139–147.

Yakovlev, P. I., & LeCours, A. R. (1967). The myelogenetic cycles of regional maturation of the brain. In A. Minkowsky (Ed.), *Regional development of the brain in early life* (pp. 3–70). Oxford: Blackwell.

5 Hemispheric Specialization in Chimpanzees: Evolution of Hand and Brain

William D. Hopkins

Hemispheric specialization refers to lateralization of motor, perceptual, and cognitive functions to the left or right cerebral hemisphere. Two of the most pronounced manifestations of hemispheric specialization in humans are handedness and language. With respect to handedness, all human cultures to date report evidence of population-level right-handedness. The archaeological records suggest that handedness can be dated back at least 2 million years. With respect to language, early studies focused on the clinical cases, such as the famous patient "Tan Tan" studied by Paul Broca, who showed pronounced and severe deficits in speech production. Postmortem analysis revealed significant damage to the left inferior frontal lobe and adjacent subcortical structures, leading Broca to conclude that the faculty of speech was localized to the left cerebral hemisphere. The initial observations of Dax, Broca, and Wernicke were affirmed later when Sperry and colleagues began their landmark work on lateralization of function in split-brain patients with epilepsy who had their corpus callosum severed as a means of isolating the seizure activity to one hemisphere (see Gazzaniga, 2000, for a review). The studies in split-brain patients corroborated the evidence that the left hemisphere was dominant for speech functions; however, the work with split-brain patients also contributed significantly to the documentation of lateralization for other cognitive processes, such as visual-spatial functions, processing of faces, and emotions (Borod, Haywood, & Koff, 1997). At the neuroanatomical level of analysis, early work focused on morphological asymmetries of the brain shape, such as petalia patterns and sylvian fissure length (Kertesz, Black, Polk, & Howell, 1986; Yeni-Komshian & Benson, 1976). One of the most significant studies was that done by Geschwind and Levitsky (1968), who documented leftward asymmetries in the planum temporale (PT) in a sample of 100 cadaver human brain specimens. The PT is the bank of tissue lying posterior to

Heschl's gyrus and includes Wernicke's area. The PT has probably been the most studied morphological asymmetry in the human brain, and modern imaging techniques have largely confirmed the original observations of Geschwind and Levitsky (see Beaton, 1997; Shapleske, Rossell, Woodruff, & David, 1999). In sum, hemispheric specialization seems to be widespread in humans at the behavioral and neuroanatomical levels of analysis.

Historically, hemispheric specialization has been considered a hallmark of human evolution. This view was largely driven by data that failed to show evidence of behavioral and neuroanatomical asymmetries in animals. For example, early attempts to characterize the handedness of animals, notably rats and primates, failed to reveal any evidence of population-level limb use (see Ettlinger, 1988; Finch, 1941; Lehman, 1993; Warren, 1980). Moreover, a number of studies were carried out on learning and memory functions in split-brain monkeys, and the findings revealed little if any evidence of specialization of one hemisphere over another for any tasks (Hamilton, 1977; Prelowski, 1979). Additionally, lesions of specific brain regions did not appear to have localized effects on motor or cognitive processes (see Warren, 1977). At the neuroanatomical level, some studies demonstrated evidence of brain asymmetries in sulcus length (Cheverud et al., 1990; Falk et al., 1990; Heilbronner & Holloway, 1988; Yeni-Komshian & Benson, 1976) and petalias (Cain & Waada, 1979; Groves & Humphrey, 1973; Holloway & De La Coste-Lareymondie, 1982; LeMay, 1985).

Recently, our views of the evolution of behavioral and brain asymmetries have begun to change owing to increasing evidence of population-level asymmetry in many vertebrates at the behavioral and neuroanatomical levels of analysis (Bradshaw & Rogers, 1993; Rogers & Andrew, 2002). Early studies were limited in a number of ways, and the procedural and methodological limitations have been mitigated in recent years. For example, regarding handedness, the focus in early work was on the measurement of hand use for simple reaching. As it turns out, simple reaching is not a particularly good measure in terms of reliability and for inducing individual hand preferences (see Hopkins, Russell, Hook, Braccini, & Schapiro, in press; Lehman, 1993). Similarly, early lesion and split-brain studies removed large brain regions and did not use what might be termed ecologically valid stimuli, such as species-specific faces or vocalizations (e.g., Dewson, 1997), but this has also improved dramatically (Hamilton & Vermiere, 1988; Heffner & Heffner, 1984; Vermiere & Hamilton, 1998). Finally, with respect to asymmetries in neuroanatomy, sample sizes were typically small and lacked adequate statistical power.

This chapter summarizes a series of studies on behavioral and neuroanatomical asymmetries in chimpanzees that have been conducted in my laboratory. The results are presented in the context of comparative work with other nonhuman primate species, and the cumulative results are discussed in relation to different evolutionary theories of hemispheric specialization in primates. Most of the research has been conducted at the Yerkes National Primate Research Center (YNPRC), in Atlanta, Georgia. Additional behavioral studies have recently been carried out at the University of Texas M.D. Anderson Cancer Center, in Bastrop, Texas, and the Alamogordo Primate Facility, in Alamogordo, New Mexico.

Behavioral Asymmetries in Captive Chimpanzees

Hand Preference

For the past 15 years, my colleagues and I have been studying whether chimpanzees show population-level handedness for specific measures of hand use. We have examined handedness for a host of tasks, including coordinated bimanual actions (Hopkins, 1995; Hopkins, Fernandez-Carriba, et al., 2001; Hopkins, Wesley, et al., 2004), manual gestures (Hopkins, Russell, Freeman, Buehler, et al., in press), throwing (Hopkins, Bard, Jones, & Bales, 1993; Hopkins, Russell, Freeman, Cantalupo, et al., in press), simple reaching (Hopkins, Cantalupo, Wesley, Hostetter, & Pilcher, 2002; Hopkins, Russell, Hook, et al., in press), tool use (Hopkins, unpublished; Hopkins & Rabinowitz, 1997), and bimanual feeding (Hopkins, 1994). In our view, these are good measures of hand preference because they show significant test-retest correlation coefficients and the majority of subjects tested show a significant bias in hand use, with the exception of simple reaching and bimanual feeding. Depicted in figure 5.1 are the mean handedness index (HI) for each of the measures described above and the sample size of subjects from which the average HI score was computed. Individual HI values are derived by subtracting the number of left-hand responses from the number of right-hand responses and dividing by the total number of responses. As can be seen, with the exception of quadrupedal reaching (Quadrupedal in the figure), the HI values are right-sided and deviate significantly from chance, as revealed by one-sample *t* tests. An HI value has also been calculated from the average HI values derived for each behavioral measure (labeled

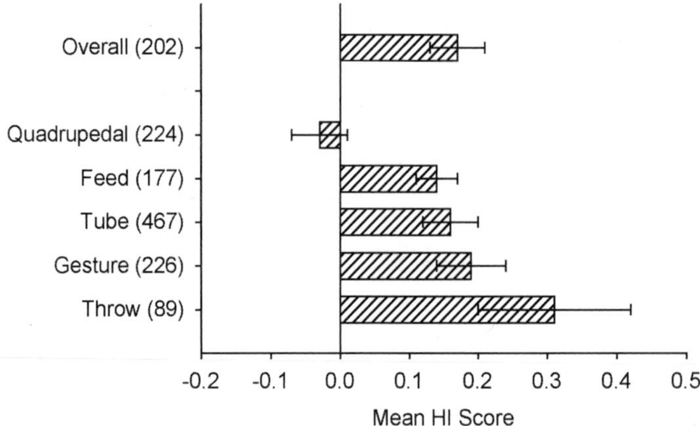

Figure 5.1
Mean handedness index (HI) and sample number for measures of handedness (*y*-axis) in chimpanzees.

Overall). This value also deviates significantly from chance, indicating that the population-level right-handedness is not necessarily task-specific, although clearly some measures are more sensitive to right-hand use than others.

One of the criticisms of our work has centered on the issue of measuring bouts of contrasted individual hand use events, particularly for a measure referred to as the TUBE task. The TUBE tasks entails the use of polyvinyl chloride (PVC) tubes (24–31 cm long, 2.5 cm wide) with peanut butter smeared on the inside edge, approximately 2–4 cm in depth. The tubes are given to the subjects in their home cage by pushing them through the cage mesh. The digit and hand used to remove the peanut butter are recorded as either the left or right each time the subjects insert a finger, remove peanut butter from the tube, and place the finger in their mouth. Observations continue until the subjects stop showing interest in the tube (usually when they have eaten all the peanut butter), dropped it for at least 10 seconds, or pushed the tube back out of their home cage through the cage mesh. Some have suggested that right- or left-hand responses for each insertion of the finger are not independent of each other and that recording each insertion as independent inflates the sample size of observations and potentially biases the characterization of hand use by the chimpanzees. To address this criticism, in two separate studies we examined bouts of hands use in conjunction with frequency for the TUBE task (Hopkins, Fernandez-Carriba, et al.,

2001, Hopkins, Cantalupo, Freeman, et al., in press). Analyses indicated that the two measures are virtually identical and the correlation between the two measures is positive and significant ($r = 0.98$, $df = 108$, $P < 0.001$). In addition, it should be pointed out that for some of our previous handedness measures, notably manual gestures and throwing, discrete responses were obtained for each response, and therefore these results cannot be accounted for by a lack of independence of data points.

Another criticism of our handedness work has been that the results have been limited to the sample of apes residing at the YNPRC (Palmer, 2003). To assess whether population-level handedness is restricted to the YNPRC chimpanzees, additional studies in two other colonies of chimpanzees, those at the University of Texas M. D. Anderson Cancer Center (BASTROP) and those at the Alamogordo Primate Facility (see Hopkins, Wesley, et al., 2004), were conducted using the same measures of hand use that were employed in the YNPRC colony. The most data for comparison with the YNPRC data were collected at BASTROP. The specific measures studied included tool use, simple reaching, manual gestures, and the TUBE task. The mean HI scores for each colony of chimpanzees and measure are shown in figure 5.2. Analysis of variance and t tests failed to reveal any evidence of significant colony differences in hand use. Moreover, we also failed to find evidence of differences in hand preference between chimpanzees raised by humans and those raised by their

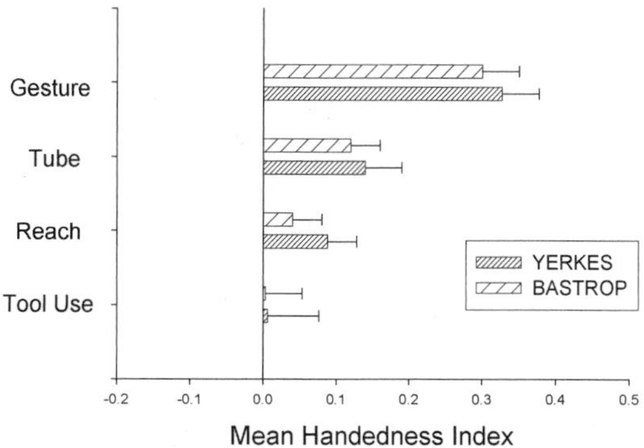

Figure 5.2
Comparison of mean HI scores among chimp populations at Yerkes National Primate Research Center and the University of Texas M. D. Anderson Cancer Center (BASTROP) on four measures of handedness. No significant between-colony differences were noted.

conspecific mothers. This suggests that the population-level handedness observed in captive chimpanzees is not due to inherent differences in how they are handled and raised by humans and chimpanzees, as has been suggested by some (McGrew & Marchant, 1997; Palmer, 2003).

Asymmetries in Motor Skill

In addition to hand preferences, we have examined differences in performance, rather than preference, of the left and right hands when grasping food items. There have been two dimensions of this work. One set of studies focused on variation in hand use in relation to grip morphology. Basically, we compared the type of grip used between the left and right hands when subjects were grasping small food items (see Hopkins, Cantalupo, et al., 2002; Hopkins & Russell, 2004; Hopkins, Russell, Hook, et al., in press). Grip morphology was characterized as thumb-index, middle-index, or single-digit responses. Whether recording hand use and grip morphology during free reaching or when systematically assessing grip morphology when an equal number of responses were obtained from each hand, the general results have been the same, with significantly greater use of thumb-index grips for the right than for the left hand (see also Christel, 1994; Tonooka & Matsuzawa, 1995).

Rates of errors in grasping were also recorded in the YNPRC chimpanzees as a direct measures of motor skill between the left and right hands (Hopkins, Cantalupo, et al., 2002; Hopkins & Russell, 2004). In these studies, subjects were required to reach and grasp an equal number of small food items with the left and right hands. The experimenters recorded the number of times the subjects made errors in grasping the food (i.e., dropped the food item). Because grip morphology varies between and within subjects, we tested subjects with and without constraints on the type of grip they used to grasp the food item. In all cases, we found that chimpanzees made more errors with the left hand than with the right hand. This effect was influenced by the handedness of the chimpanzees, with right-handed and ambiguously handed subjects making more errors with the left hand (compared with the right hand), whereas no between-hand differences were found for ambiguously handed or left-handed chimpanzees (figure 5.3).

Facial Expressions

In collaboration with Samuel Fernandez-Carriba, we assessed asymmetries in facial expressions by chimpanzees (Fernandez-Carriba, Loeches,

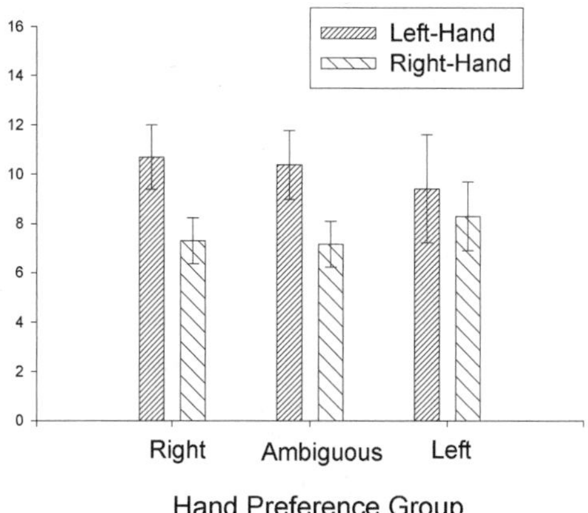

Figure 5.3
Influence of handedness on errors made with left or right hand in right-handed, left-handed, or ambiguously handed chimpanzees (Yerkes colony).

Morcillo, & Hopkins, 2002). We adopted procedures not unlike those previously used by others in humans and nonhuman primates (Hauser, 1993; Hook-Costigan & Rogers, 1998). Observations were made on a sample of 36 chimpanzees (*Pan troglodytes*) from the Yerkes colony and the colony at the Madrid Zoo-Aquarium (Madrid, Spain). Facial expressions were recorded in 10 adults, 15 subadults, and 11 juveniles. Both compounds have an outdoor area approximately 550 m^2 and five indoor rooms each about 12.6 m^2. The outdoor area is surrounded by grate walls 6.1 m high with an observation tower in one corner from which the chimpanzees were videotaped. At YNPRC and the Madrid Zoo, all observations were recorded in the outdoor portion of the chimpanzees' home cage.

At both locations, chimpanzees were observed ad libitum and all social interactions that spontaneously took place were recorded with a video camera (Sony SVHS). Videotapes were then reviewed and facial expressions were categorized according to morphological and functional criteria based on previous work in chimpanzees (Goodall, 1986). Five categories of facial expressions (*pant-hooting, play face, silent pout, silent bared-teeth display,* and *staring bared-teeth scream face*) and a neutral, nonemotional category were recorded in 9 or more subjects in our sample. Video images were analyzed using frame-by-frame procedures (24 frames/s) and then digitized in a bitmap format using a video capture card (WinView 601). The fact that chimpanzees had been recorded during their natural interactions limited the number of useful frames to usually one in each suitable sequence. If more than one frame was found in the same sequence, the best was selected. A criterion used in the selection was to obtain at least one image for each subject in each category of emotional expression, and two where possible. From the videotape, the total number of usable images of facial expressions was 183, including 39 images in the *hooting* category, 29 images in the *play* category, 14 images in the *pout* category, 31 images in the *silent bared-teeth* category, 19 images in the *scream face* category, and 51 neutral faces. The number of individual subjects represented in each of the six categories was 22, 18, 9, 20, 11, and 30, respectively.

An objective index of facial asymmetry, based on previously published procedures (Hook-Costigan & Rogers, 1998), was employed in this study. This index consisted of the length and area measures of the left and right hemimouth for each facial expression. To obtain these data, a line was drawn on the face of the focal subject between the inner corners of the two eyes using the program Adobe Photoshop 4.0.1. (Adobe Systems, San Jose, Calif.) and its midpoint was calculated. A perpendicular line that split the face into two halves was then drawn at the midpoint. The image was then vertically rotated until the midline made a perfect 90° angle with the horizontal line. For the hemimouth length, a straight line was drawn from each outer corner of the mouth to the midline (mm). To measure the area, a line surrounding the mouth perimeter was drawn using a freehand tool and the inner surface was calculated (mm^2). Both area and length were measured using the program Scion Image (Scion Corp., Frederick, Md.). Following the procedure used by Hook-Costigan and Rogers (1998), we also tried to prevent the use of portraits that were not absolutely frontal. For this, we calculated the distance between the outer corner of each eye to the midline for all the images using the program Scion Image. This served two purposes: first, based on this measure, we could assess the degree of variation in the frontal view of the subjects depicted in the image, and second, this measure allowed us to compare the relative degree of asymmetry of the mouth compared to the symmetry of the image.

For each image, the measure of the left hemimouth was subtracted from the measure of the right hemimouth and divided by the sum of right and left measures (right – left)/(right + left) to derive a facial asymmetry index (FAI). The FAI was calculated for both the area and length measures and allowed us to compare the asymmetries of images with different pixel densities and resolutions and interpret them in the same way (negative values as left asymmetry, positive values as right asymmetry,

Table 5.1
Asymmetries for Five Facial Expressions in Chimpanzees

Expression Type	N	Mean FAI	SE	*t* Value
Hooting	22	–0.0667	0.024	–2.87*
Play	18	–0.0775	0.031	–2.45*
Pout	9	–0.0427	0.063	–0.58
Silent bared-teeth	20	–0.0477	0.015	–2.83**
Scream face	11	–0.0578	0.029	–2.10***

*$P < 0.01$, **$P < 0.05$, ***$P < 0.10$.
Abbreviation: FAI, facial asymmetry index.

and 0 as symmetry). FAI values were also calculated for the measures of the distance from the outer corners of the eyes to the midline.

Table 5.1 lists the mean FAIs for each expression type and associated *t* values. Population-level biases in facial expression were tested using a one-sample *t* test, and our test value was compared to the FAI measure that was calculated from the category average for the distances from the outer corner of the eyes to the middle line. The results indicated that length measures in the *hooting*, *silent bared-teeth*, and *scream face* categories deviated significantly to the left. With respect to area measures, *hooting*, *play*, and *silent bared-teeth* also showed a significant leftward asymmetry. The correlation between the two hemimouth measures (length and area) taken in each image was positive and significant ($r = 0.718$, N = 132, $P < 0.001$).

Neuroanatomical Asymmetries in Chimpanzees

For the past 6 years, structural magnetic resonance images (MRIs) have been obtained in a sample of chimpanzees and other great apes. To date, 70 chimpanzees, 5 bonobos, 5 gorillas, and 5 orangutans have been scanned. The bulk of scans have been obtained in vivo, but a small subsample have been obtained from cadaver specimens that were either in the tissue bank or were taken after the animals died of natural causes. The specific scanning protocols and landmarks used to define each region have been described elsewhere and will not be presented in detail here (see Cantalupo & Hopkins, 2001; Cantalupo, Pilcher, & Hopkins, 2003; Hopkins, Marino, Rilling, & MacGregor, 1998).

Measurement of asymmetries has focused on cortical and subcortical brain areas. In the cortex, measures have been taken from the planum temporale (PT), inferior frontal gyrus (IFG), the motor/hand area, referred to as the KNOB (Hopkins & Pilcher, 2001), central sulcus depth (CS), posterior central gyrus (PCG), inferior parietal lobe (IPL), and sylvian fissure length (SF) (Cantalupo et al., 2003; Hopkins, Marino, et al., 1998; Hopkins, Pilcher, & MacGregor, 2000; Hopkins & Cantalupo, 2004a). Asymmetries in subcortical areas have included the anterior cingulate gyrus, hippocampus, and amygdala (Freeman, Cantalupo, & Hopkins, 2004). Torque asymmetries have been obtained from the cerebral cortex and cerebellum (Cantalupo, Freeman, & Hopkins, in press; Pilcher, Hammock, & Hopkins, 2001).

For each brain, an asymmetry quotient has been calculated using the formula (AQ = (R − L)/((R + L) * 0.5)). Positive values reflect right hemisphere biases and negative values represent left hemisphere biases. Shown in figure 5.4 are the mean AQ values for the cortical and sub-

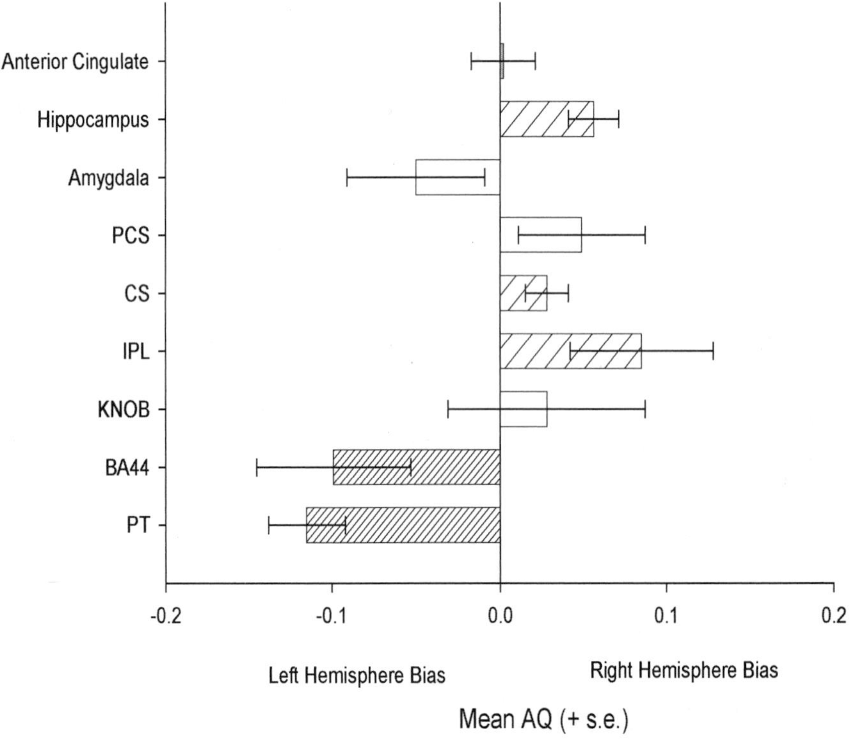

Figure 5.4
Mean asymmetry quotient (AQ, ±SE) reflecting a population-level right or left hemisphere bias in the specified cortical and neocortical brain areas, determined from in vivo and posthumous MRI measurements.

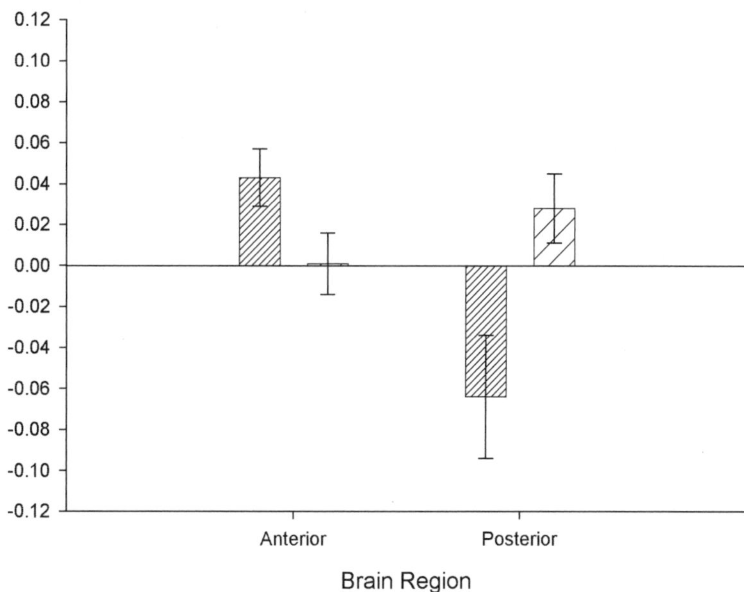

Figure 5.5
Population-level asymmetry in the cerebral cortex, determined using MRI.

cortical regions of interest. Population-level left hemisphere asymmetries have been found for the IFG, PT, and SF. Population-level right hemisphere biases have been found for the CS and IPL and hippocampus. No population-level biases have been found for the cingulate gyrus, postcentral gyrus, and amygdala. For the torque measures, a right-frontal, left-occipital asymmetry has been found for the cerebral cortex, but not the cerebellum (Cantalupo, Freeman, & Hopkins, in press) (figure 5.5).

The evidence of leftward asymmetries for the SF is consistent with one previous report in cadaver specimens (Yeni-Komshian & Benson, 1976). Similarly, the leftward asymmetries in the PT found in our sample of apes are consistent with previous reports based on cadaver measurements (Gannon, Holloway, Broadfield, & Braun, 1998) and other in vivo studies (Gilissen, 2001). The issue of asymmetries in the IFG remains somewhat controversial. Sherwood, Broadfield, Holloway, Gannon, and Hof (2003) claim that the cytoarchitectonic map used by Cantalupo and Hopkins (2001) was old (which we acknowledged in our paper) and the region we measured was not made up wholly of BA44 cells but included BA45 cells as well. Notwithstanding, this does not negate the pattern of asymmetry observed because we used consistent landmarks to define the region of interest.

Behavioral Correlates of Neuroanatomical Asymmetries

The evidence of population-level handedness and neuroanatomical asymmetries in chimpanzees naturally leads to the question of whether asymmetries in these two systems are associated with each other. We have some evidence that handedness correlates with asymmetries in the PCG of the brain of chimpanzees (Hopkins & Cantalupo, 2004a). For this study, we compared the AQ scores of left- and right-handed subjects for three separate measures, the TUBE measure, bimanual feeding, and simple reaching. We selected these three behavioral measures because we had the most complete set and because these three measures differ with respect to their sensitivity in detecting individual hand preferences and are uncorrelated with each other. With respect to their sensitivity in detecting individual differences in hand use, the absolute values of the HI scores for the TUBE task are significantly higher (mean = 0.33) than the bimanual feeding (mean = 0.22) and reaching tasks (mean = 0.11). For the TUBE task, we found significant differences in the KNOB region, with right-handed subjects having a larger left KNOB and left-handed subjects having a larger right KNOB (figure 5.6). Differences between

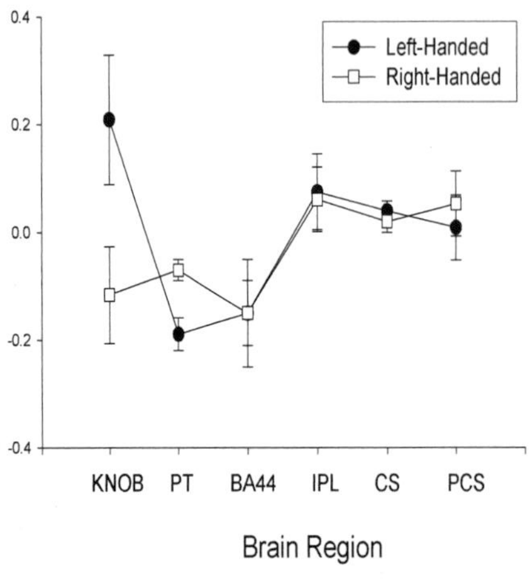

Figure 5.6
Comparison of population-level asymmetries in brain areas with right- or left-handedness. Significant differences were found in the KNOB region on the TUBE task.

other brain regions as a function of handedness were not evident. This same pattern of results was found for the bimanual feeding task but failed to reach conventional levels of statistical significance.

Factors Influencing Individual Differences in the Expression of Hemispheric Specialization

We have also been interested in the influence of genetic and nongenetic factors on the development of hemispheric specialization in chimpanzees. These analyses have primarily focused on handedness, specifically handedness as determined on the TUBE task, because much more data are available for testing various genetic and nongenetic models. The potential role of genetic factors is particularly important because some genetic models of handedness in humans presume that lateralization for speech and language is the mechanism driving the expression of right-handedness. For example, the right-shift (RS) theory of handedness (Annett, 1985) is one of the most commonly cited and accepted genetic models of human handedness and therefore is a good model to test. The RS theory proposes that individuals either inherit or do not inherit a gene that codes for left hemisphere dominance for speech. In this model, right-handedness is a consequence of the left hemisphere dominance for language, and therefore indirectly indicates the expression of the gene. According to the RS model, the RS gene (r+) is inherited from either the sire or the dam and is dominant. Therefore, genotypes made up of (r+/r+), (r+/r–), (r–/r+) would all be right-handed. Individuals inheriting the (r–/r–) genotype would be randomly distributed in their hand preferences, with half being left-handed and half being right-handed. The principal observations in support of the RS genetic model (and many others) for hand preference are that hand preference runs in families (Curt, De Agostini, Maccario, & Dellatolas, 1995; Laland, Kumm, Van Horn, & Feldman, 1995; McGee & Cozad, 1980; McManus & Bryden, 1992) and that offspring typically exhibit patterns of hand preference more similar to the patterns of their biological parents than to offspring who have been either adopted (Carter-Saltzman, 1980) or raised by step-parents (Hicks & Kinsbourne, 1978).

To assess whether handedness runs in the families of chimpanzees, we conducted similar analyses of heritability in hand use. For these analyses, we examined the association between offspring and maternal and paternal hand preference in a cohort of 467 chimpanzees for which hand preference data had been collected using the TUBE task. Subjects were classified as left- or right-handed on the basis of the sign of their

Table 5.2
Distribution of Handedness in Offspring Born to Right- and Left-Handed Dams and Sires

	Offspring Handedness	
	Left-Handed	Right-Handed
Dam		
Left-handed	21	41
Right-handed	41	95
Sire		
Left-handed	15	38
Right-handed	18	39

HI scores. Subjects with positive values were classified as right-handed and subjects with negative HI values or a value of zero were classified as left-handed. Table 5.2 shows the distribution of left- and right-handed offspring as a function of the handedness of their biological dam and sire. Chi-square tests of independence failed to reveal significant associations between maternal or paternal hand preference and the handedness of their offspring.

Some of the chimpanzees tested were raised by their biological conspecific mothers and others were raised by humans. This differential rearing of genetically related individuals allowed us to assess the influence of rearing on the potential genetic expression of handedness in chimpanzees, in much the same manner that partial cross-fostering or adoption studies are conducted in humans. Overall, concordance in handedness between mothers and their offspring was significantly greater than chance (63%, 116/184), but this was not the case for offspring and sires (49%, 54/110). Chi-square tests of independence failed to reveal a significant association between maternal-offspring handedness and whether the offspring was reared with (63%) or apart from (55%) the biological mother.

Maternal Age/Birth Order

In addition to the heritability analyses, we have also examined the influence of birth order on handedness (Hopkins & Dahl, 2000; Hopkins, Dahl, & Pilcher, 2000). Birth order has been our primary variable of interest because previous studies in humans have reported that pre- and perinatal factors, such as birth trauma and birth stress, can influence handedness. Most stressful birth events of this type covary with maternal age; thus, birth order is a simple means of considering the potential role of these perinatal events. We have classified the chimpanzees as being either first-born, middle-born (parities 2–6), or latter-born (parities of 7 or higher). Analysis of variance examining the effects of sex, rearing, and birth order revealed a significant main effect for birth order ($F(2, 333) = 3.28$, $P < 0.03$). The mean HI score for first-born chimpanzees was significantly lower (mean = 0.023) than the mean HI score for the middle-born (mean = 0.143) but not the latter-born (mean = 0.092).

Because birth order has an influence on handedness, it seems reasonable to suggest that this variable may also influence the expression of handedness in related individuals. To this end, concordance rates in hand preference between dams and offspring were compared as a function of birth order (first, second or later). A significant association was found ($\chi^2(1, N = 196) = 4.36$, $P < 0.04$). Concordance in hand preference was significantly higher between dams and offspring in the second-and-later-born cohort (64%) than in the first-born cohort (41%). Thus, the possible genetic basis for handedness in chimpanzees may be modified to some extent by the parity of the fetus.

Birth Order and Neuroanatomical Asymmetry

As noted earlier in the discussion, hand preference as observed on the TUBE task correlates with neuroanatomical asymmetries in the KNOB area of the brain. Because birth order has a significant effect on handedness for the TUBE task, it follows that birth order might also have a significant influence on the development of neuroanatomical asymmetries. To consider this possibility, the cortical areas of brain asymmetry were compared as a function of the birth order classification previously used. A significant main effect for birth order was found for the KNOB ($F(2, 46) = 4.29$, $P < 0.03$). Within the cohort of chimpanzees for which MRI scans were available, a similar main effect for birth order was found on handedness for the TUBE task ($F(2, 46) = 4.42$, $P < 0.02$). The mean HI for the TUBE task and the mean AQ for the KNOB are given in table 5.3. There is a significant proportion of left-handed first-born chimpanzees, and these subjects have strongly rightward asymmetries in the KNOB area. In contrast, middle-born chimpanzees have a right-hand bias for the TUBE task and leftward asymmetry in the KNOB area. The anomalous cohort is the latter-born chimpanzees. These individuals are somewhat right-handed but also show larger rightward asymmetries in

Table 5.3
Relation Between Birth Order and Handedness and Neuroanatomical Asymmetry of the KNOB Area of Brain

	Birth Order		
	1	2–6	7+
KNOB AQ	0.026	–0.061	0.372
	(0.12)	(0.07)	(0.14)
TUBE measure	–0.217	0.267	0.179
	(0.13)	(0.06)	(0.15)

the KNOB area. Thus, there is some dissociation between hand preferences and these neurobiological correlates.

Comment

In my opinion, there is now very good evidence of hemispheric specialization in nonhuman primates, particularly chimpanzees. Chimpanzees show population-level neuroanatomical and behavioral asymmetries, and at least some of the lateralized behaviors correlate with certain brain regions. Our preliminary studies suggest that genetic and nongenetic factors influence the expression of handedness in chimpanzees, but these variables need to be further explored.

Despite the considerable evidence of population-level laterality in vertebrates that has accumulated over the past 15 years, some caveats remain that warrant discussion. First, the evidence of population-level handedness seems fairly robust in captive chimpanzees. Evidence of population-level asymmetries in wild chimpanzees is less frequently reported, if not virtually absent (see McGrew & Marchant, 1997). Why the differences exist between wild and captive chimpanzees remains unclear; several rival hypotheses have been postulated, but the issue remains unresolved (McGrew & Marchant, 1997; Hopkins & Cantalupo, 2004a). For example, in contrast to studies in wild chimpanzees, studies in captive animals have had relatively large sample sizes and good experimental control over situational and postural factors. In addition, for at least some studies in wild chimpanzees, the behaviors of interest often fail to elicit hand preferences at the individual level (Marchant & McGrew, 1996; McGrew & Marchant, 2001). Thus, there are a number of methodological differences between studies in wild and captive chimpanzees that need to be rectified before it is possible to conclude that true differences are present between these cohorts. It should also be noted that in many studies of captive chimpanzees, distributions in hand preferences have been compared between wild-caught and captive-born individuals, and there is no evidence of a significant difference.

Second, the relative distribution of right- and left-handed subjects in chimpanzees is about 2:1 or 3:1, depending on the measure, values that are substantially lower than those reported in human samples (Raymond & Pontier, 2004). The nature of this difference also remains unclear, with some suggesting genetic factors (Corballis, 1997) and others proposing life history, cultural, or measurement factors as potential explanations (Hopkins, 2004). This issue requires much more data, particularly as it relates to the mechanisms that influence the expression of handedness in human and nonhuman primates. Our preliminary studies on the role of genetic and nongenetic factors do not suggest a genetic explanation, but there are a number of limitations to these studies, including (1) multiple representation of individual sires and dams within the same cohort and (2) limited statistical power. Much larger samples of chimpanzees will be necessary to address the potential role of genetics in the development of hemispheric specialization in chimpanzees and other primates. The virtual absence of studies in twins is also problematic for comparative studies of hemispheric specialization (but see Rogers & Kaplan, 1998). I believe that data from some of the species that have twin births relatively frequently, such as tamarins and marmosets, would be particularly useful for exploring the role of genetic and nongenetic factors in the development of hemispheric specialization.

Third, the relationship between structural and functional asymmetries has been limited to correlational analyses primarily between measures of hand use and structural MRI measures. This is a good starting point, but whether other lateralized behaviors, such as orofacial asymmetries, are linked to the structural asymmetries remains unaddressed. Moreover, structural MRI does not measure neurophysiology and connectivity between brain regions that underlie a specific behavior. Clearly, functional imaging techniques need to be further developed for nonhuman primates, including the larger and stronger great apes. PET and fMRI have been recently employed in New and Old World monkeys (e.g., Stefanacci et al., 1998), and some of these procedures and techniques work with chimpanzees, as well (Rilling et al., 2001). This remains an important avenue of future research.

Fourth, comparatively, there are some similarities between the observed behavioral and neuroanatomical asymmetries in chimpanzees and other nonhuman primate species, but there are also some interesting differences that warrant further study. For example, sylvian fissure length is probably the most common measure of neuroanatomical asymmetries in nonhuman primates. Chimpanzees and other apes show a leftward asymmetry, particularly in the posterior portion of the sylvian fissure. Asymmetries in sylvian fissure length have been reported in the genus *Macaca* by some (Heilbronner & Holloway, 1988; Hopkins, Dahl, et al., 2000) but not others (Falk et al., 1990). Moreover, asymmetries in the sylvian fissue are more pronounced in the anterior than in the posterior region, which is the case in chimpanzees. Thus, how to interpret this pattern of asymmetry remains unclear, as does the significance of these differences. Behaviorally, there are some similarities as well, but also many differences. For example, asymmetries in orofacial expressions seem to be left-sided in chimpanzees, rhesus monkeys, and common marmosets (Fernandez-Carriba et al., 2002; Hauser, 1993; Hook-Costigan & Rogers, 1998), suggesting a prolonged evolutionary history in primates. In contrast, asymmetries in handedness for identical measures reveal quite different results in various primates. For example, the TUBE task has been administered to several primate species, including chimpanzees, gorillas, orangutans, baboons, rhesus monkeys, and capuchin monkeys (Hopkins, Stoinski, et al., 2003). Figure 5.7 shows the mean HI scores for the combined data from each study and species. Chimpanzees, gorillas, baboons, and capuchin monkeys all show population-level right-handedness. Orangutans show a population-level left-hand bias, and rhesus macaques show no population-level bias. There is no easy evolutionary interpretation of these findings, and clearly a more careful consideration of the ecological, social, or morphological factors that influence behavioral and neuroanatomical asymmetries in primates needs to be undertaken.

Finally, many theories on the origins of hemispheric specialization are rooted in a neuropsychological perspective that emphasizes the role of language or other higher cognitive processes, such as tool use, as the skills that were selected for in evolution (for reviews, see Bradshaw & Rogers, 1993; Calvin, 1982; Corballis, 1997, 2002). The data from chimpanzees affirm previous claims that language is not a necessary condition for the expression of brain asymmetry; rather, brain asymmetry appears to be a fundamental attribute of the central nervous system of primates and other vertebrates (Rogers & Andrew, 2002). More recently,

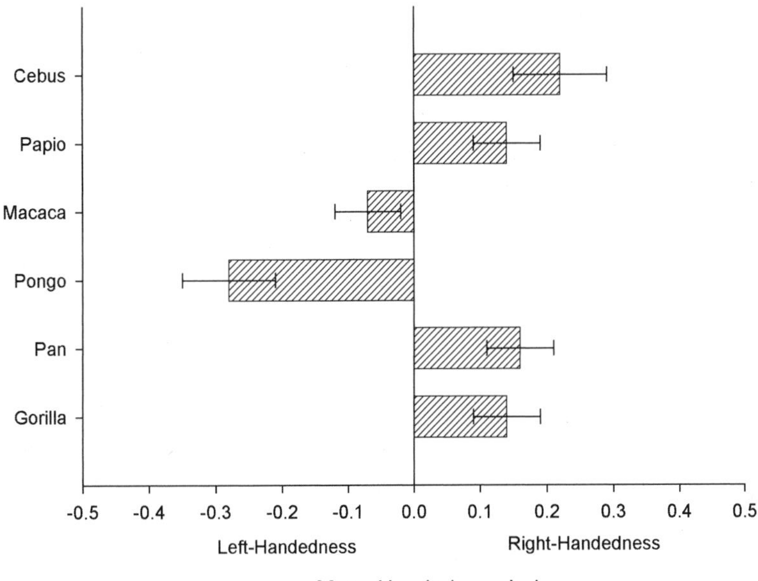

Figure 5.7
Population-level mean HI scores in several nonhuman primate species on the TUBE task.

some have suggested that the neuropsychological emphasis is far too narrow and that hemispheric specialization should be considered in a wider evolutionary framework, perhaps associated with social or developmental factors (e.g., Hopkins, 1994). One of the more intriguing theories to emerge on the evolution of hemispheric specialization has been proposed by Vallortigara and Rogers (in press). They argue that many theories on the origins of hemispheric specialization have proposed that duality of function allowed for the doubling of cognitive ability by, in essence, removing duplication in function between hemispheres. This argument can certainly be made based on the extant literature, but Vallortigara and Rogers argue that duality of function could be accomplished in the absence of population-level asymmetries. In other words, there may have been selection for more duality of function between hemispheres, but conformity to the same biases would not have been necessary to attain this level of dual functions. Rather than focus on duality of function, Vallortigara and Rogers argue that conformity in directional biases was selected for in social behaviors as an evolutionarily stable strategy to address predator and prey relations. The extent to which this theory applies to variation in primate laterality remains to be seen, but this approach certainly offers new vistas in the comparative study of hemispheric specialization in primates, including humans.

Acknowledgments Work was supported in part by NIH grants RR-00165, NS-36605, NS-42867, and HD-38051. I am grateful to Hani Feeman, Autumn Hostetter, Michael Wesley, Jamie Russell, Dawn Pilcher, Dr. Claudio Cantalupo, and Dr. Steve Schapiro for their steadfast assistance and collegiality.

References

Annett, M. (1985). *Left, right, hand, and brain: The right-shift theory.* London: Erlbaum.

Beaton, A. A. (1997). The relation of planum temporale asymmetry and morphology of the corpus callosum to handedness, gender, and dyslexia: A review of the evidence. *Brain and Language*, *15*, 255–322.

Borod, J. C., Haywood, C. S., & Koff, E. (1997). Neuropsychological aspects of facial asymmetry during emotional expression: A review of the normal adult literature. *Neuropsychology Review*, *7*, 41–60.

Bradshaw, J., & Rogers, L. J. (1993). *The evolution of lateral asymmetries, language, tool use and intellect.* San Diego: Academic Press.

Byrne, R. W., & Byrne, J. M. (1991). Hand preferences in the skilled gathering tasks of mountain gorillas (*Gorilla gorilla berengei*). *Cortex*, *27*, 521–536.

Cain, D. P., & Wada, J. A. (1979). An anatomical asymmetry in the baboon brain. *Brain, Behavior and Evolution*, *16*, 222–226.

Calvin, W. H. (1982). *The throwing Madonna: Essays on the brain.* New York: McGraw-Hill.

Cantalupo, C., Freeman, H., & Hopkins, W. D. (in press). Patterns of cerebellar asymmetry in great apes as revealed by MRI. *Brain, Behavior and Evolution.*

Cantalupo, C., & Hopkins, W. D. (2001). Asymmetric Broca's area in great apes. *Nature*, *414*, 505.

Cantalupo, C., Pilcher, D., & Hopkins, W. D. (2003). Are planum temporale and sylvian fissure asymmetries directly related? A MRI study in great apes. *Neuropsychologia*, *41*, 1975–1981.

Carter-Saltzman, L. (1980). Biological and sociocultural effects on handedness: Comparison between biological and adoptive parents. *Science*, *209*, 1263–1265.

Cheverud, J. M., Falk, D., Hildebolt, C., Moore, A. J., Helmkamp, R. C., & Vannier, J. (1990). Heritability and association of cortical petalia in rhesus monkeys (*Macaca mulatta*). *Brain, Behavior and Evolution*, *35*, 368–372.

Christel, M. I. (1994). Catarrhine primates grasping small objects: Techniques and hand preferences. In J. R. Anderson, J. J. Roeder, B. Thierry, & N. Herrenschmidt (Eds.), *Current primatology, Vol. IV, Behavioral neuro-*

science, physiology and reproduction (pp. 37–49). Strasbourg: Universite Louis Pasteur.

Corballis, M. C. (1997). The genetics and evolution of handedness. *Psychological Review*, *104*, 714–727.

Corballis, M. C. (2002). *From hand to mouth: The origins of language*. Princeton, N.J.: Princeton University Press.

Curt, F., De Agostini, M., Maccario, M., & Dellatolas, G. (1995). Parental hand preference and manual functional asymmetry in preschool children. *Behavior Genetics*, *25*, 525–536.

Damerose, E., & Vauclair, J. (2002). Posture and laterality in human and nonhuman primates: Asymmetries in maternal handling and infant's early motor asymmetries. In L. Rogers & R. Andrews (Eds.), *Comparative vertebrate lateralization* (pp. 306–362). Oxford: Oxford University Press.

Dewson, J. H. (1977). Preliminary evidence of hemispheric asymmetry of auditory function in monkeys. In S. Harnad, R. W. Doty, L. Goldstein, J. Jaynes, & G. Krauthamer (Eds.), *Lateralization in the nervous system* (pp. 63–74). New York: Academic Press.

Ettlinger, G. F. (1988). Hand preference, ability and hemispheric specialization: How far are these factors related in the monkey? *Cortex*, *24*, 389–398.

Falk, D., Hildebolt, C., Cheverud, J., Vannier, M., Helmkamp, R. C., & Konigsberg, L. (1990). Cortical asymmetries in the frontal lobe of rhesus monkeys (*Macaca mulatta*). *Brain Research*, *512*, 40–45.

Fernandez-Carriba, S., Loeches, A., Morcillo, A., & Hopkins, W. D. (2002). Asymmetry of facial expression of emotions by chimpanzees. *Neuropsychologia*, *40*, 1523–1533.

Finch, G. (1941). Chimpanzee handedness. *Science*, *94*, 117–118.

Freeman, H., Cantalupo, C., & Hopkins, W. D. (2004). Asymmetries in the hippocampus and amygdala of chimpanzees (*Pan troglodytes*). *Behavioral Neuroscience*, *118*, 1460–1465.

Gannon, P. J., Holloway, R. L., Broadfield, D. C., & Braun, A. R. (1998). Asymmetry of chimpanzee planum temporale: Humanlike pattern of Wernicke's brain language area homolog. *Science*, *279*, 220–222.

Gazzaniga, M. S. (2000). Cerebral specialization and interhemispheric communication: Does the corpus callosum enable the human condition? *Brain*, *123*, 1293–1326.

Geschwind, N., & Levitsky, W. (1968). Human brain: Left-right asymmetries in the temporal speech region. *Science*, *151*, 186–187.

Gilissen, E. (2001). Structural symmetries and asymmetries in human and chimpanzee brains. In D. Falk & K. R. Gibson (Eds.), *Evolutionary anatomy of the primate cerebral cortex* (pp. 187–215). Cambridge: Cambridge University Press.

Goodall, J. (1986). *The chimpanzees of Gombe: Patterns in adaptation*. Cambridge, MA: Harvard University Press.

Groves, C. P., & Humphrey, N. K. (1973). Asymmetry in gorilla skulls: Evidence of lateralized brain function? *Nature*, *244*, 53–54.

Hamilton, C. R. (1977). An assessment of hemispheric specialization in monkeys. *Annals of the New York Academy of Sciences*, *299*, 222–232.

Hamilton, C. R., & Vermiere, B. A. (1988). Complimentary hemispheric specialization in monkeys. *Science*, *242*, 1694–1696.

Hauser, M. C. (1993). Right hemisphere dominance in the production of facial expression in monkeys. *Science*, *261*, 475–477.

Heffner, H. E., & Heffner, R. S. (1984). Temporal lobe lesions and perception of species-specific vocalizations by macaques. *Science*, *226*, 75–76.

Heilbronner, P. L., & Holloway, R. L. (1988). Anatomical brain asymmetries in New World and Old World monkeys: Stages of temporal lobe development in primate evolution. *American Journal of Physical Anthropology*, *76*, 39–48.

Heilbronner, P. L., & Holloway, R. L. (1989). Anatomical brain asymmetry in monkeys: Frontal, temporoparietal, and limbic cortex in *Macaca*. *American Journal of Physical Anthropology*, *80*, 203–211.

Hicks, R., & Kinsbourne, M. (1976). Genetic basis for human handedness: Evidence from a partial cross-fostering study. *Science*, *192*, 908–910.

Holloway, R. L., & De La Coste-Lareymondie, M. C. (1982). Brain endocast asymmetry in pongids and hominids: Some preliminary findings on the paleontology of cerebral dominance. *American Journal of Physical Anthropology*, *58*, 101–110.

Hook-Costigan, M. A., & Rogers, L. J. (1998). Lateralized use of the mouth in production of vocalizations by marmosets. *Neuropsychologia*, *36*, 1265–1273.

Hopkins, W. D. (1994). Hand preferences for bimanual feeding in 140 captive chimpanzees (*Pan troglodytes*): Rearing and ontogenetic factors. *Developmental Psychobiology*, *27*, 395–407.

Hopkins, W. D. (1995). Hand preferences for a coordinated bimanual task in 110 chimpanzees: Cross-sectional analysis. *Journal of Comparative Psychology*, *109*, 291–297.

Hopkins, W. D. (2004). Laterality in maternal cradling and infant positional biases: Implications for the evolution and development of hand preferences inn nonhuman primates. *International Journal of Primatology*, *25*, 1243–1265.

Hopkins, W. D., Bard, K. A., Jones, A., & Bales, S. (1993). Chimpanzee hand preference for throwing and infant cradling: Implications for the origin of human handedness. *Current Anthropology*, *34*, 786–790.

Hopkins, W. D., & Cantalupo, C. (2004a). Handedness in chimpanzees is associated with asymmetries in the primary motor but not with homologous language areas. *Behavioral Neuroscience*, *118*, 1176–1183.

Hopkins, W. D., & Cantalupo, C. (2004b). Individual and setting differences in the hand preferences of chimpanzees (*Pan troglodytes*): A critical analysis and some alternative explanations. *Laterality*, *10*, 65–80.

Hopkins, W. D., Cantalupo, C., Freeman, H., Russell, J., Kachin, M., & Nelson, E. (in press). Chimpanzees are right-handed when recording bouts of hand use. *Laterality*.

Hopkins, W. D., Cantalupo, C., Wesley, M. J., Hostetter, A., & Pilcher, D. (2002). Grip morphology and hand use in chimpanzees (*Pan troglodytes*): Evidence of a left hemisphere specialization in motor skill. *Journal of Experimental Psychology: General*, *131*, 412–423.

Hopkins, W. D., & Dahl, J. F. (2000). Birth order and hand preference in chimpanzees (*Pan troglodytes*): Implications for pathological models of human handedness. *Journal of Comparative Psychology*, 114, 302–306.

Hopkins, W. D., Dahl, J. F., & Pilcher, D. (2001). Genetic influence on the expression of hand preferences in chimpanzees (*Pan troglodytes*): Evidence in support of the right shift theory and developmental instability. *Psychological Science*, *12*, 299–303.

Hopkins, W. D., Fernandez-Carriba, S., Wesley, M. J., Hostetter, A., Pilcher, D., & Poss, S. (2001). The use of bouts and frequencies in the evaluation of hand preferences for a coordinated bimanual task in chimpanzees (*Pan troglodytes*): An empirical study comparing two different indices of laterality. *Journal of Comparative Psychology*, *115*, 294–299.

Hopkins, W. D., & Marino, L. M. (2000). Cerebral width asymmetries in nonhuman primates as revealed by magnetic resonance imaging (MRI). *Neuropsychologia*, *38*, 493–499.

Hopkins, W. D., Marino, L., Rilling, J., & MacGregor, L. (1998). Planum temporale asymmetries in great apes as revealed by magnetic resonance imaging (MRI). *NeuroReport*, *9*, 2913–2918.

Hopkins, W. D., & Pilcher, D. L. (2001). Neuroanatomical localization of the motor hand area using magnetic resonance imaging: The left hemisphere is larger in great apes. *Behavioral Neuroscience*, *115*, 1159–1164.

Hopkins, W. D., Pilcher, D. L., & MacGregor, L. (2000). Sylvian fissure length asymmetries in primates revisited: A comparative MRI study. *Brain, Behavior and Evolution*, *56*, 293–299.

Hopkins, W. D., & Rabinowitz, D. M. (1997). Manual specialization and tool-use in captive chimpanzees (*Pan troglodytes*): The effect of unimanual and bimanual strategies on hand preference. *Laterality*, *2*, 267–277.

Hopkins, W. D., & Russell, J. L. (2004). Further evidence of a right hand advantage in motor skill by chimpanzees (*Pan troglodytes*). *Neuropsychologia*, *42*, 990–996.

Hopkins, W. D., Russell, J., Freeman, H., Buehler, N., Reynolds, E., & Schapiro, S. J. (in press). The distribution and development of handedness for manual gestures in captive chimpanzees (*Pan troglodytes*). *Psychological Science*.

Hopkins, W. D., Russell, J., Freeman, H., Cantalupo, C., & Schapiro, S. (in press). Chimpanzee right-handedness: Throwing out the first pitch. *Journal of Comparative Psychology*.

Hopkins, W. D., Russell, J., Hook, M., Braccini, S., & Schapiro, S. (in press). Simple reaching is not so simple: Association between hand use and grip preferences in captive chimpanzees. *International Journal of Primatology*.

Hopkins, W. D., Stoinski, T., Lukas, K., Ross, S., & Wesley, M. J. (2003). Comparative assessment of handedness for a coordinated bimanual task in chimpanzees (*Pan*), gorillas (*Gorilla*), and orangutans (*Pongo*). *Journal of Comparative Psychology*, *117*, 302–308.

Hopkins, W. D., Wesley, M. J., Izard, M. K., Hook, M., & Schapiro, S. J. (2004). Chimpanzees are predominantly right-handed: Replication in three colonies of apes. *Behavioral Neuroscience*, *118*, 659–663.

Kertesz, A., Black, S. E., Polk, M., & Howell, J. (1986). Cerebral asymmetries on magnetic resonance imaging. *Cortex*, 22, 117–127.

Laland, K. N., Kumm, J., Van Horn, J. D., & Feldman, M. W. (1995). A gene-culture model of human handedness. *Behavior Genetics, 25*, 433–445.

Lehman, R. A. W. (1993). Manual preference in prosimians, monkeys, and apes. In J. P. Ward & W. D. Hopkins (Eds.), *Primate laterality: Current behavioral evidence of primate asymmetries* (pp. 107–124). New York: Springer-Verlag.

LeMay, M. (1985). Asymmetries of the brains and skulls of nonhuman primates. In S. D. Glick (Ed.), *Cerebral lateralization in nonhuman species* (pp. 223–245). New York: Academic Press.

Manning, J. T., Heaton, R., & Chamberlain, A. T. (1994). Left-side cradling: Similarities and differences between apes and humans. *Journal of Human Evolution, 26*, 77–83.

Marchant, L. F., & McGrew, W. C. (1996). Laterality of limb function in wild chimpanzees of Gombe National Park: Comprehensive study of spontaneous activities. *Journal of Human Evolution, 30*, 427–443.

McGee, M. G., & Cozad, T. (1980). Population genetic analysis of human hand preference: Evidence for generation differences, familial resemblance and maternal effects. *Behavior Genetics, 10*, 263–275.

McGrew, W. C., & Marchant, L. F. (1997). On the other hand: Current issues in and meta-analysis of the behavioral laterality of hand function in non-human primates. *Yearbook of Physical Anthropology, 40*, 201–232.

McGrew, W. C., & Marchant, L. F. (2001). Ethological study of manual laterality in the chimpanzees of the Mahale mountains, Tanzania. *Behaviour, 138*, 329–358.

McManus, I. C., & Bryden, M. P. (1992). The genetics of handedness, cerebral dominance and lateralization. In I. Rapin & S. J. Segalowitz (Eds.), *Handbook of neuropsychology, Vol. 6, Developmental neuropsychology*, Pt. 1 (pp. 115–144). Amsterdam: Elsevier.

Palmer, A. R. (2002). Chimpanzee right-handedness reconsidered: Evaluating the evidence with funnel plots. *American Journal of Physical Anthropology, 118*, 191–199.

Palmer, A. R. (2003). Reply to Hopkins and Cantalupo: Chimpanzee right-handedness reconsidered: Sampling issues and data presentation. *American Journal of Physical Anthropology, 121*, 382–384.

Pilcher, D., Hammock, L., & Hopkins, W. D. (2001). Cerebral volume asymmetries in non-human primates as revealed by magnetic resonance imaging. *Laterality, 6*, 165–180.

Preilowski, V. (1979). Performance differences between hands and lack of transfer of finger posture skill in intact rhesus monkeys: Possible model of the origins of cerebral asymmetry. *Neuroscience Letters (Suppl 3)*, 589.

Raymond, M., & Pontier, D. (2004). Is there geographical variation in human handedness? *Laterality, 9*, 35–51.

Rilling, J. K., Winslow, J. T., O'Brien, D., Gutman, D. A., Hoffman, J. M., & Kilts, C. D. (2001). Neural correlates of maternal separation in rhesus monkeys. *Biological Psychiatry, 49*, 146–157.

Rogers, L. J., & Andrew, R. J. (2002). *Comparative vertebrate lateralization.* Cambridge: Cambridge University Press.

Rogers, L. J., & Kaplan, G. (1998). Teat preference for suckling in common marmosets: Relationship to side of being carried and hand preference. *Laterality*, 3, 269–281.

Shapleske, J., Rossell, S. L., Woodruff, P. W. R., & David, A. S. (1999). The planum temporale: A systematic, quantitative review of its structural, functional and clinical significance. *Brain Research Reviews*, *29*, 26–49.

Sherwood, C. C., Broadfield, D., Holloway, R., Gannon, P., & Hof, P. (2003). Variability in Broca's area homologue in African great apes: Implications for language evolution. *Anatomical Record*, *271A*, 276–285.

Stefanacci, L., Reber, P., Costanza, J., Wong, E., Buxton, R., Zola, S., et al. (1998). fMRI of monkey visual cortex. *Neuron*, *20*, 1051–1057.

Tonooka, R., & Matsuzawa, T. (1995). Hand preferences in captive chimpanzees (*Pan troglodytes*) in simple reaching for food. *International Journal of Primatology*, *16*, 17–34.

Vallortigara, G., & Rogers, L. J. (in press). Survival with an asymmetrical brain: Advantages and disadvantages of cerebral lateralization. *Behavioral and Brain Sciences*.

Vermiere, B. A., & Hamilton, C. R. (1998). Inversion effect for faces in split-brain monkeys. *Neuropsychologia*, *36*, 1003–1014.

Warren, J. M. (1977). Handedness and cerebral dominance in monkeys. In S. Harnard, R. W. Doty, L. Goldstein, J. Jaynes, & G. Krauthamer (Eds.), *Lateralization in the nervous system* (pp. 151–172). New York: Academic Press.

Warren, J. M. (1980). Handedness and laterality in humans and other animals. *Physiological Psychology*, *8*, 351–359.

Yeni-Komshian, G., & Benson, D. (1976). Anatomical study of cerebral asymmetry in the temporal lobe of humans, chimpanzees and monkeys. *Science*, *192*, 387–389.

6 The Evolution of Brain Size and Intelligence

J. Philippe Rushton and C. Davison Ankney

In this chapter, we update our earlier reviews of the literature (Rushton & Ankney, 1996, 1997) on the relation between whole brain size and general intelligence (IQ). In 55 samples in which IQ scores (or their proxy) were correlated with external head size measures, the mean *r* was 0.20 (N = 62,602; $P < 10^{-10}$); in 27 samples using brain imaging techniques the mean was 0.40 (N = 1,341; $P < 10^{-10}$); and in 5 samples using the method of correlated vectors to extract *g*, the general factor of mental ability from test scores, the mean was 0.57. Further, we update our review of brain size/cognitive ability correlations with age, sex, social class, and race, which provide further information about the brain-behavior relationship. Finally, we examine the evolution of brain size from a behavior genetic and life history perspective.

Throughout the nineteenth and early twentieth centuries, the relation between brain size and intelligence was almost universally accepted (Broca, 1861; Darwin, 1871; Morton, 1849; Topinard, 1878). The renowned French neurologist Paul Broca (1824–1880), for example, made major contributions to refining early techniques for estimating brain size by measuring external and internal skull dimensions and weighing wet brains at autopsy. He concluded that variation in whole brain size was related to intellectual achievement, observing that mature adults had larger brains than either children or the very elderly, skilled workers had larger brains than unskilled workers, and eminent individuals had larger brains than those less eminent.

Broca's studies were cited by Charles Darwin (1871) in support of the theory of evolution in *The Descent of Man*, where he wrote:

> No one, I presume, doubts that the large size of the brain in man, relatively to his body, in comparison with that of the gorilla or orang, is closely connected with his higher mental powers. We meet the closely analogous facts with insects,

> in which the cerebral ganglia are of extraordinary dimensions in ants; these ganglia in all the Hymenoptera being many times larger than in the less intelligent orders, such as beetles. . . .
>
> The belief that there exists in man some close relation between the size of the brain and the development of the intellectual faculties is supported by the comparison of the skulls of savage and civilized races, of ancient and modern people, and by analogy of the whole vertebrate series.

Darwin's cousin, Sir Francis Galton (1888), was the first to quantify the relation between human brain size and mental ability in living subjects. He multiplied head length by breadth by height and plotted the results against age (19–25 years) and class of degree (A, B, C) in more than 1,000 male undergraduates at Cambridge University. He reported that (1) cranial capacity continued to grow after age 19, and (2) men who obtained high honors degrees had a brain size 2%–5% greater than those who did not. Years later, Karl Pearson (1906) reanalyzed Galton's data and found a correlation of 0.11 using the Pearson coefficient he had invented for this type of analysis. Pearson (1924, p. 94), who was also Galton's disciple and biographer, reported Galton's response: "He was very unhappy about the low correlations I found between intelligence and head size, and would cite against me those 'front benches' [the people on the front benches at Royal Society meetings whom Galton perceived as having large heads]; it was one of the few instances I noticed when impressions seemed to have more weight with him than measurements."

Following World War II (1939–1945) and the revulsion evoked by Hitler's racial policies, however, craniometry became associated with extreme forms of racial prejudice. After the U.S. civil rights movement became prominent in the 1950s, along with the cold war struggle for the hearts and minds of the Third World, research on brain size and intelligence virtually ceased, and the literature underwent vigorous critiques, notably from Philip V. Tobias (1970), Leon Kamin (1974), and Stephen Jay Gould (1978, 1981). As we shall show, however, modern studies confirm many of the nineteenth-century observations.

Tables 6.1 and 6.2 update and extend several recent reviews of the brain size/IQ literature (Gignac, Vernon, & Wickett, 2003; Gray & Thompson, 2004; Jensen & Sinha, 1993; McDaniel, in press; Rushton & Ankney, 1996, 1997; Vernon, Wickett, Bazana, & Stelmack, 2000). All samples were nonclinical. To be included, the published reference had to report an actual correlation; personal communications, unpublished papers, and works merely cited were excluded. The average or most

Table 6.1
Head Size and IQ Relationships Determined in Neurologically Normal Subjects

Study	Sample	Head Size Measure	IQ Measure	Correlation
Pearl (1906)	935 German male soldiers	Perimeter	Officers' rating	0.14
Pearson (1906)	2,398 British boys aged 3–20 years, standardized to age 12	Length	Teachers' estimate	0.14
Pearson (1906)	2,188 British girls aged 3–20 years, standardized to age 12	Length	Teachers' estimate	0.08
Pearson (1906)	1,011 British male university students	Length	Grades	0.11
Murdock & Sullivan (1923)	291 American boys aged 6–17 years, standardized by age	Perimeter	Various IQ tests	0.20
Murdock & Sullivan (1923)	395 American girls aged 6–17 years, standardized by age	Perimeter	Various IQ tests	0.27
Reid & Mulligan (1923)	449 Scottish male medical students	Capacity	Grades	0.08
Sommerville (1924)	105 white male university students	Capacity	Thorndike	0.08
Estabrooks (1928)	172 white boys age 7	Capacity	Binet IQ test	0.23
Estabrooks (1928)	207 white girls age 7	Capacity	Binet IQ test	0.16
Porteus (1937)	200 white Australian children	Perimeter	Porteus Maze	0.20
Schreider (1968)	80 adult Otomi Amerindians from Mexico	Perimeter	Form board	0.39

Table 6.1 (continued)

Study	Sample	Head Size Measure	IQ Measure	Correlation
Schreider (1968)	158 French farmers of unreported sex	Perimeter	Raven's Matrices	0.23
Klein et al. (1972)	172 Guatemalan Amerindian boys aged 3–6 years	Perimeter	Knowledge tests, with age standardized	0.23
Klein et al. (1972)	170 Guatemalan Amerindian girls aged 3–6 years	Perimeter	Knowledge tests, with age standardized	0.29
Susanne & Sporcq (1973)	2,071 Belgian male conscripts	Perimeter	Raven's Matrices	0.19
Weinberg et al. (1974)	334 white boys aged 8–10 years	Perimeter	WISC	0.35
Passingham (1979)	415 English villagers (212 men, 203 women) aged 18–75 years	Capacity	WAIS	0.13
Susanne (1979)	2,071 Belgian male conscripts	Perimeter	Matrices	0.19
Pollitt et al. (1982)	91 boys and girls aged 3–6 years	Perimeter	Stanford-Binet	0.23
Majluf (1983)	120 boys and girls aged 8–20 months	Perimeter	Bayley Motor Development Test	0.35
Ounsted et al. (1984)	214 boys age 4	Perimeter	Language Test	0.06
Ounsted et al. (1984)	167 girls age 4	Perimeter	Language Test	0.07
Henneberg et al. (1985)	151 Polish male medical students aged 18–30 years	Capacity	Baley's Polish language IQ test	0.09
Henneberg et al. (1985)	151 Polish female medical students aged 18–30 years	Capacity	Baley's Polish language IQ test	0.19

Table 6.1 (continued)

Study	Sample	Head Size Measure	IQ Measure	Correlation
Sen et al. (1986)	150 16- to 18-year-old males in India	Perimeter	Raven's Matrices	0.02
Sen et al. (1986)	150 16- to 18-year-old females in India	Perimeter	Raven's Matrices	0.54
Broman et al. (1987)	18,907 black boys and girls age 7 years	Perimeter	WISC	0.19
Broman et al. (1987)	17,241 white boys and girls age 7 years	Perimeter	WISC	0.24
Ernhart et al. (1987)	257 3-year-old boys and girls	Perimeter	Stanford-Binet	0.12
Bogaert & Rushton (1989)	216 white Canadian male and female university students, adjusted for sex	Perimeter	MAB	0.14
Lynn (1989)	161 Irish boys aged 9–10 years	Perimeter	PMAT	0.15
Lynn (1989)	149 Irish girls aged 9–10 years	Perimeter	PMAT	0.23
Lynn (1990)	205 Irish children aged 9 years	Perimeter	Raven's Matrices	0.26
Lynn (1990)	91 English children aged 9 years	Perimeter	Raven's Matrices	0.26
Osborne (1992)	106 European–American boys aged 13–17 years, controls for height and weight	Capacity	Basic	0.16
Osborne (1992)	84 African–American boys aged 13–17 years, controls for height and weight	Capacity	Basic	0.34

Table 6.1 (continued)

Study	Sample	Head Size Measure	IQ Measure	Correlation
Osborne (1992)	118 European–American girls aged 13–17 years, controls for height and weight	Capacity	Basic	0.23
Osborne (1992)	168 African–American girls aged 13–17 years, controls for height and weight	Capacity	Basic	0.13
Rushton (1992b)	73 Asian–Canadian male and female university students	Perimeter	MAB	0.14
Rushton (1992b)	211 white Canadian male and female university students	Perimeter	MAB	0.21
Lynn & Jindal (1993)	100 9-year-old boys from northern India	Perimeter	Matrices	0.14
Lynn & Jindal (1993)	100 9-year-old girls from northern India	Perimeter	Matrices	0.25
Reed & Jensen (1993)	211 European–American male college students	Capacity	Raven's Matrices	0.02
Wickett et al. (1994)	40 white Canadian female university students	Perimeter	MAB	0.11
Furlow et al. (1997)	128 undergraduates, 60% female	Perimeter	CFIT	0.19
Rushton (1997)	100 East Asian–American 7-year-olds, 54% female	Perimeter	WISC	0.21

Table 6.2 (continued)

Source	Sample	IQ Measure	Correlation
Ivanovic et al. (2004)	47 male 18-year-old high school students in Chile selected from the richest and poorest counties	WAIS-R	0.55
Ivanovic et al. (2004)	49 female 18-year-old high school students in Chile selected from the richest and poorest counties	WAIS-R	0.37

Number of samples: 27
Total N: 1,341
Unweighted mean $r = 0.39$
n-weighted mean $r = 0.37$

Abbreviations: CFIT, Culture-Free Intelligence Test; MAB, Multidimensional Aptitude Battery; MRI, Magnetic Resonance Imaging; NART, New Adult Reading Test; PMAT, Primary Mental Abilities Test; WAIS-R, Wechsler Adult Intelligence Scale–Revised; WISC, Wechsler Intelligence Scale for Children.

correlations with any other variable, a procedure known as a Jensen effect (Rushton, 1998). Jensen (1994) found a simple correlation of 0.19 between head circumference and *g* on 17 cognitive tests among 286 adolescents, but when he used the method of correlated vectors he obtained a correlation of 0.64. When Wickett, Vernon, and Lee (2000) correlated brain volume by means of MRI in 68 adult subjects, they found $r = 0.38$, with *g* extracted from an extensive cognitive ability battery that also included mean and standard deviation (SD) reaction time measures, but when they used the method of correlated vectors, they found the correlation rose to 0.59. Similarly, the head perimeter measure went from 0.19 to 0.34. Schoenemann, Budinger, Sarich, and Wang (2000) obtained a simple correlation of 0.45 between brain volume and *g*, which Jensen (1998, p. 147) found to be 0.51 using the method of correlated vectors. Finally, Jensen (personal communication, August 8, 2002) carried out a vector analysis of the MRI study of MacLullich et al. (2002) in older persons and raised the correlation between *g* and cognitive ability from 0.42 to 0.78.

The evidence shows that external head size is a good proxy for brain volume. Head perimeter correlates with brain mass at autopsy from birth through childhood at correlation values of 0.80 to 0.98 (Brandt, 1978; Bray, Shields, Wolcott, & Madsen, 1969; Cooke, Lucas,

Yudkin, & Pryse-Davies, 1977). It correlates with MRI brain volume at an average value of 0.66, based on five studies (0.55 in 10 pairs of identical twins aged 24–43 years, Tramo et al., 1998; 0.66 in 34 pairs of brothers aged 20–35 years, Wickett et al., 2000; 0.74 in 103 university students of both sexes in Turkey, Tan et al., 1999; 0.56 in 83 normal controls aged 8–46 years in the United States, Aylward, Minshew, Field, Sparks, & Singh, 2002; and 0.79 in 96 high school graduates of both sexes in Chile, Ivanovic et al., 2004).

Additional findings shown in table 6.2 are of interest. For example, the brain volume–IQ correlation is equally strong in males and females (e.g., Andreasen et al., 1993; Wickett, Vernon, & Lee, 1994, 2000). It is also found for people of East Asian, East Indian, European, Turkish, African, South American, and Amerindian descent. Age, although it plays a role in brain size and intelligence, does not confound the results. Studies using a narrow age range or younger or older samples show the same magnitude of correlations (e.g., Egan et al., 1994; MacLullich et al., 2002; Reiss et al., 1996). Several studies have examined whether different regions of the brain would show differential correlations with IQ; these studies appear to show that the size effects are manifest throughout the brain and are not specific to any particular region (Andreasen et al., 1993; Egan et al., 1994; Haier, Jung, Yeo, Head, & Alkire, 2004; Reiss et al., 1996), notwithstanding a study by Duncan et al. (2000) showing it centered in the lateral frontal cortex.

A functional relation between brain size and cognitive ability has been implied in three studies showing that the correlation between brain size and IQ holds true within families as well as between families (Gignac et al., 2003; Jensen, 1994; Jensen & Johnson, 1994) (although one study that examined only sisters failed to find the within-family relation: Schoenemann et al., 2000). The within-family finding is of special interest because it controls for most of the sources of variance that distinguish families, such as social class, styles of childrearing, and general nutrition.

The number of neurons available to process information may mediate the correlation between brain size and cognitive ability. Haug (1987, p. 135) showed a correlation of $r = 0.479$ ($N = 81$, $p < 0.001$) between number of cortical neurons (based on a partial count of representative areas of the brain) and brain size, including both men and women in the sample. The regression equating the two was given as: number of cortical neurons (in billions) = 5.583 + 0.006 (cm^3 brain volume). This means that a person with a brain size of 1,400 cm^3 has, on average, 600 million fewer cortical neurons than an individual with a brain size of 1,500 cm^3. The difference between the low end of normal (1,000 cm^3) and the high end (1,700 cm^3) works out to be 4.283 billion neurons (a difference of 27% more neurons from a 41% increase in brain size). Subsequently, Pakkenberg and Gundersen (1997) found a correlation of $r = 0.56$ between brain size and number of neurons (0.56). The human brain may contain up to 100 billion (10^{11}) nerve cells classifiable into 10,000 types, resulting in 100,000 billion synapses (Kandel, 1991). Even storing information at the low average rate of one bit per synapse, which would require two levels of synaptic activity (high and low), the structure as a whole would generate 10^{14} bits. Contemporary supercomputers, by comparison, command a memory of about 10^9 bits of information.

It is also predictable, however, that correlations between IQ and overall brain size will be modest. First, much of the brain is not involved in producing what we call intelligence; thus, variation in the size or mass of that tissue will lower the magnitude of the correlation. Second, IQ, of course, is not a perfect measure of intelligence, and thus variation in IQ scores is an imperfect measure of variation in intelligence.

Brain size and IQ are also correlated with body size. Results from autopsy studies such as the one by Dekaban and Sadowsky (1978) of 2,773 men and 1,963 women, as well as the one by Ho, Roessmann, Straumfjord, and Monroe (1980) of 644 men and 617 women, suggest a correlation of about 0.20 between brain mass (grams) and stature and body mass. Similarly, MRI studies find an average correlation of about 0.20 (Pearlson et al., 1989; Wickett et al., 1994). The relationship is higher (0.30–0.40) with measures of the skull (cm^3), estimated either from endocranial volume or from external head measures. In a stratified random sample of 6,325 U.S. servicemen, cranial capacity correlated, on average, 0.38 with height and 0.41 with mass in 2,803 women and 3,522 men (Rushton, 1992a). There is also a correlation of about 0.25 between IQ and height. However, this correlation may involve no causal or intrinsic functional relation but may occur instead as a result of the common assortment of the genetic factors for both height and intelligence, which in North American society are desirable characteristics, so that there is a fairly high degree of positive assortative mating for both. The result is a *between-families* genetic correlation between height and IQ, while the best evidence is that there is no *within-family* correlation between the traits (Jensen & Sinha, 1993).

There is, however, disagreement about whether or not brain size should be corrected for body size before examining brain size/IQ correlations. As noted by Rushton and Ankney (1996), controlling for body size obviously changes the question from "Is IQ correlated with absolute brain size?" to "Is IQ correlated with relative brain size?" Although these are quite different questions, evidence shows that the answer to both is yes. Controlling for body size can be regarded to some degree as an overcorrection because head size itself is part of stature and body weight.

Group (age, sex, social class, and race) differences exist in average brain size and cognitive ability. Because group distributions overlap substantially on the variables in question, with average differences amounting to between 4% and 34%, it is impossible to generalize from group averages to individuals. Nonetheless, significant among-group variation in brain size and cognitive ability does exist, and therefore a review is required if a full understanding of the relation between brain size and IQ is to be achieved. We emphasize that enormous variability exists within each of the populations to be discussed. We also emphasize that the relationships reported are correlational.

Age Differences

Autopsy studies show that brain mass increases during childhood and adolescence and then, beginning as early as 20 years, slowly decreases through middle adulthood, and finally decreases more quickly in old age (Dekaban & Sadowsky, 1978; Ho et al., 1980; Pakkenberg & Voigt, 1964; Voigt & Pakkenberg, 1983). Broca first showed these relationships in the nineteenth century (see reanalysis by Schreider, 1966). The data of Ho et al. (1980), collated for 2,037 subjects from autopsy records, for various subgroups, 1,261 of them between the ages of 25 and 80, are shown in figure 6.1. All brains were weighed on the same balance at the Institute of Pathology at Case Western Reserve University after excluding those brains with lesions or other abnormalities. The average mass of the brain increases from 397g at birth to 1,180g at 6 years. Growth then slows, and brain mass peaks at about 1,450g before age 25 years. The mass declines slowly from age 26 to 80 at an average of 2g per year. The decrease after age 80 years is much steeper, the loss

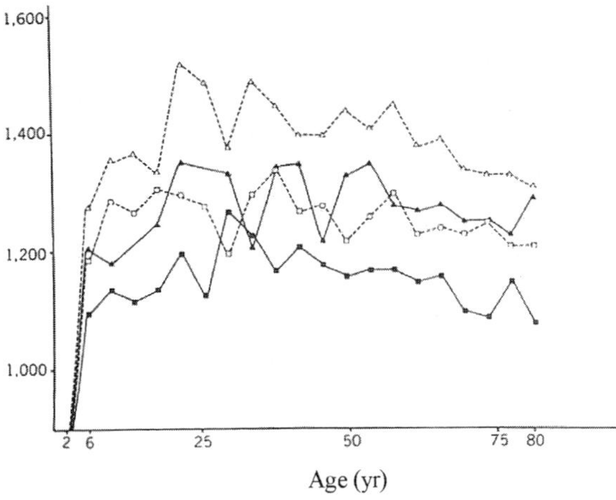

Figure 6.1
Mean brain weight for four-year age periods in various subgroups. Brain weight is plotted at the midpoint of each age period (e.g., the point at age 6 years represents the average for subjects between 4 and 8 years) (white men, open triangles; black men, solid triangles; white women, open squares; black women, solid squares). Differences in brain weights among various groups become apparent at age 6 years. (From Ho et al., 1980, p. 636, Figure 2.)

being 5 g per year. As shown in figure 6.1, although the rate of decrease varies slightly, it is essentially similar for various subgroups.

MRI investigations also show a curvilinear pattern of growth and change, with an overall decrease in brain volume following the late teens as gray matter is replaced with cerebrospinal fluid (CSF) (range of r values = −0.32 to −0.71; Gur et al., 1991; Jernigan et al., 1991). Pfefferbaum et al. (1994) demarcated cell growth, myelination, pruning, and atrophy. With a sample of 88 male and female subjects aged 3 months to 30 years, cortical gray matter volume (mainly cell bodies) peaked at around age 4 years and then declined steadily throughout the life span; cortical white matter volume (myelin sheath) increased steadily until about age 20 years and appeared stable thereafter; and the volume of cortical CSF remained stable from 3 months to 20 years. In a sample of 73 male subjects aged 21–71 years, CSF increased exponentially over the five decades of adulthood studied. Ventricular enlargement between ages 20 and 30 years suggested a possible marker for the onset of atrophy, whether it be due to cell loss or cell shrinkage. In the Baltimore Longitudinal Study of Ageing, participants aged 59–85 years showed significant annual increases of 1,526 mm^3 in ventricular volume (Resnick et al., 2000). A Danish study by Garde et al. (2000) reported significant increases in white matter hyperintensities (WMHs) in a 30-year longitudinal study of 68 healthy 50- to 80-year-olds. These WMHs were significantly related to concomitant IQ declines.

General intelligence shows concomitant increases during childhood and adolescence and then (slow) decreases between ages 25 and 45, and (faster) decreases after age 45. It once was claimed that this age-related decline in IQ was spurious because early longitudinal studies contradicted findings from cross-sectional studies; thus, the cross-sectional observations were derogated as a generation or "cohort" effect, perhaps due to "more favorable" environments for younger cohorts. However, several subsequent longitudinal studies, reviewed by Brody (1992) and Deary (2000), have corroborated results from cross-sectional studies. Brody (1992, p. 238) concluded, "Declines in fluid ability over the life span up to age 80 might well average 2 standard deviations." The 68 healthy Danes in the study by Garde et al. (cited above) similarly showed a decrease in IQ by 14 points (1 SD) from age 50 to 80.

Sex Differences

An absolute difference in average brain size between men and women has not been disputed since at least the time of Broca (1861). It is often claimed, however, that this difference disappears when corrections are made for body size or age of people sampled (Gould, 1981, 1996). However, Ankney (1992) demonstrated that the sex difference in brain size remains after correction for body size in a sample of similarly aged men and women (following tentative results by Dekaban & Sadowsky, 1978; Gur et al., 1991; Hofman & Swaab, 1991; Holloway, 1980; Swaab & Hofman, 1984; Willerman, Schultz, Rutledge, & Bigler, 1991).

Ankney (1992) suggested that the large sex difference in brain size went unnoticed for so long because earlier studies used improper statistical techniques to correct for sex differences in body size and thus incorrectly made a large difference "disappear." The serious methodological error was the use of brain mass/body size ratios instead of analysis of covariance (see Packard & Boardman, 1988). Ankney (1992) illustrated why this is erroneous by showing that, in both men and women, the ratio of brain mass to body size declines as body size increases. Thus, as

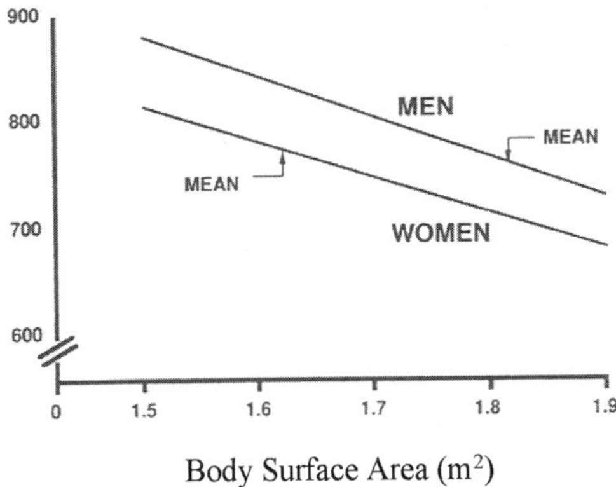

Figure 6.2
The relation between the ratio of brain mass/body surface area and body surface area in white men and women. Ankney (1992) calculated the ratios by estimating brain mass at a given body surface area using the equations in Ho et al. (1980, Table 3): *men*, brain mass = 1,077g (±56) + 173 (±31) × body surface area ($r = +0.27$, $P < 0.01$); *women*, brain mass = 949g (±52) + 188 (±32) × body surface area ($r = +0.24$, $P < 0.01$). (From Ankney, 1992, p. 331, Figure 1. Copyright 1992 by Ablex Publishing Corp. Reprinted with permission.)

can be seen in figure 6.2, larger women have a lower ratio than smaller women, and the same holds for larger men compared with smaller men. Therefore, because the average-sized man is larger than the average-sized woman, their brain mass to body size ratios are similar. Consequently, the only meaningful comparison is that of brain mass to body size ratios of men and women of equal size. Such comparisons show that at any given size, the ratio of brain mass to body size is much higher in men than in women (figure 6.2).

Ankney reexamined autopsy data on 1,261 American adults (Ho et al., 1980) and found that at any given body surface area or height, brains of white men are heavier than those of white women, and brains of black men are heavier than those of black women. For example, among whites 168cm (5′7″) tall (the approximate overall mean height for men and women combined), the brain mass of men averages about 100g heavier than that of women (figure 6.3), whereas the average difference in brain mass, uncorrected for body size, is 140g. Thus, only about 30% of the sex difference in brain size is due to differences in body size.

Ankney's results were confirmed in a study of cranial capacity in a stratified random sample of 6,325 U.S. Army personnel (Rushton,

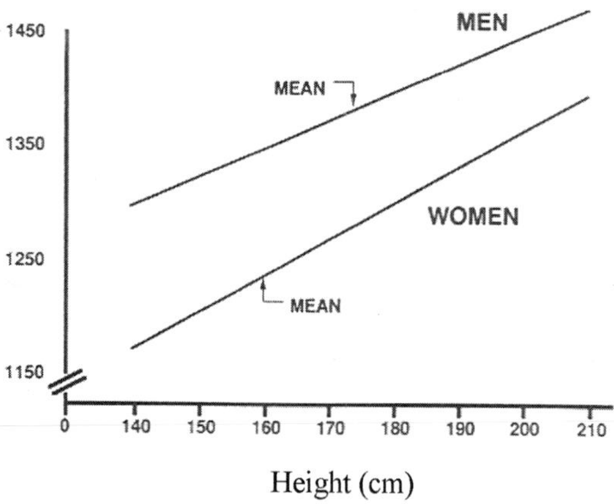

Figure 6.3
The relation between brain mass and body height in white men and women. Lines are drawn from equations in Ho et al. (1980, Table 1): *men*, brain mass = 920 g (±113) + 2.70 (±0.65) × body height ($r = 0.20$, $P < 0.01$); *women*, brain mass = 748 g (±104) + 3.10 (±0.64) × body height ($r = +0.24$, $P < 0.01$). (From Ankney, 1992, p. 333, Figure 4. Copyright 1992 by Ablex Publishing Corp. Reprinted with permission.)

1992a). After adjusting, by means of analysis of covariance, for effects of age, stature, weight, military rank, and race, men averaged 1,442 cm^3 and women 1,332 cm^3. This difference was found in all of the 20 or more separate analyses (shown in figure 6.4) conducted to rule out any body size effect. Moreover, the difference was replicated across samples of East Asians, whites, and blacks, as well as across officers and enlisted personnel. Parenthetically, in the army data, East Asian women constituted the smallest sample (N = 132), and it is probable that this caused the "instability" in estimates of their cranial size when some corrections were made for body size (figure 6.4). The sex difference of 110 cm^3 found by Rushton, from analysis of external head measurements, is remarkably similar to that (100 g) obtained by Ankney, from analysis of brain mass (1 cm^3 = 1.036 g; Hofman, 1991).

Studies using MRI have also confirmed the sex difference in adult brain size (Gur et al., 1991; Harvey, Persaud, Ron, Baker, & Murray, 1994; Reiss et al., 1996; Willerman et al., 1991). Thus, Ivanovich et al. (2004) carried out a study that controlled for body size in 96 18-year-old male and female high school graduates in Chile and found that the males averaged 1,480 cm^3 (SD = 125) before body size adjustments and 1,470 cm^3 (SD = 40) after adjustments, while the females

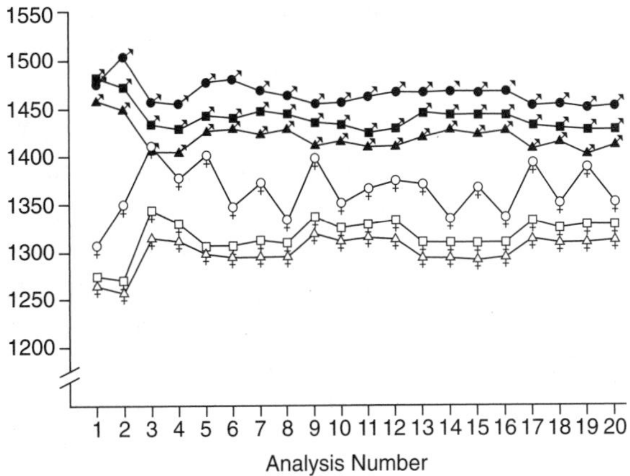

Figure 6.4
Cranial capacity for a stratified random sample of 6,325 U.S. Army personnel. The data, grouped into six sex-by-race categories, are collapsed across military rank (East Asian men, closed circles; white men, closed squares; black men, closed triangles; East Asian women, open circles; white women, open squares; black women, open triangles). The data show that, across the 19 different analyses controlling for body size, men averaged larger cranial capacities than women, and East Asians averaged larger cranial capacities than whites or blacks. Analysis 1 presents the data unadjusted for body size and shows no difference between East Asian and European men. (Adapted from Rushton, 1992a, p. 408, Figure 1. Copyright 1992 by Ablex Publishing Corp. Reprinted with permission.)

averaged 1,394 cm^3 (SD = 89) before and 1,404 cm^3 (SD = 37) after adjustments.

A stereological investigation by Pakkenberg and Gundersen (1997) found that men had about 4 billion more cortical neurons than did women, and this was not accounted for by differences in height. The average number of neocortical neurons was 19 billion in female brains and 23 billion in male brains, a 16% difference. In their study, which covered the age range from 20 years to 90 years, approximately 10% of all neocortical neurons were lost over the life span in both sexes. Sex and age were the main determinants of the total number of neurons in the human cortex, whereas body size per se had no influence on neuron number.

From birth through the early months, Rushton and Ankney (1996) found the sex difference held across several autopsy studies when, following Ankney's (1992) procedure (see figure 6.3), brain masses of boys and girls were compared after matching them for stature (Dekaban

& Sadowsky, 1978; Pakkenberg & Voigt, 1964; Voigt & Pakkenberg, 1983). From 7 to 17 years, sex differences in cranial capacity are in the range of 60–100 cm^3 (Lynn, 1993; Rushton & Osborne, 1995).

The sex differences in brain size present a paradox. Women have proportionately smaller average brains than do men but apparently have the same intelligence test scores. According to Kimura (1999), women excel in verbal ability, perceptual speed, and motor coordination within personal space, whereas men do better on various spatial tests and on tests of mathematical reasoning. A review by Voyer, Voyer, and Bryden (1995) showed that on the "purest" spatial measures, such as rotating an imaginary object or shooting at a moving rather than a stationary target, the sex difference approaches 1 SD. Ankney (1992, 1995) therefore hypothesized that the sex difference in brain size relates to those intellectual abilities at which men excel; that is, spatial and mathematical abilities require more "brain power." Analogously, whereas increasing word-processing power in a computer requires some extra capacity, increasing three-dimensional processing, as in graphics, requires a major increase in capacity.

Unfortunately for this hypothesis, what little information there is from the two MRI studies to date suggests that brain size is not significantly related to results on purely spatial tests (such as mental rotation) in either men or women (Wickett et al., 1994, 2000). Yet in the same studies, brain size did correlate significantly with IQ. However, one of these studies looked at only women and the other looked at only men. It would be more informative to know what happens in a *combined* sample of men and women, since the hypothesis that the extra brain size relates to men's better spatial scores would predict a correlation that should appear across sexes. So far, no comparison of brain size and spatial scores has been made in a mixed-sex group.

Baron-Cohen (2003) hypothesized that men on average tend to be systemizers (seeking to analyze, explore, and construct systems) while women tend to be empathizers (seeking to identify with another person's emotions and thoughts). Baron-Cohen speculates that having more brain cells allows storing of more information and greater attention to detail, which itself would lead to better systematizing.

The nineteenth-century proposition that men average slightly higher in general intelligence than women (e.g., Broca, 1861, p. 153) is not without contemporary exponents. Lynn's (1994, 1999) resolution of the brain size/sex difference paradox, which he dubbed "the Ankney-

Rushton anomaly" (1999, p. 1), was to produce evidence that contradicts the consensus view that there is no difference in general intelligence. He reviewed data from Britain, Greece, China, Israel, the Netherlands, Norway, Sweden, and Indonesia, as well as the United States, to show that men averaged about 4 IQ points higher than women on a number of published intelligence tests.

Subsequently, Lynn and Irwing (2004) carried out a meta-analysis of 57 studies of sex differences in general population samples on the Standard and Advanced Progressive Matrices. Results showed that while there is no difference among children aged 6–14 years, males do obtain higher means from the age of 15 through old age. Among adults, the male advantage is equivalent to 5 IQ points. These results disconfirm the frequent assertion that there is no average sex difference on the Progressive Matrices and support a developmental theory, namely, that a male advantage appears from the age of 15 years, around when brain size differences peak. Lynn and Irwing also carried out a meta-analysis of 15 studies of child samples on the Colored Progressive Matrices and found that among children aged 5–11 years old, boys had an advantage of 3 IQ points. They suggest that the Raven tests measure two cognitive skills, "visualization" and logical reasoning, that the Colored Matrices measure visualization even more than the Standard Matrices, and that it is this difference in what is tested that gives the younger boys their advantage over girls on this test.

Socioeconomic Differences

Nineteenth- and early twentieth-century data from Broca (1861) and others (Hooton, 1939; Sorokin, 1927; Topinard, 1878) suggested that people in higher-status occupations averaged a larger brain or head size than those in lower ones. For example, Galton collected head measurements and information on the educational and occupational background of thousands of individual visitors to the South Kensington Natural Science Museum in London. However, he had no statistical method for testing the significance of the differences in head size between various occupational or educational groups. Nearly a century later, Galton's data were analyzed by Johnson et al. (1985), who found that professional and semiprofessional groups averaged significantly larger head sizes (in both length and width) than unskilled occupational groups. Subsequently, Rushton and Ankney (1996) calculated cranial capacities from the summary by Johnson et al. (1985) of Galton's head size data and found that cranial capacity increased from unskilled to professional classes from 1,324 to 1,468 cm^3 in men and from 1,256 to 1,264 cm^3 in women. These figures are uncorrected for body size.

The relationship between head size and occupational status has also been found after correcting for body size. Jensen and Sinha (1993) reviewed much of the literature. They drew an important distinction between a person's socioeconomic status (SES) of origin (the SES attained by the person's parents) and the individual's attained SES (the SES attained by the person in adulthood). Correlations of IQ, head size, and other variables are always smaller when derived from the SES of origin than when derived from attained SES. Thus, Jensen and Sinha analyzed the head circumference data from the National Collaborative Perinatal Project (Broman, Nichols, & Kennedy, 1975) of approximately 10,000 white and 12,000 black 4-year-old children and found a small but significant correlation with social class of origin within both the white and black populations, after height was controlled for ($r = 0.10$). Jensen and Sinha also reanalyzed autopsy data reported by Passingham (1979) on 734 men and 305 women and found an overall correlation between brain mass and achieved occupational level of about 0.25, independent of body size.

Studies using brain imaging techniques have also reported significant main effects of brain size on occupational status and education level; higher-status subjects had, on average, a larger brain than lower-status subjects (Andreasen et al., 1990; Pearlson et al., 1989). Rushton (1992a) used the externally measured cranial size of 6,325 U.S. servicemen and found that officers averaged significantly larger cranial capacities than enlisted personnel either before or after adjusting for the effects of stature, weight, race, and sex (1,384 vs. 1,374 cm^3 before adjustments; 1,393 vs. 1,375 cm^3 after adjustments). The differences between officers and enlisted personnel were found for both men and women, as well as for East Asians, whites, and blacks, and in fact were in the opposite direction from predictions based on body size.

IQ test scores are significantly correlated with the socioeconomic hierarchies of modern Europe, North America, and Japan (Herrnstein & Murray, 1994; Jensen, 1998). The basic finding is that there is a difference of nearly 3 SD (45 IQ points) between average members of professional and unskilled classes. These are group mean differences with considerable overlap of distributions. Nonetheless, the overall correlation between an individual's IQ and his or her SES of origin is between

0.30 and 0.40, and the correlation between IQ and attained SES, or occupational level, is about 0.50 (Herrnstein & Murray, 1994). In studies of intergenerational social mobility, Mascie-Taylor and Gibson (1978) and Waller (1971) obtained IQ scores of fathers and their adult sons. They found that, on average, children with lower test scores than their fathers had gone down in social class as adults, but those with higher test scores had gone up. A within-family study was also conducted by Murray (1998), who found that among the 1,074 sibling pairs in the National Longitudinal Survey of Youth who had taken the Armed Forces Qualification Test, the sibling with the higher IQ achieved a higher level of education, a higher occupational status, and greater take-home pay than the sibling with the lower IQ.

Race Differences

The races differ in average brain size, and this shows up at birth. Rushton (1997) analyzed the Collaborative Perinatal Project's head circumference measurements and IQ scores from 40,000 children followed from birth to age 7 years (Broman et al., 1987). The results showed that at birth, 4 months, 1 year, and 7 years, the East Asian American children in the study averaged larger cranial volumes than the white American children, who averaged larger cranial volumes than the black American children (figure 6.5). Within each race, the children with the larger head sizes obtained higher IQ scores. By age 7, the East Asian children averaged an IQ of 110, white children an IQ of 102, and black children an IQ of 90. Moreover, the East Asian children, who averaged the largest craniums, were the shortest in stature and the lightest in weight, whereas the black children, who averaged the smallest craniums, were the tallest in stature and the heaviest in weight. Therefore, the race differences in brain size were not due to body size.

Dozens of studies from the 1840s to the 1990s, using different methods on different samples, reveal the same strong pattern. Four different methods of measuring brain size—MRI, endocranial volume measured from empty skulls, wet brain weight at autopsy, and external head size measurements—all yield the same results. Using MRI, for example, Harvey et al. (1994) found that 41 Africans and West Indians had a smaller average brain volume than did 67 Caucasians, although Harvey et al. provided no details on how, or if, the samples were matched for age, sex, or body size. In another study from the same mixed-race area of South London, Jones et al. (1994) found a (not significant) trend for

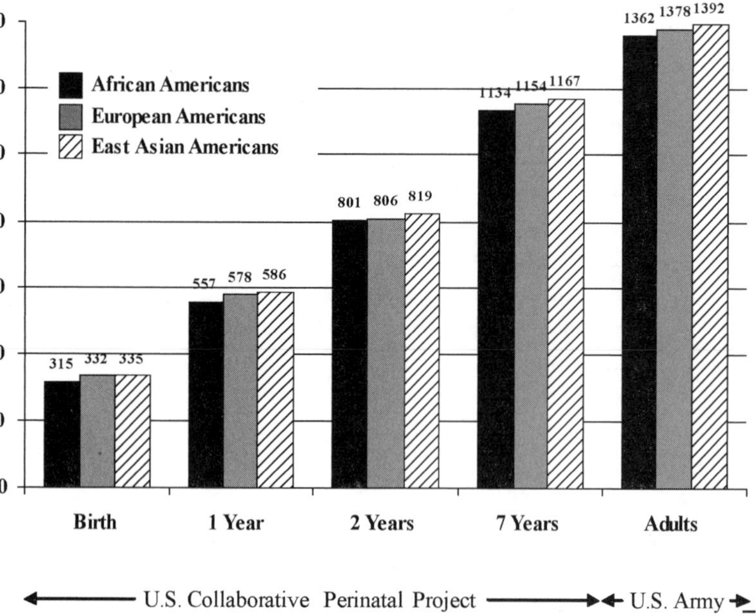

Figure 6.5
Mean cranial capacity (cm^3) for African Americans, European Americans, and East Asian Americans from birth through adulthood. Data for birth through age 7 are from the U.S. Perinatal Project; data for adults are from the U.S. Army data in figure 6.4. (From Rushton, 1997, p. 15, Figure 2. Copyright 1997 by Ablex Publishing Corp. Reprinted with permission.)

whites to have a 30 cm^3 larger intracranial volume but smaller ventricles than Afro-Caribbeans.

Using the method of measuring endocranial volume, the American anthropologist Samuel George Morton (1849) filled over 1,000 skulls with packing material and found that blacks averaged about 5 cubic inches less cranial capacity than whites. These results have stood the test of time (Todd, 1923; Gordon, 1934; Simmons, 1942). Subsequently, Beals, Smith, and Dodd (1984) carried out the largest study of race differences in endocranial volume to date, with measurements of up to 20,000 skulls from around the world. Their study found that East Asians, Europeans, and Africans averaged cranial volumes of 1,415, 1,362, and 1,268 cm^3, respectively. The skulls from East Asia were 3 cubic inches larger than those from Europe, which in turn were 5 cubic inches larger than those from Africa.

Using the method of weighing brains at autopsy, Broca (1873) found that whites averaged heavier brains than blacks, and had more complex convolutions and larger frontal lobes. (He corroborated the black-white difference using endocranial volume and also found that East Asians averaged larger cranial capacities than whites.) Subsequent studies have found an average black-white difference of about 100g (Bean, 1906; Mall, 1909; Pearl, 1934; Vint, 1934). Some studies have found that the more white admixture (judged independently from skin color), the greater the average brain weight in blacks (Bean, 1906; Pearl, 1934). In their autopsy study of 1,261 American adults, Ho et al. (1980) found that 811 white Americans averaged 1,323g and 450 black Americans averaged 1,223g—a difference of 100g. Since the blacks and whites in the study were similar in body size, differences in body size cannot explain away the differences in brain weight.

As yet unpublished, the largest cross-racial autopsy study carried out to date, at Columbia University Medical School, is by physical anthropologist Ralph Holloway (personal communications, February 21, 1997, March 16, 2002, and August 26, 2004). Holloway found that black and white men between ages 18 and 65 years differed by about 80g in brain weight, the samples being of very similar body size. The amount was less for women, about 40g. The data showed that 615 blacks, 153 Hispanics, and 1,391 whites averaged brain weights of 1,222, 1,253, and 1,285g, respectively. There were also a very large number (N = 5,731) of autopsied brain weights from 15- to 50-year-old Chinese from Hong Kong and Singapore that averaged 1,290g. Holloway himself remains agnostic as to the cause of these differences and whether they are related to general intelligence.

A final means of estimating brain size is by cranial volume calculated from external head size measurements (length, width, height). The results again confirm the racial differences. Rushton (1991, 1992a, 1993, 1994; Rushton & Osborne, 1995) carried out a series of studies estimating brain size this way from five large archival data sets. In the first of these studies, Rushton (1991) examined head size measures in 24 international military samples collated by the U.S. National Aeronautics and Space Administration (NASA) and, after adjusting for the effects of body height, weight, and surface area, found the cranial capacity for East Asians was 1,460 cm^3 and for Europeans was 1,446 cm^3. In the most comprehensive of these studies, Rushton (1992a) calculated average cranial capacities for East Asians, whites, and blacks from a stratified random sample of more than 6,000 U.S. Army personnel. The East Asians, whites, and blacks averaged 1,416, 1,380, and 1,359 cm^3, respectively. The East Asians averaged 36 cm^3 more capacity than the whites, and the whites averaged 21 cm^3 more capacity than the blacks. This study allowed precise adjustments for all kinds of body size measures. Yet adjusting for these or other variables did not erase the average racial differences in cranial capacity.

No exact solution is possible, of course, to the question of how large the racial differences are in brain size. There is much variability from sample to sample, with a clear overlap of distributions. Nonetheless, the consistency of results found even with the use of different procedures is noteworthy. Rushton (1995) reviewed the world database from (1) autopsies, (2) endocranial volume measurements, (3) head measurements, and (4) head measurements corrected for body size. The results in cm^3 or equivalents were: East Asians = 1,351, 1,415, 1,335, 1,356 (mean = 1,364); whites = 1,356, 1,362, 1,341, 1,329 (mean = 1,347); and blacks = 1,223, 1,268, 1,284, and 1,294 (mean = 1,267). The overall mean for East Asians was 17 cm^3 more than that for Europeans and 97 cm^3 more than that for Africans. Within-race differences, due to method of estimation, averaged 31 cm^3.

To reduce the uncertainty about race differences in brain size still further, Rushton and Rushton (2003) extended the parameters of the debate by examining race differences in 37 musculoskeletal variables shown in standard evolutionary textbooks to change systematically with increments in brain size in the hominoid line from chimpanzees to australopithecenes to *Homo erectus* to modern humans. The 37 variables included cranial traits (such as jaw size and shape, tooth size and shape, muscle attachment sites on the head, and indentations in the skull for muscles to run along), and postcranial traits (such as pelvic width, thighbone curvature, and knee joint surface area). Across the three populations, the correlations between brain size and the 37 morphological traits averaged a remarkable $r = 0.94$. It is noteworthy that the correlation for 12 lower limb traits was as high ($r = 0.98$) as the correlation for the 11 cranial traits ($r = 0.91$). If the races did not differ in brain size, these correlations with the concomitant musculoskeletal traits could not have been found. It must be concluded that the race differences in average brain size are securely established. They were acknowledged by Ulric Neisser, chair of the American Psychological Society's Task Force on Intelligence, who noted that, with respect to "racial differences in the mean measured sizes of skulls and brains (with East Asians having the largest, followed by whites and then blacks) . . . there is indeed a small overall trend" (Neisser, 1997, p. 80).

Racial differences in measured intelligence around the world parallel those found in brain size (Jensen, 1998; Lynn & Vanhanen, 2002; Rushton, 2000). In the United States and around the world, East Asians, measured in North America and in Pacific Rim countries, typically average IQs in the range of 101–111. Caucasoid populations in North America, Europe, and Australasia typically average IQs of 85–115, with an overall mean of 100. African populations living south of the Sahara, in North America, in the Caribbean, and in Britain typically have mean IQs of 70–90.

Serious questions have been raised about the validity of using IQ tests for racial comparisons. However, because the tests show similar patterns of internal item consistency and predictive validity for all groups, and because the same differences are found on relatively culture-free tests, many psychometricians have concluded that the tests *are* valid measures of racial differences, at least among people sharing the culture of the authors of the test (Herrnstein & Murray, 1994; Jensen, 1998). This conclusion was endorsed by an American Psychological Association Task Force's statement: "Considered as predictors of future performance, the tests do not seem to be biased against African Americans" (Neisser et al., 1996, p. 93).

Subsequent work has been carried out on the construct validity of IQ tests in Africa. For example, the study by Sternberg et al. (2001) of Kenyan 12- to 15-year-olds found that IQ scores predicted school grades with a mean $r = 0.40$, $P < 0.001$ (and continued to do so after controlling for age and SES, $r = 0.28$, $P < 0.01$) just as they do for white children in Europe and America. Similarly, Rushton, Skuy, and Bons (2004) found that among engineering students at the University of the Witwatersrand, the test items "behave" in the same way for African students as they do for non-African students, thereby indicating the test's internal validity, while concurrent validity was demonstrated by finding that the test scores correlated as highly with other test scores (an English Comprehension test, the Similarities subscale from the South African Wechsler Adult Intelligence Scale, end-of-year university grades, and high school grade point average) for Africans as they do for non-Africans.

The same three-way pattern of race differences has been found using the simplest culture-free cognitive measures such as reaction time tasks, which are so easy that 9- to 12-year-old children can perform them in less than 1 second. On these simple tests, children with higher IQ scores perform faster than children with lower scores, perhaps because reaction time measures the neurophysiological efficiency of the brain's capacity to process information accurately—the same ability measured by intelligence tests (Deary, 2000; Jensen, 1998). Children are not trained to perform well on reaction time tasks (as they are on certain paper-and-pencil tests), so the advantage of those with higher IQ scores on these tasks cannot arise from practice, familiarity, education, or training. Lynn and Vanhanen (2002) found that East Asian children from Hong Kong and Japan were faster than European children from Britain and Ireland, who in turn were faster than African children from South Africa. Using similar tasks, this pattern of racial differences was also found in California (Jensen, 1998).

Behavioral Genetics and Evolution

Heritabilities for mental ability range from 50% to 80% and have been established in numerous adoption, twin, and family studies (Bouchard & McGue, 2003). Noteworthy are the 80% heritabilities found in adult twins raised apart. Genetic influence is also found in studies of non-whites, including African Americans (Osborne, 1980; Scarr, Weinberg, & Waldman, 1993) and Japanese (Lynn & Hattori, 1990).

Both brain size and its relation to general intelligence are also highly heritable—80% or higher (Pennington et al., 2000; Posthuma et al., 2002; Thompson et al., 2001). In the largest and most recent of these studies, Posthuma et al. (2002) scanned the brains of 258 Dutch adults from 112 extended twin families using MRI and found high heritability for whole-brain gray matter volume (82%), whole-brain white matter volume (87%), and general intelligence (86%). The high heritability of gray matter implies that interindividual variation in cell-body volume is not modified by experience. Similarly, the high heritability of white matter, which reflects the degree of interconnections between different neurons and might be expected to be more influenced by experience, suggests that either experience barely contributes to interindividual variation therein or, alternatively, exposure to relevant environmental experience is under strong genetic control. Posthuma et al. also found r values = 0.25 (P values < 0.05) between gray matter volume, white matter volume, and g. The genetic correlations (the cross-trait/cross-twin correlations) showed that the relation between both measures of brain volume and g was mediated entirely by genetic factors.

These results on heritability may or may not pertain to race and other group differences because heritability studies have typically undersampled people from the most deprived segments of society, where lower heritabilities might be expected due to harmful environmental effects damaging brains and lowering IQs. Thus, in a study of cranial capacity in 236 pairs of black and white adolescent twins aged 13–17 years, Rushton and Osborne (1995) found a lower heritability for blacks (12%–31%) than for whites (47%–56%) and a higher within-family environmental effect for blacks than for whites (42%–46% vs. 28%–32%).

Nonetheless, transracial adoption studies do show some genetic contribution to the between-group differences in IQ. Studies of Korean and Vietnamese children adopted into white American, Belgian, and Dutch homes have shown that, although as babies many had been hospitalized for malnutrition, they grew to excel in academic ability, with IQs 10 points or more higher than their adoptive national homes (Clark & Hanisee, 1982; Frydman & Lynn, 1989; Stams, Juffer, Rispens, & Hoksbergen, 2000; Winick, Meyer, & Harris, 1975). By contrast, black and mixed-race children adopted into white middle-class families performed at a lower level than the white siblings with whom they had been raised (Scarr et al., 1993). Multifarious other sources of evidence suggest that racial differences in intelligence are partly genetic (Jensen, 1998; see Rushton & Jensen, 2005, for a full review).

It is reasonable to hypothesize that bigger brains evolved based on natural selection for increased intelligence (Jerison, 1973). Over the last 575 million years of evolutionary history, neural complexity and brain size have increased in vertebrates and invertebrates alike (figure 6.6), little of which can be explained by body size increases. Russell (1983) calculated encephalization quotients, or EQs, a measure of actual brain size to expected brain size for an animal of that body weight (following Jerison, 1973; EQ = Cranial capacity (cm^3)/(0.12) (body weight in grams)$^{0.67}$). Russell found that the mean EQ was only about 0.30 for mammals living 65 million years ago, compared to the average of 1.00 today. EQs for molluscs varied between 0.043 and 0.31, and for insects between 0.008 and 0.045, with the less encephalized species resembling forms that appeared early in the geological record and the more encephalized species resembling those that appeared later. Russell (1989) also demonstrated how, over 140 million years, dinosaurs showed increasing encephalization before going extinct 65 million years ago

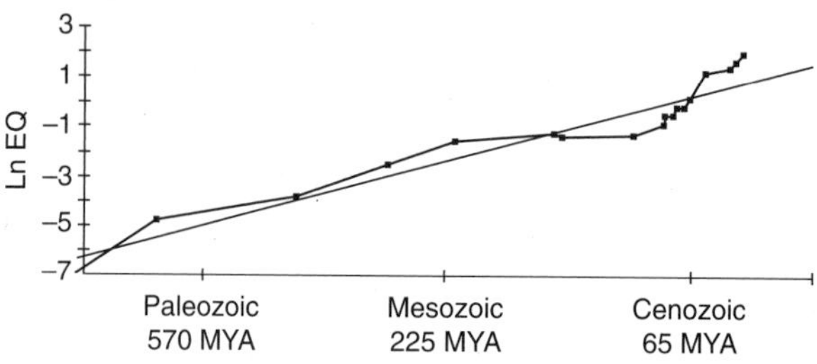

Figure 6.6
Average encephalization quotient (EQ; natural log), a measure of neural tissue corrected by body size, plotted against elapsed geological time in millions of years (After Russell, 1983.)

(probably because of an asteroid impact or other catastrophic event). He extrapolated the data to suggest that if dinosaurs had continued on, they would have progressed to a large-brained, bipedal descendant. The tripling in size of the hominoid brain over the last 5 million years (chimpanzees ≈ 380 cm^3, australopithecenes ≈ 450 cm^3, *Homo erectus* ≈ 1,000 cm^3, and *Homo sapiens* ≈ 1,350 cm^3) may be a special case of the more general trend to larger brains.

Others have also shown the value of an evolutionary perspective on brain size relations. Bonner (1980, 1988) reviewed naturalistic data and found that the more recently an animal species had evolved, the larger was its brain and the more complex was its culture. Passingham (1982) reviewed experimental studies of "visual discrimination learning" that measured the speed with which children and other mammals abstracted such rules as "pick the same object each time to get food." More intelligent children, assessed by standardized IQ tests, learned faster than did those with lower IQ scores, and mammals with larger brains learned faster than did those with smaller brains (i.e., chimpanzees > rhesus monkeys > spider monkeys > squirrel monkeys > marmosets > cats > gerbils > rats > squirrels). Madden (2001) found that species of bowerbirds that build more complex bowers have larger brains than species that build less complex ones.

Metabolically, the human brain is an expensive organ. Representing only 2% of body mass, the brain uses about 5% of basal metabolic rate in rats, cats, and dogs, about 10% in rhesus monkeys and other primates, and about 20% in humans (Armstrong, 1990). Moreover, as large brains evolved, they required more prolonged and complex life histories to sustain them. For example, across 21 primate species, Smith (1989) found that brain size correlates 0.80–0.90 with life span, length of gestation, age of weaning, age of eruption of first molar, age at complete dentition, age at sexual maturity, interbirth interval, and body weight. Similarly, Rushton (2004) found that across 234 mammalian species, brain weight correlated with longevity (0.70), gestation time (0.67), birth weight (0.46), litter size (–0.22), age at first mating (0.50), duration of lactation (0.54), body weight (0.61), and body length (0.63). Remarkably, even after the effects of body weight and body length were controlled for, brain weight still correlated with longevity (0.59), gestation time (0.66), birth weight (0.16), litter size (–0.18), age at first mating (0.63), and duration of lactation (0.61). From an adaptationist perspective, unless large brains substantially contributed to evolutionary fitness (defined as increased survival of genes through successive generations), they would not have evolved.

The sexual dimorphism in cranial size and cognitive ability likely originated partly through evolutionary selection of men's hunting ability (Ankney, 1992; Kolakowski & Malina, 1974) and partly through the reproductive success socially dominant men have traditionally enjoyed (Lynn, 1994). Race differences in cranial capacity may have originated from evolutionary pressures in colder climates for greater intelligence (Rushton, 1995). Of course, brain size and intellectual performance are also affected by nutrition and experience (Sternberg, 2004).

Conclusion

The preponderance of evidence demonstrates that brain size is correlated positively with intelligence and that both brain size and cognitive ability are correlated with age, sex, social class, and race. Correlation does not prove cause and effect, but, just as zero correlations provide no support for a hypothesis of cause and effect, non-zero correlations do provide support. We are convinced that the brain size/cognitive ability correlations that we have reported are in fact due to cause and effect. This is because we are unaware of any variable, other than the brain, that can directly mediate cognitive ability.

Numerous issues still require research, and several paradoxes require resolution. For example, the average brain size of white women is equal to or less than the average brain size of black men (see figures 6.1 and 6.4), but white females obtain a higher average mental test score than do black males. We hypothesize that, within race, at least some of the additional brain tissue/neurons that men have, as compared with women, are related to the average male advantage in dynamic spatial abilities (not measured on standard IQ tests), such as in throwing balls and the like at stationary or moving targets. If so, that could untangle the aforementioned paradox. Additional research using MRI with a wider array of cognitive tasks may shed light on this puzzle.

Although it is established that the correlation between brain volume and *g* is mediated by common genetic factors, this is only the first step in unveiling the relation between them. One important next step will be to identify specific genes that influence both brain volume and *g*. Since genes have been identified that regulate brain size during development, particularly in the ape lineage leading from mammals to humans (Evans, Anderson, Vallender, Choi, & Lahn, 2004), these might be useful candidates for examining the underlying process.

ACKNOWLEDGMENTS We thank B. Cox and E. Hanna for helpful comments, and R. Holloway for sharing his data.

References

Andreasen, N. C., Ehrhardt, J. C., Swayze, V. W., Alliger, R. J., Yuh, W. T. C., Cohen, G., et al. (1990). Magnetic resonance imaging of the brain in schizophrenia. *Archives of General Psychiatry*, *47*, 35–44.

Andreasen, N. C., Flaum, M., Swayze, V., O'Leary, D. S., Alliger, R., Cohen, G., et al. (1993). Intelligence and brain structure in normal individuals. *American Journal of Psychiatry*, *150*, 130–134.

Ankney, C. D. (1992). Sex differences in relative brain size: The mismeasure of woman, too? *Intelligence*, *16*, 329–336.

Ankney, C. D. (1995). Sex differences in brain size and mental ability: Comments on R. Lynn and D. Kimura. *Personality and Individual Differences*, *18*, 423–424.

Armstrong, E. (1990). Brains, bodies and metabolism. *Brain, Behavior and Evolution*, *36*, 166–176.

Aylward, E. H., Minshew, N. J., Field, K., Sparks, B. F., & Singh, N. (2002). Effects of age on brain volume and head circumference in autism. *Neurology*, *59*, 175–183.

Baron-Cohen, S. (2003). *The essential difference*. New York: Basic Books.

Beals, K. L., Smith, C. L., & Dodd, S. M. (1984). Brain size, cranial morphology, climate, and time machines. *Current Anthropology*, *25*, 301–330.

Bean, R. B. (1906). Some racial peculiarities of the Negro brain. *American Journal of Anatomy*, *5*, 353–432.

Bogaert, A. F., & Rushton, J. P. (1989). Sexuality, delinquency and *r/K* reproductive strategies: Data from a Canadian university sample. *Personality and Individual Differences*, *10*, 1071–1077.

Bonner, J. T. (1980). *The evolution of culture in animals*. Princeton, NJ: Princeton University Press.

Bonner, J. T. (1988). *The evolution of complexity*. Princeton, NJ: Princeton University Press.

Bouchard, T. J., Jr., & McGue, M. (2003). Genetic and environmental influences on human psychological differences. *Journal of Neurobiology*, *54*, 4–45.

Brandt, I. (1978). Growth dynamics of low-birth weight infants with emphasis on the perinatal period. In F. Falkner & J. M. Tanner (Eds.), *Human growth: Vol. 2* (pp. 557–617). New York: Plenum Press.

Bray, P. F., Shields, W. D., Wolcott, G. J., & Madsen, J. A. (1969). Occipitofrontal head circumference: An accurate measure of intracranial volume. *Journal of Pediatrics*, *75*, 303–305.

Broca, P. (1861). Sur le volume et la forme du cerveau suivant les individus et suivant les races. *Bulletin Société d'Anthropologie Paris*, *2*, 139–207, 301–321, 441–446.

Broca, P. (1873). Sur les crânes de la caverne de l'Homme Mort (Loere). *Revue d'Anthropologie*, *2*, 1–53.

Brody, N. (1992). *Intelligence*. New York: Academic Press.

Broman, S. H., Nichols, P. L., & Kennedy, W. A. (1975). *Preschool IQ: Prenatal and early development correlates*. Hillsdale, NJ: Erlbaum.

Broman, S. H., Nichols, P. L., Shaughnessy, P., & Kennedy, W. (1987). *Retardation in young children*. Hillsdale, NJ: Erlbaum.

Castellanos, F. X., Giedd, J. N., Eckburg, P., Marsh, W. L., Vaituzis, A. C., Kaysen, D., et al. (1994). Quantitative morphology of the caudate nucleus in attention deficit hyperactivity disorder. *American Journal of Psychiatry*, *151*, 1791–1796.

Cattell, R. B. (1982). *The inheritance of personality and intelligence*. New York: Academic Press.

Clark, E. A., & Hanisee, J. (1982). Intellectual and adaptive performance of Asian children in adoptive American settings. *Developmental Psychology*, *18*, 595–599.

Cooke, R. W. I., Lucas, A., Yudkin, P. L. N., & Pryse-Davies, J. (1977). Head circumference as an index of brain weight in the fetus and new born. *Early Human Development*, *112*, 145–149.

Darwin, C. (1871). *The descent of man*. London: Murray.

Deary, I. J. (2000). *Looking down on human intelligence: From psychometrics to the brain*. Oxford: Oxford University Press.

Dekaban, A. S., & Sadowsky, D. (1978). Changes in brain weights during the span of human life: Relation of brain weights to body heights and body weights. *Annals of Neurology*, *4*, 345–356.

Duncan, J., Seitz, R. J., Kolodny, J., Bor, D., Herzog, H., Ahmed, A., et al. (2000). A neural basis for general intelligence. *Science*, *289*, 457–460.

Egan, V., Chiswick, A., Santosh, C., Naidu, K., Rimmington, J. E., & Best, J. J. K. (1994). Size isn't everything: A study of brain volume, intelligence and auditory evoked potentials. *Personality and Individual Differences*, *17*, 357–367.

Egan, V., Wickett, J. C., & Vernon, P. A. (1995). Brain size and intelligence: Erratum, addendum, and correction. *Personality and Individual Differences*, *19*, 113–115.

Ernhart, C. B., Marler, M. R., & Morrow-Tlucak, M. (1987). Size and cognitive development in the early preschool years. *Psychological Reports*, *61*, 103–106.

Estabrooks, G. H. (1928). The relation between cranial capacity, relative cranial capacity and intelligence in school children. *Journal of Applied Psychology*, *12*, 524–529.

Evans, P. D., Anderson, J. R., Vallender, E. J., Choi, S. S., & Lahn, B. T. (2004). Reconstructing the evolutionary history of microcephalin, a gene controlling human brain size. *Human Molecular Genetics*, *13*, 1139–1145.

Fisher, R. A. (1970). *Statistical methods for research workers* (14th ed.). New York: Hafner Press.

Flashman, L. A., Andreasen, N. C., Flaum, M., & Swayze II, V. W. (1998). Intelligence and regional brain volumes in normal controls. *Intelligence*, *25*, 149–160.

Frydman, M., & Lynn, R. (1989). The intelligence of Korean children adopted in Belgium. *Personality and Individual Differences*, *10*, 1323–1326.

Furlow, F. B., Armijo-Prewitt, T., Gangestad, S. W., & Thornhill, R. (1997). Fluctuating asymmetry and psychometric intelligence. *Proceedings of the Royal Society of London, B, Biological Sciences*, *264*, 823–829.

Galton, F. (1888). Head growth in students at the University of Cambridge. *Nature*, *38*, 14–15.

Garde, E., Mortensen, E. L., Krabbe, K., Rostrup, E., & Larsson, H. B. W. (2000). Relation between age-related decline in intelligence and cerebral white-matter hyperintensities in healthy octogenarians: A longitudinal study. *Lancet*, *356*, 628–634.

Gignac, G., Vernon, P. A., & Wickett, J. C. (2003). Factors influencing the relationship between brain size and intelligence. In H. Nyborg (Ed.), *The scientific study of general intelligence: Tribute to Arthur R. Jensen* (pp. 93–106). London: Elsevier.

Gordon, H. L. (1934). Amentia in the East African. *Eugenics Review*, *25*, 223–235.

Gould, S. J. (1978). Morton's ranking of races by cranial capacity. *Science*, *200*, 503–509.

Gould, S. J. (1981). *The mismeasure of man*. New York: Norton.

Gould, S. J. (1996). *The mismeasure of man* (2nd ed.). New York: Norton.

Gray, J. M., & Thompson, P. M. (2004). Neurobiology of intelligence: Science and ethics. *Nature Reviews: Neuroscience*, *5*, 471–482.

Gur, R. C., Mozley, P. D., Resnick, S. M., Gottlieb, G. L., Kohn, M., Zimmerman, R., et al. (1991). Gender differences in age effect on brain atrophy measured by magnetic resonance imaging. *Proceedings of the National Academy of Sciences, U.S.A.*, *88*, 2845–2849.

Gur, R. C., Turetsky, B. I., Matsui, M., Yan, M., Bilkur, W., Hughett, P., et al. (1999). Sex differences in brain gray and white matter in healthy young adults: Correlations with cognitive performance. *Journal of Neuroscience, 19*, 4065–4072.

Haier, R., Jung, R. E., Yeo, R. A., Head, K., & Alkire, M. T. (2004). Structural brain variation and general intelligence. *NeuroImage, 23*, 425–433.

Harvey, I., Persaud, R., Ron, M. A., Baker, G., & Murray, R. M. (1994). Volumetric MRI measurements in bipolars compared with schizophrenics and healthy controls. *Psychological Medicine, 24*, 689–699.

Haug, H. (1987). Brain sizes, surfaces, and neuronal sizes of the cortex cerebri. *American Journal of Anatomy, 180*, 126–142.

Henneberg, M., Budnik, A., Pezacka, M., & Puch, A. E. (1985). Head size, body size and intelligence: Intraspecific correlations in *Homo sapiens sapiens*. *Homo, 36*, 207–218.

Herrnstein, R. J., & Murray, C. (1994). *The bell curve.* New York: Free Press.

Ho, K. C., Roessmann, U., Straumfjord, J. V., & Monroe, G. (1980). Analysis of brain weight: I and II. *Archives of Pathology and Laboratory Medicine, 104*, 635–645.

Hofman, M. A. (1991). The fractal geometry of convoluted brains. *Journal für Hirnforschung, 32*, 103–111.

Hofman, M. A., & Swaab, D. F. (1991). Sexual dimorphism of the human brain: Myth and reality. *Experimental Clinical Endocrinology, 98*, 161–170.

Holloway, R. L. (1980). Within-species brain-body weight variability: A reexamination of the Danish data and other primate species. *American Journal of Physical Anthropology, 53*, 109–121.

Hooton, E. A. (1939). *The American criminal*: *Vol. 1*. Cambridge, MA: Harvard University Press.

Ivanovic, D. M., Leiva, B. P., Castro, C. G., Olivares, M. G., Jansana, J. M. M., Castro, V. G., et al. (2004). Brain development parameters and intelligence in Chilean high school graduates. *Intelligence, 32*, 461–479.

Ivanovic, D. M., Olivares, M. G., Castro, C. G., & Ivanovic, R. M. (1996). Nutrition and learning in Chilean school age children: Chile's Metropolitan Region Survey 1986–1987. *Nutrition, 12*, 321–328.

Ivanovic, R. M.; Forno, H. S., Castro, C. G., & Ivanovic, D. M. (2000). Intellectual ability and nutritional status assessed through anthropometric measurements of Chilean school-age children from different socioeconomic status. *Ecology of Food and Nutrition, 39*, 35–59.

Jensen, A. R. (1994). Psychometric g related to differences in head size. *Personality and Individual Differences, 17*, 597–606.

Jensen, A. R. (1998). *The g factor.* Westport, CT: Praeger.

Jensen, A. R., & Johnson, F. W. (1994). Race and sex differences in head size and IQ. *Intelligence, 18*, 309–333.

Jensen, A. R., & Sinha, S. N. (1993). Physical correlates of human intelligence. In P. A. Vernon (Ed.), *Biological approaches to the study of human intelligence* (pp. 139–242). Norwood, NJ: Ablex.

Jerison, H. J. (1973). *Evolution of the brain and intelligence.* New York: Academic Press.

Jernigan, T. L., Archibald, S. L., Berhow, M. T., Sowell, E. R., Foster, D. S., & Hesselink, J. R. (1991). Cerebral structure on MRI: Part 1. Localization of age-related changes. *Biological Psychiatry*, *29*, 55–67.

Johnson, R. C., McClearn, G. E., Yuen, S., Nagoshi, C. T., Ahern, F. M., & Cole, R. E. (1985). Galton's data a century later. *American Psychologist*, *40*, 875–892.

Jones, P. B., Harvey, I., Lewis, S. W., Toone, B. K. Van Os, J., Williams, M., & Murray, R. M. (1994). Cerebral ventricle dimensions as risk factors for schizophrenia and affective psychosis: an epidemiological approach to analysis. *Psychological Medicine*, *24*, 995–1011.

Kamin, L. J. (1974). *The science and politics of IQ*. Hillsdale, NJ: Erlbaum.

Kandel, E. R. (1991). Nerve cells and behavior. In E. R. Kandel, J. H. Schwartz, & T. M. Jessell (Eds.), *Principles of neural selection* (3rd ed.). New York: Elsevier.

Kareken, D. A., Gur, R. C., Mozley, P. D., Mozley, L. H., Saykin, A. J., Shtasel, D. L., et al. (1995). Cognitive functioning and neuroanatomic volume measures in schizophrenia. *Neuropsychology*, *9*, 211–219.

Kimura, D. (1999). *Sex and cognition*. Cambridge, MA: MIT Press.

Klein, R. E., Freeman, H. E., Kagan, J., Yarborough, C., & Habicht, J. P. (1972). Is big smart? The relation of growth to cognition. *Journal of Health and Social Behavior*, *13*, 219–250.

Kolakowski, D., & Malina, R. M. (1974). Spatial ability, throwing accuracy, and man's humting heritage. *Nature*, *251*, 410–412.

Lynn, R. (1989). A nutrition theory of the secular increases in intelligence: Positive correlations between height, head size and IQ. *British Journal of Educational Psychology*, *59*, 372–377.

Lynn, R. (1990). New evidence on brain size and intelligence: A comment on Rushton and Cain and Vanderwolf. *Personality and Individual Differences*, *11*, 795–797.

Lynn, R. (1993). Further evidence for the existence of race and sex differences in cranial capacity. *Social Behavior and Personality*, *21*, 89–92.

Lynn, R. (1994). Sex differences in intelligence and brain size: A paradox resolved. *Personality and Individual Differences*, *17*, 257–271.

Lynn, R. (1999). Sex differences in intelligence and brain size: A developmental theory. *Intelligence*, *27*, 1–12.

Lynn, R., & Hattori, K. (1990). The heritability of intelligence in Japan. *Behavior Genetics*, *20*, 545–546.

Lynn, R., & Irwing, P. (2004). Sex differences on the Progressive Matrices: A meta-analysis. *Intelligence*, *32*, 481–498.

Lynn, R., & Jindal, S. (1993). Positive correlations between brain size and intelligence: Further evidence from India. *Mankind Quarterly*, *34*, 109–123.

Lynn, R., & Vanhanen, T. (2002). *IQ and the wealth of nations*. Westport, CT: Praeger.

MacLullich, A. M. J., Ferguson, K. J., Deary, I. J., Seckl, J. R., Starr, J. M., & Wardlaw, J. M. (2002). Intracranial capacity and brain volumes are associated with cognition in healthy elderly men. *Neurology*, *59*, 169–174.

Madden, J. (2001). Sex, bowers and brains. *Proceedings of the Royal Society of London, Series B, Biological Sciences, 268*, 833–838.

Mall, F. P. (1909). On several anatomical characters of the human brain, said to vary according to race and sex, with special reference to the weight of the frontal lobe. *American Journal of Anatomy, 9*, 1–32.

Mascie-Taylor, C. G. N., & Gibson, J. B. (1978). Social mobility and IQ components. *Journal of Biosocial Science, 10*, 263–276.

McDaniel, M. A. (in press). Big-brained people are smarter: A meta-analysis of the relationship between in vivo brain volume and intelligence. *Intelligence*.

Morton, S. G. (1849). Observations on the size of the brain in various races and families of man. *Proceedings of the Academy of Natural Sciences, Philadelphia, 4*, 221–224.

Murdock, J., & Sullivan, L. R. (1923). A contribution to the study of mental and physical measurements in normal school children. *American Physical Education Review, 28*, 209–330.

Murray, C. (1998). *Income inequality and IQ*. Washington, DC: American Enterprise Institute.

Neisser, U. (1997). Never a dull moment. *American Psychologist, 52*, 79–81.

Neisser, U., Boodoo, G., Bouchard, T. J., Jr., Boykin, A. W., Brody, N., Ceci, S. J., et al. (1996). Intelligence: Knowns and unknowns. *American Psychologist, 51*, 77–101.

Osborne, R. T. (1980). *Twins: Black and white*. Athens, GA: Foundation for Human Understanding.

Osborne, R. T. (1992). Cranial capacity and IQ. *Mankind Quarterly, 32*, 275–280.

Ounsted, M., Moar, V. A., & Scott, A. (1984). Associations between size and development at four years among children who were small-for-dates and large-for-dates at birth. *Early Human Development, 9*, 259–268.

Packard, G. C., & Boardman, T. J. (1988). The misuse of ratios, indices, and percentages in ecophysiological research. *Physiological Zoology, 61*, 1–9.

Pakkenberg, B., & Gundersen, H. J. G. (1997). Neocortical neuron number in humans: Effects of sex and age. *Journal of Comparative Neurology, 384*, 312–320.

Pakkenberg, H., & Voigt, J. (1964). Brain weight of the Danes. *Acta Anatomica, 56*, 297–307.

Passingham, R. E. (1979). Brain size and intelligence in man. *Brain, Behavior and Evolution, 16*, 253–270.

Passingham, R. E. (1982). *The human primate*. San Francisco, CA: Freeman.

Pearl, R. (1906). On the correlation between intelligence and the size of the head. *Journal of Comparative Neurology and Psychology, 16*, 189–199.

Pearl, R. (1934). The weight of the Negro brain. *Science, 80*, 431–434.

Pearlson, G. D., Kim, W. S., Kubos, K. L., Moberg, P. J., Jayaram, G., Bascom, M. J., et al. (1989). Ventricle-brain ratio, computed tomographic density, and brain area in 50 schizophrenics. *Archives of General Psychiatry, 46*, 690–697.

Pearson, K. (1906). On the relationship of intelligence to size and shape of head, and to other physical and mental characters. *Biometrika*, *5*, 105–146.

Pearson, K. (1924). *The life, letters and labours of Francis Galton: Vol. 2.* London: Cambridge University Press.

Pennington, B. F., Filipek, P. A., Lefly, D., Chhabildas, N., Kennedy, D. N., Simon, J. H., et al. (2000). A twin MRI study of size variations in the human brain. *Journal of Cognitive Neuroscience*, *12*, 223–232.

Pfefferbaum, A., Mathalon, D. H., Sullivan, E. V., Rawles, J. M., Zipursky, R. B., & Lim, K.O. (1994). A quantitative magnetic resonance imaging study of changes in brain morphology from infancy to late adulthood. *Archives of Neurology*, *51*, 874–887.

Pollitt, E., Mueller, W., & Liebel, R. L. (1982). The relation of growth to cognition in a well-nourished preschool population. *Child Development*, *53*, 1157–1163.

Porteus, S. D. (1937). *Primitive intelligence and environment.* New York: Macmillan.

Posthuma, D., De Geus, E. J. C., Baare, W. F. C., Pol, H. E. H., Kahn, R. S., & Boomsma, D. I. (2002). The association between brain volume and intelligence is of genetic origin. *Nature Neuroscience*, *5*, 83–84.

Raz, N., Torres, I. J., Spencer, W. D., Millman, D., Baertschi, J. C., & Sarpel, G. (1993). Neuroanatomical correlates of age-sensitive and age-invariant cognitive abilities: An *in vivo* MRI investigation. *Intelligence*, *17*, 407–422.

Reed, T. E., & Jensen, A. R. (1993). Cranial capacity: New Caucasian data and comments on Rushton's claimed Mongoloid-Caucasoid brain size differences. *Intelligence*, *17*, 423–431.

Reid, R. W., & Mulligan, J. H. (1923). Relation of cranial capacity to intelligence. *Journal of the Royal Anthropological Institute*, *53*, 322–331.

Reiss, A. R., Abrams, M. T., Singer, H. S., Ross, J. R., & Denckla, M. B. (1996). Brain development, gender and IQ in children: A volumetric study. *Brain*, *119*, 1763–1774.

Resnick, S. M., Goldszal, A. F., Davatzikos, C., Golski, S., Kraut, M. A., Metter, E. J., et al. (2000). One-year age changes in MRI brain volumes in older adults. *Cerebral Cortex*, *10*, 464–472.

Rushton, J. P. (1991). Mongoloid-Caucasoid differences in brain size from military samples. *Intelligence*, *15*, 351–359.

Rushton, J. P. (1992a). Cranial capacity related to sex, rank, and race in a stratified random sample of 6,325 U.S. military personnel. *Intelligence*, *16*, 401–413.

Rushton, J. P. (1992b). Life history comparisons between Orientals and whites at a Canadian university. *Personality and Individual Differences*, *13*, 439–442.

Rushton, J. P. (1993). Corrections to a paper on race and sex differences in brain size and intelligence. *Personality and Individual Differences*, *15*, 229–231.

Rushton, J. P. (1994). Sex and race differences in cranial capacity from International Labour Office data. *Intelligence*, *19*, 281–294.

Rushton, J. P. (1995). *Race, evolution and behavior: A life-history perspective.* New Brunswick, NJ: Transaction.

7 The Evolution of the Brain and Cognition in Cetaceans

Lori Marino

Cetaceans, the order of fully aquatic mammals known as dolphins, whales, and porpoises, have been one of the most captivating groups of animals throughout history. Not only have they held a particular fascination for informal observers, they have also intrigued scientists for years. The reasons for this interest no doubt relate to the behavioral qualities of dolphins and other cetaceans. They are described as playful, intentional, clever, and cooperative. These broad characteristics reflect the fact that observers perceive a great deal of intelligence in these animals even though their physical expressions, body form, communication method, and behavioral milieu are so unlike our own. That is, given the less anthropomorphic nature of cetaceans, it is even more striking that they have also been able to display to us a keen intelligence that stands out through these differences.

However, the real measure of apparent intelligence must be grounded in scientific investigation into both the neuroanatomical complexity of the animal and its cognitive-behavioral capacities tested in controlled situations. Within these scientific domains, cetaceans have substantiated our perceptions of complex intelligence. Cetaceans possess some of the largest brains on the planet, a highly convoluted and elaborated cerebral cortex, and an encephalization level or EQ (in many species) matched by no other animal except modern humans.[1] Under laboratory testing situations, cetaceans have demonstrated complex cognitive abilities on par with those of great apes and humans.

---

1 Encephalization is typically expressed as an encephalization quotient (EQ). EQ is an index that quantifies how much larger or smaller a given animal's brain is relative to the expected brain size for an animal at that body size (Jerison, 1973). Brains with EQs larger than 1 are larger than the expected size, while those with EQs less than 1 are smaller than the expected size.

In this chapter I describe the scientific evidence for these assertions by addressing the question of how cetacean brains evolved over time and what the inferred cognitive consequences were of those changes. Modern cetacean cognitive characteristics are compared with those of other animals, especially great apes and humans.

Cetacean Evolution and Phylogeny

The order Cetacea consists of one extinct and two modern suborders. The Eocene suborder, Archaeoceti, contained approximately 25 (known) genera (Thewissen, 1998) and survived from the early Eocene, around 52 million years ago (mya), until the late Eocene, around 38 mya (Bajpai & Gingerich, 1998; Barnes, Domning, & Ray, 1985; Uhen, 1998). The modern suborders, Mysticeti (comprising 13 species of baleen and rorqual whales) and Odontoceti (comprising 67 species of toothed whales, dolphins, and porpoises), are first found in the fossil record in the early Oligocene, about 38 mya (Barnes et al., 1985). Table 7.1 presents a taxonomy of modern cetaceans for reference.

The evolutionary history of Cetacea (dolphins, whales, and porpoises) represents one of the most dramatic transformations in the mammalian fossil record.

Cetacean ancestry is tied closely to that of the Ungulata (the order of hooved mammals) and specifically Artiodactyla (the suborder of even-toed ungulates). Morphological evidence from early Eocene whales confirms the cetacean-artiodactyl link (Geisler & Uhen, 2003; Gingerich, ul Haq, Zalmout, Khan, & Sadiq, 2001; Thewissen, Williams, Roe, & Hussain, 2001). Molecular evidence shows a sister-taxon relationship between extant cetaceans and the artiodactyl family Hippopotamidae (Gatesy, 1998; Milinkovitch, Berube, & Palsboll, 1998; Nikaido, Rooney, & Okada, 1996; Shimamura et al., 1997) but an early divergence from hippopotamids at least 52 mya (Gingerich & Uhen, 1998).

Methods of Studying Cetacean Brains and Cognition

A vast amount remains unknown about most cetacean species. This statement speaks to the fact that there are certain difficulties in studying cetaceans because of their size, speed, and aquatic habitat. Despite these difficulties, marine mammal researchers have been able to collect important information on the evolution of cetacean brains and cognition. Relevant data come from three general domains of research—paleobiology,

Table 7.1
Taxonomy of Modern Cetaceans

Suborder Odontoceti (toothed whales)	
Family Delphinidae ("oceanic" dolphins)	
Steno bredanensis	Rough-toothed dolphin
Sousa chinensis	Indo-Pacific hump-backed dolphin
Sousa teuszii	Atlantic hump-backed dolphin
Sotalia fluviatilis	Tucuxi
Tursiops truncatus	Bottlenose dolphin
Stenella longirostris	Spinner dolphin
Stenella clymene	Clymene dolphin
Stenella frontalis	Alantic spotted dolphin
Stenella attenuata	Pantropical spotted dolphin
Stenella coeruleoalba	Striped dolphin
Delphinis delphis	Common dolphin
Lagenodelphis hosei	Fraser's dolphin
Lagenorhynchus albirostris	White beaked dolphin
Lagenorhynchus acutus	Atlantic white-sided dolphin
Lagenorhynchus obliquidens	Pacific white-sided dolphin
Lagenorhynchus obscurus	Dusky dolphin
Lagenorhynchus australis	Peale's dolphin
Lagenorhynchus cruciger	Hourglass dolphin
Cephalorhynchus commersonii	Commerson's dolphin
Cephalorhynchus heavisidii	Heaviside's dolphin
Cephalorhynchus eutropia	Black dolphin
Cephalorhynchus hectori	Hector's dolphin
Lissodelphis borealis	Northern right whale dolphin
Lissodelphis peronii	Southern right whale dolphin
Grampus griseus	Risso's dolphin
Peponocephala electra	Melon-headed whale
Feresa attenuata	Pygmy killer whale
Pseudorca crassidens	False killer whale
Globicephala melas	Long-finned pilot whale
Globicephala macrorhynchus	Short-finned pilot whale
Orcinus orca	Killer whale
Orcaella brevirostris	Irrawaddy dolphin
Family Phocoenidae (porpoises)	
Phocoena phocoena	Harbor porpoise
Phocoena sinus	Vaquita
Phocoena dioptrica	Spectacled porpoise
Phocoena spinnipinnis	Burmeister's porpoise
Neophocaena phocaenoides	Finless porpoise
Phocoenoides dalli	Dall's porpoise
Family Monodontidae (belugas or white whales, and narwhals)	
Delphinapterus leucas	Beluga, White whale
Monodon monoceros	Narwhal

Table 7.1 (continued)

Family Platanistidae	
Platanista gangetica	Ganges and Indus river dolphin
Family Iniidae	
Inia geoffrensis	Amazon river dolphin, boto
Family Lipotidae	
Lipotes vexillifer	Yangtze river dolphin, Baiji
Family Pontoporiidae	
Pontoporia blainvillei	La Plata dolphin, Franciscana
Family Ziphiidae (beaked whales)	
Berardius bairdii	Baird's beaked whale
Berardius arnuxii	Arnoux's beaked whale
Tasmacetus shepherdi	Sheperd's beaked whale
Ziphius cavirostris	Cuvier's beaked whale
Hyperoodon ampullatus	Northern bottlenose whale
Indopacetus pacificus	Longman's beaked whale
Mesoplodon hectori	Hector's beaked whale
Mesoplodon mirus	True's beaked whale
Mesoplodon europaeus	Gervais' beaked whale
Mesoplodon ginkgodens	Ginkgo-toothed beaked whale
Mesoplodon carlhubbsi	Hubb's beaked whale
Mesoplodon stejnegeri	Stejneger's beaked whale
Mesoplodon bowdoini	Andrew's beaked whale
Mesoplodon bidens	Sowerby's beaked whale
Mesoplodon layardii	Strap-toothed whale
Mesoplodon densirostris	Blainville's beaked whale
Mesoplodon peruvianus	Pygmy beaked whale
Mesoplodon traversii	Spade-toothed whale
Mesoplodon perrini	Perrin's beaked whale
Family Physeteridae (sperm whales)	
Physeter macrocephalus	Sperm whale
Family Kogiidae	
Kogia breviceps	Pygmy sperm whale
Kogia simus	Dwarf sperm whale
Suborder Mysticeti (baleen whales)	
Family Balaenidae (right whales)	
Eubalaena glacialis	Atlantic northern right whale
Eubalaena japonica	Pacific northern right whale
Eubalaena australis	Southern right whale
Balaena mysticetus	Bowhead whale
Family Neobalaenidae	
Caperea marginata	Pygmy right whale
Family Balaenopteridae (rorquals)	
Balaenoptera musculus	Blue whale
Balaenoptera physalus	Fin whale

Table 7.1 (continued)

Balaenoptera borealis	Sei whale
Balaenoptera edeni	Bryde's whale
Balaenoptera acutorostrata	Common minke whale
Balaenoptera bonaerensis	Antarctic minke whale
Megaptera novaeangliae	Humpback whale
Family Eschrichtiidae	
Eschrichtius robustus	Gray whale

Note: This taxonomy is based largely on Rice (1998).

neuroanatomy, and experimental laboratory studies. All of these approaches have contributed to an evolutionary cognitive neuroscience of cetaceans.

Paleobiology

Until recently, the fossil record of cetacean postcranial and cranial morphology has relied on a slow trickle of data from investigations of intact endocasts to estimate size and gross morphology of the brain. Not surprisingly, relatively little was known about the brains of extinct cetaceans because of difficulty in accessing the hardened matrix that fills the endocranial cavity of fossil specimens. In the past few years, however, computed tomography (CT) has emerged as a revolutionary tool in the study of fossil endocrania. CT allows nondestructive imaging and measurement of endocranial features and gives unprecedented views into the previously largely inaccessible world of fossil cetacean endocranial morphology. The result has been an increase in our knowledge of cetacean brain evolution as derived from fossil specimens. This chapter examines very recent findings on cetacean brain evolution from CT-based studies of fossils.

Neuroanatomy

Studies of neuroanatomical structure and function are crucial for understanding the neurobiological basis of cognition and behavior in cetaceans. Relatively little work has been done on the neuroanatomy and neurobiology of cetacean brains compared with what is known about other mammal brains, such as those of primates and rodents. However, the literature on cetacean neuroanatomy and neurobiology is growing. A number of morphological studies of adult, juvenile, and fetal cetacean brains exist. Additionally, much work on cortical architecture has been done using histological and immunocytochemical techniques to investigate the cellular and chemical organization of cetacean cortex. In addition to CT, another imaging modality, magnetic resonance imaging (MRI), has opened up whole new avenues of research into the neuroanatomical structure of cetacean brains. Whereas morphological descriptions based on standard techniques are difficult to obtain because of the bulky size of most cetacean brains, MRI has proved to be a very efficient way to obtain large amounts of data on cetacean brain morphology. MRI offers a means of examining the internal structures of the brain in their precise anatomical positions because the whole fixed brain is kept intact during the scanning, therefore minimizing the spatial distortions associated with traditional methods.

In addition to brain structure, studies of brain function are necessary for a complete understanding of how the brain gives rise to behavior and cognition. Studies done mainly in Russia have used electrophysiological recording methods to map functional areas of the dolphin cortex. This literature remains our primary source of information about cortical sensorimotor organization in cetaceans. Most of these methods involve invasive surface recording from dolphin cortex in awake animals. These kinds of studies are not allowed in the United States because of animal welfare regulations. However, the advent of functional imaging techniques such as positron emission tomography (PET) and functional MRI (fMRI) has provided a new, noninvasive avenue for examining the function of cetacean brains. In this chapter all of these methods are discussed as they contribute to our understanding of cetacean brain structure and function.

Experimental Cognition

Laboratory-based studies under well-controlled conditions are critical for testing hypothesized cognitive abilities. An overwhelming preponderance of the experimental research on cognition and behavior in cetaceans has focused on the bottlenose dolphin, *Tursiops truncatus*, with a relatively smaller proportion of studies on other odontocetes such as the killer whale, *Orcinus orca*, and the beluga whale, *Delphinapterus leucas*, and a still smaller number of studies conducted on other odontocetes. Therefore, as in any taxonomic group, one should expect cognitive and behavioral differences across cetacean species. However, the sizable literature on the bottlenose dolphin can, at the very least, serve as a database of "existence proofs" for cognitive and behavioral capacities in cetaceans in general. This chapter explores the domains in which modern dolphins and other cetaceans have demonstrated their cognitive capacities. These findings are then used to form hypotheses about how cognition and behavior changed with the brain throughout cetacean evolution.

Cetacean Brain and Cognitive Evolution

This section explores the pattern of evolutionary change in cetacean brains and the possible relationship between these changes and cognition by putting together observations in the three research areas noted above: paleobiology, neuroanatomy, and experimental cognition studies.

The First Cetaceans

Over the course of approximately 13 million years, from around 52 mya to 39 mya, cetaceans evolved from fully terrestrial to semiaquatic to, eventually, obligate aquatic animals (Gingerich et al., 2001; Thewissen, Williams, Roe, Hussain, 2001; Uhen, 1998). The early terrestrial cetaceans had five toes ending in short hooves and typical heterodont (differentiated) mammalian dentition (Uhen, 2004). The earliest specimens have been found in Indo-Pakistani and Eurasian regions that once surrounded the subtropical Tethys Sea (Gingerich & Russell, 1990; Kumar & Sahni, 1986; Thewissen, Madar, & Hussain, 1996). It is thought that terrestrial archaeocete behavioral ecology mainly consisted of near-shore piscivory (Thewissen & Bajpai, 2001; Uhen, 2004). Later archaeocetes moved from the Tethys to other oceans. These fully aquatic whales were predators with elongated jaws and an intimidating dentition (Uhen, 2004).

Archaeocete brains (figure 7.1A) were characterized by small, elongated cerebral hemispheres ending rostrally in large olfactory peduncles and bulbs (Edinger, 1955). There is no evidence for echolocation abilities in archaeocetes (Uhen, 2004). Importantly, archaeocete brains changed very little over their 13-million-year period of adaptation to a fully aquatic existence. Archaeocete EQ levels remained very low (i.e., <1) over the entire period of aquatic adaptation. Even the most recent archaeocetes from the later Eocene possessed significantly lower EQs

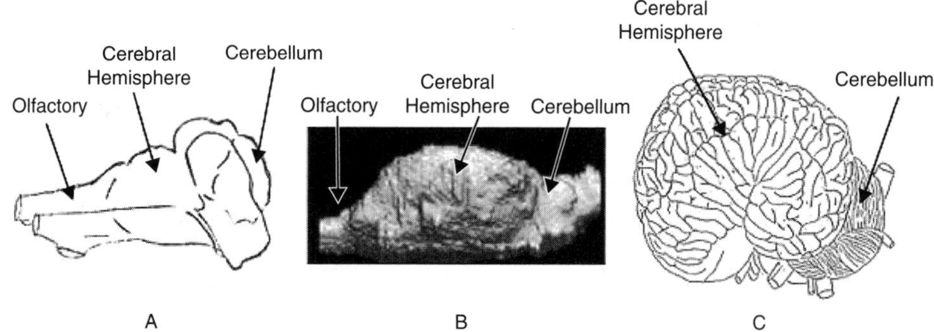

Figure 7.1
Sagittal images of cetacean brains at three stages of evolution: (A) drawing of an archaeocetes brain from a natural endocast; (B) three-dimensional reconstruction of a 27-million-year-old fossil odontocete brain from CT images; (C) drawing of a modern bottlenose dolphin brain from a photograph. Representations are not to scale.

than their successors in the Oligocene (Marino, McShea, & Uhen, 2004). Therefore there is no support for the hypothesis that cetacean brains were selected to be large because of adaptation to the aquatic environment (Marino, McShea, et al., 2004). All of the morphological evidence taken together suggests that the cognitive abilities of archaeocetes were probably the aquatic version of the typical predatory Cretaceous mammal and not nearly as cognitively or behaviorally complex as their descendants.

The Early Modern Suborders

By the beginning of the Oligocene, about 39 mya, archaeocetes went extinct and their descendants, early modern odontocetes and mysticetes, appeared in the fossil record (Barnes et al., 1985). These early forms were very different from archaeocetes. Over the next 24 million years of cetacean evolution, cetacean brains underwent distinct patterns of change (figure 7.1B). The signature occurrences were the substantial hyperproliferation of the cerebral hemispheres and the extreme enlargement of auditory processing areas. Cerebral enlargement mainly occurred in the parietal, temporal and occipital regions with little or no elaboration of frontal areas. Olfactory structures were still in evidence but were proportionately much smaller in relation to the hemispheres.

is unsurpassed among mammals, including humans (2,275 cm^2) (Elias & Schwartz, 1969; Ridgway & Brownson, 1984). This exceptional surface area may be due to the extreme replication of vertical or "cross-laminar" functional units (Glezer, Jacobs, & Morgane, 1988; Morgane, Glazer, & Jacobs, 1988, 1990). Moreover, the cetacean neocortex is relatively thin, with a width between 1.3 and 1.8 mm, as compared with the 3.0-mm-thick human neocortex (Haug, 1969; Kesarev, 1971; Ridgway & Brownson, 1984).

Because of the early divergence of cetaceans from other mammalian lineages, their neocortex evinces elaboration on themes that are recognizably phylogenetically old or conserved in mammals. However, it would be a mistake to interpret this remark as implying that the cetacean neocortex is more primitive or simpler than primate brains and the like. On the contrary, despite building on an apparently conservative laminar theme, cetacean neocortex is highly derived and elaborated.

The neocortex of primates and many other mammals consists of six well-defined layers. Main input from the thalamus is to layer 4, which is topographically organized and consists of small stellate cells that give it a granular appearance. Layer 6 sends feedback to the input. Layers 2 and 3 (the external layers) send output to other parts of the cortex, and layer 5 (the internal pyramidal layer) sends output to subcortical structures. Pyramidal cells with apical dendrites oriented perpendicular to the neocortical surface span much of the cortex and end in layer 1 (Allman, 1999). Although stellate and pyramidal cells dominate, other kinds of cellular morphotypes are clearly distinguishable, and there is considerable heterogeneity (Swanson, 2003). These general features are considered the basic plan of the mammalian neocortex.

In contrast, cetacean neocortex possesses mainly five layers. It is characterized by a very thick layer 1 that contains apical dendrites of extraverted pyramidal cells from a highly accentuated layer 2 (Glezer et al., 1988; Morgane et al., 1988). The strong pyramidalization of layer 2 is also a key feature of cetacean neocortex (Morgane et al., 1988). It has been suggested that, in cetaceans, all of the thalamocortical afferents feed into the thick layer 1 and through the extraverted neurons of layer 2 to deeper levels (Glezer et al., 1988; Morgane et al., 1990). One of the most salient features of cetacean neocortex is the general lack of granularity, which is due primarily to the absence of (or, at best, barely identifiable) granular layer 4. Morgane et al. (1988) identified two types of visual cortex in the bottlenose dolphin. Heterolaminar cortex appears to contain a very meager layer 4. In homolaminar cortex, however, layer 4 is entirely absent. To most investigators, the general dysgranularity of the cetacean neocortex is viewed as evidence that cetaceans diverged from the mammalian line prior to the neocortical granularization trend evinced in other mammals. Furthermore, the general lack of layer 4 in cetaceans has important implications for afferentation patterns. In primates and other mammals only *some* afferent connections come through layer 1 to dendritic connections from layer 2 neurons, while other, specialized thalamocortical afferents synapse directly on layer 4. In cetaceans the *majority* of afferents appear to go through the very thick layer 1 to synapse en passage on extraverted neurons of layer 2 (Glezer et al., 1988). A small portion of afferents go to layers 3 and 5 as well (Garey & Revishchin, 1989; Revishchin & Garey, 1990). Some investigators view the segregation of afferents to layers 4 and 1 to be a later evolutionary development than the pattern evinced in cetacean neocortex (Glezer et al., 1988; Kesarev, 1975; Morgane, Jacobs, & Galaburda, 1986; Morgane et al., 1990). Therefore, according to this view, the cetacean neocortex has expanded on a highly conserved theme that essentially bypasses an entire stage of cortical evolution found in many other mammals and takes an alternative route to complexity.

In addition to striking differences in input-output and integrative organization between the cetacean brain and the primate brain, there are also major differences in chemoarchitecture (Hof et al., 1995, 1999, 2000) and the range of cellular morphotypes (Morgane et al., 1988, 1990). Also, the presence of numerous "transitional" types of neurons has been noted (Garey, Winkelmann, & Brauer, 1985; Kesarev, Malofeyeva, & Trykova, 1977; Morgane et al., 1986).

To summarize, the cetacean cortex, with its extreme surface area and different architecture, represents an alternative to the evolutionary route toward cortical expansion taken by other large mammals, including primates. The cetacean brain is arguably the most highly developed version of this particular kind of cortex.

Neuroanatomy and Cognitive Function in Modern Cetaceans

Relatively little is known about the functional morphology of cetacean brains. In this section we examine available data in order to address the question, what cognitive abilities do cetaceans display that can be related to known features of their brain?

Johnson et al., 2005). Interestingly, in terms of the relationship between olfaction and the hippocampus, cetaceans and humans represent opposite trends. The loss of olfaction in cetaceans has resulted in a reduced hippocampus and related structures. Contrarily, reduction in olfaction in humans is not paralleled by a loss of hippocampal structures. Quite the opposite is true. The human hippocampus is probably the most highly developed of all mammals (Jacobs et al., 1979).

The hippocampus serves a primary role in memory consolidation (Gazzaniga, Ivry, & Mangun, 1998) and spatial learning (Morris, Pickering, Abraham, & Feigenbaum, 1996; O'Keefe & Nadel, 1978; Sherry, Jacobs, & Gaulin, 1992). This makes the extremely reduced hippocampus in cetaceans all the more puzzling, in light of the evidence for highly sophisticated ranging and distribution patterns in cetaceans that depend heavily on spatial memory skills (Baird, 2000) and superb performance on direct memory tasks (Mercado III et al., 1998, 1999; Thompson & Herman, 1977) and cognitive tests that rely on memory consolidation (see Herman, 2002, for a review). These include studies in which dolphins are required to mentally access their long-term memory and "declare" (through explicit actions) their knowledge of whether objects are present or absent (Herman & Forestell, 1985).

An interesting corollary feature to the diminished hippocampal formation in cetaceans is the extremely well-developed cortical limbic lobe (particularly cingulate cortex and entorhinal cortex) in cetaceans (Jacobs et al., 1979; Marino, Sherwood, et al., 2004; Morgane et al., 1980). This juxtaposition of a vastly reduced hippocampus and a highly elaborated periarchicortical/entorhinal zone in the face of complex memory and spatial skills suggests the possibility that there was a transfer of hippocampus-like functions from the olfactory-based hippocampal domain to other cortical, including periarchicortical and entorhinal, regions (see Jacobs et al., 1979, for a review) during cetacean evolution.

The Possible Role of the Insular Cortex in Communication

Another notably elaborated area in cetacean brains is the insular cortex and surrounding temporal operculum (Jacobs, Galaburda, McFarland, & Morgane, 1984; Marino, Sherwood, et al., 2004). The insula mediates viscerosensation, gustation, and some somatosensation in most mammals. Although homology between the cetacean insula and that of other mammals has yet to be established, Morgane et al. (1980) suggest that, on the basis of architectonic evidence, the operculum may cortically represent trigeminal (rostrum) and glossopharyngeal (nasal respiratory tract) innervation. Given that various sounds are modified by structures associated with the control of air flow through the nasal region, it is a speculative but not altogether unreasonable possibility that the operculum could serve a similar function as the speech-related opercular cortex in humans. Others have suggested that the insular region surrounded by the operculum is related to specializations of the auditory cortex (Manger, Sum, Szymanski, Ridgway, & Krubitzer, 1998), a function closely tied to communication.

No Frontal Lobes, Yet Self-Awareness

The cetacean hemispheres evince hyperexpansion of temporal, parietal, and occipital regions. The notable exception is the so-called frontal lobe. In fact, there is no identifiable neocortical region in the cetacean brain that is homologous with primate frontal lobe or prefrontal cortex. This situation would normally not be puzzling except for one interesting fact. Despite the lack of frontal lobes, dolphins are capable of cognition that is typically attributed to the frontal lobes in primates. These cognitive capacities include executive functions such as foreplanning and, intriguingly, aspects of self-awareness (see Gazzaniga et al., 1998, for a review).

Bottlenose dolphins demonstrate evidence for anticipating, organizing, monitoring, and modifying goal-directed behavior on the basis of contingencies. For instance, McCowan, Marino, Vance, Walke, & Reiss (2000) applied statistical tests to observations of bubble ring play in captive bottlenose dolphins and revealed findings consistent with the hypothesis that the dolphins were monitoring the quality of their bubble rings and planning future bubble production actions based on this information.

Bottlenose dolphins also more directly demonstrate capacities in the domain of self-awareness. Self-awareness is a multidimensional phenomenon that, at its core, refers to the ability to think about one's own mental states (metacognition) and identity. Wheeler, Stuss, and Tulving (1997) proposed the similar term "autonoetic consciousness" to encapsulate the "capacity that allows adult humans to mentally represent and to become aware of their protracted existence across subjective time" (p. 335). An individual's capacity for self-awareness can be probed experimentally in a number of ways. One of those ways is through the mirror self-recognition paradigm. Mirror self-recognition requires self-awareness (or self-identity) in the physical domain. It is an exceed-

Figure 7.2
Bottlenose dolphins use mirrors to investigate marks on their bodies. (Photograph by Diana Reiss, Wildlife Conservation Society.)

ingly rare capacity in the animal kingdom that, until recently, had been demonstrated only in humans and great apes (see Povinelli et al., 1997, for a review of this literature), but did not extend as far phylogenetically as monkeys or lesser apes (Anderson & Roeder, 1989; Bayart & Anderson, 1985; Hyatt, 1998; Shaffer & Renner, 2000; Suarez & Gallup, 1986). However, in 2001 Reiss and Marino reported conclusive evidence of mirror self-recognition in a nonprimate species, the bottlenose dolphin. In a series of controlled variations of the procedures used with primates, both dolphins in the study consistently used a mirror to investigate marked parts of their bodies (figure 7.2). Bottlenose dolphins have also demonstrated related capacities on other kinds of cognitive tasks in the domains of awareness of one's own body parts and one's own behavior (Herman, Matus, Herman, Ivancic, & Pack, 2001; Mercado et al., 1998, 1999).

To add to the growing evidence for self-awareness in dolphins, bottlenose dolphins placed in a difficult auditory discrimination task with the option of making an "Uncertain Response" do so in exactly the same manner as humans and monkeys under the same experimental conditions (Smith et al., 1995, 2003). In other words, dolphins *know* when they do not know something. This kind of response requires the kinds of metacognitive levels of processing attributed to human frontal lobe function.

There is an interesting literature showing a relationship between right prefrontal cortical function and self-recognition in humans (Keenan, Wheeler, Gallup, & Pascual-Leone, 2000; Keenan, Nelson, & Pascual-Leone, 2001; Miller et al., 2001; Platek et al., 2004; Stuss, Gallup, & Alexander, 2001), supporting the hypothesis that the cortical circuitry of the prefrontal cortex in humans and great apes is the necessary neuroanatomical substrate for self-recognition and other dimensions of self-awareness. The Reiss and Marino findings show that dolphins are capable of mirror self-recognition despite possession of unelaborated frontal lobes (or absence of homologous frontal lobe structures). These results suggest that there is more than one way to evolve self-awareness and that there are parallel neurobiological substrates operating in dolphins and humans/great apes. These results in no way call into question the validity of the neurobiological human literature. However, they do show that the emergence of self-recognition (and likely other dimensions of self-awareness) is not a byproduct of factors exclusive to great apes and humans.

Self-recognition is one of the most striking examples of cognitive convergence between dolphins and primates. It is not unreasonable to suggest that the high encephalization levels of both humans and great apes, on the one hand, and bottlenose dolphins on the other are related to their shared capacities. However, the size of the brain and its structures is but one correlate of cognition. The other is architecture, and specifically neocortical architecture. Cognitive complexity derives not only from having a sufficient mass of tissue but also from possessing an organizational structure that allows for complex computations—that is, one with many intricate connections, various functional units, and so on. However, when we compare neocortical architecture in primates (including humans) with that in dolphins, the differences between the two groups are only furthered. As described above, the underlying cytoarchitectural and organizational scheme of the dolphin neocortex is unique and highly different from that in primates. These differences further support the notion that the same cognitive capacities in primates and dolphins are underwritten by different neurobiological "themes" resulting in convergent cognition.

Approaches on the Horizon

Evidence from various domains of research, including paleobiology, experimental cognition, and neuroanatomy, demonstrates that cetacean brains underwent extreme elaboration during their evolution, with resulting hyperexpansion of the neocortex and some subcortical regions associated with auditory processing. Cortical evolution, however, proceeded along very different lines in cetaceans than in primates and other large mammals. Yet many cetaceans evince some of the most sophisticated cognitive abilities seen among mammals and exhibit striking cognitive evolutionary convergences with primates, including humans. In order to bring our understanding of cetacean cognitive neuroscience to the next level, not only do these structural and behavioral research domains need to be vastly extended, they need to be joined by critical research approaches to *linking* anatomy and cognitive performance. These are functional anatomy and computational modeling.

There is a critical need for knowledge in cetacean functional anatomy. Because of the inappropriateness of invasive neurobiological investigations in cetaceans, noninvasive methods are crucial. The advent of functional imaging techniques such as positron emission tomography (PET), single-proton emission computed tomography (SPECT), and functional MRI has opened up a new, noninvasive methodological avenue to examining the neuroanatomical basis of cognitive processing in awake cetaceans. The use of these vanguard techniques is part of a nascent research program being undertaken by Sam H. Ridgway and his colleagues at the Navy Marine Mammal Center. They have succeeded in producing the first functional images of the awake dolphin brain by using SPECT (Houser et al., 2004; Ridgway et al., 2003). Through these studies they have demonstrated that functional imaging can be employed safely and productively in awake dolphins to obtain valuable information about brain structure and function. This very new and exciting research approach is bound to expand and develop further in the near future and promises to provide an unprecedented window of knowledge on cetacean functional anatomy.

Ultimately, anatomical and functional information must be examined within the context of how these features combine to arrive at complexity in information processing. Cetaceans evince complex cognitive abilities that must be manifestations of neurobiological complexity. The term *complexity* has many meanings but can be operationalized in a number of ways, such as by assessing the depth of hierarchical structure in a system or by using concepts drawn from information theory that express the degree of interaction between elements of a system. In order to understand, at the deepest level, how, despite fundamental structural differences, the cetacean neocortex provides the same level of cognitive complexity as the primate neocortex, neurobiological or computational modeling must be initiated. By using our increasing knowledge of how cetacean brains are put together, we can start applying quantitative models to test hypotheses about what aspects of cetacean brains provide the information-processing complexity that forms the basis of the notable cetacean intelligence.

References

Anderson, J. R., & Roeder, J. J. (1989). Responses of capuchin monkeys (*Cebus apella*) to different conditions of mirror-image stimulation. *Primates, 30*, 581–587.

Allman, J. M. (1999). *Evolving brains*. New York: Scientific American.

Baird, R. W. (2000). The killer whale: Foraging specializations and group hunting. In J. Mann, R. C. Connor, P. L. Tyack, & H. Whitehead (Eds.), *Cetacean societies: Field studies of dolphins and whales* (pp. 127–153). Chicago: University of Chicago Press.

Bajpai, S., & Gingerich, P. D. (1998). A new Eocene archaeocete (Mammalia, Cetacea) from India and the time of origin of whales. *Proceedings of the National Academy of Sciences, U.S.A., 95*, 15464–15468.

Balonov, L. Ia., Deglin, V. L., Kaufman, D. A., & Nikolaenko, N. N. (1981). [Functional asymmetry of the animal brain]. *Zhurnal Evoliutsionnoi Biokhimii i Fiziologii, 17*, 225–233. (In Russian, English translation 1982 by Plenum)

Barnes, L. G., Domning, D. P., & Ray, C. E. (1985). Status of studies on fossil marine mammals. *Marine Mammal Science, 1*, 15–53.

Bayart, F., & Anderson, J. R. (1985). Mirror-image reactions in a tool-using adult male *Macaca tonkeana. Behavioural Processes, 10*, 219–227.

Breathnach, A. S. (1960). The cetacean central nervous system. *Biological Review, 35*, 187–230.

Buhl, E. H., & Oelschlager, H. A. (1988). Morphogenesis of the brain in the harbour porpoise. *Journal of Comparative Neurology, 277*, 109–125.

Edinger, T. (1955). Hearing and smell in cetacean history. *Monatsschrift für Psychiatrie and Neurologie, 129*, 37–58.

Elias, H., & Schwartz, D. (1969). Surface areas of the cerebral cortex of mammals determined by stereological methods. *Science, 166*, 111–113.

Fauteck, J. D., Lerchl, A., Bergmann, M., Moller, M., Fraschini, F., Wittkowski, W., et al. (1994). The adult human cerebellum is a target of the neuroendocrine system involved in circadian timing. *Neuroscience Letters, 179*, 60–64.

Flanigan, W. F. (1974). Nocturnal behavior of captive small cetaceans. I. The bottlenosed porpoise, *Tursiops truncatus. Sleep Research, 3*, 84.

Flanigan, W. F. (1975). More nocturnal observations of small captive cetaceans. I. The killer whale, *Orcinus orca. Sleep Research, 4*, 139.

tions of the telencephalon of the bottlenose dolphin with comparative anatomical observations in four other cetacean species. *Brain Research Bulletin*, *5*, 1–107.

Morris, R., Pickering, A., Abraham, S., & Feigenbaum, J. D. (1996). Space and the hippocampal formation in humans. *Brain Research Bulletin*, *40*, 487–490.

Mukhametov, L. M. (1984). Sleep in marine mammals. *Experimental Brain Research, Supplement 8*, 227–238.

Mukhametov, L. M., Supin, A. Y., & Polyakova, I. G. (1977). Interhemispheric asymmetry of the electroencephalographic sleep patterns in dolphins. *Brain Research*, *134*, 581–584.

Nikaido, M., Rooney, A. P., & Okada, N. (1996). Phylogenetic relationships among cetartiodactyls based on insertions of short and long interspersed elements: Hippopotamuses are the closest extant relatives of whales. *Proceedings of the National Academy of Sciences, U.S.A.*, *96*, 10261–10266.

Oelschlager, H. A., & Oelschlager, J. S. (2002). Brains. In W. F. Perrin, B. Wursig, & H. Thewissen (Eds.), *Encyclopedia of marine mammals* (pp. 133–158). San Diego: Academic Press.

O'Keefe, J., & Nadel, L. (1978). *The hippocampus as a cognitive map*. Oxford: Oxford University Press.

Pack, A. A., & Herman, L. M. (1995). Sensory integration in the bottlenosed dolphin: Immediate recognition of complex shapes across the senses of echolocation and vision. *Journal of the Acoustical Society of America*, *98*, 722–723.

Paulin, M. G. (1993). The role of the cerebellum in motor control and perception. *Brain, Behavior and Evolution*, *41*, 39–50.

Platek, S. M., Kenan, J. P., Gallup Jr., G. G., & Mohamed, F. B. (2004). Where am I? The neurological correlates of self and other. *Cognitive Brain Research*, *19*, 114–122.

Povinelli, D. J., Gallup, G. G., Eddy, T. J., Bierschwale, D., Engstrom, M. C., Perilloux, H. K., et al. (1997). Chimpanzees recognize themselves in mirrors. *Animal Behavior*, *53*, 1083–1088.

Reiss, D., & Marino, L. (2001). Mirror self-recognition in the bottlenose dolphin: A case of cognitive convergence. *Proceedings of the National Academy of Sciences, U.S.A.*, *98*, 5937–5942.

Revishchin, A. V., & Garey, L. J. (1990). The thalamic projection to the sensory neocortex of the porpoise, *Phocoena phocoena*. *Journal of Anatomy*, *169*, 85–102.

Rice, D. W. (1998). *Marine mammals of the world: Systematics and distribution* (pp. 1–231). Lawrence, KS: Allen Press.

Ridgway, S. H. (1972). Homeostasis in the aquatic environment. In S. H. Ridgway (Ed.), *Mammals of the sea: Biology and medicine* (p. 715). Springfield, IL: Charles C. Thomas.

Ridgway, S. H. (1986). Physiological observations on dolphin brains. In J. A. Thomas & F. G. Wood (Eds.), *Dolphin cognition and behavior* (pp. 31–59). Hillsdale, NJ: Erlbaum.

Ridgway, S. H. (1990). The central nervous system of the bottlenose dolphin. In S. Leatherwood & R. R. Reeves (Eds.), *The bottlenose dolphin* (pp. 69–100). San Diego: Academic Press.

Ridgway, S. H. (2000). The auditory central nervous system of dolphins. In W. Au, A. Popper, & R. Fay (Eds.), *Hearing in whales and dolphins* (pp. 273–293). New York: Springer-Verlag.

Ridgway, S. H., & Brownson, R. H. (1984). Relative brain sizes and cortical surface areas of odontocetes. *Acta Zoologica Fennica, 172*, 149–152.

Ridgway, S. H., Houser, D., Finneran, J., Carder, D., Van Bonn, W., Smith, C., et al. (2003). Functional brain imaging for bottlenose dolphins, *Tursiops truncatus*. In *Proceedings of the 15th Biennial Conference on the Biology of Marine Mammals* (p. 138). Greensboro, NC.

Schwerdtfeger, W. K., Oelschlager, H. A., & Stephan, H. (1984). Quantitative neuroanatomy of the brain of the La Plata dolphin, *Pontoporia blainvillei*. *Anatomy and Embryology, 170*, 11–19.

Shaffer, V. A., & Renner, M. J. (2000). Black-and-white colobus monkeys (*Colobus guereza*) do not show mirror self-recognition. *International Journal of Comparative Psychology, 13*, 154–160.

Sherry, D. F., Jacobs, L. F., & Gaulin, S. J. C. (1992). Spatial memory and adaptive specialization of the hippocampus. *Trends in Neuroscience, 15*, 298–303.

Shimamura, M., Yasue, H., Ohshima, K., Abe, H., Kato, H., Kishiro, T., et al. (1997). Molecular evidence from retroposons that whales form a clade within even-toed ungulates. *Nature, 388*, 666–671.

Shurley, J., Serafetinides, E., Brooks, R., Elsner, R., & Kenney, D. (1969). Sleep in cetaceans. I. The pilot whale, *Globicephala scammoni*. *Psychophysiology, 6*, 230.

Smith, J. D., Schull, J., Strote, J., McGee, K., Egnor, R., & Erb, L. (1995). The uncertain response in the bottlenose dolphin (*Tursiops truncatus*). *Journal of Experimental Psychology: General, 124*, 391–408.

Smith, J. D., Shields, W. E., & Washburn, D. A. (1995). The comparative psychology of uncertainty monitoring and metacognition. *Behavioral and Brain Sciences, 26*, 317–373.

Sobel, N., Supin, A. Ya., & Myslobodsky, M. S. (1994). Rotational swimming tendencies in the dolphin (*Tursiops truncatus*). *Behavioural Brain Research, 65*, 183–186.

Stuss, D. T., Gallup, G. G., Jr., & Alexander, M. P. (2001). The frontal lobes are necessary for "theory of mind." *Brain, 124*, 279–286.

Suarez, S. D., & Gallup, G. G., Jr. (1986). Social responding to mirrors in rhesus macaques (*Macaca mulatta*): Effects of changing mirror location. *American Journal of Primatology, 11*, 239–244.

Supin, A. Y., Mukhametov, L. M., Ladygina, T. F., Popov, V. V., Mass, A. M., & Poliakova, I. G. (1978). [*Electrophysiological studies of the dolphin's brain*]. Moscow: Izdatel'ato Nauka. (In Russian)

Swanson, L. W. (2003). *Brain architecture*. Oxford: Oxford University Press.

Tarpley, R. J., & Ridgway, S. H. (1994). Corpus callosum size in delphinid cetaceans. *Brain, Behavior and Evolution, 44*, 156–165.

III Reproduction and Kin Selection

Reproduction and selection are the keys to the evolutionary process. Put simply, the brain and the cognitive repertoire that we have today throughout the animal kingdom are the result of many generations of reproduction. Pressure is placed on successful reproduction at both the individual and species levels. This basic tenet of evolution has shaped the nervous system of animals, as well as the cognitive and behavioral abilities tied to those nervous systems.

At first it seems odd and quite challenging to imagine that modern humans are born with a brain that is more attracted to some entities (animate or inanimate) than to others. However, it is a logical conclusion based on basic evolutionary understanding. At a fundamental level, we can expect that those individuals attracted to X will outreproduce individuals attracted to Y. Because the X will change over subsequent generations as a result of selection pressures, the mapping of attraction and selection is difficult to do. For example, X and Y represent honest signals to underlying genetic qualities or adaptations that could be passed to offspring.

The central and peripheral nervous systems in humans are tied directly to our reproductive system at both a functional and an anatomical level. The two interact efficiently and almost seamlessly. Cognitive function is tied directly to hormonal secretions, for example, and hormonal secretions are tied to our cognitions. Thus, both top-down and bottom-up considerations (to borrow from traditional cognitive science) are in play in terms of the brain, its related systems, and reproduction. In nonhuman animals, it is simple to demonstrate both effects, as comparative psychology and biopsychology have done for over a century by manipulations at numerous levels of the cognitive-reproductive chain. Now, with advances in methodological techniques, we can further elucidate these relationships in animals, as well as directly investigate them in humans.

There are a number of cognitive mechanisms that traditional cognitive psychologists do not typically investigate but that are important for understanding the results of evolutionary processes. One is the identification of kin and nonkin. Cognitive neuroscience has demonstrated the existence of mechanisms and structures associated with the recognition of objects, words, places, and faces, for example. However, are there such mechanisms associated with kin recognition in animals, let alone humans? Further, if such a neural substrate exists, is it genetically mediated, and if so, in what manner? Is such recognition, as one might suspect in humans, primarily visual?

Studies in humans advance as our methodological tools reach a greater degree of sophistication. Much attention, even within this volume, is directed to the advancement of neuroimaging. In Part III we find that techniques such as MRI and SPECT are even being employed with chimpanzees and dolphins to advance our understanding of the evolution of the brain. An interesting manipulation found in this section is the elevation of serotonin via SSRIs in humans through the use of psychopharmacological agents. This work reminds us that with the neuroimaging revolution at hand, there is a largely ignored level of information that contributes to brain function at the neurotransmitter level. One surprising trend has been the lack of single-neurotransmitter studies in the field of neuroscience, which we hope will change (see Northoff et al., 1999). That said, Part III examines evolutionary cognitive neuroscience through multiple channels, from basic animal and human development to imaging and neurotransmitter manipulation. The chapters demonstrate the flexibility and creativity that is typical of researchers in this field, as well as the diversity of the areas of research. It is clear that one of the exciting features of evolutionary cognitive neuroscience is the richness and variety of research that encompasses the discipline.

Reference

Northoff, G., Steinke, R., Czcervenka, C., Krause, R., Ulrich, S., Danos, P., Kropf, D., Otto, H., & Bogerts, B. (1999). Decreased density of GABA-A receptors in the left sensorimotor cortex in akinetic catatonia: Investigation of in vivo benzodiazepine receptor binding. *Journal of Neurology, Neurosurgery, and Psychiatry*, 67, 445–450.

8 The Social Control of Reproduction: Physiological, Cellular, and Molecular Consequences of Social Status

Russell D. Fernald

Animals of every species experience life somewhat differently because their sensory and motor capacities provide a unique perceptual world. Von Uexküll (1921) first described this perception as the species's *Umwelt*, and the variety of *Umwelts* is evident in extreme cases, such as bats that use echoes from ultrasound waves they emit to probe the darkness, forming images from sound reflected from surroundings. Bats can function in total darkness using a sensory channel unavailable to most other animals. However, bats can also use vision in the ultraviolet wavelength, possibly for nectar foraging (Winter, Lopez, & von Helverson, 2003), giving some bat species two unique windows onto the world. Such sensory capacities limit what can be sensed, both enabling and constraining behavioral responses of an animal. Writing at the beginning of the twentieth century, Von Uexküll could not possibly have anticipated the discovery of magnetic, electric, pressure, temperature, and other senses, nor could he have imagined the "visual" sense extending from the infrared into the ultraviolet. Konrad Lorenz (1932) expanded the idea of *Umwelt* to include the detection of stimuli not just from physical surroundings but also from other animals. His influential article, "Companions as Factors in the Bird's Environment," showed that behavioral scientists needed to enlarge their notion of an animal's perceptual world to include other individuals and especially their social context. Behavior is the ultimate arbiter of animal survival, so the responses of animals during their interactions with others and with their environment shape the phenotype. Yet behavior in turn depends on intricate physiological, cellular, and ultimately molecular adaptations forged during evolution and modified during development.

A major challenge in cognitive science is to discover how behavior is controlled by physiological processes and, correspondingly, how behavior influences physiological, cellular, and molecular events. Though ambitious, this goal seems realistic with the availability of several new techniques. Addressing ultimate questions about the evolution and control of behavior and especially cognitive interactions requires understanding causal mechanisms in animals as they interact with one another, preferably in a reasonably natural setting. Yet the vast majority of experiments are performed on isolated individuals, many of them domesticated species. Clearly, little can be learned about how evolution has shaped social behavior by analyzing the behavior of individual animals in isolation, since social interactions are not possible. In this chapter, I describe results from our research program in which multiple techniques are used to study individuals in a seminatural social context. The experiments are focused on discovering how the social context of reproductive behavior shapes the brain and in turn alters the behavior of animals as they interact.

Ethologists transcended descriptive analysis by providing a framework for understanding order in animal behavior. By studying important life events such as feeding and reproduction in species with less complex behavioral interactions, Konrad Lorenz (1981) and Nikolaas Tinbergen (1951) were able to articulate what are now the central tenets of classical ethology, and brought rational discussion to understanding behavior. Discoveries about how animals respond to stimuli from conspecifics have provided significant insights into both the proximate and ultimate factors responsible for animal behavior. However, behavior can only be understood in the natural context of the animal, and in real life, animals behave and interact continuously, seamlessly integrating what they see with what they do. Given the importance of complex social interactions, scientists have sought model systems suitable for investigation of social interactions.

In 1950, Baerends and Baerends van Roon published a landmark monograph describing the behavior of numerous cichlid fish species, describing this vast collection as excellent models for ethological analysis. Cichlids are among the most successful fish species on the planet, having extended their range through the great lakes of Africa and South America. Cichlids were deemed excellent choices for ethological analysis for several reasons. First, within reasonable limits, cichlid fish could be studied in the laboratory without compromising their natural behavioral context or the ecological validity of the results. Second, cichlids are active, making the collection of quantitative data a realistic goal. Third, some cichlid species evolved facing different environmental constraints, making comparisons among closely related species a possible strategy for identifying potential selective forces. Indeed, cichlids have played an important role in understanding the rate of evolution (Verheyen, Salzburger, Snoeks, & Meyer, 2003) and in discovering how environmental conditions can compromise sexual selection (Seehausen, Van Alphen, & Witte, 1997). Thus, cichlids offered unique opportunities to examine both proximate mechanisms and ultimate functions of animal behavior in the same model system. To be sure, other fish model systems have been used to good effect to analyze aspects of social behavior, and indeed, there are claims for a rather remarkable suite of behavioral adaptations (e.g., Bshary, Wickler, & Fricke, 2002). Several fish species have been useful for the analysis of sound communication (Bass, Bodnar, & Marchaterre, 2000), the role of behavior in sex change (e.g., Godwin, Luckenbach, & Borski, 2003), and the social modulation of androgen levels, which has been studied in teleosts (e.g., Oliveria, Almada, & Caniaro, 1996).

Fishes make up the largest group of vertebrates, with more than 27,000 known species on the planet, more than all other vertebrates combined. As the most ancient vertebrates, fish occupy nearly every available aquatic niche, from hot sulfur springs to ponds that are frozen part of each year. This remarkable radiation into disparate ecological habitats has been possible because of the extraordinary range of physiological adaptations. But fish species also are successful because individuals have evolved a variety of behavioral adaptations for living, whether alone or in social groups with variously sophisticated social interactions. Indeed, almost every kind of vertebrate social system can be found among fish species, and these specialized behavioral adaptations often have resulted in corresponding physiological adaptations such as social regulation of reproduction, including sex change. Because the evolution of similar social systems in response to common ecological pressures among disparate vertebrate species might rest on common adaptations, it will be an interesting challenge to discover whether there are common neural substrates for specialized social systems.

Over the past decades during the emergence of the field of neuroscience, fishes proved to be valuable model systems for understanding the neural basis of vertebrate sensory systems, brain organization, and motor outputs. Yet there has been resistance to the idea of using fishes, or for that matter other cold-blooded vertebrates, to study the biological roots of cognition. Why is this? One obvious reason is the common conception that the psychological and cognitive skills of fishes rank them near plants in level of sophistication attributed to their behavior. Yet recent reviews have argued that fish might exploit Machiavellian strategies (e.g., Bshary & Würth, 2001), might cooperate (e.g., Pitcher, Green, & Magurran, 1986), learn in a social context (Laland, Brown, & Krause, 2003), eavesdrop on interactions among conspecifics to gain social insights (McGregor, 1993), and generally display a wide range of social activities comparable in some instances with those of primates (e.g., Bshary et al., 2002).

One emerging focus for cognitive science is understanding the mechanistic bases of cognitive skills. We expect the basis of cognition will be rooted in the evolutionary history of the species, but will comparisons across species yield common neurobiological mechanisms? The history of the comparative approach suggests that the answer is yes for many of the basic neural processes found in all animals. Indeed, the overwhelming message from the explosion of molecular details to emerge from the past 25 years is how many common properties there are at all levels of analysis. Programmed cell death in nematodes yielded the responsible genes whose identity and function have been largely conserved through to humans. So for molecular biologists, the mantra has become understanding how similar structures and their constituent molecules are across great phylogenetic divides. The flip side of the astounding similarities at the genetic level is the question of what gives rise to the many important differences biologists encounter among organisms at a macroscopic level. The glib answer usually invokes natural selection as a tinkerer, with post hoc inferences about the likely selective processes involved. But are there common cognitive strategies across taxa, and do these strategies use the same neural structures to support similar processes? To date, the possibility of answering these questions has been restricted because behavioral studies have been interpreted in functional rather than mechanistic terms, although that focus is beginning to change (e.g., Renn, Aubin-Horth, & Hofmann, 2004). However, detailed measurements of what happens in the parts of the brain responsible for social interactions are unavailable.

Why is it likely that analysis of the social interactions of fishes might yield general insights into brain mechanisms that support social interactions? Basically, stable social systems in any animal species require that individuals behave predictably. However, what an individual does at any moment in time may depend on its status relative to others, its reproductive state, and its recent behavioral interactions. Moreover, an assessment of environmental factors (e.g., predators, prey, or resource competitors) also needs to be incorporated into any plan for action. So, to be successful in a social group, individuals must be aware of the immediate behavior of others and use that information to regulate their own activity. But what exactly does an individual need to know, and how do individuals acquire the knowledge that lets them act appropriately? It is possible that apparently subtle social interactions can be explained and understood in terms of contingencies. That is, a set of if-then rules with associated probabilities might suffice in many situations to explain the behavior of animals in social groups. Because it has been argued that species in the fish taxon have demonstrated many but not all the social skills that arguably led to the evolution of complex brain structures in primates (Bshary et al., 2002), how might we proceed to exploit these social skills in understanding neural mechanisms? In this chapter I discuss the range of social competence and requirements in a particular species of fish and what we know about how these social interactions require particular perceptual and motor skills to be supported by the brain. Of note, this model system provides evidence of how social behavior shapes the brain in ways that depend on the developmental stage, social circumstance, and environmental context.

The social system of the African cichlid fish, *Astatotilapia* (*Haplochromis*) *burtoni* (Günther), has two kinds of adult males: those with and those without territories (Fernald, 1977). Territorial (T) males are brightly colored, with basic blue or yellow body coloration, a dark black stripe through the eye, a black spot on the tip of the gill cover, and a large red humeral patch just behind it. In contrast, nonterritorial (NT) males are cryptically colored, making them difficult to distinguish from the background and from females that are similarly camouflaged (figure 8.1). In their natural habitat, the shallow shorepools and river estuaries of Lake Tanganyika (Coulter, 1991), *A. burtoni* live in a leklike social system in which T males vigorously defend contiguous territories (Fernald & Hirata, 1977a, 1977b). Social communication among these fish appears to depend primarily on visual signals (Fernald, 1984; see also discussion following).

A. burtoni territorial males perform 19 distinct behavioral patterns in social interactions (Fernald, 1977). T males dig a pit in their territory, exchange threat displays with neighboring territorial males, chase NT animals from their territories, and solicit and court females. When soliciting and courting females, T males display bright coloration patterns toward the female being courted. A T male will lead a female toward his territory, typically using large movements of his tail, and he courts by quivering his opened, brightly colored anal fin in front of the female.

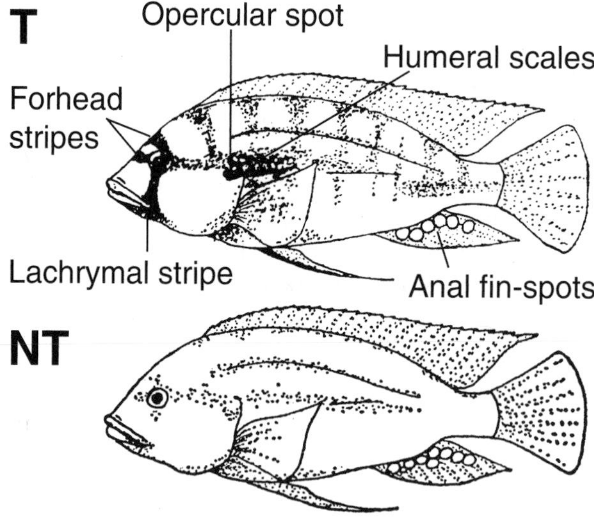

Figure 8.1
Illustration of the body patterns for typical territorial (T) and nonterritorial (NT) males. *Top*: nonterritorial males are camouflage colored without the robust markings of their territorial counterparts. *Bottom*: territorial males are brightly colored, including orange humeral scales, and have distinctive anal fin spots and dark forehead and lachrymal stripes. The overall body color may be either yellow or blue.

When a T male manages to lure a female into his territory, she will normally eat by sifting the substrate in the territory. NT males mimic female behavior sufficiently well that the T male allows NTs to enter the territories and feed before the deception is discovered. This NT male behavior occurs because only sites defended as territories contain food, so NT males need to enter to eat. Normally, however, the NT female impersonator is quickly chased off. If a female responds to male courtship, the T male leads her to his pit and continues courtship movements. T males swim vigorously in front of the female, quivering their entire body with spread anal fins. If appropriately stimulated, the female will lay her eggs in the pit and collect them in her mouth immediately. After she has deposited several eggs, the male will swim in front of her, displaying the egglike spots on his anal fin (ocelli). T males display this fin because the spots may seem to the female like eggs not yet collected (Wickler, 1962). Thus, while attempting to "collect" the egg-spots, the female ingests milt ejected near them by the male and ensures fertilization. The spawning male may repeatedly interrupt his courtship and mating to chase off intruders into his territory. After several bouts of egg laying and fertilization, the female departs with fertilized eggs, which she broods in her mouth (Fernald, 1984).

Even this abbreviated description of the natural behavior of *A. burtoni* shows the important role visual signals play in mediating social behavior. As is typical for this kind of rapid social interactions, each behavioral act influences the next, both in the individual and in other animals involved in the encounter. What do animals attend to during aggressive social interactions? Using ethological methods, early workers identified several fixed action patterns and key stimuli that mediate social signaling in *A. burtoni*. Specifically, Leong (1969) analyzed the role of the black eyebar by testing how T males responded to *A. burtoni* dummies painted with various configurations of the distinctive body patterns. When the eyebar was presented alone, T males increased their readiness to attack targets, while presentation of the orange-red patch of the humeral scales alone decreased attack readiness. Subsequent experiments tested the importance of the orientation of the eyebar relative to the body and other visual stimuli (Heiligenberg & Kramer, 1972; Heiligenberg, Kramer, & Schulz, 1972). All the work supported the hypothesis that the black eyebar and the red humeral patch influence the aggressiveness of T males in opposite directions. Males reared from hatching in complete isolation, showed the same response to the presentation of dummies as did normal animals, suggesting that the response to these key stimuli is innate (Fernald, 1980).

In *A. burtoni*, the visual system has remarkable adaptations to the behavioral signals of the species. In the primary habitat, shorepools and river estuaries along Lake Tanganyika, color patterns on the body match the filtering properties of the water, maximizing the visibility of crucial visual signals (Fernald & Hirata, 1977a). The *A. burtoni* retina has three types of cone photoreceptors and one type of rod characteristically sensitive to different wavelengths of light, implying that they could have trichromatic vision (Fernald & Liebman, 1980). The cone photoreceptors are arranged in a square array that is optimal for trichromatic vision (Fernald, 1981), and spectral sensitivity measured behaviorally (Allen & Fernald, 1985) shows that *A. burtoni* can distinguish colors, as predicted from the morphological measurements. Since the eye continues to grow through adding new neurons, the visual system of *A. burtoni* has been useful for understanding how retinal development is controlled (e.g., Fernald, 2000a, 2000b).

One of the most remarkable features of vertebrates with indeterminate growth is how ongoing sensory and motor functions are maintained during changes in body size. For example, the growth of the eye in *A. burtoni* is so fast that the body of a newly released fry could fit in the eye of a 1-year-old T male (Zygar, Lee, & Fernald, 1999). Growth is achieved by adding new cells to the lens and retina without compromising vision (Fernald, 1983, 1989; Fernald & Wright, 1983; Johns & Fernald, 1981). Through observing the animals it is evident that the growth rate is not uniform and depends critically on the social situation (Fraley & Fernald, 1982). Growth rate depending on social situation has been reported for other fish species (e.g., Berglund, 1991; Borowsky, 1973; Francis, 1988; Schultz et al., 1991), but the mechanisms by which such control is exerted are not understood. Using the *A. burtoni* social system, we are beginning to discover mechanisms through which social behavior is controlled by and also regulates the physiology of *A. burtoni*.

Following release of the young by the mother, the growth, behavioral, and gonadal development of the fry depend critically on the social environment (Fraley & Fernald, 1982). Rearing animals either physically isolated with visual contact or in groups of broodmates showed no difference in growth based on standard length and weight for the first 10 weeks (figure 8.2). Group-reared males that become NT gain less weight than those that become T, though this difference is no longer evident at 20 weeks (figure 8.2A). Gonads also develop more rapidly in T males than in NT males, though more slowly than in isolated males at 14 weeks (figure 8.2A). Physically isolated males effectively become T males and display all the behaviors associated with that status. Possibly they develop larger size and gonads because they face no actual physical competition. When the onset of behavioral attributes is compared, group-reared T males exhibit characteristic agonistic behaviors (chase, tailbeat, fin spread) and coloration (eyebar, opercular spot) more than 2 weeks earlier than animals reared in physical isolation (figure 8.3). Of note, these aggressive behaviors are fully suppressed in NT males reared in groups. In the *A. burtoni* social system, where territorial space is a limiting factor, this robust regulation of maturation in early development seems to be an adaptive solution to a limited resource.

Being reared with broodmates can suppress early social and physical development, but in *A. burtoni* even more effective social regulation can occur when older animals are kept with younger animals. Davis and Fernald (1990) raised animals from hatching in the presence of adult males and showed that these fish have suppressed gonadal maturation

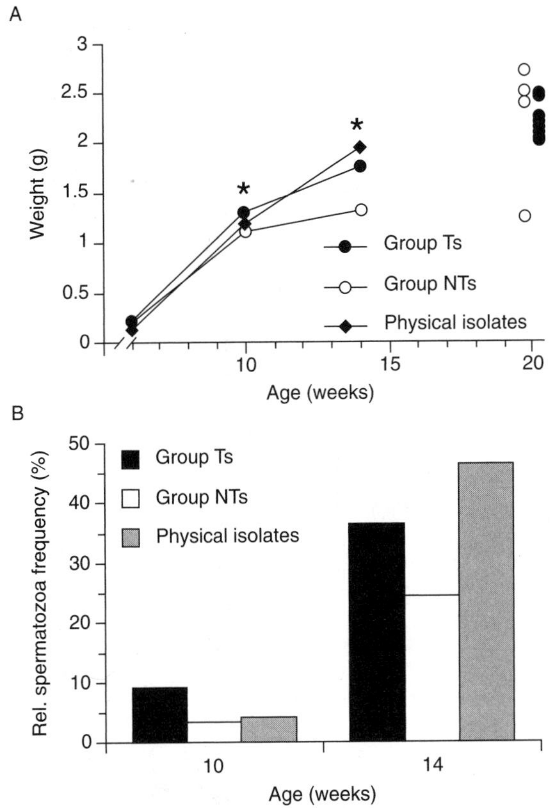

Figure 8.2
Development and maturation of *A. burtoni* fry reared either in groups (open and solid circles) or physically isolated (diamonds). (A) Body weight shown as a function of time. Asterisks signify that group-reared territorial fish (Ts, solid circles) weighed significantly more after 10 and 14 weeks than NT tankmates. After 20 weeks, size differences were no longer evident. (B) Relative amount of mature spermatozoa in cross-sections of the central testicular lobule. (After Fraley & Fernald, 1982.)

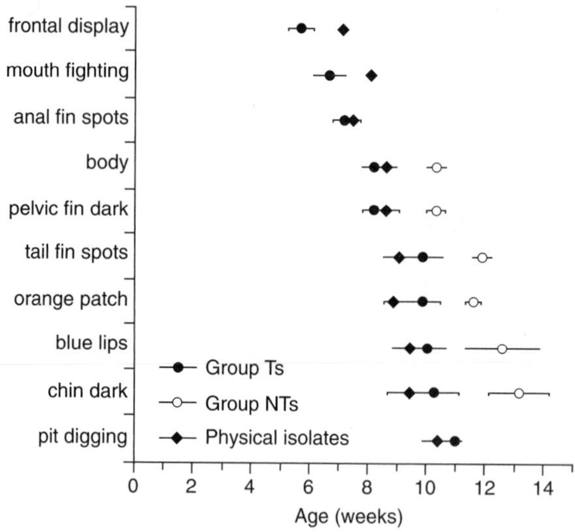

Figure 8.3
Ontogeny of color patterns and agonistic behavioral patterns in *A. burtoni* fry reared in groups or isolated (see figure 8.2). Symbols representing social conditions show means ± SD in days when each pattern was first observed. The origin is when the fry were released from the mouth. (Modified from Fraley & Fernald, 1982.)

relative to fish reared in the absence of adults (figure 8.4). This experiment showed that the suppressed animals had not only hypogonadal testes but also smaller gonadotropin-releasing hormone (GnRH)–containing neurons in the preoptic area (POA). GnRH neurons are the key point in the brain-pituitary-gonadal axis that controls reproduction in all vertebrates. In *A. burtoni*, as in all vertebrates, the GnRH neurons project to the pituitary (Bushnik & Fernald, 1995), where they release GnRH, the signaling peptide sent from the brain to the pituitary to trigger the release of gonadotropins and ultimately testes growth. Davis and Fernald (1990) showed that the GnRH-containing cells in the brain are eightfold larger in T than in NT males. Thus, the social control of maturation in *A. burtoni* is effected by changing the GnRH-containing cells in the brain.

Social status can regulate the physiology of the reproductive state, even in adult *A. burtoni*, as shown by switching males from T→NT or NT→T by moving them to new communities. Specifically, when T males were moved to communities with larger T males, they became NT (e.g., T→NT), and similarly, when NT males were moved to communities with smaller conspecifics, they became T (e.g., NT→T). After 4 weeks in the altered social setting, GnRH cell size was measured (figure 8.5), and it

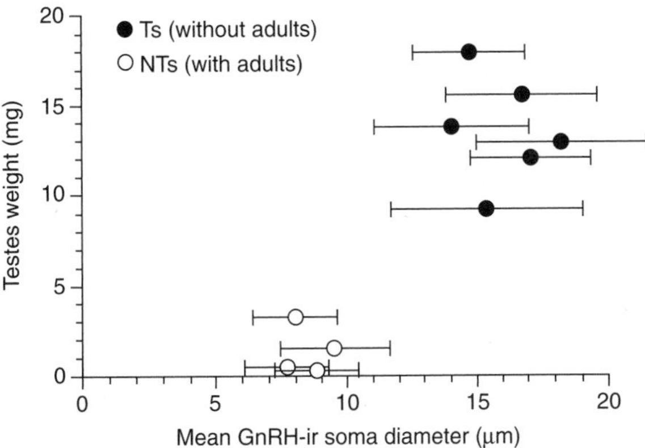

Figure 8.4
Gonad weight plotted as a function of average GnRH-immunoreactive soma size (± SD) for largest 30% of cells. Data were obtained for animals reared without adults present (Ts; filled circles) or with adults present (NTs; empty circles) at 20 weeks. Note that GnRH neuron sizes are independent of body size in this experiment; also note the large differences in cell size and testes weight between the T and NT males. (After Davis & Fernald, 1990.)

was found that changing the social status alone was sufficient to change GnRH neuron size in the brain. As expected, the gonadosomatic index (GSI) was changed correspondingly (Francis, Soma, & Fernald, 1993). Thus, adults as well as juveniles are subject to the social control of reproduction via changes in the GnRH neurons in the brain.

Although causing a change in brain structure by changing social status is quite remarkable, the time scale of this initial experiment did not reflect how rapidly behavioral and neural changes could occur. Indeed, the 4-week interval tested was substantially longer than any observed changes in behavior following status switches, which can occur in minutes. Analyzing socially induced changes in neural structures on a significantly shorter time scale revealed another surprise.

Using a paradigm of changing social status by moving animals similar to that described above, White, Nguyen, and Fernald (2002) discovered several important new aspects of the social control of the reproductive axis. First, on social ascent from NT to T status, the change in cell size was quite rapid, with substantial growth in a single day, and the T male GnRH cell size was reached in 1 week. The GnRH neurons continued to grow so that at 2 weeks they were significantly larger than normal T male size before returning to the size appropriate for a T male (figure 8.6). This massive upregulation of GnRH production very likely allows

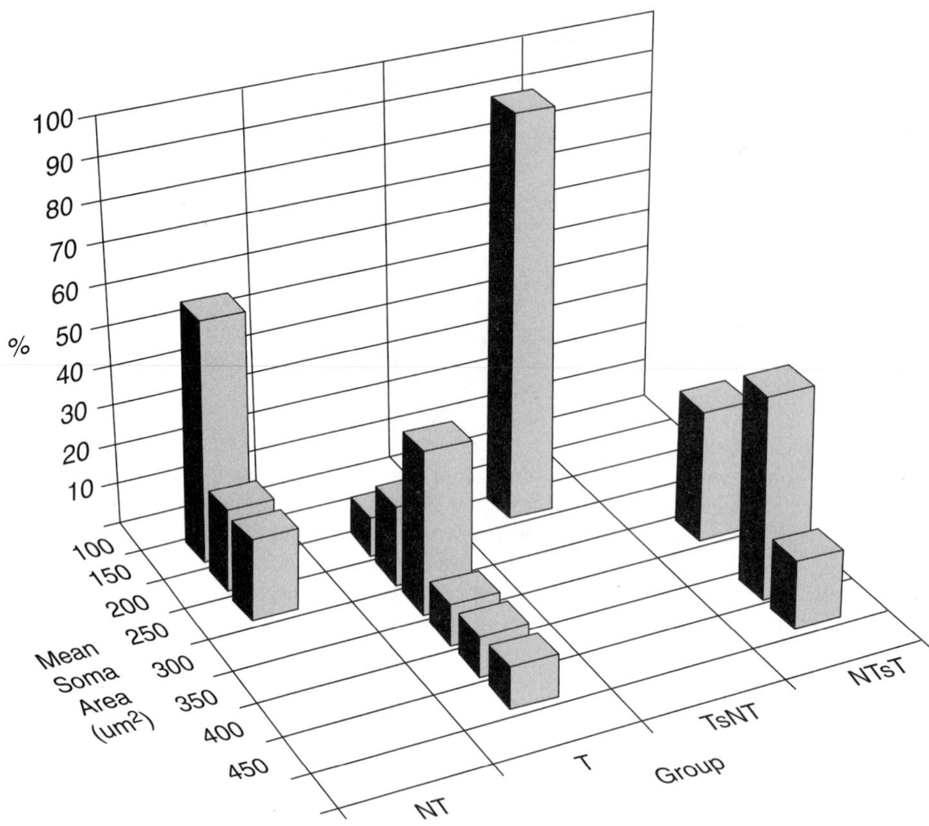

Figure 8.5
Three-dimensional presentation of the mean soma sizes of GnRH-containing neurons in the POA as a function of experimental group. Note significant differences between T and T→NT males and between NT and NT→T males. The vertical axis shows the percentage of individuals with the mean soma size in a given bin. (Modified from Francis et al., 1993.)

the socially ascending animal to achieve reproductive competence rapidly and was obviously not observed in the 4-week experiment described above. The behavioral switch from NT to T, although immediately evident as a change from nonaggressive to aggressive behavior, does not fully match that of a stable T male for ca. 1.5 weeks. The second discovery in this experiment was that the change between T and NT is remarkably asymmetrical. Fishes of descending social status (T→NT) stop displaying aggressive behaviors immediately, but the GnRH-containing neurons in the POA do not reduce to NT size until ca. 3 weeks after defeat, whereas the NT→T ascent takes less than a week. The significance

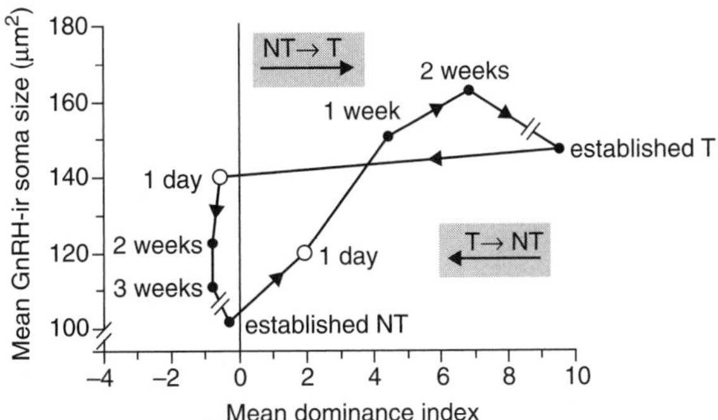

Figure 8.6
Mean two-dimensional GnRH-ir neuronal soma size plotted against the frequency of aggressive and submissive behaviors expressed as a dominance index (sum of aggressive acts minus sum of escape events/3-minute observation interval). Note the hysteresis-like function, as social status changes are asymmetrical in regard to behavior and soma size. Although the behavioral change in T®NT males is significantly faster (ca. 1 day) than in NT®T males (ca. 2 weeks), the latter achieve soma sizes equivalent to those of Ts in 1 week, while T®NTs require 3 weeks for their neurons to shrink to NT sizes. Empty circles indicate cases where soma size is predicted. Variances are not shown, for clarity. (Data from White et al., 2002.)

of this hysteresis in neural and behavioral changes between T and NT males may be explained as a consequence of life in an unstable world, where reproductive opportunities may arise quickly for NTs (see below). After a defeat, switching to subordinate behaviors rapidly likely reduces the chances of injuries to the loser. However, given that the chance to establish a territory could arise soon, maintaining an active reproductive system for a bit longer may be adaptive. Social status sets both soma size of preoptic area (POA) GnRH-containing neurons and GSI, and these effects are reversible. The relatively large testes and GnRH neurons characteristic of T males are a consequence of their social dominance, and when this dominance advantage is lost, both neurons and testes shrink, although, as seen here (White et al., 2002), there is striking asymmetry in the physiological responses. Social information about status causes the changes in the brain, but how this is achieved is not known.

White et al. (2002) also showed that the socially induced changes in status resulted in significant changes in gene expression. Measuring changes in mRNA from all 3 forms of GnRH, they found that only the POA GnRH mRNA was regulated corresponding to a change in social status. The change in mRNA in the POA form of GnRH was evident at

3 days after a change in social status. Such social regulation demonstrates that key social information is used to control cellular and molecular processes in the brain.

It is important to note that the effect of social status on GnRH cell size and GnRH mRNA expression is limited to the GnRH-containing neurons of the preoptic area. As we have shown, *A. burtoni* has three distinct genes that code for three distinct GnRH-like molecules (White & Fernald, 1993; White et al., 1994) expressed at three distinct sites in the brain (White, Kasten, Bond, Adelman, & Fernald, 1995). The GnRH forms not found in the POA are expressed in two other distinct cell groups, one located in the terminal nerve region, the other in the mesencephalon (see White et al., 1995, for details). Neither of these other GnRH-containing cells showed any change in size as a function of social status (Davis & Fernald, 1990), nor did their mRNA change with status change. Thus, the status-linked variation in soma size is not a general property of GnRH-containing neurons but rather is confined to the POA population (Davis & Fernald, 1990). The same result has been shown for GnRH mRNAs using in situ hybridization (White et al., 1995). Males and females share the brain-pituitary-gonadal axis used to control reproduction, but female *A. burtoni* have a strikingly different system that regulates reproduction. GnRH-containing cells in the POA of females also change size, but do so depending on their reproductive status alone (White & Fernald, 1993) and there is no effect of social status.

As expected, social control of the reproductive axis via GnRH also influences important endocrine factors. Androgen released from the gonads depends on social status. Castrated *A. burtoni* T males have hypertrophied GnRH neurons (Francis, Jacobsen, Wingfield, & Fernald, 1992a; Soma, Francis, Wingfield, & Fernald, 1996), showing that androgen has a feedback effect on GnRH cell size (figure 8.7). The important point is that setpoint for this feedback is social status, since T males have larger GnRH neurons despite having higher androgen levels (Soma et al., 1996). T males that are castrated are able to maintain their rank despite having lowered androgen levels (Francis, Jacobsen, Wingfield, & Fernald, 1992b). Possibly prior dominance experience on the part of the T male and the size difference among animals contribute to this result. It is possible but less likely that individual recognition could also play a role.

Social status regulates the production and release of GnRH in the pituitary. Another potential site for regulation is the GnRH receptor in the pituitary. Recent work in our laboratory has shown that *A. burtoni* has genes that encode two distinct GnRH receptors (Robison et al., 2001). Using real-time polymerase chain reaction (PCR) technology, we

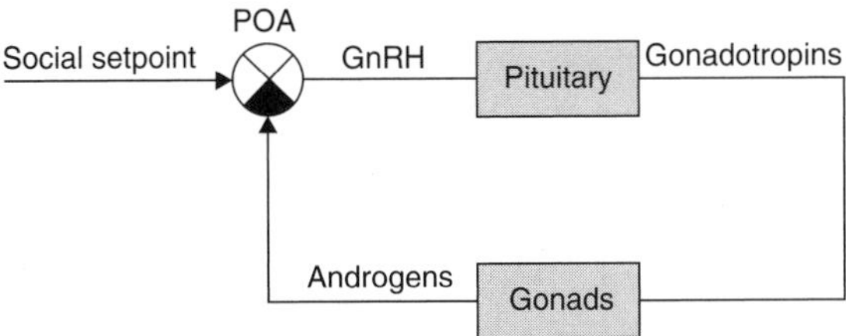

Figure 8.7
Feedback control model of GnRH regulation in male *A. burtoni*. Neurons in the preoptic area (POA) integrate both social and hormonal signals to regulate GnRH release. In this model, the setpoint for the GnRH level is determined by social signals and the maintenance of the GnRH level at this setpoint is achieved by negative feedback from gonadal androgens. (Modified from Soma et al., 1996.)

have been able to show that the mRNA of one of these receptor types is upregulated rapidly and dramatically in the pituitary of T males as compared with NT males (Au, Illing, & Fernald, 2003). It remains to be discovered whether this receptor regulation results solely from a change in social status or if other factors are also involved.

An interesting feature of *A. burtoni* and other cichlid species is their bright coloration and the critical role of vision in social interactions. The recent demonstration that visibility affects intraspecies communication in cichlids of lake Victoria, such that turbidity can cause loss of species, underscores this point (Seehausen et al., 1997). The production and detection of visual signals have been subjected to natural and sexual selection, as evidenced by the neural control of the black eye stripe in *A. burtoni*. The eyebar is controlled by a small branch of cranial nerve VI, which controls the migration of melanin granules to change the color from clear to black (Muske & Fernald, 1987a). The eyebars of T males are much more sensitive to the neurotransmitter norepinephrine than are NT males. In addition, over the longer term, the eyebar of T males inserts iridiphores behind the black pigment, enhancing contrast and efficacy of the eyebar signal (Muske & Fernald, 1978b). This means that at the cellular and molecular level, both the control and efficacy of this visual signal also depend on social state. At all levels examined, there are socially induced changes in the physiology underlying the T-NT differences.

When Von Uexküll (1921) first described the *Umwelt* of an animal, he recognized that habitat was important for animals and was likely to be viewed differently from that of a human observer. The elaboration of *Umwelt* to include the social world implies that there can be direct effects of habitat on social structure (e.g., Lott, 1982). In *A. burtoni*, habitat complexity influences the fraction of T males able to maintain a territory, and the stability of that habitat influences the duration of territorial tenure (Hofmann, Benson, & Fernald, 1999). Because the habitat near Lake Tanganyika is subject to high daily winds and hence to disruption, the social regulation of reproduction, growth, and development appears adaptive. Not all males can be T males and hence breed at any given time (ca. 10%–30%), and these animals appear to be more vulnerable to avian predators, making their territorial ownership relatively brief (Fernald & Hirata, 1977b). Brightly colored animals are attractive to predators, as has been shown for several fish species, with the consequence being differential selection on those individuals (e.g., Brick, 1998; Endler, 1988, 1991; Godin & McDonough, 2003; Haas, 1976; O'Steen, Cullum, & Bennett, 2002). It is easy to imagine that reproductive opportunities might come and go rapidly, possibly explaining the asymmetric response of GnRH neurons to changes in social status (e.g., figure 8.7).

Our analysis of the role of habitat in social change led to a number of interesting conclusions. First, there is an intrinsic instability in the maintenance of territories (Hofmann & Fernald, 2000). The instability is due to differential growth rates. The growth rates measured in adults is quite different from that observed in young animals described above. At early ages (e.g., from 0 to 14–21 weeks), animals subjected to social influence from conspecifics can have their growth slowed and their reproductive development retarded (see above). This early form of social influence is different from that experienced by adult animals of similar size ranges living in social colonies. We attribute this difference to a number of factors that distinguish early suppression from social interaction among older animals. In nearly size-matched animals, there seem to be behavioral strategies that allow animals to function successfully among larger conspecifics and escape the regulation of body size but not that of gonadal regulation. As a result, NTs and NT→Ts grow faster than Ts and T→NTs (figure 8.8). Observations suggest that the T males, though they may have a growth spurt upon gaining T status, then begin to expend energy at a much higher rate than NT males. This heightened energy cost results in T male growth slowing (Hofmann & Fernald, 2000). The second discovery is that the social regulation of growth among adults may depend on somatostatin release in the pituitary, where this neurohormone inhibits the release of growth hormone (GH; Brazeau et al., 1973; Gillies, 1997; Lin, Otto, Cardenas, & Peter, 2000; Very,

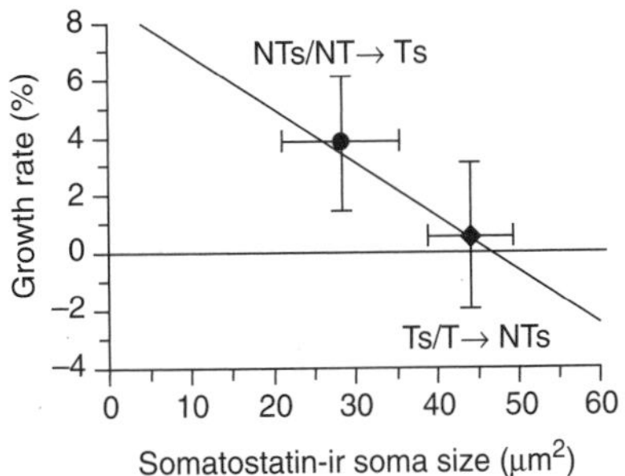

Figure 8.8.
Growth rates plotted as a function of the mean somatostatin-immunoreactive soma size in *A. burtoni*. NTs and NT®T males (solid circle; mean ± SD) have smaller soma cross-sectional areas and grow faster than Ts and T®NTs (solid diamond; mean ± SD). A linear regression analysis results in $y = 0.19 * x + 8.95$, with $r^2 = 0.4163$ ($P < 0.001$; n = 18). (Modified from Hofmann & Fernald, 2000.)

Knutson, Kittilson, & Sheridan, 2001). Supporting this idea is the fact that somatostatin-containing neurons in the POA change size when social status and, consequently, growth rate change (Hofmann & Fernald, 2000, and figure 8.8).

Animals lose territories because their growth rates have diminished, and in some cases those T males even shrink. As noted above, it seems likely that behavioral stress may play a role. As shown by Fox, White, Kao, and Fernald (1997), in *A. burtoni*, status switches in both directions can be accompanied by elevated levels of the major stress hormone cortisol, with the T→NT change showing the most pronounced increase. NT→T fish with increased cortisol levels usually did not maintain territoriality. T→NT males consistently had high cortisol levels. As has been shown in another cichlid, the tilapia *Oreochromis mossambicus*, chronic administration of cortisol leads to a reduction in body weight and reproductive parameters such as gamete size and levels of sex steroids (Foo & Lam, 1993). Although the regulatory interactions between GH and cortisol are very complex (Thakore & Dinan, 1994; van Weerd & Komen, 1998, for critical reviews), in vivo experiments have demonstrated an inhibitory effect of glucocorticoids on somatic growth in many vertebrates, including fish (e.g., Pickering, 1990).

Fox et al. (1997) showed that cortisol levels in Ts and NTs do not differ as long as the fish community is unstable, but when stability is achieved, T males have low cortisol levels and NT males have high cortisol levels. Because NT males can grow faster than T males, their growth may not be effectively inhibited by cortisol, but other factors may become important. Recently, we identified, cloned, and characterized the cortisol receptors in *A. burtoni* (Greenwood et al., 2003). Interestingly, there are four forms of cortisol receptors in *A. burtoni*, and quantitative PCR revealed differential distribution of their expression. The selective binding of cortisol to these receptors showed quite different levels of response, suggesting that the animal could regulate its responsiveness to cortisol by modifying the receptor subtype expressed. Given the social modulation of the GnRH receptor, this finding might not be unexpected.

Although we have focused our work on the role of males in the social system described here, females are important as well. Recently we showed that differences in female reproductive state that are due to differences in hormone levels correspond to changes in females' affiliation preference with males (Clement, Grens, & Fernald, 2005). Gravid females preferentially associated with T males, whereas nongravid females showed no preference, and this preference did not depend on male size. These data suggest that females use a hierarchy of cues in decision making.

The important and difficult question that remains is how social information causes cellular and molecular changes in the brain and nervous system. *A. burtoni* have stable social interactions requiring that they follow rules in their behavior relative to others. To do this, they use information about other animals based on social and reproductive state and recent behavioral encounters. All this behavior is supported by physiological, cellular, and ultimately molecular mechanisms. Understanding how such control occurs depends on evaluating many animals simultaneously in an ecologically realistic context. Our recent work suggests that *A. burtoni* attend to their surroundings and respond appropriately in ways we did not anticipate. To test their response in reliable social contexts, we are developing a virtual fish that will allow us to present repeated stimuli in a social context (Rosenthal, 1999) to observe animals in social situations that can be accurately replicated. Reducing the variance that is a central part of many animals interacting will help us discern the important interactions from the rest. In addition, we are developing neuroanatomical marking techniques that trace circuits active when the animals are experiencing social change. In this way we will be able to understand where and when particular brain regions play a role in the social response. Finally, a new project analyzing gene expression globally within animals that have experienced different social situations should give us glimpses of what collections of genes might be important for successful social interactions (Hofmann, 2003; Hofmann et al., 1999).

The modulation of the brain by behavior makes sense in an evolutionary framework in which the behavioral phenotype is the locus of selective pressure. Phenotypic plasticity allows *A. burtoni* to adapt its behavior and physiology reversibly to changing social opportunities, thus allocating resources between reproduction and growth (Williams, 1966). Given the limited territorial space in their natural habitat, the selective advantage to animals that can modify behavior and physiology quickly seems obvious. The evolution of this life history strategy shaped the *A. burtoni* brain and nervous system and offers a chance to understand the mechanisms that support this flexibility. The remarkable diversity of cichlids in Africa and South America offers the chance to discover general principles of the selective pressures of habitat, behavior, and the brain.

Acknowledgments Work was supported by the Jacob Javits Award (NIH-NS 34950) and by grant NIH-EY-05051 to R.D.F. I thank my many colleagues over the years for invaluable discussions and assistance with the work described in this chapter.

References

Allen, E. E., & Fernald, R. D. (1985). Spectral sensitivity of the African cichlid fish, *Haplochromis burtoni*. *Journal of Comparative Physiology, A*, *157*, 247–253.

Au, T. M., Illing, N., & Fernald, R. D. (2003). Social regulation of gonadotropin-releasing hormone receptor in the cichlid fish, *Haplochromis burtoni*. *Society for Neuroscience, Abstracts*, abstn. 611.6.

Baerends, G. P., & Baerends van Roon, J. M. (1950). An introduction to the study of the ethology of cichlid fishes. *Behavior, Supplement 1*, 1–235.

Bass, A. H., Bodnar, D. A., & Marchaterre, M. A. (2000). Midbrain acoustic circuitry in a vocalizing fish. *Journal of Comparative Neurology*, *419*, 505–531.

Berglund, A. (1991). Egg competition in a sex-role reversed pipefish: Subdominant females trade reproduction for growth. *Evolution*, *45*, 770–774.

Borowsky, R. (1973). Social control of adult size in males of *Xiphophorus variatus*. *Nature*, *245*, 332–335.

Brazeau, P., Vale, W., Burgus, R., Ling, N., Butcher, M., Rivier, J., et al. (1973). Hypothalamic polypeptide that inhibits the secretion of immunoreactive pituitary growth hormone. *Science*, *179*, 77–79.

Brick, O. (1998). Fighting behavior, vigilance and predation risk in the cichlid fish *Nannacara anomala*. *Animal Behavior*, *56*, 309–317.

Bshary, R., Wickler, W., & Fricke, H. (2002). Fish cognition: A primate's eye view. *Animal Cognition*, *5*, 1–13.

Bshary, R., & Würth, M. (2001). Cleaner fish *Labroides dimidiatus* manipulate client reef fish by providing tactile stimulation. *Procedings of the Royal Society of London, Series B, Biological Sciences*, *268*, 1495–1501.

Bushnik, T. L., & Fernald, R. D. (1995). The population of GnRH-containing neurons showing socially mediated size changes project to the pituitary in a teleost, *Haplochromis burtoni*. *Brain, Behavior and Evolution*, *46*, 371–377.

Clement, T. S., Grens, K. E., & Fernald, R. D. (2005). Female association preference depends on reproductive state in the African cichlid fish, *Astatotilapia burtoni*. *Behavioral Ecology*, *16*, 83–88.

Coulter, G. W. (1991). *Lake Tanganyika and its life*. Oxford: Oxford University Press.

Davis, M. R., & Fernald, R. D. (1990). Social control of neuronal soma size. *Journal of Neurobiology*, *21*, 1180–1188.

Endler, J. A. (1988). Frequency-dependent predation, crypsis and aposematic coloration. *Philosophical Transactions of the Royal Society of London, Series B, Biological Sciences*, *319*, 505–523.

Endler, J. A. (1991). Variation in the appearance of guppy color patterns to guppies and their predators under different visual conditions. *Vision Research*, *31*, 587–608.

Fernald, R. D. (1977). Quantitative observations of *Haplochromis burtoni* under semi-natural conditions. *Animal Behavior*, *25*, 643–653.

Fernald, R. D. (1980). Response of male cichlid fish, *Haplochromis burtoni*, reared in isolation to models of conspecifics. *Zeitschrift für Tierpsychologie*, *54*, 85–93.

Fernald, R. D. (1981). Chromic organization of a cichlid fish retina. *Vision Research*, *21*, 1749–1753.

Fernald, R. D. (1983). Neural basis of visual pattern recognition in fish. In J. P. Ewert, R. R. Capranica, & D. J. Ingle (Eds.), *Advances in vertebrate neuroethology* (pp. 569–580). New York: Plenum Press.

Fernald, R. D. (1984). Vision and behavior in an African cichlid fish. *American Scientist*, *72*, 58–65.

Fernald, R. D. (1989). Retinal rod neurogenesis. In B. L. Finley, & D. R. Sengelaub (Eds.), *Development of the vertebrate retina* (pp. 31–42). New York: Plenum Press.

Fernald, R. D. (2000a). Sensory Systems 15.1. Vision—Gross functional anatomy. In G. K. Ostrander (Ed.), *Laboratory fish* (pp. 225–235). London: Academic Press.

Fernald, R. D. (2000b). Sensory Systems 27.1. Vision—Microscopic functional anatomy. In G. K. Ostrander (Ed.), *Laboratory fish* (pp. 451–462). London: Academic Press.

Fernald, R. D., & Hirata, N. R. (1977a). Field study of *Haplochromis burtoni*: Quantitative behavioural observations. *Animal Behavior*, *25*, 964–975.

Fernald, R. D., & Hirata, N. R. (1977b). Field study of *Haplochromis burtoni*: Habitats and cohabitants. *Environmental Biology of Fishes*, *2*, 299–308.

Fernald, R. D., & Liebman, P. A. (1980). Visual receptor pigments in the African cichlid fish, *Haplochromis burtoni*. *Vision Research*, *20*, 857–864.

Fernald, R. D., & Wright, S. E. (1983). Maintenance of optical quality during crystalline lens growth. *Nature*, *301*, 618–620.

Fernald, R. D., & White, R. B. (1999). Gonadotropin-releasing hormone genes: Phylogeny, structure, and functions. *Frontiers in Neuroendocrinology*, *20*, 224–240.

Foo, J. T. W., & Lam, T. J. (1993). Retardation of ovarian growth and depression of serum steroid levels in the tilapia *Oreochromis mossambicus* by cortisol implantation. *Aquaculture*, *115*, 133–143.

Fox, H. E., White, S. A., Kao, M. H. F., & Fernald, R. D. (1997). Stress and dominance in a social fish. *Journal of Neuroscience*, *17*, 6463–6469.

Fraley, N. B., & Fernald, R. D. (1982). Social control of developmental rate in the African cichlid, *Haplochromis burtoni*. *Zeitschrift für Tierpsychologie*, *60*, 66–82.

Francis, R. C. (1988). Socially mediated variation in growth rate of the midas cichlid: The primacy of early size differences. *Animal Behaviour*, *36*, 1844–1845.

Francis, R. C., Jacobsen, B., Wingfield, J. C., & Fernald, R. D. (1992a). Hypertrophy of gonadotropin releasing hormone–containing neurons after castration in the teleost, *Haplochromis burtoni*. *Journal of Neurobiology*, *23*, 1084–1093.

Francis, R. C., Jacobsen, B., Wingfield, J. C., & Fernald, R. D. (1992b). Castration lowers aggression but not social dominance in male *Haplochromis burtoni* (Cichlidae). *Ethology*, *90*, 247–255.

Francis, R. C., Soma, K., & Fernald, R. D. (1993). Social regulation of the brain-pituitary-gonadal axis. *Proceedings of the National Academy of Sciences, U.S.A.*, *90*, 7794–7798.

Gillies, G. (1997). Somatostatin: The neuroendocrine story. *Trends in Pharmacological Sciences*, *18*, 87–95.

Godin, J. G. J., & McDonough, H. E. (2003). Predator preference for brightly colored males in the guppy: A viability cost for a sexually selected trait. *Behavioral Ecology*, *14*, 194–200.

Godwin, J., Luckenbach, J. A., & Borski, R. J. (2003). Ecology meets endocrinology: Environmental sex determination in fishes. *Evolution & Development*, *5*, 40–49.

Greenwood, A. K., Butler, P., DeMarco, U., White, R. C., Pearce, D., & Fernald, R. D. (2003). Multiple corticosteroid receptors in a teleost fish: Distinct sequences, expression patterns, and transcriptional activities. *Endocrinology*, *144*, 4226–4236.

Haas, R. (1976). Sexual selection in *Nothobranchius-Guentheri pisces cyprinodontidae*. *Evolution*, *30*, 614–622

Heiligenberg, W., & Kramer, U. (1972). Aggressiveness as a function of external stimulation. *Journal of Comparative Physiology and Psychology*, *77*, 332–340.

Heiligenberg, W., Kramer, U., & Schulz, V. (1972). The angular orientation of the black eye-bar in *Haplochromis burtoni* (Cichlidae, Pisces) and its relevance to aggressivity. *Zeitschrift für vergleichende Physiologie, 76,* 168–176.

Hofmann, H. A. (2003). Functional genomics of neural and behavioral plasticity. *Journal of Neurobiology, 54,* 272–282.

Hofmann, H. A., Benson, M. E., & Fernald, R. D. (1999). Social status regulates growth rate: Consequences for life-history strategies. *Proceedings of the National Academy of Sciences, U.S.A., 96,* 14171–14176.

Hofmann, H. A., & Fernald, R. D. (2000). Social status controls somatostatin-neuron size and growth. *Journal of Neuroscience, 20,* 4740–4744.

Johns, P. R., & Fernald, R. D. (1981). Genesis of rods in teleost fish retina. *Nature, 293,* 141–142.

Laland, K. N., Brown, C., & Krause, J. (2003). Learning in fishes: From three-second memory to culture. *Fish and Fisheries, 4,* 199–202.

Lin, X. W., Otto, C. J., Cardenas, R., & Peter, R. E. (2000). Somatostatin family of peptides and its receptors in fish. *Canadian Journal of Physiology and Pharmacology, 78,* 1053–1066.

Leong, C.-Y. (1969). The quantitative effect of releasers on the attack readiness of the fish *Haplochromis burtoni. Zeitschrift für vergleichende Physiologie, 65,* 29–50.

Lorenz, K. (1932). Der Kumpan in der Umwelt des Vogels. *Journal für Ornithologie,* 80, Heft 1.

Lorenz, K. (1981). *The foundations of ethology (Vergleichende Verhaltensforschung).* New York: Springer-Verlag.

Lott, D. F. (1982). *Intraspecific variation in the social systems of wild vertebrates.* Cambridge Studies in Behavioural Biology. Cambridge: Cambridge University Press.

McGregor, P. K. (1993). Signalling in territorial systems: A context for individual identification, ranging and eavesdropping. *Philosophical Transactions of the Royal Society of London, Series B, Biological Sciences, 340,* 237–244.

Muske, L. E., & Fernald, R. D. (1987a). Control of a teleost social signal. I. Neural basis for differential expression of a color pattern. *Journal of Comparative Physiology, A, Sensory, Neural, and Behavioral Physiology, 160,* 89–97.

Muske, L. E., & Fernald, R. D. (1987b). Control of a teleost social signal. II. Anatomical and physiological specializations of chromatophores. *Journal of Comparative Physiology, A, Sensory, Neural, and Behavioral Physiology, 160,* 99–107.

Oliveira, R. F., Almada, V. C., & Caniaro, V. (1996). Social modulation of sex steroid concentrations in the urine of male cichlid fish *Orechromis mossambicus. Hormones and Behavior, 30,* 2–12.

O'Steen, S., Cullum, A. J., & Bennett, A. F. (2002). Rapid evolution of escape ability in Trinidadian guppies (*Poecilia reticulata*). *International Journal of Organismal Evolution, 56,* 776–84.

Pickering, A. D. (1990). Stress and the suppression of somatic growth in teleost fish. In A. Epple, C. G. Scanes, & M. H. Stetson (Eds.), *Progress in comparative endocrinology. Progress in Clinical and Biological Research, 432,* 473–479.

2000). In other words, approximately 1 in 10 children are the product of female infidelity. This asymmetry in parental certainty has produced an asymmetry in human parental investment (Bjorklund & Shackelford, 1999; Geary, 2000). As a consequence of having to carry a child to term, females, by default, invest more in and provision more for children than do males. Additionally, if a female nurses her offspring, she could be bound to a minimum of 1.5–2 years of further parental investment.

Males, however, are not bound by their biology to provide care for offspring and instead tend to provide care proportional to their confidence or certainty of paternity (Burch & Gallup, 2000; Daly & Wilson, 1996, 1998). The risk of cuckoldry appears to have driven the evolution of male anticuckoldry tactics, or tactics to limit and control female infidelity in an attempt to reduce the likelihood of extrapair paternity (Buss, 1988, 1994, 1999; Buss & Shackelford, 1997; Davis & Gallup, submitted; Gallup & Burch, in press; Gallup, Burch, Zappierri, Parvez, & Davis, in press; Goetz et al., in press; Platek, 2002; Platek et al., 2002, 2003; Shackelford, et al., 2002).

We observe a similar pattern among many other mammals. For example, paternal care usually manifests only in those mammalian species with relatively high paternal certainty, whereas in most species males provide little or no direct investment in their offspring. Among those few species that do engage in paternal provisioning, it appears that the males have evolved several anticuckoldry tactics that increase the certainty that they are the source of paternity (e.g., Lacy & Sherman, 1983; see also Platek & Shackelford, in press). In an attempt to limit provisioning for offspring that are the consequence of female extrapair copulations, the males of some species are driven to what may appear to be extreme behaviors. For example, when a male Langur monkey overthrows another male and gains dominance within a troop, he will systematically kill young infants that were fathered by the previous alpha male. By resorting to infanticide when the paternity of an offspring is ostensibly foreign (e.g., Hrdy, 1974), his behavior serves two adaptive functions: it eliminates the possibility that he will invest valuable resources in unrelated offspring, and it induces menstrual cycling (i.e., sexual receptivity) in those females whose offspring he killed. This allows the new dominant male to use the females to his own reproductive advantage. Additionally, male baboons appear to invest resources in offspring proportional to the degree to which they monopolized the female prior to insemination (Buchan, Alberts, Silk, & Altman, 2003).

There is growing evidence that human males are similarly affected by these evolutionary pressures to invest in offspring as a function of paternal certainty. It is well known that men differentially invest resources in children to whom they are genetically related. For example, it is not uncommon for unrelated or otherwise stepchildren to be treated significantly worse than biological children (e.g., Anderson, Kaplan, Lam, & Lancaster, 1999). Burch and Gallup (2000) have shown that males spend less time with, invest fewer resources in, and are more likely to abuse ostensibly unrelated children than children they assume to be their genetic offspring. Daly and Wilson (1988a; and see Daly, Wilson, & Weghorst, 1982) estimated the incidence of abuse that results in infanticide among stepchildren to be 100 times that directed toward genetically related children. In Daly and Wilson's (1988b) landmark book, *Homicide*, they interpret spousal homicide (uxoricide) as a byproduct of cuckoldry fear and sexual jealousy among men. These data suggest a strong link between paternity uncertainty and family violence.

Detecting Paternity Without DNA Paternity Tests

If males selectively invest in offspring as a function of shared genetic information, one question that arises is how males selectively detect paternity in the absence of DNA paternity tests. As a way of elucidating the importance of paternity for males, Daly and Wilson (1982) and Regalski and Gaulin (1993) observed families in maternity wards. They measured the number of times people remarked whether the infant looked more like the mother or the father. Both studies found that people were more likely to ascribe resemblance to putative fathers than they were to say the child resembled the mother. Both studies noted that the mother and her family were more likely to attribute resemblance than were other people. They interpreted this behavior as an attempt on the part of the female and her family to convince the putative father that he sired the child.

Burch and Gallup (2000) found that the less a male thinks a child looks like him, regardless of actual genetic relationship to the child, the worse he treats the child and the worse the relationship with that child is.

As a result of paternal uncertainty, male anticuckoldry tactics designed to limit and control the incidence of female infidelity have developed; the design features of such mechanisms are to increase the likelihood that the offspring they provide for are genetically related to them.

Emerging data, and an emerging model (Platek & Shackelford, in press), suggest that males have evolved at least three stages of anticuckoldry defense tactics: (1) monitoring/mate guarding his partner during the fertile period (Daly & Wilson, 1998), (2) intravaginal intersexual competition, such as sperm competition (Birkhead, 1995, 1996; Shackelford et al., 2002) and semen displacement (Gallup et al., in press), and (3) assessment of paternity post parturition (for a review, see Platek & Shackelford, in press). The first and third of these three anticuckoldry tactics actually represent solutions to information-processing problems faced by males, and presumably problems that our male ancestors recurrently faced during evolutionary history. The fact that a number of problem-solving mechanisms revolve around the detection of paternity or the prevention of extrapair copulation suggests that strong positive selection for anticuckoldry tactics was prevalent during our ancestry.

Similar to the way in which mate-guarding strategies and sexual jealousy are likely the result of perception of subtle behavioral changes in one's mate (Buss, 1992), the detection of paternity appears to be the result of perception of subtle facial features expressed by offspring. This hypothesis, that males utilize physical (i.e., facial) resemblance as a cue toward relatedness, was first put forth by Daly and Wilson (1998) and extended by others (Platek, 2002; Platek et al., 2002, 2003; Volk & Quinsey, 2002).

Assessment of Paternal Resemblance

Because paternity detection was a recurrent adaptive problem during human evolutionary history, perceptual mechanisms of direct and indirect detection of resemblance have evolved. The indirect way that resemblance might affect male behavior toward children is by way of a social mirror (Burch & Gallup, 2000): other people can ascribe paternal resemblance to the child as a means of swaying a male to act more positively toward the child. Daly and Wilson (1982) and Regalski and Gaulin (1993) have shown that mothers and family members actively ascribe paternal resemblance to children and that when males express doubt, they are quick to reassure them of the resemblance. Males may be predisposed to take into account social mirror information because of the importance of paternity (see Hauber & Sherman, 2001, and Neff & Sherman, 2002, for a model of parentage).

Burch and Gallup (2000) found that the more a sample of convicted spouse abusers felt that their children looked like them, the better the children were treated. The childhoods of the abusive males themselves were also affected by how much they thought they resembled their fathers. Perceptions of paternal resemblance were negatively correlated with the incidence of physical and sexual abuse they experienced as children, as well as with feelings of closeness to their father. How often others had told them that the child physically resembled the adult male also correlated with a male's ratings of his relationship to the child.

A male, however, may also assess the degree to which a child actually resembles him by a direct perceptual detection mechanism. Platek et al. (2002) morphed the faces of participants (figure 9.1) with the face of

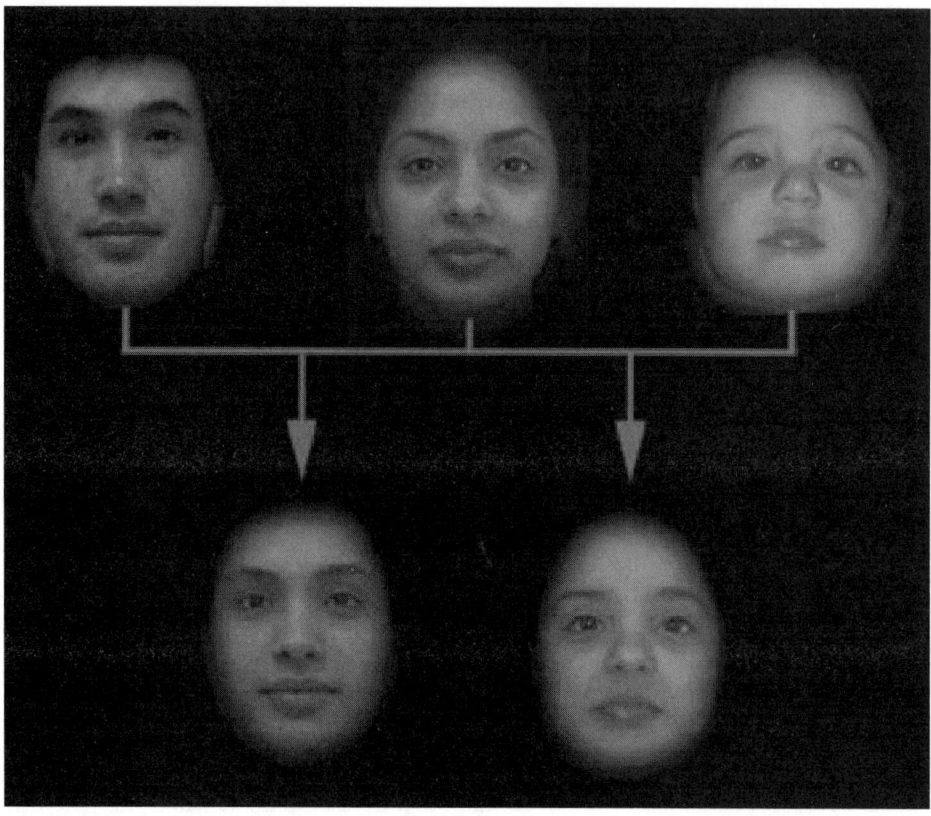

Figure 9.1
Example of morphing paradigm. An unknown adult face (upper left face) is morphed with a subject's face (upper center face) to create a composite adult morph, and an unknown child's face (bottom left face) is morphed with a subject's face to create a composite child morph (bottom right face).

toddlers and measured reactions to hypothetical investment questions (e.g., Which one of these children would you spend the most time with? Which one would you adopt?). Males were more likely than females and more likely than chance to select a face theirs had been morphed with when asked to react positively toward the faces. Thus, it seems that actual resemblance also plays a role in a male's reactions toward children's faces, and this might be modulated by a cortical mechanism or module (e.g., self-referent phenotype matching; see Neff & Sherman, 2002) dedicated to controlling the affective nature of males' reactions toward children.

In a test of how actual resemblance and social mirror–mediated resemblance interact to affect reactions toward children's faces, Platek (2002) provided participants with social mirror information about children's faces, some of which were morphed to resemble the subject, and found that social mirror information affected both male and female reactions similarly. However, unlike females, males were still more affected by actual resemblance: males selected a face primarily based on whether the face resembled theirs. This study replicated our previous findings that males use actual resemblance in their reactions toward hypothetical children (Platek et al., 2002, 2003) and supports the idea that social perceptions of resemblance affect father-child relationships (Burch & Gallup, 2000), but the degree to which social perceptions are as important or more important when it comes to resemblance is not completely understood.

It is obvious that convincing a male of paternity and securing his investment would be in the evolutionary best interests of females. However, because the incidence of cuckoldry is appreciable (1%–20%, as noted earlier), it is hardly in the best interests of males to be easily convinced. If ascriptions of resemblance were completely persuasive throughout evolutionary history, males might have been deceived in investing in children that were the byproduct of cuckoldry. Those who remained wary and used their own perceptions of resemblance and invested accordingly might have stood a better chance of passing on their genetic material and maximizing their fitness through their reproductive and parental efforts.

McLain, Setters, Moulton, and Pratt (2000) investigated this idea by objectively comparing new mothers' ascriptions of paternal resemblance to the ability of independent raters to match photographs of children to their putative fathers. Maternal ascriptions of resemblance could not be verified by the objective, unrelated raters. In no case were the infants' pictures matched to the fathers' photographs more often than chance; the mothers' opinions, although highly reliable, held no validity. These findings could in turn explain the reluctance of males to prematurely agree with their partner's assertions of paternity (Daly & Wilson, 1982; Regalski & Gaulin, 1993). In some cases males would agree only after several maternal attempts to persuade them. Interesting as these data are, they are still flawed in that actual paternity was never determined, which could have masked or obscured independent raters' ability to match children to the males (see also Bredart & French, 1999; Christenfeld & Hill, 1995; Nesse, Silverman, & Bortz, 1990).

In addition, a male may have adopted a strategy of comparing his offspring to his kin in order to assess resemblance, and might also choose to believe information provided only by those that also shared genes in common with him, but this hypothesis has not yet been tested. Platek et al. (2003) and DeBruine (2002) have provided indirect evidence that parentage and trust may also be mediated by facial resemblance. Platek et al. (2003; but see DeBruine, 2002) found that males reacted favorably toward children's faces that shared 25% of their characteristics, which is approximately the proportion of genes shared in common with kin one step removed—grandchildren, nieces and nephews, aunts and uncles, and half-siblings. DeBruine (2002) found that participants tended to trust faces that resembled them more than those that did not, and has since shown a self-resemblance attractiveness bias (DeBruine, 2004).

In all of our studies to date (Platek, 2002; Platek et al., 2002, 2003; Platek, unpublished data), none of the subjects were aware of the effect resemblance had on their choices. In a re-analysis of existing data from our previous morphing studies (Platek, 2002; Platek et al., 2002, 2003) and data not yet published, no subject reported using resemblance to choose a child's face. When queried about their choices, none of the subjects identified resemblance as a factor in how they chose which child to select, nor did they realize that their faces had been morphed with the child's. In fact, during debriefing, when subjects were told the hypothesis and shown the morphing procedure, most subjects responded with surprise and asked to see the faces again in an attempt to identify more consciously which face it was that theirs had been morphed with. Even under these conditions, subjects still had difficulty selecting the face their face had been morphed with. It was not until their real picture was available for comparison with the child morphs that subjects could tell which face theirs had been morphed with. This suggests that males possess a mechanism that processes information about resemblance at largely unconscious levels. Therefore, it has been hypothesized that male brains may support neural architecture for a resemblance detection module that is situation (child care)–specific.

Neurobiological Correlates of Facial Resemblance

If the sex difference in reaction to children's faces is driven by evolutionary pressures and is occurring at levels below conscious awareness, one might expect the coevolution of (sex)-specialized modules (or programs) for processing such information. Questions about the neurobiological correlates of presumed evolved, adaptive information-processing mechanisms have only recently begun to be asked (e.g., Cosmides & Tooby, 1992, 2005), but are presumed to be part of a hierarchical adaptive domain specificity along the spectrum of general face recognition to self-face recognition.

Face recognition is particularly well developed in primate species and apparently is processed by unique neural architecture, the fusiform face area (Kanwisher et al., 1997). Additionally, humans have evolved the capacity for self-face recognition (Platek et al., 2004, 2005; see also Chapter 16, this volume). Because of humans' reliance on information from the face domain for social interaction (e.g., the communication of emotional information and identity information), it seems plausible to assume that selection for self-referent phenotype matching in humans might have occurred in the face perception domain. This might also be the case among chimpanzees (Parr & de Waal, 1999).

Self-face referential phenotype matching must be housed within a larger model for self-face recognition as part of the evolution of a general face processing mechanism (figure 9.2). Within the context of face recognition/self-face recognition, there are hypotheses that can be generated from a domain specificity model that further support the position that facial resemblance was a means for processing kin-based information. The model suggests that self-referent phenotype matching occurs at subconscious levels but taps into what we have called self-face identity nodes (the correlate to face identity nodes as presented in the face recognition models of Bruce & Young, 1986, and Breen et al., 2001). The information-processing connection between self-face identity nodes and self-reference phenotype matching is also recursively linked through an affective labeling node, which we presume drives the subconscious reactions to facial resemblance. This processing loop is predicted to produce

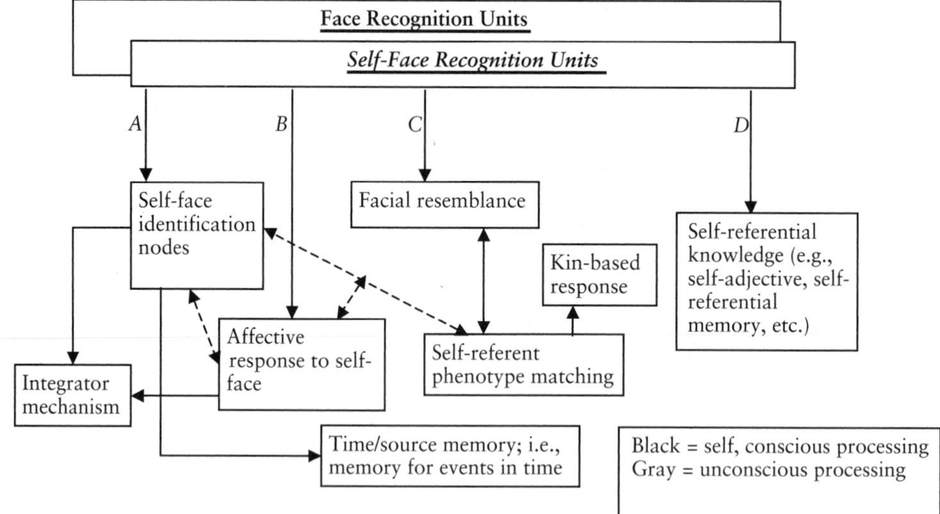

Figure 9.2
Model for self-face recognition units (Platek, in prep.) from within a broader model for face recognition (Breen et al., 2001). The figure shows a possible link between face recognition and self-face recognition, and the possible processing for self-referent phenotype matching from self-face recognition units. Gray boxes and lines indicate subconscious processing. Note the connections between self-face identification nodes and self-referent phenotype matching and the link through the affective response node. (The units are described in Breen et al., 2001.) See text for further discussion.

approach-avoidance-based behavioral responses as a function of kin/nonkin decisions.

Neural Correlates of Facial Resemblance Perception

Platek et al. (2004, 2005) have conducted the only functional neuroimaging studies of facial resemblance to date. In a series of functional magnetic resonance imaging (fMRI) studies, they investigated the neural correlates of face resemblance using adult and child faces morphed to resemble subjects. When subjects saw a face that was morphed to resemble their face, primarily posteromedial (e.g., precuneus and cingulate) and limbic structures (anterior cingulate) were active. This suggests a link between self-referential processing (medial structures) and affective labeling (limbic structures) of the information (figure 9.3).

These data are the first to show that the neural substrates that become activated for the perception of facial resemblance are possibly

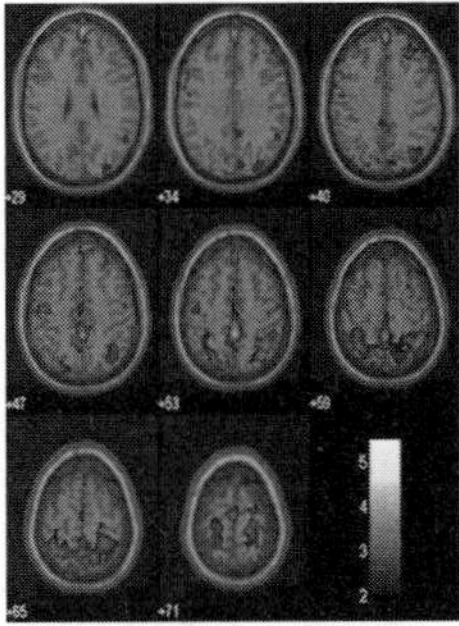

Figure 9.3
Spatial parametric map of activation associated with facial resemblance across sex. (Bar = value of *t* statistic, $P < 0.005$.)

unique, and at least different from both general face recognition and self-face recognition.

Findings for Sex Differences in Brain Activation: Neuroimaging Support for Parental Investment Theory

Two recent studies by our research group have provided initial evidence in favor of a sex-specific modular response to facial resemblance as a function of the age of the face. In the first of these studies, Platek et al. (2004) asked nine subjects to look at images of children, some of which had been morphed with the subject's image. Two interesting findings emerged from this initial study: first, males showed activation in the anterior left prefrontal lobe and anterior cingulate gyrus (figure 9.4 and Table 9.1A) when viewing children's faces that had been morphed to resemble theirs, but females did not, and second, females activated a set of substrates in the right and medial prefrontal cortices when viewing all children's faces, irrespective of whether the face had been morphed to resemble theirs (figures 9.5, & 9.6, and table 9.1B).

To extend and improve on this existing study, we (Platek et al., 2005) designed a more rigorous study that included a control condition for adult faces. Not only did we control for exposure to adult faces, we also used a more sensitive experimental paradigm (known as event-related fMRI) and asked subjects to actively respond to the faces as they were presented in the scanner. This study revealed consistent results. We found that males' brains were much more active when responding to children's faces that resembled the male subjects than when responding to other children's faces

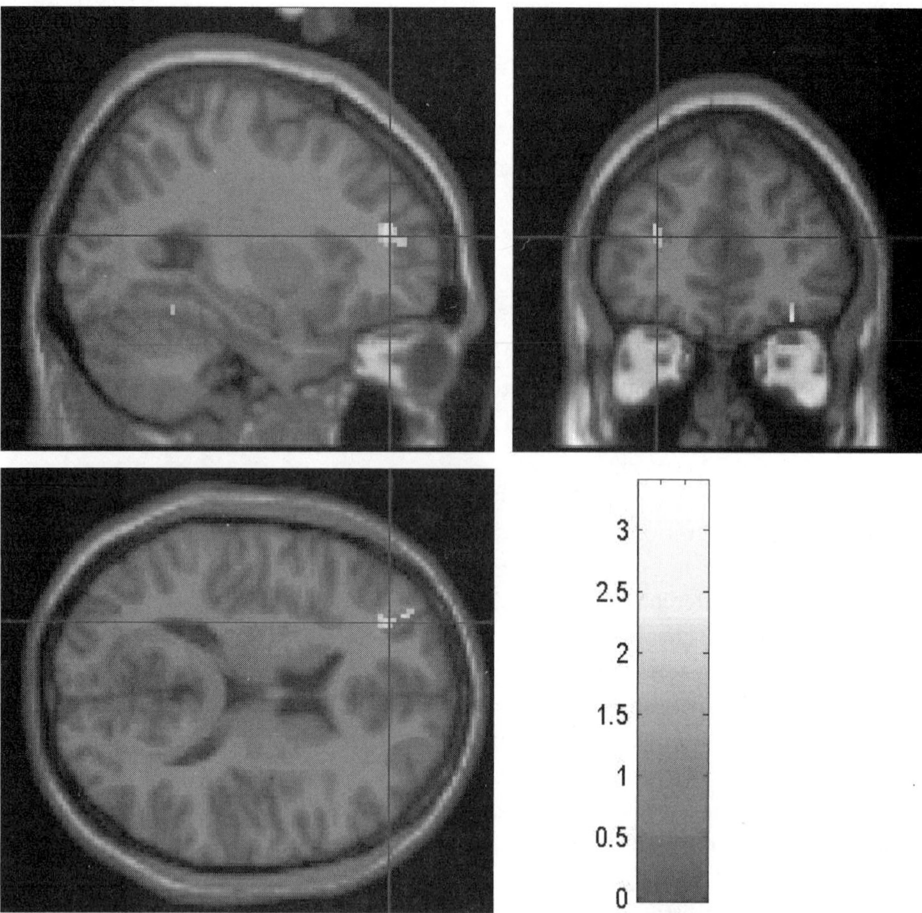

Figure 9.4
Left superior and middle frontal, and frontal subgyral activation in males in a comparison of activation between self-child morph and non-self-child morph conditions. (Bar = value of *t* statistic, $P < 0.005$.)

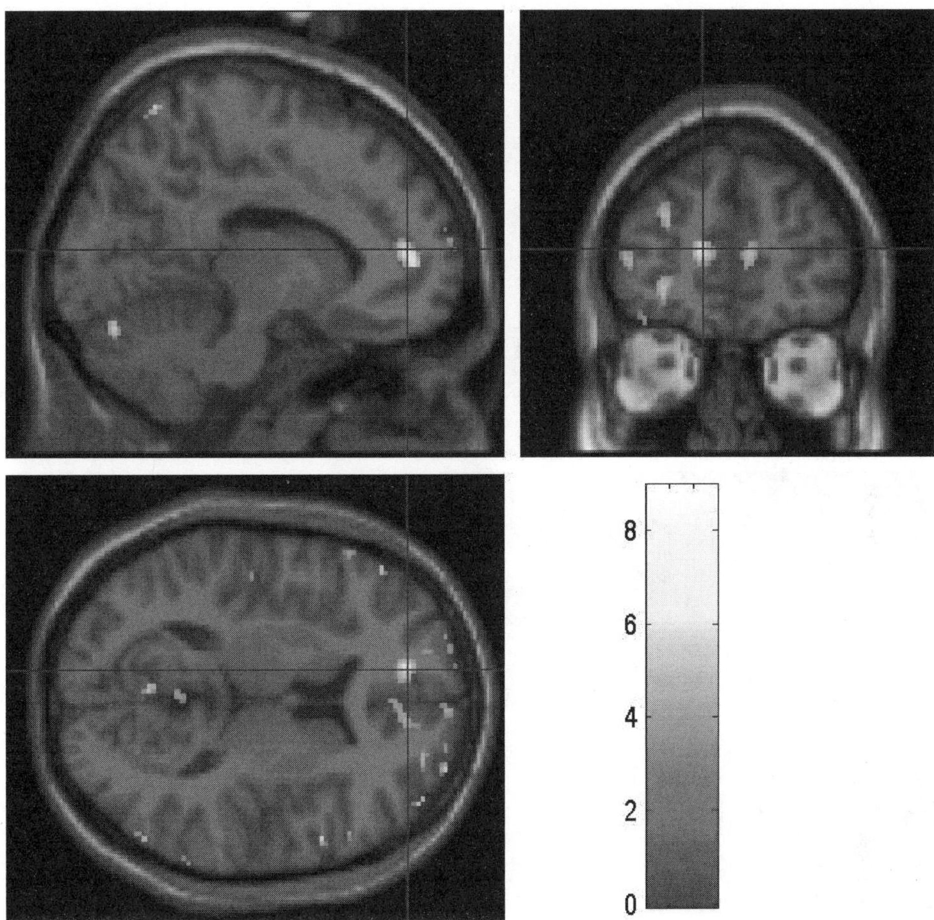

Figure 9.5
Left medial and medial superior frontal gyrus activation in females in a comparison of activation between self-child morph and non-self-child morph conditions. (Bar = value of t statistic, $P < 0.005$.)

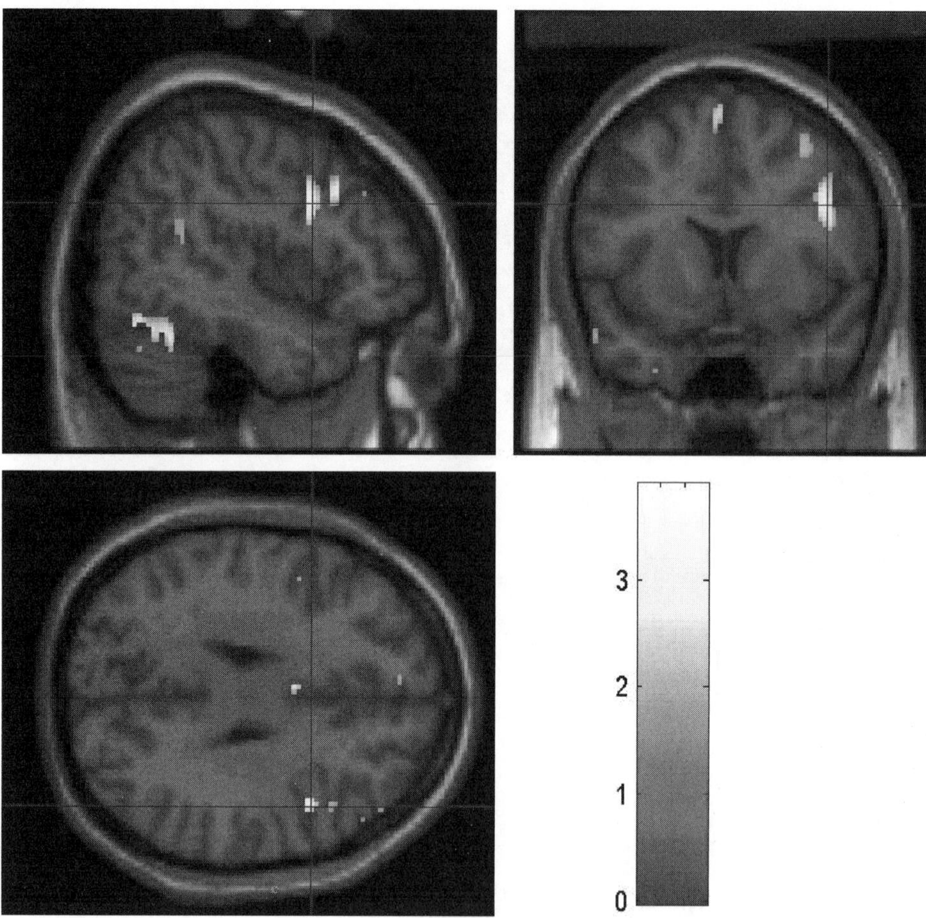

Figure 9.6
Right middle and inferior frontal lobe activation in females in a comparison of activation between self-child morph and non-self child morph conditions. (Bar = value of *t* statistic, $P < 0.005$.)

Table 9.1
Brain Activation Regions (Talairach Coordinates) in a Comparison of Activation Between Self-Child Morphs and Non-Self-Child Morphs in Males and Females

	Coordinates				
Right Hemisphere	*x*	*y*	*z*	BA	*P*
A. Areas of Activation in Males					
Middle frontal gyrus	–28	44	18	10	<0.05
Superior frontal gyrus	–30	54	14	10	<0.05
B. Areas of Activation in Females					
Inferior frontal gyrus	60	14	16	44/45	<0.01
Middle frontal gyrus	46	10	30	9	<0.01
Left medial frontal lobe					
Medial frontal gyrus	–12	50	10	10	<0.01
Medial superior frontal gyrus	–18	60	8	—	<0.01

Abbreviation: BA, Brodmann's area.

or to adult faces who did and did not resemble them. The male subjects' brains also showed more activation than female subjects' brains to child faces that resembled theirs. When comparing activation between responding to children's faces that resembled the subjects' face and children's faces that did not resemble the subjects' face, we found significant activation in the anterior cingulate and anterior left prefrontal lobe in males and in the caudate nucleus in females (figure 9.7 and table 9.2).

Discussion

These data suggest that unique neural substrates are involved in the detection of facial resemblance and that the behavioral difference males and females show in reactions toward children's faces (Burch & Gallup, 2000; Daly & Wilson, 1998; Platek, 2002; Platek et al., 2002, 2003, 2004) is driven by differences in neural processing (Platek et al., 2004, 2005) that might be related to an affective labeling recursive loop designed to attribute affective information to self-referent phenotype matches or nonmatches.

Unlike females, males showed significant neural activation in the left frontal cortex and anterior cingulate region, which has been hypothesized to be involved in the inhibition of negative responses (Collette et al., 2001; Davidson, 1997; Harmon-Jones & Sigelman, 2001). Applying this

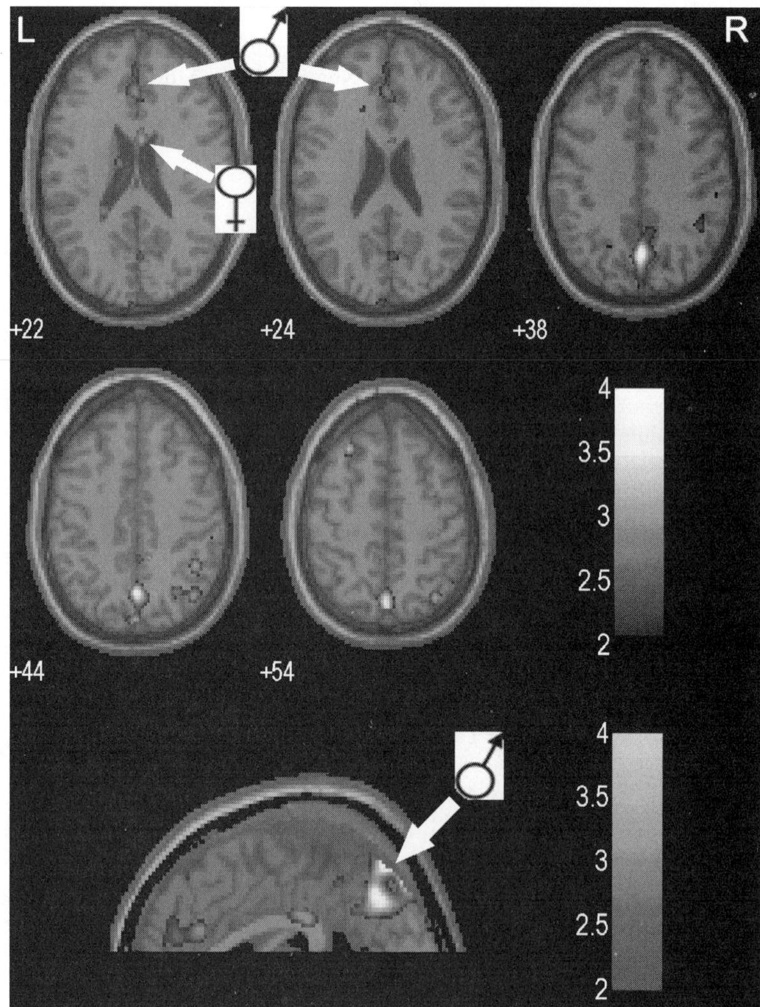

Figure 9.7
Activations. Male activation = anterior cingulate, posterior cingulate, precuneus, right superior temporal gyrus, and left middle frontal gyrus. Female activation = left insulas and anterior cingulate gyrus.

Table 9.2
Brain Activation Regions (Talairach Coordinates) in a Comparison of Activation Between Self-Child Morphs and Non-Self-Child Morphs in Males and Females, Using an Event-Related fMRI Experimental Study Design

Region	Hemisphere	Coordinates			Z-score
		x	*y*	*z*	
Male > Female					
Superior temporal gyrus	R	28	14	–24	4.54
Cingulate gyrus	L	–18	26	23	4.27
Inferior parietal lobe	R	46	–45	41	4.03
Precuneus	R	6	–62	45	4.01
Anterior cingulate gyrus	L	0	34	17	3.88
Anterior cingulate gyrus	L	–2	–18	32	3.49
Female > Male					
Insula	L	–40	–5	13	3.52
Cingulate gyrus	L	0	25	37	3.46

hypothesis to these data, it would appear that males may possess a generalized skepticism about children that is inhibited when (1) the child resembles the subject and (2) the male is faced with the adaptive problem of provisioning for offspring. Thus, the left frontal activation associated with viewing self-child morphs may be a situation-specific, evolutionarily adaptive response in males, which supports the hypothesis put forth by Daly and Wilson (1998) and Platek and colleagues (Platek, 2002; Platek et al., 2002, 2003) that males use self-resemblance to assess paternity.

The data from female subjects are harder to explain. However, in light of anecdotal behavioral evidence, it might be the case that the right lateralized frontal and medial frontal activity may be part of a mentalizing (e.g., theory of mind/mental state attribution) module. In each of our behavioral studies (e.g., Platek et al., 2002) participants were queried as to how they made their choices. Whereas males usually reported "going with a gut feeling" or no strategy, females often attributed specific personality characteristics to the children's faces. Female subjects also reported that they tried to find the "nicest" child to give money to, spend time with, or adopt and the "meanest" or "brattiest" child to punish, not spend time with, and not spend money on. In other words, females may be making decisions to invest in or discipline children based on inferences about the psychological characteristics of the child.

This hypothesis is supported by recent neuropsychological and neuroimaging data on mentalizing. For example, Fletcher et al. (1995) found medial prefrontal cortex (MPFC) activation when comparing activation associated with reading mentalizing stories with activation associated with reading physical stories. Gallagher et al. (2000) and Vogely et al. (2001) found similar MPFC as well as right hemisphere activation associated with mental state processing. In a test similar to that devised by Povinelli, Rulf, and Bierschwale (1994) for understanding intention in chimpanzees, Berthoz et al. (2002) reported MPFC activation associated when subjects read accounts of social transgressions that were both deliberate and accidental. These data extend those of Castelli, Happe, Frith, and Frith (2000) and Klin, Jones, Schultz, and Volkmar (2003) which demonstrated activation in the temporal pole and MPFC when subjects observed the motion of inanimate objects that could be interpreted as indicating intention or desire. Stuss and colleagues (Ishii et al., 2002; Stuss, Gallup, & Alexander, 2001) showed that patients with damage to the right frontal lobes, but not other parts of the cortex, were deficient in understanding visual perspective taking, deception, and emotional mental states. Using fMRI, Platek et al. (submitted) showed activation in the medial and right frontal lobes when participants were asked to think about the mental states of others when seeing only the person's eyes (Eye in the Mind–Revised Test; Baron-Cohen, Wheelright, Hill, & Raste, 2001). These data support the idea that females may utilize a mentalizing approach when thinking about how to invest resources in children.

There are obvious limitations to these studies. For example, the sample sizes are small. However, large sample sizes typically are not needed to achieve adequate statistical power in functional imaging studies. Additionally, a boxcar design was used in our first study, which is not as sensitive to subtle changes in neural activations as event-related fMRI, and one runs the risk of habituation effects. However, in order to account for the possibility of low activity levels and the possibility of habituation, we used six blocks, which has been shown to be a reliable number of epochs to produce maximum activation associated with stimulus exposure while limiting the likelihood of habituation effects (Mohamed, personal communication). Additionally, we used an event-related fMRI design in our second study to account for these possible effects, and found similar activation patterns.

Outstanding Questions

If facial resemblance is filtered through a mechanisms for self-referential processing (i.e., medial cortical structures), then the model assumes a positive valence for such recursion when appetitive kin social responses are enacted. What of individuals who do not have a positive self-referential bias? Would their activation be exactly opposite of what we found? Does selection work against negative self-reference and parental investment? What is the relationship between negative/positive self-referential processing and parental effort and success? Clinically interesting questions also arise. For example, are patient populations that are deficient in processing self-referential information also deficient in processing self-resemblance (e.g., autism, schizophrenia)?

Acknowledgments The authors thank Feroze Mohamed and Scott Faro for their assistance with the functional magnetic resonance imaging. We also thank Gordon Gallup, Rebecca Burch, Julian Keenan, Todd Shackelford, Jennifer Davis, Thomas Myers, Samuel Critton, Ivan Panyavin, Brett Wasserman, Gordon Bear, David Smith, and Robert Haskell for helpful discussions concerning this line of investigation.

References

Anderson, K., Kaplan, H., Lam, D., & Lancaster, J. (1999). Paternal care of genetic fathers and stepfathers. II. Reports by Xhosa high school students. *Evolution and Human Behavior*, *20*, 433–451.

Ashburber, J., & Friston, K. J. (1999). Nonlinear spatial normalization using basis functions. *Human Brain Mapping*, *7*, 254–266.

Baker, R. R., & Bellis, M. A. (1995). *Human sperm competition: Copulation, masturbation, and infidelity*. London: Chapman and Hall.

Baron-Cohen, S., Wheelright, S., Hill, J., Raste, Y., & Plumb, I. (2001). The "Reading the Mind in the Eyes" Test revised version: A study with normal adults, and adults with Asperger syndrome or high-functioning autism. *Journal of Child Psychology and Psychiatry*, *42*, 241–251.

Berthoz, S., Armony, J., Blair, R., & Dolan, R. (2002). An fMRI study of intentional and unintentional (embarrassing) violations of social norms. *Brain*, *125*, 1696–1708.

Birkhead, T. R. (1995). Sperm competition: Evolutionary causes and consequences. *Reproduction, Fertility, and Development*, *7*, 755–775.

Birkhead, T. R. (1996). Sperm competition: Evolution and mechanisms. *Current Topics in Developmental Biology*, *33*, 103–158.

Bjorklund, D. F., & Shackelford, T. K. (1999). Differences in parental investment contribute to important differences between men and women. *Current Directions in Psychological Science*, *8*(3), 86–89.

Brédart, S., & French, R. (1999). Do babies resemble their fathers more than their mothers? A failure to replicate Christenfeld and Hill. *Evolution and Human Behavior*, *20*, 129–135.

Breen, N., Caine, D., & Coltheart, M. (2001). Mirrored-self misidentification: Two cases of focal onset dementia. *Neurocase*, *7*, 239–254.

Bruce, V., & Young, A. (1986). Understanding face recognition. *British Journal of Psychology*, *77*, 305–327.

Buchan, J. C., Alberts, S. C., Silk, J. B., & Altmann, J. (2003). True paternal care in a multi-male primate society. *Nature*, *425*(6954), 179–181.

Burch, R. L., & Gallup, G. G., Jr. (2000). Perceptions of paternal resemblance predict family violence. *Evolution and Human Behavior*, *21*, 429–435.

Buss, D. M. (1988). From vigilance to violence: Tactics of mate retention in American undergraduates. *Ethology and Sociobiology*, *9*, 291–317.

Buss, D. M. (1992). Mate preference mechanisms: Consequences for partner choice and intrasexual competition. In J. H. Barkow, L. Cosmides, & J. Tooby (Eds.), *The adapted mind: Evolutionary psychology and the generation of culture* (pp. 249–266). Oxford: Oxford University Press.

Buss, D. M. (1994). *The evolution of desire*. New York: Basic Books.

Buss, D. M. (1999). *Evolutionary psychology*. Needham Heights, MA: Allyn & Bacon.

Buss, D. M. (2000). *The dangerous passion*. New York: Free Press.

Buss, D. M., & Shackelford, T. K. (1997). From vigilance to violence: Mate retention tactics in married couples. *Journal of Personality and Social Psychology*, *72*, 346–361.

Castelli, F., Happe, F., Frith, U., & Frith, C. D. (2000). Movement and mind: A functional imaging study of perception and interpretation of complex intentional movement patterns. *NeuroImage, 12*, 314–325.

Cerda-Flores, R. M., Barton, S. A., Marty-Gonzales, L. F., Rivas, F., & Chakraborty, R. (1999). Estimation of nonpaternity in the Mexican population of Nuevo Leon: A validation study with blood group markers. *American Journal of Physical Anthropology*, *109*, 281–293.

Christenfeld, N., & Hill, E. (1995). Whose baby are you? *Nature*, *378*, 669.

Collette, F., Van der Linden, M., Delfiore, G., Degueldre, C., Luxen, A., & Salmon, E. (2001). The functional anatomy of inhibition processes investigated with the Hayling Task. *NeuroImage*, *14*, 258–267.

Cosmides, L., & Tooby, J. (1992). Cognitive adaptations for social exchange. In J. Barkow, L. Cosmides, & J. Tooby (Eds.). *The adapted mind*. New York: Oxford University Press.

Cosmides, L., & Tooby, J. (2005). Neurocognitive adaptations designed for social exchange. In D. M. Buss (Ed.), *Evolutionary psychology handbook*. New York: Wiley.

Daly, M., & Wilson, M. (1982). Whom are newborn babies said to resemble? *Ethology and Sociobiology*, *3*, 69–78.

Daly, M., & Wilson, M. (1984). A sociobiological analysis of human infanticide. In G. Hausfater, & S. B. Hrdy (Eds.), *Infanticide: comparative and evolutionary perspectives* (pp. 487–502). New York: Aldine.

Daly, M., & Wilson, M. (1988a). Evolutionary social psychology and family homicide. *Science*, 242, 519–524.

Daly, M., & Wilson, M. (1988b). *Homicide*. Hawthorne, NY: Aldine de Gruyter.

Daly, M., & Wilson, M. (1996). Violence against stepchildren. *Current Directions in Psychological Science*, *5*, 77–81.

Daly, M., & Wilson, M. (1998). *The truth about Cinderella: A Darwinian view of parental love.* New Haven, CT: Yale University Press.

Daly, M., Wilson, M., & Weghorst, S. J. (1982). Male sexual jealousy. *Ethology and Sociobiology, 3*(1), 11–27.

Davidson, R. J. (1997). Emotion and affective style: physiological substrates. *Electroencephalography and Clinical Neurophysiology,* 102, 1P. Abstracts of the American Encephalographic Society.

Davis, J. A., & Gallup, G. G., Jr. (in press). Preeclampsia and other pregnancy complications as an adaptive response to unfamiliar semen. In S. M. Platek & T. K. Shackelford (Eds.), *Female infidelity & parental uncertainty.* New York: Cambridge University Press.

DeBruine, L. M. (2002). Facial resemblance enhances trust. *Proceedings of the Royal Society of London, B, Biological Sciences, 269,* 1307–1312.

DeBruine, L. M. (2004). Resemblance to self increases the appeal of child faces to both men and women. *Evolution and Human Behavior, 25,* 142–154.

DeBruine, L. M. (2005). Trustworthy but not lust-worthy: Context-specific effects of facial resemblance. *Proceedings of the Royal Society of London B, Biological Sciences, 272,* 919–922.

Fletcher, F. C., Happe, F., Frith, U., Dolan, R. J., Fracowiak, R. S., & Frith, C. D. (1995). Other minds in the brain: A functional imaging study of "theory of mind" in story comprehension. *Cognition, 57,* 109–128.

Gallagher, H. L., Happe, F., Brunswick, N., Fletcher, P. C., Frith, U., & Frith, C. D. (2000). Reading the mind in cartoons and stories: An fMRI study of "theory of mind" in verbal and nonverbal tasks. *Neuropsychologia, 38,* 11–21.

Gallup, G. G., Jr., & Burch, R. L. (in press). The semen displacement hypothesis. In T. Shackelford & N. Poune (Eds.), *Sperm competition: New and old readings.*

Gallup, G. G., Jr., Burch, R. L., Zappierri, M. L., Parvez, J., & Davis, J. (in press). Semen displacement: Behavioral and morphological evidence. *Evolution and Human Behavior.*

Geary, D. C. (2000). Evolution and proximate expression of human paternal investment. *Psychological Bulletin, 126,* 55–77.

Goetz, A. T., Shackelford, T. K., Weekes-Shackelford, V. A., Euler, H. A., Hoier, S., & Schmitt, D. P. (in press). Mate retention, semen displacement, and human sperm competition: Tactics to prevent and correct female infidelity.

Hamilton, W. D. (1964a). The evolution of social behavior I. *Journal of Theoretical Biology, 7,* 1–16.

Hamilton, W. D. (1964b). The genetical evolution of social behaviour I, II. *Journal of Theoretical Biology, 7,* 17–52.

Harmon-Jones, E., & Sigelman, J. (2001). State anger and prefrontal brain activity: Evidence that insult-related relative left-prefrontal activation is associated with experienced anger and aggression. *Interpersonal Relations and Group Processes, 5,* 797–803.

Hauber, M. E., & Sherman, P. W. (2001). Self-referent phenotype matching: Theoretical considerations and empirical evidence. *Trends in Cognitive Sciences, 10,* 609–616.

Henson, R. N., Shallice, T., Gorno-Tempini, M. L., & Dolan, R. J. (2002). Face repetition effects in implicit and explicit memory tests as measured by fMRI. *Cerebral Cortex*, *12*, 178–186.

Hrdy, S. (1974). Male-male competition and infanticide among the langurs (*Presbytis entellus*) of Abu, Rajasthan. *Folia Primatologica (Basel)*, *22*(1), 19–58.

Ishii, R., Stuss, D. T., Gorjmerac, C., Gallup, G. G., Jr., Alexander, M. P., & Pantev, C. (2002). Neural correlates of "theory of mind" in emotional vignettes comprehension studied with spatially filtered magnetoencephalography. In H. Nowak, J. Haueisen, F. Giessler, & R. Hounker (Eds.), *Biomag 2002: Proceedings of the 13th International Conference on Biomagnetism* (pp. 291–293). Berlin: VDE Verlag.

Kanwisher, N., McDermott, J., & Chun, M. (1997). The fusiform face area: A module in human extrastriate cortex specialized for face perception. *Journal of Neuroscience*, *17*, 4302–4311.

Klin, A., Jones, W., Schultz, R., & Volkmar, F. (2003). The enactive mind, or from actions to cognition: Lessons from autism. *Philosophical Transactions of the Royal Society of London, B, Biological Sciences*, *358*, 345–360.

Lacy, R. C., & Sherman, P. W. (1983). Kin recognition by phenotype matching. *American Naturalist*, *121*, 489–512.

McLain, D. K., Setters, D., Moulton, M. P., & Pratt, A. E. (2000). Ascription of resemblance of newborns by parents and nonrelatives. *Evolution and Human Behavior*, *21*, 11–23.

Neale, M. C., Neale, B. M., & Sullivan, P. F. (2002). Nonpaternity in linkage studies of extremely discordant sib pairs. *American Journal of Human Genetics*, *70*, 526–529.

Neff, B. D., & Sherman, P. W. (2002). Decision making and recognition mechanisms. *Proceedings of the Royal Society of London, B, Biological Sciences*, *269*, 1435–1441.

Nesse, R., Silverman, A., & Bortz, A. (1990). Sex differences in ability to recognize family resemblance. *Ethology and Sociobiology*, *11*, 11–21.

Ogawa, S., Tank, D. W., Menon, R., et al. (1992). Intrinsic signal changes accompanying sensory stimulation: Functional brain mapping with magnetic resonance imaging. *Proceedings of the National Academy of Sciences, U.S.A.*, *89*, 5951–5955.

Oldfield, R. C. (1971). The assessment and analysis of handedness: The Edinburgh inventory. *Neuropsychologia*, *9*, 97–113.

Parr, L. A., & de Waal, F. B. M. (1999). Visual kin recognition in chimpanzees. *Nature*, *399*(6737), 647–648.

Platek, S. M. (2002). Unconscious reactions to children's faces: The effect of resemblance. *Evolution and Cognition*, *8*, 207–214.

Platek, S. M., Burch, R. L., Panyavin, I. S., Wasserman, B. H., & Gallup, G. G., Jr. (2002). Reactions towards children's faces: Resemblance matters more for males than females. *Evolution and Human Behavior*, *23*, 159–166.

Platek, S. M., Critton, S. R., Burch, R. L., Frederick, D. A., Myers, T. E., & Gallup, G. G., Jr. (2003). How much paternal resemblance is enough? Sex differences in the reaction to resemblance but not in ability to detect resemblance. *Evolution and Human Behavior*, *24*, 81–87.

Platek, S. M., Keenan, J. P., Gallup, G. G. Jr., & Mohamed, F. B. (submitted). Where am I? Neural correlates of self and other.

Platek, S. M., Keenan, J. P., & Mohamed, F. B. (2005). Neural correlates of facial resemblance. *NeuroImage*, *25*, 1336–1344.

Platek, S. M., & Shackelford, T. K. (in press). *Female infidelity and parental uncertainty*. New York: Cambridge University Press.

Povinelli, D. J., Rulf, A. B., & Bierschwale, D. T. (1994). Absence of knowledge attribution and self-recognition in young chimpanzees (*Pan troglodytes*). *Journal of Comparative Psychology*, *108*, 74–80.

Regalski, J., & Gaulin, S. (1993). Whom are Mexican infants said to resemble? Monitoring and fostering paternal confidence in the Yucatan. *Ethology and Sociobiology*, *14*, 97–113.

Sasse, G., Muller, H., Chakraborty, R., & Ott, J. (1994). Estimating the frequency of nonpaternity in Switzerland. *Human Heredity*, *44*, 337–343.

Shackelford, T. K., LeBlanc, G. J., Weekes-Shackelford, V. A., Bleske-Rechek, A. L., Euler, H. A., & Hoier, S. (2002). Psychological adaptations to human sperm competition. *Evolution and Human Behavior*, *23*, 123–138.

Stuss, D. T., Gallup, G. G., Jr., & Alexander, M. P. (2001). The frontal lobes are necessary for theory of mind. *Brain*, *124*, 279–286.

Sykes, B., & Irven, C. (2000). Surnames and the Y chromosome. *American Journal of Human Genetics*, *66*, 1417–1419.

Talairach, J., & Tournoux, P. (1988). *Co-planar stereotaxic atlas of the human brain: 3-dimensional proportional system. An approach to cerebral mapping*. New York: Thieme.

Trivers, R. L. (1972). Parental investment and sexual selection. In B. Campbell (Ed.), *Sexual selection and the descent of man* (pp. 1871–1971). Chicago: Aldine.

Vogely, K., Bussfeld, P., Newen, A., Happe, F., Falkai, P., Maier, W., et al. (2001). Mind reading: Neural mechanisms of theory of mind and self-perspective. *NeuroImage*, *14*, 170–181.

Volk, A. & Quinsey, V. L. (2002). The influence of infant facial cues on adoption preferences. *Human Nature*, *13*(4), 437–455.

Wedekind, C., Seebeck, T., & Paepke, A. J. (1995). MHC dependent mate preferences in humans. *Proceedings of the Royal Society of London*, *260*, 245–252.

Wedekind, C., and Furi, S. (1997). Body odour preferences in men and women: Do they aim for specific MHC combinations or simply heterozygosity? *Proceedings of the Royal Society of London B, Biological Sciences*, *264*, 1471–1479.

10 Lust, Romance, Attachment: Do the Side Effects of Serotonin-Enhancing Antidepressants Jeopardize Romantic Love, Marriage, and Fertility?

Helen E. Fisher and J. Anderson Thomson, Jr.

Today, millions of people of reproductive age take selective serotonin-reuptake inhibitors (SSRIs) and other serotonin-enhancing antidepressants. Approximately 80% of these drugs are prescribed by nonpsychiatric physicians, including internists, general practitioners, pediatricians, and gynecologists, who disseminate them to a wide array of men and women. In the first five months of 2004, American doctors wrote 46 million prescriptions for antidepressants, largely for these drugs. In the United States alone, antidepressants account for $14 billion a year in wholesale revenues (Morais, 2004).

These medications effectively treat a wide range of serious conditions, including major depression, posttraumatic stress disorder, generalized anxiety disorders, panic disorders, obsessive-compulsive disorder, social phobias, eating disorders, Asperger's syndrome, irritable bowel syndrome, and chronic pain syndromes. But they also produce various side effects. In both men and women, these antidepressants can cause emotional blunting, weight gain, and several types of sexual dysfunction, interfering with sexual desire, sexual arousal, genital sensation, lubrication, erection, ejaculation, and orgasm (Montejo, Lorca, Izquierdo, & Rico-Villademoros, 2001; Rosen, Lane, & Menza, 1999). The number of men and women affected by these forms of sexual dysfunction vary; some studies report that as many as 73% of patients taking serotonin-enhancing antidepressants experience one or more of these sexual side effects (Montejo et al., 2001).

We propose that serotonin-enhancing antidepressants can have far more serious psychological, social, and genetic consequences through their effects on several other neural mechanisms that evolved to enable mate assessment, mate choice, mate pursuit, feelings of romantic love, and expressions of attachment to a long-term partner.

This chapter discusses the neural correlates of the three primary brain systems for courtship, mating, pair formation, and reproduction: the sex drive, romantic love, and male-female attachment (companionate love). It explores the neurochemical relationships between these three neural systems to show how serotonin-enhancing antidepressants can potentially jeopardize the ability to fall in love and maintain a stable, long-term partnership. It discusses the potential effects of the long-term use of serotonin-enhancing medications on other brain-body mechanisms that evolved to foster courtship and pair-bond stability, including penile erection and female orgasm. Finally, the discussion considers how serotonin-enhancing antidepressants can adversely affect fertility and one's genetic future.

Three Neural Systems for Mating and Reproduction

Neuroscientists currently believe that the basic human emotions and motivations arise from distinct systems of neural activity, that these brain systems derive from mammalian precursors, and that these brain mechanisms evolved to enable survival and reproduction (Davidson, 1994; Panksepp, 1998). Among these primary neural systems are three discrete, interrelated motivation/emotional systems for mating, reproduction, and parenting: the sex drive, romantic love, and male-female attachment. Each of these motivation/emotional systems is associated with a different behavioral repertoire, each is associated with a different and dynamic constellation of neural correlates, and each evolved to direct a different aspect of reproduction (Fisher, 1998).

The sex drive is characterized by the craving for sexual gratification. In nonprimate mammalian species, it is associated primarily with the estrogens and androgens. In humans and other higher primates, the estrogens have little direct influence on sexual desire (Meston & Frolich, 2000); instead, the androgens, particularly testosterone, are crucial to sexual desire in both sexes (Edwards & Booth, 1994; Sherwin, 1994; Van Goozen, Wiegant, Endert, Helmont, & Van de Poll, 1997). The sex drive evolved principally to motivate individuals to seek sexual union with a *range* of reproductive partners.

Romantic love (also known as obsessive love, passionate love, or being in love) is characterized by intense energy, focused courtship attention, ecstasy, mood swings, sexual possessiveness, emotional dependency, obsessive thinking about the beloved, craving for emotional union with the beloved, and intense motivation to win this *preferred* mating partner

(Fisher, 1998; Gonzaga, Keltner, Londahl, & Smith, 2001; Harris, 1995; Hatfield, 1988; Hatfield & Sprecher, 1986; Shaver, Schwartz, Kirson, & O'Connor, 1987; Tennov, 1979). Evidence suggests that romantic love is primarily associated with elevated activity in dopaminergic pathways of the reward system of the brain (Aron et al., 2005; Bartels & Zeki, 2000, 2004), and data suggest that other mammals share central biological and behavioral aspects of this brain system (Fabre-Nys et al., 1997; Gingrich, Liu, Cascio, Wang, & Insel, 2000; Liu & Wang, 2003; Wang et al., 1999). The neural system associated with romantic love evolved to motivate individuals to prefer a *specific* mating partner, thereby conserving courtship time and energy.

Partner attachment in humans is associated with feelings of calm, security, social comfort, and emotional union with a long-term mating partner, as well as with some of the traits of mammalian attachment, including mutual territory defense and nest (home) building, mutual feeding and grooming, maintenance of close proximity, separation anxiety, shared parental chores, and affiliative gestures (Carter et al., 1997; Lim, Murphy, & Young, 2004; Lim & Young, 2004; Young, Wang, & Insel, 1998). Animal studies suggest that this brain system is associated primarily with the neuropeptides oxytocin and vasopressin (Carter, 1992; Lim, Murphy et al., 2004; Lim & Young, 2004; Winslow, Hastings, Carter, Harbaugh, & Insel, 1993). Adult male-female partner attachment evolved primarily to motivate individuals to sustain an affiliative connection with a reproductive partner at least long enough to complete species-specific parental duties (Fisher, 1992).

We propose that when individuals use serotonin-enhancing antidepressants, they can potentially jeopardize not only their sex drive but also these related neural mechanisms for romantic love and partner attachment.

The Sex Drive

The androgens, particularly testosterone, are central to sexual desire in both men and women (Edwards & Booth, 1994; Sherman, 1994; Van Goozen, Wiegant, Endert, Helmond, & Van de Poll, 1997). Individuals with higher circulating levels of testosterone tend to engage in more sexual activity (Edwards & Booth, 1994; Sherman, 1994). Male athletes who use testosterone and other anabolic steroids to increase their strength and stamina have more sexual thoughts, more morning erections, more sexual encounters, and more orgasms. Middle-aged women who inject or apply testosterone cream to the skin boost their sexual desire. The male libido peaks in the early twenties, when the activity of testosterone is highest. Many women feel more sexual desire around ovulation, when testosterone increases (Van Goozen et al., 1997). Both sexes also have fewer sexual fantasies, masturbate less regularly, and engage in less frequent intercourse as they age and testosterone levels decline (Edwards & Booth, 1994). People vary in their degree and frequency of sexual desire, in part because levels of testosterone are inherited (Meikle, Stringham, Bishop, & West, 1988). Moreover, the balance between testosterone, estrogen, and other bodily systems, as well as social circumstances, childhood experiences, and a host of other factors, play a role in determining when, where, and how often one feels lust (Nyborg, 1994). Nevertheless, testosterone is central to the sex drive.

The sex drive is also associated with a specific range of neural correlates. Using functional magnetic resonance imaging (fMRI), Arnow and colleagues reported that when young male heterosexual subjects viewed erotic video material while wearing a custom-built pneumatic pressure cuff around the penis, they showed strong activations in the right subinsular region, including the claustrum, the left caudate and putamen, the right middle occipital/middle temporal gyri, the bilateral cingulate gyrus and right sensorimotor and premotor regions, and the right hypothalamus (Arnow et al., 2002). Beauregard, Levesque, and Bourgouin (2001) measured brain activation (using fMRI) in men as the subjects viewed erotic film excerpts. Activations occurred in limbic and paralimbic structures, including the right amygdala, right anterior temporal pole, and hypothalamus.

Using fMRI, Karama and colleagues (2002) also recorded brain activity while men and women viewed erotic film excerpts. Activity increased in the anterior cingulate, medial prefrontal cortex, orbitofrontal cortex, insula, and occipitotemporal cortices, as well as in the amygdala and the ventral striatum. Men showed activation in the thalamus and significantly greater activation than women in the hypothalamus, specifically in a sexually dimorphic area associated with sexual arousal and behavior. In another experiment, researchers measured brain activity among eight men as these subjects experienced orgasm. Blood flow decreased in all regions of the cortex except one region of the prefrontal cortex, where it increased (Tiihonen et al., 1994). Animal studies also indicate that several brain structures are associated with the sex drive and sexual expression, including the medial amygdala, medial preoptic area, paraventricular nucleus, and periaqueductal gray (Heaton,

2000), and the septum and ventromedial hypothalamus (Dixson, 1998).

These data indicate that the constellation of neural correlates associated with the sex drive are dynamic yet specific. Moreover, data on the neural correlates associated with romantic love indicate that the sex drive and romantic love are overlapping yet distinct neural systems.

The Neural Correlates of Romantic Love

Intense courtship attraction, commonly known as romantic love, is recorded in all human societies for which data are available (Jankowiak & Fischer, 1992), and despite the varied ways that this phenomenon is expressed cross-culturally, this multipartite experience is associated with a specific constellation of motivations and emotions (Fisher, 1998; Gonzaga et al., 2001; Harris, 1995; Hatfield, 1988; Hatfield & Sprecher, 1986; Shaver et al., 1987; Tennov, 1979).

Romantic love begins as a person starts to regard another as special, unique. The lover focuses his or her attention on the beloved, doting on the beloved's worthy traits and overlooking or minimizing that person's flaws. The lover expresses increased energy, ecstasy when the love affair is going well, and mood swings into despair during times of adversity. Barriers heighten romantic passion, in what has been referred to as "frustration attraction" (Fisher, 2004). The lover suffers separation anxiety when apart from the beloved and often a host of sympathetic nervous system reactions when with the beloved, including sweating and a pounding heart. Lovers are emotionally dependent; they tend to change their priorities and daily habits to remain in contact with or to impress the beloved. They exhibit empathy for the beloved; many are willing to sacrifice, even die for this special other. The lover expresses sexual desire for the beloved, as well as intense sexual possessiveness. Yet their craving for emotional union supersedes their craving for sexual union. Most characteristic, the lover thinks obsessively about the beloved. Rejected lovers generally protest and try to win the beloved back, as well as express "abandonment rage" and despair. Romantic passion is also involuntary, difficult to control, and generally impermanent.

To investigate the neural correlates of romantic love, Fisher, Brown, Aron, and colleagues used fMRI to study the neural activity of 10 women and 7 men who reported being "madly in love" (Aron et al., 2005). The participants' age range was 18–26 years (mean, 20.6; median,

21), and subjects reported being in love an average of 7.4 months (median, 7; range, 1–17 months).

A preliminary investigation had identified a photograph of the beloved as an effective stimulus for eliciting feelings of intense romantic love (Mashek et al., 2000), so the protocol employed photographs and consisted of four tasks presented in an alternating block design. For 30 seconds each participant viewed a photo of the beloved (positive stimulus); for the following 40 seconds each performed a countback distraction task; for the following 30 seconds each viewed a photograph of an emotionally neutral acquaintance (neutral stimulus); and for the following 20 seconds each performed a similar countback task. The countback task involved viewing a large number, such as 8,421, and mentally counting backward (beginning with this number) in increments of seven. The countback task was included to decrease the carryover effect after the participant viewed the positive stimulus because it is difficult to quell intense feelings of romantic love. This four-part sequence (or a counterbalanced version beginning with the neutral stimulus) was repeated six times; the total stimulus protocol was 12 minutes.

Group activation specific to the beloved occurred in the right ventral tegmental area (VTA), localized in the region of A10 dopamine cells, and the right medial and posterodorsal body of the caudate nucleus (Aron et al., 2005). The VTA is rich in cells that produce and distribute dopamine to many brain regions, including the caudate nucleus. The VTA is also a central part of the brain's "reward system" (Breiter, Aharon, Kahneman, Dale, & Shizgal, 2001; Fiorillo, Tabler, & Schultz, 2003; Martin-Soelch et al., 2001; Schultz, 2000; Schultz, Dayan, & Read Montague, 1997; Wise, 1989), the neural network associated with sensations of pleasure, general arousal, focused attention, and motivation to pursue and acquire rewards (Delgado, Nystrom, Fissel, Noll, & Fiez, 2000; Elliot, Newman, Longe, & Deakin, 2003; Gold, 2003; Schultz, 2000). The caudate nucleus is also associated with reward, motivation, and goal-oriented behaviors. It plays a role in reward detection and expectation, the representation of goals, and the integration of sensory inputs to prepare for the appropriate actions to win rewards (Lauwereyns et al., 2002; Martin-Soelch et al., 2001; O'Doherty et al., 2004; Schultz, 2000). Some 80% of receptor sites for dopamine reside in the caudate nucleus.

Using fMRI, Bartels and Zeki also investigated brain activity in 6 men and 11 women who reported being "truly, deeply, and madly in love" (Bartels & Zeki, 2000). Participants looked at a photograph of the beloved, as well as photographs of three friends of similar age, sex, and length of friendship. Individuals reported being in love an average of 28.8 months, longer than the love relationships studied by Aron et al. (2005), who were in love an average of 7.4 months. Those in the Bartels and Zeki study also were less intensely in love. In spite of these differences, Bartels and Zeki (2000, 2004) found that romantic love also activated regions of the caudate nucleus and the VTA, as well as several different brain areas. These combined data support the hypothesis that dopaminergic pathways in the reward system of the brain play a central role in the focused attention and motivation associated with romantic love (Fisher, 1998).

Elevated activity of central dopamine is also associated with ecstasy, intense energy, hyperactivity, sleeplessness, mood swings, emotional dependence, and craving (Abbott, 2002; Colle & Wise, 1988; Kiyatkin, 1995; Post, Weiss, & Pert, 1988; Robbins & Everitt, 1996; Salamone, 1996; Schultz et al., 1997; Wise, 1988, 1996), more central traits of romantic love. The addictive behaviors associated with romantic love are most likely related to dopamine activity as well (Fisher, 2004), because acute cocaine injection has been shown to activate the VTA in fMRI studies of humans (Breiter et al., 1997); animal studies of cocaine addiction also implicate mesolimbic dopamine pathways (David, Segu, Buhot, Ichaye, & Cazala, 2004; Kalivas & Duffy, 1998; McBride, Murphy, & Ikemoto, 1999; Wise & Hoffman, 1992).

Norepinephrine also may be associated with human romantic love (Fisher, 1998), although this has not yet been recorded by neuroimaging. Increased activity of norepinephrine generally produces alertness, energy, sleeplessness, loss of appetite (Coull, 1998; Robbins et al., 1998), and increased attention (Marracco & Davidson, 1996; Posner & Petersen, 1990), some of the basic characteristics of romantic love (Fisher, 2004; Hatfield & Sprecher, 1986; Tennov, 1979). Elevated activity of central norepinephrine also increases memory for new stimuli (Griffin & Taylor, 1995), so this neurotransmitter may also contribute to the lover's ability to remember the smallest details of the beloved's actions and cherished moments spent together. Because norepinephrine is also associated with sympathetic nervous system responses, including increased heart rate and blood pressure, and these responses often occur in early stage, intense romantic love, norepinephrine may contribute to these aspects of romantic love as well.

Low activity of central serotonin also may be involved in feelings of intense romantic love (Fisher, 1998; Marazziti, Akiskal, Rossi, &

Cassano, 1999). This is hypothesized because a striking sympton of romantic love is incessant, obsessive thinking about the beloved (Fisher, 1998, 2004; Hatfield & Sprecher, 1986; Tennov, 1979), and low activity of central serotonin is associated with obsessive-compulsive disorder (OCD) (Insel, Mueller et al., 1985; Insel, Zohar et al., 1990). In fact, most forms of OCD are treated with antidepressants that elevate the activity of central serotonin (Flament, Rapoport, & Berg, 1985; Hollander et al., 1988; Thoren, Asberg, & Bertilsson, 1980).

A recent study supports the hypothesis that romantic love is associated with low levels of central serotonin. In this experiment, 20 men and women who had fallen in love in the previous 6 months, 20 patients with unmedicated OCD, and 20 normal (control) individuals who were not in love were all tested for plasma levels of serotonin (Marazziti et al., 1999). Both the in-love participants and those with OCD showed significantly lower concentrations of the platelet serotonin transporter (Marazziti et al., 1999). Although bodily activities of serotonin do not necessarily correlate with serotonin activities in the brain (Kendrick, Keverne, Baldwin, & Sharman, 1986), decreased activity of central serotonin may contribute to the lover's obsessive thinking. Because impulsivity is also associated with low activity of central serotonin (Tiihonen et al., 1997), decreased activity of this neurotransmitter may also produce the impulsivity associated with romantic love.

These data suggest that the constellation of neural correlates associated with romantic love are largely distinct from those of the sex drive. Moreover, both neural systems are fundamental human drives (Fisher, 2004).

The Drive to Love

Psychologists distinguish between emotions, affective states of feeling, and motivations, brain systems oriented around the planning and pursuit of a specific want or need; and Aron has proposed that romantic love is not primarily an emotion but a motivation system designed to enable suitors to build and maintain an intimate relationship with a preferred mating partner (Aron & Aron, 1991; Aron, Paris, & Aron, 1995). Because the experiments described in the previous section indicate that romantic love is associated with activity in the VTA and caudate nucleus, Aron's hypothesis is supported: motivation and goal-oriented behaviors are central to the experience of intense, early-stage romantic love. These data suggest that romantic love is a *primary* motivation system, a fundamental human mating drive (Fisher, 2004).

Pfaff defines a drive as a neural state that energizes and directs behavior to acquire a particular biological need to survive or reproduce (Pfaff, 1999). Like drives, romantic love is tenacious; emotions come and go. Like drives, romantic love is focused on a specific reward, in this case the beloved; emotions, such as fear, are associated with a wider range of objects and ideas. Like drives, romantic love is not associated with any particular facial expression; all of the primary emotions have stereotypic facial poses. Like drives, romantic love is difficult to control; it is harder to curb thirst, for example, than to control anger. Finally, like all of the basic drives (Pfaff, 1999), romantic love is associated with elevated activity in the dopaminergic reward system in the brain.

Drives lie along a continuum (Fisher, 2004). Some, like thirst and the need for warmth, cannot be extinguished until satisfied. The sex drive, hunger, the craving for salt, and the maternal instinct can often be redirected, even quelled. Falling in love is evidently stronger than the sex drive because when one's sexual advances are rejected, people do not kill themselves or someone else, whereas rejected lovers sometimes commit suicide or homicide (Meloy & Fisher, in press).

Mammalian Courtship Attraction

Not only are romantic love and the sex drive distinct neural systems, but evidence suggests that they may have been distinct since the proliferation of mammalian species some 70 million years ago. All mammals have mate preferences; none will copulate with *any* conspecific (Fisher, Aron, Masher, Strong et al., 2002). The drive to pursue a *preferred* mating partner is so common that the ethological literature regularly uses several terms to describe it, including "mate choice," "female choice," "individual preference," "favoritism," "sexual choice," and "selective proceptivity" (Andersson, 1994).

This mate preference in mammals, referred to as *courtship attraction*, is associated with many of the same characteristics as human romantic love, including heightened energy, focused attention, obsessive following, sleeplessness, loss of appetite, possessive "mate guarding," affiliative gestures, goal-oriented courtship behaviors, and intense motivation to win a specific mating partner (Fisher, 2004). Moreover, animal studies indicate that elevated activities of dopaminergic reward pathways play a primary role in mammalian mate preference, data that correlate with the previously presented evidence for the role of dopaminergic pathways in human romantic love.

For example, when a female laboratory-maintained prairie vole (*Microtus ochrogaster*) is mated with a male, she forms a distinct preference for him, associated with a 50% increase of dopamine in the nucleus accumbens (Gingrich, Liu, Cascio, & Insel, 2000). When a dopamine antagonist is injected into the accumbens, the female no longer prefers this partner; and when a female is injected with a dopamine agonist, she begins to prefer the conspecific who is present at the time of infusion, even if she has not mated with this male (Gingrich et al., 2000; Liu & Wang, 2003; Wang et al., 1999). An increase in central dopamine is associated with courtship attraction in female sheep (Fabre-Nys et al., 1998). In male rats, increased striatal dopamine release has also been shown in response to the presence of a receptive female rat (Montague et al., 2004; Robinson, Heien, & Wightman, 2002).

In most species, this excitatory state is brief (Fisher, 2004); among humans, romantic love can last 12 months or more (Marazziti, 1999). Nevertheless, mammalian courtship attraction and human romantic love have much in common, including behavior patterns and neural mechanisms. It is parsimonious to hypothesize that the neural correlates of courtship attraction developed into those for human romantic love some time during hominid evolution, perhaps along with the development of the hominid brain some 2 million years ago (Fisher, 2004). Moreover, it is likely that this neural mechanism serves the same purpose in all mammalian species: to enable individuals to discriminate between the courtship displays of an array of suitors, prefer those that advertise superior genes, better resources, or more parental investment, and motivate males and females to focus their courtship attention on these preferred individuals, thereby conserving mating time and energy (Fisher, 2004; Fisher, Aron, Mashek, Strong et al., 2002).

Despite the biological distinctions between romantic love and the sex drive, and despite what is likely their long evolutionary history, the brain systems for the sex drive and romantic love interact in many ways, suggesting that serotonin-enhancing antidepressants can potentially suppress feelings of romantic love.

Interactions Between the Sex Drive and Romantic Love

Men and women in Western societies do not confuse the ecstasy, focused attention, and obsessive thinking associated with romantic love with the mere appetite for sexual release (Hatfield & Rapson, 1996; Tennov, 1979). Men and women in an array of traditional societies also make this distinction (Jankowiak, 1995). On the Polynesian island of Mangaia, "real love" is called *inangaro kino*, a state of romantic passion distinct from one's sexual desires (Harris, 1995). The Taita of Kenya call lust *ashiki*, whereas they refer to love as *pendo* (Bell, 1995). In Caruaru, northeastern Brazil, locals say, "*Amor* is when you feel a desire to always be with her, you breathe her, eat her, drink her, you are always thinking of her, you don't manage to live without her" (Rebhun, 1995, p. 253). *Paixao*, on the other hand, is "horniness," and *tesao* is "a very strong sexual attraction for a person" (Rebhun, 1995, p. 254).

Despite people's ability to distinguish between feelings of passionate romantic love and feelings of sexual desire, those who fall in love regularly begin to find their beloved enormously sexually attractive; sexual desire is a central trait of human romantic love. This positive association between romantic love and the sex drive may be due in part to the biological link between these two brain systems. Dopamine can stimulate a cascade of reactions, including the release of testosterone and estrogen (Hull, Du, Lorrain, & Matuszewich, 1995, 1997; Kawashima & Takagi, 1994; Szezypka, Zhou, & Palmiter, 1998; Wenkstern, Pfaus, & Fibiger, 1993; Wersinger & Rissman, 2000), and the increasing activity of testosterone and estrogen can promote dopamine release (Appararundaram, Huller, Lakhlani, & Jennes, 2002; Auger, Meredith, Snyder, & Blaustein, 2001; Becker, Rudnick et al., 2001; Creutz & Kritzer, 2002; Hull et al., 1999; Pfaff, 2005).

Animal studies confirm this positive correlation between the sex drive and the dopaminergic arousal system. When a male laboratory rat is placed in an adjacent cage where he can see or smell an estrous female, his levels of central dopamine increase and elevate sexual arousal and pursuit of the female (Hull et al., 1995, 1997; Hull, Meisel, & Sachs, 2002; Wenkstern et al., 1993; West, Clancy, & Michael, 1992). When the barrier is removed and the male is allowed to copulate, levels of dopamine continue to rise (Hull et al., 1995). When dopamine is injected into specific regions of the brain in male rats, the infusion stimulates copulatory behavior (Ferrari & Giuliani, 1995). Conversely, blocking the activities of central dopamine in rats diminishes several proceptive sexual behaviors, including hopping and darting (Herbert, 1996).

Pfaff (2005) reports that in male rats, dopamine increases male sexual behavior through at least three functional roles. It increases sexual arousal and courtship behavior, it potentiates the motor acts of mounting, and it faciliates genital responses to stimulation.

This positive correlation between central dopamine, the sex steroids, and sexual arousal and performance is not only common in animals (Herbert, 1996; Liu, Sachs, & Salamone, 1998; Pfaff, 2005); it also occurs in humans (Clayton, McGarvey, Warnock, et al., 2000; Heaton, 2000; Walker, Cole, Gardner, et al., 1993). When individuals who suffer from hypoactive sexual desire disorder are treated with dopamine-enhancing medications, their libido improves (Segraves, Goft, Kavoossi, et al., 2001). When patients with depression take drugs that elevate the activity of dopamine, their sex drive often improves as well (Ascher et al., 1995; Coleman et al., 1999; Walker et al., 1993). In fact, some patients who currently take serotonin-enhancing antidepressants supplement their therapy with medications that elevate the activity of dopamine (and norepinephrine) solely to maintain or elevate sexual arousal (Ascher et al., 1995; Coleman et al., 1999; Rosen et al., 1999; Walker et al., 1993).

Norepineprhine is also positively linked with sexual motivation and sexual arousal (Clayton et al., 2002; Etgen & Morales, 2002; Fraley, 2002; Pfaff, 2005; Van Bockstaele, Pieribone, & Aston-Jones, 1989). When a female prairie vole is exposed to a drop of male urine on the upper lip, norepinephrine is released in parts of the olfactory bulb, contributing to the release of estrogen and concomitant proceptive behavior (Dluzen, Ramirez, Carter, & Getz, 1981), and in rats, estradiol and progesterone result in the release of norepinephrine in the hypothalamus to produce lordosis (Etgen et al., 1999). Last, when ovariectomized, sexually receptive female rats receive injections of estrogen and are then permitted to mate, copulation results in the release of norepinephrine in the lateral ventromedial hypothalamus (Etgen & Morales, 2002).

This positive relationship between norepinephrine and the sex drive may be due in part to its interaction with the androgens. Norepinephrine, like dopamine, stimulates the production of testosterone (Cardinali, Nagle, Gomez, & Rosner, 1975; Fernandez, Vidal, & Dominguez, 1975; Mayerhofer, Steger, Gow, & Bartke, 1992), and increasing levels of testosterone can elevate the activity of norepinephrine (Jones, Dunphy, Milsted, & Ely, 1998) and dopamine (Becker, 2001; Hull et al., 1999; Pfaff, 2005). Drug users attest to this positive chemical connection between norepinephrine and the sex drive. In the right oral dose, amphetamines (norepinephrine agonists) enhance sexual desire (Buffum, Moser, & Smith, 1988).

These data indicate that romantic love is associated with elevated activity of dopamine (and most likely also norepinephrine) in general arousal systems in the brain. Moreover, these catecholamines are positively correlated with sexual motivation and sexual arousal. Most important to this discussion, elevated serotonin activity can directly suppress all pathways for dopamine (Meston & Frohlic, 2000; Stahl, 2000) and norepinephrine (Done & Sharp, 1992), as well as suppress testosterone activity (Gonzalez, Farabollini, Albonetti, & Wilson, 1994; Netter, Hennig, Meier, & Rohrmann, 1998; Sundblad & Eriksson, 1997). Hence, serotonin-enhancing antidepressants that negatively affect the sex drive and sexual arousal are also likely to adversely affect feelings of romantic love.

Case study: A 20-year-old, single, white, female undergraduate patient with an eating disorder, recurrent depressions, and attention-deficit disorder was administered an SSRI at relatively high doses for her eating disorder. When asked about side effects, she said she had none. When asked specifically about sexual side effects, she wasn't certain and asked that they be explained. Once they were explained, she acknowledged that she did have sexual side effects but that she had attributed them to problems in her relationship. "I have not been as much in love with my boyfriend," she reported. "I am not as interested in intimate time with him. I find myself wanting more space." At the time she reported this, the dose of the SSRI had just been increased.

Emotional Blunting and Romantic Love

Serotonin-enhancing medications can also jeopardize feelings of romantic love indirectly, by affecting the emotions. A striking characteristic of romantic love is obsessive thinking about a beloved. As discussed above, this intrusive thinking is most likely associated with a low activity of central serotonin. Hence, individuals taking serotonin-enhancing antidepressants are likely to suppress the obsessive thinking characteristic of romantic love. Elation is another primary feature of romantic love, and individuals who take serotonin-enhancing antidepressants are likely to suppress this ecstasy as well.

Serotonin-enhancing medications are well known to blunt the emotions. An unsolicited letter to *The New York Times* in response to our ideas (Fisher & Thomson, 2004; O'Connor, 2004) illustrates the impact that an SSRI had on Dr. Jerry Frankel, of Plano, Texas:

> After two bouts of depression in 10 years, my therapist recommended I stay on serotonin-enhancing antidepressants indefinitely. As appreciative as I was to have regained my health, I found that my usual enthusiasm for life was replaced with blandness. My romantic feelings for my wife declined drastically. With the approval of my therapist, I gradually discontinued my medication. My enthusiasm returned and our romance is now as strong as ever. I am prepared to deal with another bout of depression if need be, but in my case the long-term side effects of antidepressants render them off limits. (Frankel, 2004)

The Drive to Attach

Love changes over time. The ecstasy, energy, focused attention, obsessive thinking, yearning, and intense motivation to win the beloved gradually diminish, often transforming into feelings of comfort, calm, and emotional union with one's partner. This male-female partner attachment system is characterized in birds and mammals by mutual territory defense and nest building, mutual feeding and grooming, the maintenance of close proximity, separation anxiety, shared parental chores, and other affiliative behaviors. In humans, partner attachment is also characterized by feelings of calm, security, social comfort, and emotional union with a partner. Hatfield refers to this feeling of attachment as "companionate love," defining it as "a feeling of happy togetherness with someone whose life has become deeply entwined with yours" (Hatfield, 1988, p. 191).

Just as men and women distinguish between feelings of romantic love and the sex drive, people distinguish between feelings of romance and those of attachment to a long-term partner. Nisa, a !Kung Bushman woman of the Kalahari Desert, Botswana, explained the feeling of man-woman attachment this way: "When two people are first together, their hearts are on fire and their passion is very great. After a while, the fire cools and that's how it stays. They continue to love each other, but it's in a different way—warm and dependable" (Shostak, 1981, p. 268). The Taita of Kenya report that love comes in two forms, an irresistible longing, a "kind of sickness," and a deep, enduring affection for another (Bell, 1995, p. 158). Brazilians have a poetic proverb that distinguishes between these feelings: "Love is born in a glance and matures in a smile"

(Rebhun, 1995, p. 252). For Koreans, *sarang* is a word close to the Western concept of romantic love, while *chong* is more like feelings of long-term attachment. Abigail Adams described these feelings, writing to John Adams in 1793, "Years subdue the ardor of passion, but in lieu thereof friendship and affection deep-rooted subsists, which defies the ravages of time, and whilst the vital flame exists" (McCullough, 2001).

Bowlby (1969, 1973) and Ainsworth, Blehar, Waters, and Wall (1978) proposed that, to promote survival of the young, primates have evolved an innate attachment system designed to motivate infants to seek comfort and safety from their primary caregiver, generally their mother. More recently, researchers have emphasized that this attachment system remains active throughout life and serves as a foundation for attachment between spouses as they raise children (Hazan & Diamond, 2000; Hazan & Shaver, 1987).

This parental attachment system has been associated with the activity of two neuropeptides, oxytocin in the nucleus accumbens and arginine vasopressin in the ventral pallidum (Carter, 1992; Lim, Murphy, et al., 2004; Lim & Young, 2004; Wang, Ferris, & De Vries, 1994; Winslow et al., 1993; Young et al., 1998), although the brain's opioid system (Moles, Kieffer, & D'Amato, 2004) and other neural systems are most likely also involved (Kendrick, 2000). When vasopressin was injected intracerebroventricularly into virgin, laboratory-raised male prairie voles, they began to defend the space around them from other males, an aspect of pair formation among prairie voles. When each was introduced to a female, he became instantly possessive of her as well (Wang et al., 1994; Winslow et al., 1993). Moreover, arginine vasopressin antagonists infused into the ventral pallidum prevented partner preference formation among male prairie voles, suggesting that V1a receptor activation in this region is necessary for their pair-bond formation (Lim & Young, 2004, p. 1).

This distinct distribution of vasopressin receptors in the ventral forebrain seen in monogamous male prairie voles is also seen in monogamous California mice and monogamous marmoset monkeys, whereas promiscuous white-footed mice and promiscuous rhesus monkeys do not express this distribution of V1a receptors in the ventral pallidum (Bester-Meredith, Young, & Marler, 1999; Wang et al., 1997; Young, 1999; Young, Winslow, Nilsen, & Insel, 1997), further suggesting that vasopressin activity in this region of the brain's reward system is directly associated with pair bonding and attachment behaviors (Lim, Murphy, et al., 2004).

Oxytocin also stimulates the bonding process between a mother and her offspring (Carter, 1992; Pedersen, Caldwell, Walker, Ayers, & Mason, 1994) and between mating partners (Lim, Murphy, et al., 2004). When oxytocin is administered intracerebroventricularly, ovariectomized female prairie voles preferred the partner that was present at the time of infusion and formed a pair bond with him (Williams, Insel, Harbaugh, & Carter, 1994). When an oxytocin receptor antagonist is infused directly into the nucleus accumbens of a female prairie vole, it blocks partner preference and pair-bond formation (Lim, Murphy, et al., 2004; Young, Lim, Gingrich, & Insel, 2001).

A specific gene also has been associated with attachment behaviors and pair bonding. When this gene was manipulated to increase V1a receptors in the ventral pallidum, male prairie voles with increased V1aR expression exhibited heightened levels of social affiliation, formed a preference for a specific female, and began to cohabit with her, even though they had not mated with her (Pitkow et al., 2001). When Lim and colleagues introduced this gene into a male meadow vole (a promiscuous species), vasopressin receptors upregulated and the vole began to fixate on a particular female and mate exclusively with her, even though other females were available (Lim, Wang, et al., 2004).

Oxytocin and vasopressin appear to be associated with both partner preference and attachment/pair bonding, whereas dopamine and perhaps other monoamines are related only to partner preference. Thus, Young maintains that when monogamous prairie voles and individuals of other monogamous species engage in sex, they trigger the activity of vasopressin and oxytocin in specific reward centers of the brain; then dopamine in these reward centers enable males and females to prefer their current mating partner, thereby initiating attachment and pair bonding (Lim, Murphy, et al., 2004). Moreover, males of promiscuous species, which lack one link in this chain (V1a receptors in the ventral pallidum), may feel attraction to but do not associate this pleasurable feeling with a specific female and do not initiate an attachment to her.

Data from the Demographic Yearbooks of the United Nations on 97 societies suggest the prevalence of this attachment system in humans: approximately 93.1% of women and 91.8% of men marry by age 49 (Fisher, 1992). Moreover, when Fisher and colleagues examined a subset of their fMRI subjects who were in longer relationships, specifically those who were in love between 8 and 17 months, they found activation in the ventral pallidum, the brain region where activity has been linked with

to motivate ancestral females to copulate with multiple partners, thereby confusing the identity of the biological father of a forthcoming child and obliging each male to contribute to the survival of the infant (Hrdy, 1999). The sperm retention hypothesis proposes that female orgasm evolved to transport sperm through the cervix, enhancing the probability of conception (Fox, Wolfs, & Baker, 1970).

The above data and theories suggest that female orgasm is a multipurpose mechanism designed to promote pair bonding with appropriate males, promote "extra pair copulations" to increase female fecundity, and enable a single woman to identify and win the best possible partner when she seeks a new relationship. All of these functions of female orgasm are jeopardized by serotonin-enhancing antidepressants.

Chemical Clitoridectomy

Women who take serotonin-enhancing antidepressants also disrupt related evolutionary mechanisms for mate selection, pair formation, and pair maintenance. The ring of nerves around the vaginal opening measures penis width and, by distending surrounding muscles, elevates sexual excitement. The clitoris also responds to minor variations in touch and angle, thereby measuring a partner's skill, patience, determination, and sensitivity to her needs (Miller, 2000). By creating a chemical clitoridectomy, serotonin-enhancing antidepressants dull the responses of these devices (Frolich & Meston, 2000), contribute to anorgasmia, and diminish a woman's ability to discern appropriate mating and marital partners. Anorgasmia may also motivate a woman to look beyond her primary relationship, even though this male may have superior genes, resources, and parenting capabilities (Small, 1995).

Serotonin-enhancing antidepressants may affect other subtle female mechanisms for courtship, mating, and reproduction. At midcycle, ovulating women tend to have more erotic fantasies, initiate more sexual activity, and experience a lower threshold for orgasm. They have a better sense of smell (Doty, 1986) and are better able to discriminate healthy from unhealthy available males. At midcycle, women are also more likely to prefer men with higher bodily and facial symmetry and men who are creative, humorous, and display other signs of good genes (Grammer et al., 2003; Miller, 2000; Thornhill et al., 1995). Attraction to individuals with MHC histocompatibility or other immunological profiles may be linked to sex drive and sexual arousal, too. These and many other courtship mechanisms evolved to aid mate assessment, mate choice, and pair formation, and any and all of these brain responses could potentially be altered by serotonin-enhancing antidepressants.

Like drugs that blur vision, serotonin-enhancing medications may impair myriad female adaptive mechanisms, obscuring a woman's ability to make appropriate mating choices, fall in love, or sustain appropriate long-term reproductive relationships.

Penile Erection, Seminal Fluid, and Antidepressants

Men who take serotonin-enhancing antidepressants also inhibit an array of adaptive mechanisms that evolved to promote mate selection and partnership formation. For example, the penis may function as an internal courtship device (Miller, 2000). With its width, length, and turgidity, it stimulates the vagina to give pleasure; it also advertises psychological and physical fitness (Miller, 2000). When men take antidepressants that produce impotence, they cripple these courtship functions.

The penis also deposits seminal fluid, which contains dopamine and norepinephrine, as well as tyrosine, a building block of these catecholamines (Burch & Gallup, in press). These compounds do not pass through the blood-brain barrier. Nevertheless, when a man taking a serotonin-enhancing antidepressant fails to ejaculate, he fails to deposit these catecholamines in the vaginal tract, neurotransmitters that could contribute to his partner's feelings of romantic attraction to him.

Seminal fluid also contains several other mood-altering hormones, including testosterone, estrogen, follicle-stimulating hormone (FSH), and luteinizing hormone (LH), chemicals that can also affect sexual desire and function (Clayton, 2003). Gallup and colleagues have demonstrated that these and other chemicals in seminal fluid have antidepressant effects on women (Gallup, Burch, & Platek, 2002). When a man fails to ejaculate, he suppresses his ability to stimulate in his partner a positive mood that could potentially change her threshold for romantic attraction or deep attachment to him.

SSRIs and Psychological Barriers to Romance and Marriage

Serotonin-induced sexual dysfunction can adversely affect feelings of romantic love and partner attachment in psychological ways as well. For example, some men and women taking these medications shy away from a liaison that could become romantic because they are afraid of their own poor performance in bed.

Case study: A 26-year-old man had panic attacks that required high doses of a serotonin-enhancing antidepressant. He soon experienced diminished libido and impotence. A handsome, personable, intelligent man, he was readily sought after by women. However, he ended several relationships because he was too embarrassed about his inability to perform sexually. Although he tried several other medications, he was able to control his panic disorder only with high doses of serotonin enhancers. He eventually retreated into a social life in which he avoided serious dating. When last evaluated, he still confined himself to nonsexual relationships with women.

Due to low libido, other patients on serotonin-enhancing antidepressants fail to become sexually attracted to a potential partner and incorrectly attribute their lack of sexual (and romantic) interest to personality deficits in this potential mate, thereby misappraising the viability of the relationship.

Still others fail to notice potential partners.

Case study: A patient in her late twenties had recurrent major depressions that were being controlled with an SSRI. She reported sexual side effects, including diminished sexual interest and absent orgasm. However, 3–4 weeks after the SSRI medication was reduced and an antidepressant with fewer sexual side effects was added, she noticed an increase in her sexual interest. When asked if she had noticed any change in her feelings of attraction to men, she said, "I notice someone who is attractive now which I hadn't before."

SSRIs and Fertility

SSRI medications can also influence one's genetic future.

Case study: A 35-year-old married woman with recurrent depression and generalized anxiety disorder was placed on an SSRI. She was not told about the potential negative sexual side effects of this medication. The drug relieved her depression and anxiety. However, she soon developed diminished libido and absent orgasm. This led her to conclude that she no longer loved her husband. She decided to divorce him but kept her feelings to herself for several years, planning for the appropriate time to make this major life change. She eventually switched to an antidepressant with a low frequency of sexual side effects. On this new medication, her sexual desire and orgasmic function returned. She decided not to divorce her spouse. Soon after this, she conceived. Now she and her husband have a child. A serotonin-enhancing medication had affected not only her social life but her fertility.

These medications can also influence one's genetic future in specific biological ways. Serotonin increases prolactin levels by inhibiting dopamine activity and stimulating prolactin-releasing factors. Prolactin can impair fertility through several mechanisms, including suppressing hypothalamic gonadotropin-releasing hormone release, suppressing pituitary FSH and LH release, and suppressing ovarian hormone production (Hendrick, Gitlin, Altshuler, & Korenman, 2000). Also, clomipramine, a strong serotonin-enhancing antidepressant, adversely affects sperm volume and motility (Maier & Koinig, 1994).

The number and range of unconscious psychobiological mechanisms that have evolved to enable men and women to signal mating fitness, assess appropriate mating partners, pursue specific preferred individuals, and form and sustain a pair bond are largely unknown. But it is likely that many of these neural mechanisms are altered by serotonin-enhancing medications.

Conclusion

Homo sapiens has inherited three distinct yet interrelated brain systems for courtship, mating, reproduction and parenting: the sex drive, romantic love, and partner attachment. These neural systems can become active in any sequence. An individual may begin a casual sexual liaison with someone for whom he or she feels only sexual desire, then one evening falls in love with the sex partner, then gradually begins to feel deep attachment to this partner. Some couples begin their relationship with feelings of attachment instead: the man and woman become friends and achieve emotional union in the college dorm, at the office, or in their social circle. With time, this attachment metamorphoses into romantic passion, which then triggers lust. Still others fall in love with someone they hardly know, then they experience lust, and finally they experience feelings of attachment. These three neural systems can also operate independently. An individual can feel deep attachment for a long-term spouse *while* they feel romantic passion for someone else *while* they feel the sex drive for an array of other individuals.

The flexible nature of these three brain mechanisms for reproduction and their complex, dynamic interactions suggest that any medication that changes the chemical checks and balances is likely to alter an individual's courting, mating, and parenting tactics, ultimately affecting that person's fertility and genetic future.

Serotonin is the oldest known monoamine neurotransmitter; it has numerous receptors and many subtle functions. For example, activation of serotonin type 1a (5-HT1a) receptors enhances sexual desire and lowers the threshold for ejaculation; activation of serotonin type 1b (5-HT1b) and 1c (5-HT1c) receptors decreases sexual desire and inhibits orgasm; and activation of serotonin type 2 (5-HT2) and type 3 (5-HT3) receptors impairs all stages of sexual response in both men and women (Meston & Frolich, 2000). Some 90% of these serotonin receptors are located in the body, where serotonin affects the smooth muscle of the vascular system, including the smooth muscle of the genitals.

Individuals vary in the sensitivity of these serotonin receptors (Saks, 2000), as well as in many other aspects of serotonin production, synthesis, and interaction with other bodily systems. Childhood experiences and current circumstances also affect the expression of this monoamine neurotransmitter. Thus, individuals taking serotonin-enhancing antidepressants vary in their response to these medications, including their sexual side effects. In fact, data indicate that under the right circumstances, serotonin-enhancing antidepressants can considerably improve several mental and physical disorders, including disorders that affect one's romantic and marital relationships.

Nevertheless, the Food and Drug Administration has warned Americans that these medications can have potentially harmful side effects, including severe restlessness, anxiety, hostility, insomnia, and/or suicidal thinking, as well as emotional blunting and sexual dysfunction.

Because there is a positive relationship between dopamine (associated with romantic love) and testosterone (linked to sexual desire and arousal) and because there is a negative relationship between serotonin and these catecholamines and the androgens, serotonin-enhancing antidepressants can also inhibit feelings of romantic love. Moreover, because serotonin-enhancing antidepressants have a negative impact on penile erection, sexual arousal, orgasm, and other evolved psychobiological courtship mechanisms, these drugs can also negatively affect one's ability to signal genetic and psychological fitness, assess and select potential mating partners, pursue preferred individuals, and maintain stable pair bonds.

Harvard Medical School psychiatrist Joseph Glenmullen estimates that 75% of all patients on antidepressants, largely SSRIs, are "needlessly on these drugs"(cited in Morais, 2004, p. 120). Physicians who prescribe serotonin-enhancing antidepressants and individuals who plan to use these drugs should bear in mind the broad, largely unrecognized, and possibly deleterious effects of these medications.

References

Abbott, A. (2002). Addicted. *Nature, 419*(6910), 872–874.

Ainsworth, M. D. S., Blehar, M. C., Waters, E., & Wall, S. (1978). *Patterns of attachment: A psychological study of the strange situation.* Hillsdale, NJ: Erlbaum.

Andersson, M. (1994). *Sexual selection.* Princeton, NJ: Princeton University Press.

Appararundaram, S., Huller, J., Lakhlani, S., & Jennes, L. (2002). Ovariectomy-induced alterations of choline and dopamine transporter activity in the rat brain. *Society for Neuroscience Abstracts*, abstn. 368.20.

Arnow, B. A., Desmond, J. E., Banner, L. L., Glover, G. H., Solomon, A., Polan, M. L., et al. (2002). Brain activation and sexual arousal in healthy, heterosexual males. *Brain, 125*(pt. 5), 1014–1023.

Aron, A., & Aron, E. (1991). Love and sexuality. In K. McKinney & S. Sprecher (Eds.), *Sexuality in close relationships.* Hillsdale, NJ: Erlbaum.

Aron, A., Fisher, H. E., Mashek, D. J., Strong, G., Li, H. F., & Brown, L. L. (2005). Reward, motivation and emotion systems associated with early-stage intense romantic love: An fMRI study. *Journal of Neurophysiology, 94*, 327–337.

Aron, A., Paris, M., & Aron, E. N. (1995). Falling in love: Prospective studies of self-concept change. *Journal of Personality and Social Psychology, 69*, 1102–1112.

Arsenijevic, Y., & Tribollet, E. (1998). Region-specific effect of testosterone on oxytocin receptor binding in the brain of the aged rat. *Brain Research, 785*(1), 167–170.

Ascher, J. A., Cole, J. O., Colin, J. N., Feighner, J. P., Ferris, R. M., Fibiger, H. C., et al. (1995). Bupropion: A review of its mechanism of antidepressant activity. *Journal of Clinical Psychiatry, 56*(9), 396–402.

Auger, A. P., Meredith, J. M., Snyder, G. L., & Blaustein, J. D. (2001). Oestradiol increases phosphorylation of a dopamine- and cyclic AMP-regulated phosphoprotein (DARPP-32) in female rat brain. *Journal of Neuroendocrinology, 13*(9), 761–768.

Bartels, A., & Zeki, S. (2000). The neural basis of romantic love. *Neuroreport, 11*, 1–6.

Bartels, A., & Zeki, S. (2004). The neural correlates of maternal and romantic love. *NeuroImage*, *21*, 1155–1166.

Beauregard, M., Levesque, J., & Bourgouin, P. (2001). Neural correlates of conscious self-regulation of emotion. *Journal of Neuroscience*, *21*(18), RC165.

Becker, J. B., Rudick, C. N., et al. (2001). The role of dopamine in the nucleus accumbens and striatum during sexual behavior in the female rat. *Journal of Neuroscience*, *21*(9), 3236–3241.

Bell, J. (1995). Notions of love and romance among the Taita of Kenya. In W. Jankowiak (Ed.), *Romantic passion: A universal experience?* New York: Columbia University Press.

Berridge, C. W., Stratford, T. L., Foote, S. L., & Kelley, A. E. (1997). Distribution of dopamine beta-hydroxylase-like immunoreactive fibers within the shell subregion of the nucleus accumbens. *Synapse*, *27*(3), 230–241.

Bester-Meredith, J. K., Young, L. J., & Marler, C. A. (1999). Species differences in paternal behavior and aggression in *Peromyscus* and their associations with vasopressin immunoreactivity and receptors. *Hormones and Behavior*, *36*, 25–38.

Booth, A., & Dabbs, J. M. (1993). Testosterone and men's marriages. *Social Forces*, *72*(2), 463–477.

Bowlby, J. (1969). *Attachment and loss: Vol. 1. Attachment.* New York: Basic Books

Bowlby, J. (1973). *Attachment and loss: Vol. 2. Separation.* New York: Basic Books.

Breiter, H. C., Gollub, R. L., Weisskoff, R. M., Kennedy, D. N., Makris, N., Berke, J. D. (1997). Acute effects of cocaine on human brain activity and emotion. *Neuron*, *19*, 591–611.

Breiter, H. C., Aharon, I., Kahneman, D., Dale, A., & Shizgal, P. (2001). Functional imaging of neural responses to expectancy and experience of monetary gains and losses. *Neuron*, *30*, 619–639.

Buffum, J., Moser, C., & Smith, D. (1988). Street drugs and sexual function. In J. M. A. Sitsen (Vol. Ed.), *Handbook of sexology: Vol 6. The pharmacology and endocrinology of sexual function*. New York: Elsevier.

Burch, R. L., & Gallup, G. G., Jr. (in press). The psychobiology of human semen. In S. Platek & T. Shackelford (Eds.), *Female infidelity and paternal uncertainty*. Cambridge: Cambridge University Press.

Buss, D. M. (2003). *The evolution of desire: Strategies of human mating*. Rev. ed. New York: Basic Books.

Cardinali, D. P., Nagle, C. A., Gomez, E., & Rosner, J. M. (1975). Norepinephrine turnover in the rat pineal gland: Acceleration by estradiol and testosterone. *Life Sciences*, *16*, 1717–1724.

Carmichael, M. S., Humbert, R., Dixen, J., Palmisano, G., Greenleaf, W., & Davidson, J. M. (1987). Plasma oxytocin increases in the human sexual response. *Journal of Clinical Endocrinology and Metabolism*, *64*(1), 27–31.

Carter, C. S. (1992). Oxytocin and sexual behavior. *Neuroscience and Biobehavioral Reviews*, *1*(16), 131–144.

Carter, C. S., DeVries, A., Taymans, S. E., Roberts, R. L., Williams, J. R., & Getz, L. L. (1997). Peptides, steroids, and pair bonding. In C. S. Carter, I. I. Lederhendler, & B. Kirkpatrick (Eds.), *The integrative neurobiology of affiliation. Annals of the New York Academy of Sciences*, *807*, 260–272.

Clayton, A. H. (2003). Sexual function and dysfunction in women. *Psychiatric Clinics of North America*, *26*, 673–682.

Clayton, A. H., McGarvey, E. D., Warnock, J., et al. (2000). *Bupropion as an antidote to SSRI-induced sexual dysfunction.* Poster session presented at the New Clinical Drug Evaluation Unit Program, Boca Raton, FL.

Clayton, A. H., Pradko, J. F., Croft, H. A., Montano, B., et al. (2002). Prevalence of sexual dysfunction among newer antidepressants. *Journal of Clinical Psychiatry*, *63*, 357–366.

Coleman, C. C., Cunningham, L. A., Foster, V. J., Batey, S. R., Donahue, R. M. J., Houser, T. L., et al. (1999). Sexual dysfunction associated with the treatment of depression: A placebo-controlled comparison of buproprion sustained release and sertraline treatment. *Annals of Clinical Psychiatry*, *11*, 205–215.

Colle, L. M., & Wise, R. A. (1988). Facilitory and inhibitory effects of nucleus accumbens amphetamine on feeding. In P. W. Kalivas, & C. B. Nemeroff (Eds.), *The mesocorticolimbic dopamine system. Annals of the New York Academy of Sciences*, *537*, 491–492.

Coull, J. (1998). Neural correlates of attention and arousal: Insights from electrophysiology, functional neuroimaging and psychopharmacology. *Progress in Neurobiology*, *55*, 343–361.

Curtis, J. T., & Wang, Z. (2003). Forebrain c-*fos* expression under conditions conducive to pair bonding in female prairie voles (*Microtus ochrogaster*). *Physiology and Behavior*, *80*, 95–101.

Creutz, L. M., & Kritzer, M. F. (2002). Estrogen receptor-beta immunoreactivity in the midbrain of adult rats: Regional, subregional, and cellular localization in the A10, A9, and A8 dopamine cell groups. *Journal of Comparative Neurology*, *446*(3), 288–300.

David, V., Segu, L., Buhot, M. C., Ichaye, M., & Cazala, P. (2004). Rewarding effects elicited by cocaine microinjections into the ventral tegmental area of C57BL/6 mice: Involvement of dopamine D(1) and serotonin (1B) receptors. *Psychopharmacology (Berlin)*, *174*, 367–375.

Davidson, R. J. (1994). Complexities in the search for emotion-specific physiology. In P. Ekman & R. J. Davidson (Eds.), *The nature of emotion: Fundamental questions* (pp. 237–242). New York: Oxford University Press.

De Ridder, Pinxten, E., & Eens, M. (2000). Experimental evidence of a testosterone-induced shift from paternal to mating behaviour in a facultatively polygynous songbird. *Behavioral Ecology and Sociobiology*, *49*(1), 24–30.

Delgado, M. R., Nystrom, L. E., Fissel, C., Noll, D. C., & Fiez, J. A. (2000). Tracking the hemodynamic responses to reward and punishment in the striatum. *Journal of Neurophysiology*, *84*, 3072–3077.

Delville, Y., & Ferris, C. F. (1995). Sexual differences in vasopressin receptor binding within the ventrolateral hypothalamus in golden hamsters. *Brain Research*, *68*(1), 91–96.

Delville, Y., Mansour, K. M., & Ferris, C. F. (1996). Testosterone facilitates aggression by modulating vasopressin receptors in the hypothalamus. *Physiology and Behavior*, *60*(1), 25–29.

Dixson, A. F. (1998). *Primate sexuality.* Oxford: Oxford University Press.

Dluzen, D. E., Ramirez, V. D., Carter, C. S., & Getz, L. L. (1981). Male vole urine changes luteinizing hormone-releasing hormone and norepinephrine in female olfactory bulb. *Science*, *212*, 573–575.

Done, C. J., & Sharp, T. (1992). Evidence that 5-HT2 receptor activation decreased noradrenaline release in rat hippocampus in vivo. *British Journal of Pharmacology*, *107*, 240–245.

Doty, R. L. (1986). Gender and endocrine-related influences on human olfactory perception. In H. L. Meiselman & R. S. Ravlin (Eds.), *Clinical measurement of taste and smell* (pp. 377–413). New York: Macmillan.

Edwards, J. N., & Booth, A. (1994). Sexuality, marriage, and well-being: The middle years. In A. S. Rossi (Ed.), *Sexuality across the life course.* Chicago: University of Chicago Press.

Elliott, R., Newman, J. L., Longe, O. A., & Deakin, J. F. W. (2003). Differential response patterns in the striatum and orbitofrontal cortex to financial reward in humans: A parametric functional magnetic resonance imaging study. *Journal of Neuroscience*, *23*(1), 303–307.

Etgen, A. M. (2002). Estrogen regulation of neurotransmitter and growth factor signaling in the brain. In D. W. Pfaff, A. Arnold, A. Etgen, S. Fahrbach, & R. Rubin (Eds.), *Hormones, brain and behavior* (pp. 381–440). New York: Academic Press.

Etgen, A. M., Chu, H. P., Fiber, J. M., Karkanias, G. B., & Morales, J. M. (1999). Hormonal integration of neurochemical and sensory signals governing female reproductive behavior. *Behavioral Brain Research*, *105*(1), 93–103.

Etgen, A., & Morales, J. C. (2002). Somatosensory stimuli evoke norepinephrine release in the anterior ventromedial hypothalamus of sexually receptive female rats. *Journal of Neuroendocrinology*, *14*(3), 213–218.

Fabre-Nys, C., et al. (1997). Male faces and odors evoke differential patterns of neurochemical release in the mediobasal hypothalamus of the ewe during estrus: An insight into sexual motivation. *European Journal of Neuroscience*, *9*, 1666–1677.

Fabre-Nys, C. (1998). Steroid control of monoamines in relation to sexual behaviour. *Reviews of Reproduction*, *3*(1), 31–41.

Fernandez, B. E., Vidal, N. A., & Dominguez, A. E. (1975). Action of the sexual hormones on the endogenous norepinephrine of the central nervous system. *Revista Espanola de Fisiologia*, *31*(4), 305–307.

Ferrari, F., & Giuliani, D. (1995). Sexual attraction and copulation in male rats: Effects of the dopamine agonist SND 919. *Pharmacology Biochemistry and Behavior*, *50*(1), 29–34.

Ferris, C. F., & Deville, Y. (1994). Vasopressin and serotonin interactions in the control of agonistic behavior. *Psychoneuroendocrinology*, *19*(5–7), 593–601.

Fiorillo, C. D., Tobler, P. N., & Schultz, W. (2003). Discrete coding of reward probability and uncertainty by dopamine neurons. *Science*, *299*, 1898–1901.

Fisher, H. E. (1992). *Anatomy of love: The natural history of monogamy, adultery and divorce.* New York: Norton.

Fisher, H. E. (1998). Lust, attraction, and attachment in mammalian reproduction. *Human Nature*, *9*(1), 23–52.

Fisher, H. E. (1999). *The first sex: The natural talents of women and how they are changing the world.* New York: Random House.

Fisher, H. E. (2004). *Why we love: The nature and chemistry of romantic love.* New York: Henry Holt.

Fisher, H. E., Aron, A., Mashek, D., Li, H., Strong, G., & Brown, L. L. (2002). The neural mechanisms of mate choice: A hypothesis. *Neuroendocrinology Letters, Supplement 4*, *23*, 92–97.

Fisher, H. E., Aron, A., Mashek, D., Strong, G., Li, H., & Brown, L. L. (2002). Defining the brain systems of lust, romantic attraction and attachment. *Archives of Sexual Behavior*, *31*(5), 413–419.

Fisher, H. E., Aron, A., Mashek, D., Strong, G., Li, H., & Brown, L. L. (2003, January 11). *Early stage intense romantic love activates cortical-basal-ganglia reward/motivation, emotion and attention systems: An fMRI study of a dynamic network that varies with relationship length, passion intensity and gender.* Poster session presented at the annual meeting of the Society for Neuroscience, New Orleans.

Fisher, H., Aron, A., Mashek, D., Strong, G., Li, H., & Brown, L. L. (2005, November 15). *Motivation and emotion systems associated with romantic love following rejection: An fMRI study.* Poster session presented at the annual meeting of the Society for Neuroscience, Washington, DC.

Fisher, H. E., & Thomson, J. A., Jr. (2004, May 1). Do the sexual side effects of most antidepressants jeopardize romantic love and marriage? In *Sex, sexuality, and serotonin.* Symposium conducted at the annual meeting of the American Psychiatric Association.

Flament, M. F., Rapoport, J. L., & Berg, C. L. (1985). Clomipramine treatment of childhood obsessive-compulsive disorder: A double-blind controlled study. *Archives of General Psychiatry*, *42*, 977–986.

Fox, C. A., Wolfs, H. S., & Baker, J. A. (1970). Measurement of intra-vagina and intra-uterine pressures during human coitus by radio-telementry. *Journal of Reproduction and Fertility*, *22*, 243–251.

Fraley, G. S. (2002). Immunolesion of hindbrain catecholaminergic projections to the medial hypothalamus attenuates penile reflexive erections and alters hypothalamic peptide mRNA. *Journal of Neuroendocrinology*, *14*(5), 345–348.

Frankel, J. (2004, May 11). Reviving romance. *The New York Times*, p. F4, Letters.

Frohlich, P. F., & Meston, C. M. (2000). Evidence that serotonin affects female sexual functioning via peripheral mechanisms. *Physiology and Behavior*, *71*, 383–393.

Galfi, M., Janaky, T., Toth, R., Prohaszka, G., Juhasz, A., Varga, C., et al. (2001). Effects of dopamine and dopamine-active compounds on oxytocin and vasopressin production in rat neurohypophyseal tissue cultures. *Regulatory Peptides*, *98*(1–2), 49–54.

Gallup, G. G., Jr., Burch, R. L., & Platek, S. M. (2002). Does semen have antidepressant properties? *Archives of Sexual Behavior*, *13*(26), 289–293.

Gangestad, S. W., & Thornhill, R. (1997). The evolutionary psychology of extrapair sex: The role of fluctuating asymmetry. *Evolution and Human Behavior*, *18*(2), 69–88.

Gibbs, R. B. (2000). Effects of gonadal hormone replacement on measures of basal forebrain cholinergic function. *Neuroscience*, *101*(4), 931–938.

Gingrich, B., Liu, Y., Cascio, C. Z., & Insel, T. R. (2000). Dopamine D2 receptors in the nucleus accumbens are important for social attachment in female prairie voles (*Microtus ochrogaster*). *Behavioral Neuroscience*, *114*, 173–183.

Ginsberg, S. D., Hof, P. R., Young, W. G., & Morrison, J. H. (1994). Noradrenergic innervation of vasopressin- and oxytocin-containing neurons in the hypothalamic paraventricular nucleus of the macaque monkey: Quantitative analysis using double-label immunohistochemistry and confocal laser microscopy. *Journal of Comparative Neurology*, *341*(4), 476–491.

Gold, J. I. (2003). Linking reward expectations of behavior in the basal ganglia. *Trends in Neurosciences*, *26*(1), 12–14.

Gonzaga, G. C., Keltner, D., Londahl, E. A., & Smith, M. D. (2001). Love and the commitment problem in romantic relations and friendship. *Journal of Personality and Social Psychology*, *81*, 247–262.

Gonzalez, M. I., Farabollini, F., Albonetti, E., & Wilson, C. A. (1994). Interactions between 5-hydroxytryptamine (5-HT) and testosterone in the control of sexual and nonsexual behaviour in male and female rats. *Pharmacology, Biochemistry, and Behavior*, *47*(3), 591–601.

Grammer, K., Fink, B., Moller, A. P., & Thornhill, R. (2000). Darwinian aesthetics: Sexual selection and the biology of beauty. *Biological Reviews of the Cambridge Philosophical Society*, *78*, 385–407.

Griffin, M. G., & Taylor, G. T. (1995). Norepinephrine modulation of social memory: Evidence for a time-dependent functional recovery of behavior. *Behavioral Neuroscience*, *109*(3), 466–473.

Harris, H. (1995). Rethinking heterosexual relationships in Polynesia: A case study of Mangaia, Cook Island. In W. Jankowiak (Ed.), *Romantic passion: A universal experience?* New York: Columbia University Press.

Harvey, P. H., & Harcourt, A. H. (1984). Sperm competition, testes size, and breeding systems in primates. In R. Smith (Ed.), *Sperm competition and the evolution of animal mating systems* (pp. 589–659). New York: Academic Press.

Hatfield, E. (1988). Passionate and companionate love. In R. J. Sternberg & M. S. L. Barnes (Eds.), *The psychology of love*. New Haven, CT: Yale University Press.

Hatfield, E., & Rapson, R. L. (1996). *Love and sex: Cross-cultural perspectives.* Needham Heights, MA: Allyn & Bacon.

Hatfield, E., & Sprecher, S. (1986). Measuring passionate love in intimate relationships. *Journal of Adolescence*, *9*, 383–410.

Hazan, C., & Diamond, L. M. (2000). The place of attachment in human mating. *Review of General Psychology*, *4*, 186–204.

Hazan, C., & Shaver, P. R. (1987). Romantic love conceptualized as an attachment process. *Journal of Personality and Social Psychology*, *52*, 511–524.

Heaton, J. P. (2000). Central neuropharmacological agents and mechanisms in erectile dysfunction: The role of dopamine. *Neuroscience and Biobehavioral Reviews*, *24*(5), 561–569.

Hendrick, V., Gitlin, M., Altshuler, L., & Korenman, S. (2000). Antidepressant medications, mood and male fertility. *Psychoneuroendocrinology*, *25*(1), 37–51.

Herbert, J. (1996). Sexuality, stress and the chemical architecture of the brain. *Annual Review of Sex Research*, *7*, 1–44.

Hollander, E., Fay, M., Cohen, B., Campeas, R., Gorman, J. M., & Liebowitz, M. R. (1988). Serotonergic and noradrenergic sensitivity in obsessive-compulsive disorder: Behavioral findings. *American Journal of Psychiatry*, *145*(8), 1015–1017.

Hrdy, S. B. (1999). *Mother nature: A history of mothers, infants and natural selection*. New York: Pantheon.

Homeida, A. M., & Khalafalla, A. E. (1990). Effects of oxytocin and an oxytocin antagonist on testosterone secretion during the oestrous cycle of the goat (*Capra hircus*). *Journal of Reproduction and Fertility*, *89*(1), 347–350.

Hull, E. M., Du, J., Lorrain, D. S., & Matuszewich, L. (1995). Extracellular dopamine in the medial preoptic area: Implicatons for sexual motivation and hormonal control of copulation. *Journal of Neuroscience*, *15*(11), 7465–7471.

Hull, E., Du, J., Lorrain, D., & Matuszewich, L. (1997). Testosterone, preoptic dopamine, and copulation in male rats. *Brain Research Bulletin*, *44*, 327–333.

Hull, E. M., Lorrain, D. S., Du, J., Matuszewick, L., Lumley, L. A., Putnam, S. K., et al. (1999). Hormone-neurotransmitter interactions in the control of sexual behavior. *Behavioural Brain Research*, *105*(1), 105–116.

Hull, E., Meisel, R., & Sachs, B. D. (2002). Male sexual behavior. In D. W. Pfaff et al. (Eds.), *Hormones, brain, and behavior* (pp. 1–139). New York: Academic Press.

Insel, T. R., Mueller, E. A., Alterman, I., Linnoila, M., & Murphy, D. L. (1985). Obsessive-compulsive disorder and serotonin: Is there a connection? *Biological Psychiatry*, *20*, 1174–1188.

Insel, T. R., Zohar, J., Benkelfat, C., & Murphy, D. L. (1990). Serotonin in obsessions, compulsions, and the control of aggressive impulses. *Annals of the New York Academy of Sciences*, *600*, 574–586.

Jankowiak, W. (1995). Introduction. In W. Jankowiak (Ed.), *Romantic passion: A universal experience?* New York: Columbia University Press.

Jankowiak, W. R., & Fischer, E. F. (1992). A cross-cultural perspective on romantic love. *Ethnology*, *31*(2), 149.

Johnson, A. E., Coirine, H., Insel, T. R., & McEwen, B. S. (1991). The regulation of oxytocin receptor binding in the ventromedial hypothalamic nucleus by testosterone and its metabolites. *Endocrinology*, *128*(2), 891–896.

Stahl, S. M. (2000). *Essential psychopharmacology.* 2nd ed. New York: Cambridge University Press.

Sundblad, C., & Eriksson, E. (1997). Reduced extracellular levels of serotonin in the amygdala of androgenized female rats. *European Neuropsychopharmacology*, *7*(4), 253–259.

Szezypka, M. S., Zhou, Q. Y., & Palmiter, R. D. (1998). Dopamine-stimulated sexual behavior is testosterone dependent in mice. *Behavioral Neuroscience*, *112*(5), 1229–1235.

Tennov, D. (1979). *Love and limerence: The experience of being in love.* New York: Stein & Day.

Thomson, J. A., Jr., & Fisher, H. E. (2004, July 21). *Do the sexual side-effects of antidepressants jeopardize romantic love and marriage?* Paper presented at the annual meeting of the Human Behavior and Evolution Society, Berlin.

Thomas, A., Kim, N. B., & Amico, J. A. (1996). Sequential exposure to estrogen and testosterone (T) and subsequent withdrawal of T increases the level of arginine vasopressin messenger ribonucleic acid in the hypothalamic paraventricular nucleus of the female rat. *Journal of Neuroendocrinology*, *8*(10), 793–800.

Thoren, P., Asberg, M., & Bertilsson, L. (1980). Clomipramine treatment of obsessive disorder: Biochemical and clinical aspects. *Archives of General Psychiatry*, *37*, 1289–1294.

Thornhill, R., Gangestad, S. W., & Comer, R. (1995). Human female orgasm and mate fluctuating asymmetry. *Animal Behavior*, *50*, 1601–1615.

Tiihonen, J., Kuikka, J. T., Bergstrom, K. A., Karhu, J., Viinamiki, H., Lehtonen, J., et al. (1997). Single-photon emission tomography imaging of monoamine transporters in impulsive violent behaviour. *European Journal of Nuclear Medicine*, *24*(10), 1253–1260.

Tiihonen, J., Kuikka, J., Kupila, J., Partanen, K., Vainio, P., Airaksinen, J., et al. (1994). Increase in cerebral blood flow of right prefrontal cortex in men during orgasm. *Neuroscience Letters*, *170*, 241–243.

Van Bockstaele, E. J., Peoples, J., & Telegan, P. (1999). Efferent projections of the nucleus of the solitary tract to peri-locus coeruleus dendrites in rat brain: Evidence for a monosynaptic pathway. *Journal of Comparative Neurology*, *412*(3), 410–428.

Van Bockstaele, E. J., Pieribone, V. A., & Aston-Jones, G. (1989). Diverse afferents converge on the nucleus paragigantocellularis in the rat ventrolateral medulla: Retrograde and anterograde tracing studies. *Journal of Comparative Neurology*, *290*(4), 561–584.

Van Bockstaele, E. J., Saunders, A., Telegan, P., & Page, M. E. (1999). Localization of mu-opioid receptors to locus coeruleus-projecting neurons in the rostral medulla: Morphological substrates and synaptic organization. *Synapse*, *34*(2), 154–167.

Van de Kar, L. D., Levy, A. D., Li, Q., & Brownfield, M. S. (1998). A comparison of the oxytocin and vasopressin responses to the 5-HT1A agonist and potential anxiolytic drug alnespirone (S-20499). *Pharmacology, Biochemistry, and Behavior*, *60*(3), 677–683.

Van Goozen, S., Wiegant, V. M., Endert, E., Helmond, F. A., & Van de Poll, N. E. (1997). Psychoendocrinological assessment of the menstrual cycle: The relationship between hormones, sexuality, and mood. *Archives of Sexual Behavior*, *26*(4), 359–382.

Villalba, D., Auger, C. J., & De Vries, G. J. (1999). Androstenedione effects on the vasopressin innervation of the rat brain. *Endocrinology*, *140*(7), 3383–3386.

Vizi, E. S., & Volbekas, V. (1980). Inhibition by dopamine of oxytocin release from isolated posterior lobe of the hypophysis of the rat: Disinhibitory effect of beta-endorphin/enkephalin. *Neuroendocrinology*, *31*(1), 46–52.

Walker, P. W., Cole, J. O., Gardner, E. A., et al. (1993). Improvement in fluoxetine-associated sexual dysfunction in patients switched to bupropion. *Jounal of Clinical Psychiatry*, *54*, 459–465.

Wang, Z., & De Vries, G. J. (1993). Testosterone effects on paternal behavior and vasopressin immunoreactive projections in prairie voles (*Microtus ochrogaster*). *Brain Research*, *63*(1), 156–160.

Wang, Z., & De Vries, G. J. (1995). Androgen and estrogen effects on vasopressin messenger RNA expression in the medial amygdaloid nucleus in male and female rats. *Journal of Neuroendocrinology*, *7*(1), 827–831.

Wang, Z. X., Ferris, C. F., & De Vries, G. J. (1994). The role of septal vasopressin innervation in paternal behavior in prairie voles (*Microtus ochrogaster*). *Proceedings of the National Academy of Sciences, U.S.A.*, *91*, 400–404.

Wang, Z., Toloczko, D., Young, L. J., Moody, K., Newman, J. D., & Insel, T. R. (1997). Vasopressin in the forebrain of common marmosets (*Calithrix jacchus*): Studies with in situ hybridization, immunocytochemistry and receptor autoradiography. *Brain Research*, *768*, 147–156.

Wang, Z., Yu, G., Cascio, C., Liu, Y., Gingrich, B., & Insel, T. R. (1999). Dopamine D2 receptor-mediated regulation of partner preferences in female prairie voles (*Microtus ochrogaster*): A mechanism for pair bonding? *Behavioral Neuroscience*, *113*(3), 602–611.

Wenkstern, D., Pfaus, J. G., & Fibiger, H. C. (1993). Dopamine transmission increases in the nucleus accumbens of male rats during their first exposure to sexually receptive female rats. *Brain Research*, *618*, 41–46.

Wersinger, S. R., & Rissman, E. F. (2000). Dopamine activates masculine sexual behavior independent of the estrogen receptor alpha. *Journal of Neuroscience*, *20*(11), 4248–4254.

West, C. H. K., Clancy, A. N., & Michael, R. P. (1992). Enhanced responses of nucleus accumbens neurons in male rats to novel odors associated with sexually receptive females. *Brain Research*, *585*, 49–55.

Williams, J. R., Insel, T. R., Harbaugh, C. R., & Carter, C. S. (1994). Oxytocin administered centrally facilitates formation of a partner preference in female prairie voles (*Microtus ochrogaster*) *Journal of Neuroendocrinology*, *6*(3), 247–250.

Wingfield, J. C. (1994). Hormone-behavior interactions and mating systems in male and female birds. In R. V. Short & E. Balaban (Eds.), *The differences between the sexes* (pp. 303–330). Cambridge: Cambridge University Press.

Winslow, J. T., Hastings, N., Carter, C. S., Harbaugh, C. R., & Insel, T. R. (1993). A role for central vasopressin in pair bonding in monogamous prairie voles. *Nature 365*, 545–548.

Winslow, J. T., & Insel, T. R. (1991). Social status in pairs of male squirrel monkeys determines the behavioral response to central oxytocin administration. *Journal of Neuroscience 11*(7), 203–208.

Wise, R. A. (1996). Neurobiology of addiction. *Current Opinion in Neurobiology*, *6*(2), 243–251.

Wise, R. A. (1988). Psychomotor stimulant properties of addictive drugs. In P. W. Kalivas & C. B. Nemeroff (Eds.), *The mesocorticolimbic dopamine system. Annals of the New York Academy of Sciences*, *537*, 228–234.

Wise, R. A. (1989). Brain dopamine and reward. *Annual Review of Psychology*, *40*, 191–225.

Wise, R. A., & Hoffman, D. C. (1992). Localization of drug reward mechanisms by intracranial injections. *Synapse*, *10*, 247–263.

Young, L. J. (1999). Oxytocin and vasopressin receptors and species-typical social behaviors. *Hormones and Behavior*, *36*, 212–221.

Young, L. J., Lim, M. M., Gingrich, B., & Insel, T. R. (2001). Cellular mechanisms of social attachment. *Hormones and Behavior*, *40*, 133–138.

Young, L. J., Wang, Z., & Insel, T. R. (1998). Neuroendocrine bases of monogamy. *Trends in Neurological Sciences*, *21*(2), 71–75.

Young, L. J., Winslow, J. T., Nilsen, R., & Insel, T. R. (1997). Species differences in V1a receptor gene expression in monogamous and nonmonogamous voles: Behavioral consequences. *Behavioral Neuroscience*, *111*, 599–605.

11 Self-Perceived Survival and Reproductive Fitness Theory: Substance Use Disorders, Evolutionary Game Theory, and the Brain

David B. Newlin

Current evidence supports the conclusion that substance use disorders (SUDs), including alcohol and other drug abuse and dependence, have a complex, multifactorial etiology. Family (Cotton, 1987), twin (Pickens et al., 1991; Prescott & Kendler, 1999), and adoption studies (Goodwin, Schulsinger, Hermanson, Guze, & Winokur, 1973) implicate genetic factors in the development of alcoholism, although the precise genetic mechanisms are only now being identified (Long et al., 1998; Reich et al., 1998). There are also major environmental components to risk for SUD, such as peer influence, drug availability, and the environmental effects of parental alcoholism (Newlin et al., 2000), although the research tools to study environmental effects have not had the precision of genetic tools. Moreover, the interaction between genetic and environmental factors has seldom been explored, despite theories of alcoholism that emphasize the interactional nature of complex causes (e.g., Cloninger, 1987; Cloninger Bohman, & Sigvardsson, 1981). Therefore, theories of SUD that integrate diverse etiological pathways are needed to organize and synthesize the large body of data on the causes of SUD. Rigorous experimental tests are also needed for these new theories to fulfill their heuristic promise and to provide foundations that are empirically sound.

This chapter presents and discusses a recent theory about the etiology of SUD. This theory depends heavily on three concepts that, taken together, define an emerging theoretical model of SUD: (1) self-perceived survival and reproductive fitness (SPFit) (see also Newlin, 1999); (2) evolutionary game theory—the SPFit game; and (3) the conclusion that the proposed brain substrate of SPFit, the corticomesolimbic dopamine

---

Portions of this chapter appeared in different form in D. B. Newlin, "The self-perceived survival ability and reproductive fitness (SPFit) theory of substance abuse disorders," *Addiction*, 97, 427–446.

(CMDA) system, is not a "reward center" or "reward pathway," as the addictions field has often assumed, but a basic survival and reproductive motivation system that is activated both by drugs of abuse and by perceived threats to survival and reproductive fitness (i.e., stressful and novel stimuli). These concepts provide unifying systems for this new theory of SUD and suggest empirical tests that can falsify or support the theory.

The field of substance abuse was originally dominated by the question of why people would choose to take drugs that are clearly harmful to them and are addictive. After the behavioral revolution and the rise of behavioral pharmacology, the question changed to one of why abused drugs are reinforcing to animals and humans, which is a more limited and theoretically constrained question. In this chapter, we return to the original question, now recast in the framework of evolutionary biology.

Self-Perceived Survival and Reproductive Fitness

Definition

The first concept is self-perceived survival and reproductive fitness, or SPFit—a new psychological construct based on the fundamental mammalian motivations to enhance and protect survival and reproductive fitness.[1] In humans, SPFit represents an internalized, self-perceived model of survival and reproductive functioning. SPFit is embodied in such basic psychological characteristics as feelings of personal power, control, and omnipotence—all related to survival ability—and to feelings of personal sexiness (i.e., that relevant others find them sexually attractive), physical and behavioral attractiveness, and social desirability—all related to reproductive fitness. SPFit organizes and prioritizes behavior in a complex world. Moreover, these evolutionarily conserved mechanisms are not viewed as limited to rare circumstances, such as the fight-or-flight response to direct threat but instead are pervasive in human functioning and are tonically active.

Power motivation (McClelland, 1974) is directly relevant to SPFit. Specifically, the desire to acquire and enhance personal power is understood in current theory as fundamental to the perceived ability to survive. The powerful person is better able to overcome obstacles to survival than

1 As with IQ, SPFit is not designed or intended to account for or to measure racial and ethnic differences, and should not be used for this purpose. The SPFit construct would be construed and measured very differently if it were explicitly designed to perform this function.

the less powerful individual, and personal empowerment is thought to promote one's life relatively directly. Reproductive fitness is a basic concept in evolutionary biology and can also be considered important in human behavior (without the assumption of exclusively genetic control). Humans go to extraordinary lengths to make themselves more physically, socially, and sexually attractive, and sexuality is integral to much human functioning. In terms of natural (Darwinian) selection, survival and reproductive functions are under the most direct selective pressure. Miller (2001) has argued that in hominid evolution, natural selection for reproductive fitness may have played a greater role than survival fitness. He has suggested that "runaway" evolution occurred for characteristics related to sexuality and reproductive fitness and, for example, may have accounted for the development of the human big brain.

In SPFit theory, natural selection is proposed to operate on characteristics that are transmitted over generations through both Mendelian (genetic) mechanisms and, much more rapidly, through sociocultural evolution (learning). In the lifetime of the individual, SPFit evolves through interactions with the physical and social world, and puberty is viewed as a critical period in which SPFit develops full expression and becomes much less plastic.

SPFit is schematized in figure 11.1, which relates the psychological construct of SPFit to its proposed brain substrate, CMDA, and to subjective states that reflect basic survival and reproductive functioning. As figure 11.1 shows, prevailing hedonistic concepts concerning the acute effects of drugs, such as euphoria and pleasure, and mechanistic concepts

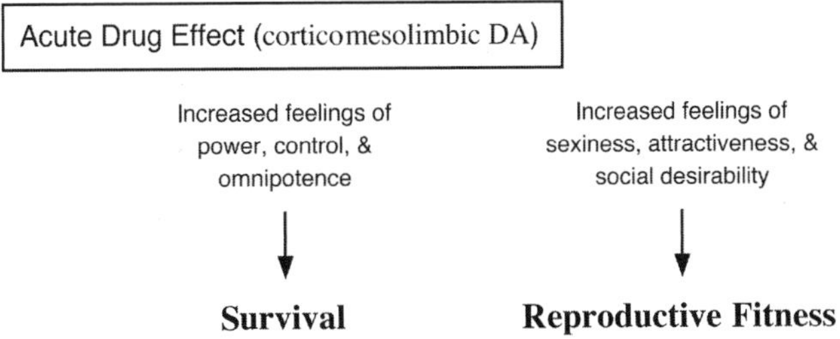

Figure 11.1
Schematic of SPFit. Note that SPFit does not employ reinforcement or reward or even euphoria as an explanatory construct. The acute drug effect involves increased SPFit, activation of the corticomesolimbic dopamine circuitry (its proposed biological substrate), and enhanced feelings of power and sexual attractiveness. The acute drug effect is activating—a characteristic of motivation—in this case to survive and to be reproductively fit.

such as reinforcement and reward, do not enter into the definition or the heuristics of SPFit. The current theory is not a hedonistic theory. It is instead a teleological model based on goal-directed motivations and behaviors to survive and to be reproductively fit. In the hierarchy of motivations, survival and reproductive fitness are vastly more basic to the animal and to the human than is pleasure seeking. A mammal that lacks motivation will die, but one that lacks pleasure will not. Pleasurable sensations or euphoria from drugs are considered incidental or epiphenomenal in the current theory. They can result from artificially increased SPFit (such as from taking an abused drug) rather than affecting it. At the same time, negative affective states such as fear, anxiety, and anger are as likely to result from activation of SPFit as are pleasurable states.

In humans, SPFit bridges the gap between the biological imperatives to survive and reproduce, on the one hand, and behavioral and physiological adaptation to a complex world on the other. SPFit is proposed as an internalized representation of these biological primitives (to survive and to reproduce) based on self-perception of personal power and sexual attractiveness. SPFit is not proposed as a measure of actual biological fitness, which can be measured only in nonhuman animals (i.e., fecundity). Instead, it is viewed as a psychological construct that is fueled by power motivation, which enhances survival, and sexuality, which enhances reproduction, but is itself an internalized assessment of these capacities. As such, it is strongly influenced by social and cultural factors that impinge on a person's self-perception.

SPFit is directly relevant to substance abuse because drugs of abuse artificially inflate feelings of personal power and sexual attractiveness. For example, abusers report that cocaine produces feelings of omnipotence and power (Sherer, Kumor, & Jaffe, 1989), as well as heightened feelings of sexual attractiveness. The "coasting" and "absolute contentment" of an opiate-induced high may reflect artificial feelings of satisfaction that survival and reproductive fitness are assured. Other drugs of abuse produce similar feelings to a greater or lesser extent. Power motivation has been studied in relation to alcohol intoxication, and enhanced feelings of power and masculinity in men (McClelland, 1974) and feminine characteristics (e.g., nurturance) in women (Wilsnack, 1974) have been reported with an alcohol-induced high.

The central paradox in the drug abuse field is the question of why people use drugs that are clearly harmful to them and in fact reduce their biological fitness. SPFit theory suggests that this temporary artificial inflation of SPFit, which can be quite dramatic (such as in a drug rush

Evidence Relating to SPFit

Power Motivation

Power Imagery McClelland performed a series of studies in the 1960s concerning power motivation, or the desire to maintain and increase one's power over self and others. McClelland (1974) proposed that individuals drink alcohol to artificially enhance feelings of power. He measured power motivation using written fantasy material expressed by male volunteers who were administered the Thematic Apperception Test (TAT), a standard projective test. Subjects took the TAT during various cocktail parties and discussion groups when alcohol was served and when alcohol was absent, and the responses were compared. TATs were administered before drinking and then twice after drinking, first at a moderate dose and then at a high dose of alcohol.

Wilsnack (1974), a colleague of McClelland, summarized her mentor's research on alcohol as follows:

> Small to moderate amounts of alcohol were found to increase thoughts of social power (s Power), power for the good of others or a cause. Larger amounts of alcohol increased thoughts of personal power (p Power), power in the interest of self-aggrandizement, without regard for others. In two studies of working class men, men with histories of heavy drinking had higher p Power scores when not drinking than men with histories of light drinking. (p. 43)

Gender Differences Wilsnack (1974) determined whether the power theory of alcohol applied only to men or characterized women's drinking as well. Using similar methodology, she found that women's TAT fantasies after consuming alcohol were unlike those of the men in the earlier studies (Kalin, McClelland, & Kahn, 1965). Among female drinkers, personal power (p Power) actually decreased after drinking relative to the condition without alcohol. Moreover, "Being Orientation," a psychological measure that in previous studies had characterized nurturant women, markedly increased after alcohol. Although it is likely that women showed more traditionally feminine characteristics after drinking, Wilsnack (1974) observed that the effect might represent only a decrease in masculine fantasy. In either case, there was strong evidence of gender differences in the response to alcohol in relation to power motivation.

Wilsnack (1974) also found that TAT images that had often been associated with men's responses were more frequent in women who had a history of heavy drinking than among light-drinking women. "It appears that high p Power men and high p Power women both drink more than their low p Power peers, yet drinking seems to have opposite effects on the power fantasies of men and women" (p. 57).

Self-Confidence

Although power motivation is fundamental to the SPFit concept, it is not a contemporary measure. Konovsky and Wilsnack (1982) measured self-confidence using the Tennessee Self-Concept Scale administered at cocktail parties with married couples. Again, gender differences were striking; men scored higher in self-confidence after drinking, but women scored lower after alcohol.

Self-Efficacy and Euphoria

Self-efficacy (Bandura, 1977) is a psychological construct that may have some overlap with SPFit. Self-efficacy is usually defined as a relatively specific belief that one can deal effectively with a specific stimulus or situation (contextual stimulus). It is a much less global construct than "self-confidence" or "self-esteem." Bandura (1977) stated that "Psychological procedures, whatever their form, serve as means of creating and strengthening expectations of personal efficacy" (p. 193). Therefore, self-efficacy is a cognitive theory that concerns expectancies about personal behavior and its outcomes.

SPFit has some similarity to all these constructs in that it is based on self-perception, but it differs in that it emphasizes survival and reproductive fitness rather than a specific stimulus such as a snake (self-efficacy) or life in general (self-confidence). Much of daily behavior is irrelevant or only indirectly relevant to survival and reproduction, so that SPFit would not be so actively engaged as when, for example, someone were held underwater or were ridiculed by peers. Moreover, SPFit might be very prominent in situations in which self-efficacy was low, such as when there was a direct threat to SPFit with which the individual felt poorly able to cope.

Therefore, it is essential in the articulated thoughts about simulated situations (ATSS) assessment procedure that self-efficacy be measured concurrently so that it can be covaried with SPFit. Of particular importance to the theory is variation in SPFit that cannot be predicted by self-efficacy beliefs in the expected success of a specific behavior in a specific situation. This residual might then be lawfully related to pharmacological manipulations (increased SPFit) and to organismic state, such as decreased SPFit during drug withdrawal or influenza infection.

Researchers often measure pleasurable affective states ("euphoria," "high," "coasting") during intoxication from drugs such as cocaine, heroin, marijuana, and alcohol, as surrogate measures of drug reward or reinforcement. This follows from prevailing hedonistic models of SUD, whether behavioral or psychobiological. In contrast, heightened SPFit should be associated with both positive (e.g., omnipotence) and negative (e.g., stress responses) emotional states. This sensitivity to negative emotional states is characteristic of SPFit, but not of hedonistic mechanisms such as reward and reinforcement. A corollary of this is that abused drugs administered acutely would be expected in SPFit theory to produce a wide range of positive and negative affective states, not just euphoria or reward. Again, the variation in SPFit that cannot be accounted for by changes in positive affect is particularly important in testing the theory (table 11.2).

Increased SPFit

The most important prediction of the model is that intoxication from alcohol and other abused drugs will inflate SPFit. Also, the perceived sphere of influence and the location in the perceived dominance hierarchy should increase during intoxication relative to a sober state. During the acute drug effect, SPFit is artificially elevated at the same time that actual fitness may be seriously compromised by the drug. This potential discrepancy between self-perception and reality illustrates the fundamental nature of SPFit. The construct serves as a flexible buffer between the individual's ecology (e.g., survival and social demands) and his or her behavioral and physiological adaptation to that environment (coping with those demands). SPFit allows for clearly maladaptive behavior precisely because of the enhanced flexibility and greater adaptive capacity of an internalized system. It is inherent in the evolutionary design of the adaptive systems of SPFit and the CMDA that the substance abuser responds to drugs as if they boosted survival and reproductive fitness when they do not actually do so. Although Nesse and Berridge (1997) were discussing nonhuman animals when they referred to delivery of a "huge fitness benefit," the same argument applies to humans. In this case, the artificial inflation of SPFit is the subjective counterpart in humans to the brain response in animals (including humans) to which these authors referred.

Table 11.2
Major Components of SPFit Theory

1. SPFit is a new psychobiological construct that mediates between (a) the primary motivation to survive and the secondary motivation to reproduce, and (b) behavioral interactions with the social environment and the internal world of abused drugs. The biological substrate of SPFit is the corticomesolimbic dopamine system and its neurophysiological connections.
2. *Fixation/Completion.* The process of fixation/completion of the motivational system during and after puberty by the abuse of drugs reflects the operation of tolerance and sensitization processes from escalating use of these drugs. It is in part a learning process that supplants the more typical fixation/completion of this motivational system by more culturally accepted processes such as mating rituals and schooling.
3. *Autoshaping/Sign Tracking/Feature Positivity.* The learning mechanism by which these processes occur is autoshaping/sign tracking. This involves self-sustaining behavioral and physiological processes by which the organism becomes oriented toward, attended to, and fixated on arbitrary stimuli that are highly predictive of activating biological events such as drug effects. The biological substrate of SPFit becomes sensitized during this process.
4. *The SPFit Game.* The SPFit game represents an evolutionary game theory implementation of the current theory. The animal or human being is organized in such a manner that the acute effects of abused drugs are countered by organismic responses that produce sensitization to the activating effects and tolerance to the deactivating effects of the drugs. In this way, the organism accentuates the positive (activating) aspects of the drug response and attenuates the aversive (deactivating) components of the drug effect. The primary motivations of the SPFit game are (a) to survive and (b) to protect and increase SPFit.
5. *High Risk.* Individuals at elevated risk for substance abuse show exaggerated counterresponses to abused drugs that perpetuate the processes noted above. Two high-risk pathways to substance abuse, an antisocial pathway and a negative affect pathway, have been identified that are predicted to have different strategies in the SPFit game.
6. *Short-circuiting the Regulatory System.* Abused drugs hijack the motivational system that evolved to regulate survival and reproductive functions. This short-circuiting occurs because drugs activate this system *as if* they were relevant to survival and reproduction. Stressful stimuli also activate this motivational system because they *are* biologically relevant, although these stimuli do not produce feelings of reward and are not positively reinforcing.

A corollary of the prediction that drugs of abuse will artificially boost SPFit is that states of drug withdrawal will be associated with SPFit that is depressed below baseline levels. Since SPFit is viewed as pervasive and paramount in human functioning, it is tonically active and can decrease at any time. In addition to measuring SPFit during withdrawal from drugs of abuse, one might also assess it in individuals who have influenza or who have just learned they have a major disease, or in individuals who are experiencing divorce or bereavement.

Biphasic Responses

The effect of alcohol (and some other drugs) on SPFit may be biphasic, such that SPFit increases in the rising blood alcohol curve, followed by decreased SPFit (below baseline levels) in the falling curve (Newlin & Thomson, 1990). Brain responses to alcohol differ in the rising and falling blood alcohol limbs (Lyons, Whitlow, & Porrino, 1998). This effect can be biphasic in terms of dosage, as well (Pohorecky, 1977). Low doses of alcohol and other abused drugs that produce behavioral activation (locomotor and psychostimulant activation) in animals should be associated with temporary enhancement of SPFit in humans. In contrast, very high doses that deactivate behavior should depress SPFit acutely.

Stress

Hobfoll (1989) proposed a conceptual model of stress based on conservation of resources. He proposed that "people strive to retain, project, and build resources and that what is threatening to them is the potential or actual loss of these valued resources" (p. 513). If one simply replaces the word "resources" with "SPFit," the result is a new definition of stress that is integral to the current theory. When people are exposed to stressful stimuli (i.e., those that threaten SPFit), they seek to boost or to maintain SPFit in the face of potential losses. Humans marshal their resources to cope with the stressor in a manner that protects or enhances SPFit. Therefore, both abused drugs and stressful stimuli can lead to temporarily elevated SPFit despite opposing affective valences of these two types of stimuli. It should be noted that prolonged, uncontrollable stressors can lead to sharply depressed SPFit if the person is unable to cope with these stimuli.

An important aspect of SPFit is that both positive and negative stimuli tend to be activating. Psychologically, this activation reflects the mobilization of resources to survive or to be reproductively fit, and physiologically it is associated with engagement of the CMDA system. This aspect of SPFit is emphasized in the SPFit game (discussed in the next section).

"Saving face" represents a similar phenomenon that is important in folk psychology and cultural anthropology. In SPFit theory, "saving face" is the successful protection of SPFit in socially difficult situations and "losing face" represents decreased SPFit.

Personality

SPFit may provide a useful link between personality theory and biological responses to stimuli that produce positive and negative affect. For example, low SPFit should be associated with negative affectivity, a highly stable and recurrent personality characteristic, and greater right frontal electroencephalographic (EEG) activation. Laboratory induction of positive mood should inflate SPFit and produce left frontal EEG activation. A prediction that is unique to this model is that controllable stressful stimuli should also increase SPFit because they activate similar brain systems (i.e., the CMDA). However, uncontrollable or prolonged stressful stimuli should decrease SPFit. SPFit theory has obvious implications for other types of psychopathology. For example, depression and anxiety should be associated with lower SPFit. Antisocial personality disorder may involve very high SPFit, and obsessive-compulsive disorder may demonstrate low SPFit.

Risk-Taking Behavior

SPFit theory predicts increased risk-taking behavior in individuals under the influence of abused drugs because the enhanced sense of empowerment (survival ability) would tend to diminish their expectancy of adverse outcomes from risky behavior.[2] For example, driving while intoxicated may be associated with enhanced SPFit and feelings of invulnerability. It is worth noting that drug taking is itself risky behavior, the perception of risk from which would also be diminished by the temporary boost in SPFit. This could lead to rapid, repeated dosing and to very high blood levels of the drug that are strongly associated with serious

2 The extension of SPFit theory to risk-taking behavior is from Mark Fillmore.

adverse consequences. This is also a potential mechanism of "loss of control" drug use.

Summary SPFit is new to the fields of psychology and biology. It captures and clarifies many characteristics of the acute response to alcohol and to other drugs, which provides a basis for its heuristic value. However, it requires substantiation using sophisticated measurement techniques such as ATSS (Davison, Vogel, & Coffman, 1997). The current theory makes strong, directional predictions concerning the empirical effects of acute intoxication and drug withdrawal on SPFit. It must be demonstrated too that SPFit is not merely redundant with self-efficacy or euphoric mood states. Therefore, SPFit meets the critical criterion of falsifiability. The degree of genetic and environmental contributions to SPFit can be estimated in classic twin studies or in other behavior genetic studies. The current theory predicts that SPFit will have a significant genetic component, although environmental influences are predicted as well. SPFit theory does not assume that SUD or SPFit (as measured by ATSS and other techniques) are under exclusive genetic control. In fact, the current theory emphasizes environmental and genetic-environmental interactions in the development and expression of these constructs, particularly during puberty.

Drug Craving

Newlin (1992) proposed an autoshaping/sign-tracking model of drug craving, with an emphasis on the persistent orientation toward and skeletal behavior directed to drugs and to cues for drugs in drug abusers and addicts. This may parallel the powerful effects of Pavlovian conditioning, with skeletal responses oriented toward the predictive sign of impending reinforcement. Newlin's (1992) autoshaping model of drug craving was integrated (Newlin, 1999) into SPFit theory as a remarkable orientation of the substance abuser toward stimuli associated with drugs and to drugs themselves as a means to artificially enhance SPFit.

Approach versus avoidance is a fundamental dimension of mammalian responses to environmental stimuli. For example, when presented with another animal or human, the organism immediately categorizes it as friend or foe, predator or prey, or, if it is a novel environment, as opportunity or threat. Davidson (1992; Davidson & Irwin, 1999) summarized electrophysiological data indicating that approach is associated with left frontal EEG activation and avoidance with right frontal activation. This frontal asymmetry appears not to reflect the dimension of positive versus negative affect because anger, a negative emotion, also activates left frontal areas (Harmon-Jones & Allen, 1998). Therefore, the approach versus avoidance dimension is more fundamental than affective tone. Moreover, this frontal asymmetry has been a highly reliable empirical finding in the psychophysiology literature.

SPFit theory predicts that drugs of abuse and drug stimuli, such as paraphernalia, drinking buddies, advertisements for tobacco products, and situations frequently associated with drug use, will produce left frontal brain activation and approach functioning only in drug abusers. Moreover, this approach dimension reflects sign tracking, or orientation toward stimuli associated with drug effects and signals of impending drug availability and drug delivery. It also reflects the "feature-positive effect" (Newman, Wolff, & Hearst, 1980) that is conceptually related to sign tracking, or an overemphasis on features that are positively related to or correlated with biologically relevant stimuli such as drugs. The drug user develops a more or less pervasive frame of reference in which there is approach or orientation toward drugs and drug stimuli as effective means to enhance SPFit. Therefore, sign tracking is a basic learning mechanism for drug craving, which is approach functioning toward signs predictive of drug effects (and their artificial enhancement of SPFit).

Zinser et al. (1999) measured right and left frontal alpha power of the EEG in smokers deprived of cigarettes for 24 hours, after exposure to smoking cues, and during actual smoking. Anticipation of smoking was associated with increased left frontal activation, and actual smoking increased right frontal brain activity. Of different models of craving and addiction that were considered, the results were most consistent with Robinson and Berridge's (1993) incentive-sensitization model because the frontal brain response indicative of approach functioning occurred to anticipation and to exposure to smoking cues rather than to actual smoking.

In SPFit theory, drug stimuli are viewed as cues for the opportunity to inflate SPFit. Many cue exposure studies have been performed in which established drug users are compared with nonusers or recreational users in terms of their autonomic and subjective response (i.e., craving) to drug cues. For example, with alcoholics, the cue might be the sight and smell of their favorite alcoholic beverage (compared to their favorite nonalcoholic drink), and the autonomic response is increased salivation (Newlin, Hotchkiss, Cox, Rausches, & Li, 1989). The current theory predicts that addicted individuals would show a short-term elevation in SPFit in response to such a cue, but only if the drug were available to them. If the drug were unavailable, there could be a transient decrease in SPFit. In autoshaping/sign tracking, the cue is the conditioned stimulus that signals impending delivery of an appetitive stimulus, in this case an abused drug. It is interesting to note that Tomie, Aguado, Pohorecky, and Benjamin (1998) have argued that autoshaping represents impulsive behavior. Drug craving is viewed in SPFit theory as impulsive responding. This contrasts markedly with Tiffany's (1990) proposal that drug craving represents controlled (effortful, deliberate, and nonautomatic) rather than automatic cognitive processes and with Cox and Klinger's (1988) description of craving as a cognitive decision-making process.

There is evidence (Earleywine, 1994; Weingardt, Stacy, & Leigh, 1996) that heavy users of alcohol have a cognitive bias toward alcohol-related stimuli. For example, Earleywine found that the tendency to understand words with double meanings one of which is alcohol related (e.g., "bar"), in terms of the alcohol-related meaning was greater in individuals who drank alcohol the most. SPFit theory would interpret these results as due to sign tracking of stimuli associated with alcohol.

The chronic smoker or drug abuser "always has something to look forward to"—their next cigarette or injection of cocaine.[3] They are always looking forward to or anticipating boosting SPFit, and this expectation is an important component of drug craving. Moreover, when they are attempting to quit using the drug, they "have nothing to look forward to." These considerations emphasize the degree to which drugs capture and control SPFit through the development of expectancies and craving.

Drug Outcome Expectancies

There is now a large literature (Goldman, Brown, Christiansen, & Smith, 1991; Leigh, 1989) concerning beliefs and attitudes in humans toward drugs (many of them false beliefs) which are particularly strong and positive among those who abuse and become dependent on drugs. These attitudes and beliefs are thought to play an important causal role in the development of drug abuse and addiction.

---

3 This observation is from Kenneth Sher.

SPFit theory posits that autoshaping/sign tracking/feature positivity is a basic mechanism by which these positive beliefs and attitudes toward drugs develop during early use and abuse of these substances. Specifically, drug users learn that taking these drugs is an easy, highly reliable way to inflate SPFit, although they may be unaware that the inflation of SPFit is false in the sense that the feelings do not correspond with reality. Moreover, drug outcome expectancies reflect the underlying belief that drugs increase personal power, sexual appeal, and social desirability—that is, they enhance SPFit. Therefore, drug outcome expectancies reflect the basic dimension of approach (as opposed to avoidance) toward drugs and drug stimuli. Recall, too, that Bolles (1972) emphasized that expectancies are what are learned in autoshaping and avoidance learning.

Distortion of the Frame of Reference

In humans, SPFit theory posits that a cognitive map or frame of reference develops concerning situations and behaviors that boost versus impair SPFit. In Western culture, this map or frame is likely viewed by the individual in terms of situations or behaviors that increase (or impair) empowerment, sexual desirability, or personal attractiveness rather than in the technical terms of survival or reproductive fitness. Drug outcome expectancies develop as part of this motivational map or frame of reference. Moreover, drug experiences distort this map because they produce new anchors and points of reference for the limits of subjective experience. For example, young teenagers may find that kissing or sexual petting strongly enhances feelings of sexual desirability, until they feel the artificial boost in SPFit from taking a strong drug of abuse such as cocaine or alcohol. Learning to drive a car is another example of an empowering experience that may pale in comparison with drug effects on SPFit. As a result, the frame of reference becomes distorted in drug abusers but not in nonusers, and in such a manner that culturally sanctioned activities that bring a sense of control and personal attractiveness are diminished relative to drug abuse. In addition, cues for drugs or positive predictors of drug availability or use become highly salient stimuli that exert increasingly strong effects on drug expectancies and drug-seeking behavior. The motivational map also reflects expectancies concerning events or behaviors that diminish SPFit, and these anchors, too, can be influenced by drug use. These considerations suggest empirical tests of an important component of SPFit theory, or the distortion of the SPFit frame of reference through drug experiences and autoshaping/sign tracking of drug cues.

This cognitive map or frame of reference for SPFit bears some similarity to the idea of Goldman et al. (1991; Dunn & Goldman, 1998) of a distributed memory system for alcohol expectancies that differs as a function of degree of use. They share in common the idea that expectancies are powerful mechanisms that control behavior, although SPFit theory is couched in terms of the construct of SPFit and autoshaping/sign tracking/feature positivity as a learning mechanism for these expectancies rather than memory per se. Also, SPFit theory emphasizes a feed-forward model of expectancies in which the animal or human responds to drug stimuli as if the stimuli were drugs themselves.

The SPFit Game

The second major concept in this new theory is an evolutionary game—the SPFit game. This game is schematized in figure 11.2. An alternative way of illustrating the game, the extensive form, is shown in figure 11.3. In this game users of alcohol or other drugs seek to artificially inflate their SPFit by abusing drugs. In animals, it is a game of survival.

Evolutionary Game Theory

Evolutionary game theory (Maynard Smith, 1973; Sigmund, 1995; Weibull, 1996) has played an important role in understanding how animal behavior evolved. The most common evolutionary game in this field is Prisoner's Dilemma, a two-player game. This game has attracted a great deal of attention among evolutionary biologists because it deals with the difficult issue of competitive versus cooperative behavior in selfish species such as mammals. However, other evolutionary games have been developed, such as the rock-paper-scissors children's game played by male side-blotched lizards (Sinervo & Lively, 1996). Newlin (1999) developed an evolutionary game of multiple chemical sensitivity, a disorder characterized by aversive responses to chemicals in the environment that have little or no effect on individuals without the disorder.

Game theory continues to be a vital area of research. For example, new strategies continue to be developed for Prisoner's Dilemma, such as "win-stay, lose-shift" (Nowak & Sigmund, 1993), which improves upon "tit for tat," as well as "negotiated" strategies in which the players interact to contract over the outcomes (McNamara, Gasson, & Houston, 1999). Prisoner's Dilemma has been played by starlings (Reboreda & Kacelnik, 1993) and even by viruses that attack bacteria (Turner & Chao, 1999). On the horizon is quantum game theory, in which a player's

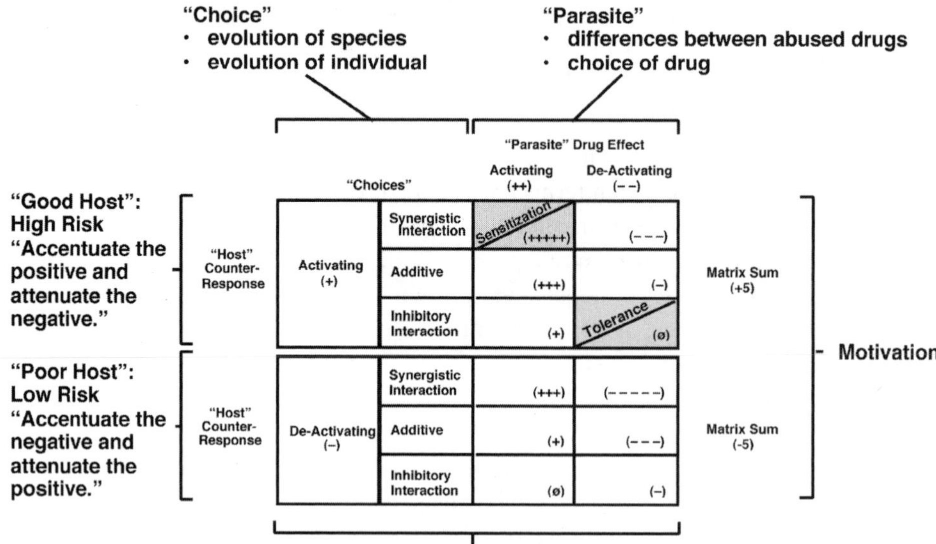

Figure 11.2
The SPFit game. The game involves drug effects that have both activating and deactivating components and the organismic counterresponse to the drug effects. The counterresponse can be activating, which leads to sensitization of activating drug effects and tolerance to deactivating effects, or it can be deactivating, most often in the case of individuals who are at low risk for substance use disorders. This "choice" (no conscious effort is implied) of counterresponse, which is a result of the way in which the organism is constructed, is part of the addictive process. The act of accentuating the positive (activating) effects of drugs and attenuating the negative (deactivating) drug effects is viewed as a function that is normally biologically adaptive with natural stimuli but leads to addiction with drugs of abuse. The host is the drug abuser and the parasite is the drug (see text). The motivation is to boost SPFit as much as possible and to avoid strong negatives, such as overdose. The payoff matrix defines these outcomes and forms the basis for motivated behavior such as substance abuse. Since SPFit is defined only in humans, the game is one of survival in animals.

move has characteristics of both possible plays at the same time (Collins, 2000).

Purpose

The SPFit game was developed to model the interaction between a drug abuser (the game player) and the drug effect. The "behavior" of the

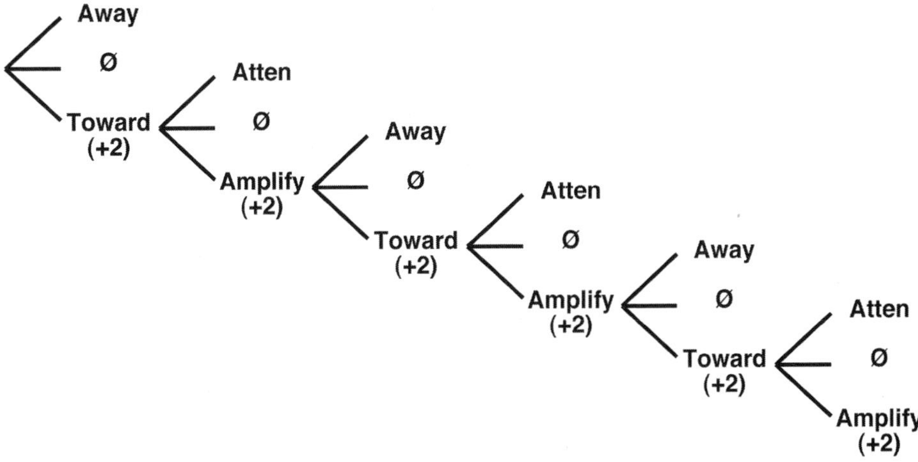

Figure 11.3
The extensive form of the SPFit game. There is a series of decision points as time moves from left to right. The addicted individual moves toward drug cues and drugs themselves, and amplifies the drug effects through activating counterresponses to the drugs.

player consists primarily of physiological counterresponses to drug stimuli. This is an unusual game because it really has only one and one-half players, not the two or more players in most evolutionary games. The half-player is the drug effect itself, although it is relatively constant in its plays (hence, one-half). The player's (drug abuser's) behavior is viewed in the SPFit game as relatively consistent once it reaches asymptote, although there is considerable variation between the one-half plays of different drugs. Only the player varies his or her physiological behavior in relation to the drug effect.

The player has evolved in geological time (i.e., through Mendelian genetics) in such a manner as to promote survival and reproductive functions. The physiological behavior of the player reflects both this organization as it relates to his or her evolutionary history of dealing with biologically relevant stimuli (such as parasites and both appetitive and aversive stimuli) and the ways in which the individual player learns to adapt to potent drugs of abuse within his or her lifetime (i.e., Lamarckian evolution). The payoff matrix is determined by the player's evolutionary history, but the "choices" of behavior in response to an individual drug relate both to the player's evolutionary history and to learning processes in the lifetime of the individual. Another way of expressing this is that the payoff matrix is hard-wired, whereas the player's choices in the SPFit game are partially hard-wired but are also under the control of Pavlovian conditioning processes.

Motivation

The player is motivated to avoid strong negatives, such as overdose death. For example, the player would avoid combining deactivating (i.e., SPFit-reducing) properties of the drug, including sedation, with a deactivating counterresponse, such as locomotor suppression. The player also seeks to experience strong positives, such as those associated with sensitization. In this case, the player may counter the locomotor activating (i.e., SPFit-increasing) effects of the drug with an activating response such as tachycardia due to withdrawal of parasympathetic tone. This leads to strong sensitization (i.e., very high SPFit) because of the combined activating effects of the drug and the player's activating counterresponse. In practice, the drug effect and the organismic counterresponse interact either synergistically or subadditively, depending on whether the drug effect is activating (e.g., locomotor activation) or deactivating (e.g., analgesia). The two effects rarely combine additively, despite theoretical assumptions to the contrary (Siegel, 1975). The trade-off determined by the payoff matrix is that this strong positive will be accompanied by a tolerant but still present deactivating effect of the aversive components of the drug response.

Rules of the SPFit Game

The foundation of the SPFit game is a specific pattern of empirical findings in the drug conditioning literature concerning the dynamics of conditioned tolerance and sensitization. These are summarized in table 11.3 and illustrated in figure 11.4. The rules of the game are summarized in

Table 11.3
Foundation of the SPFit Game in Consistent Characteristics of Conditioned Tolerance and Sensitization in the Animal Drug Conditioning Literature

Drug effect	Deactivating	Activating
Counterresponse	Drug-opposite	Druglike
Counterresponse	Activating	Activating
Temporal trend	Tolerance	Sensitization
Interaction of effect and response	Inhibitory (subadditive)	Synergistic (superadditive)

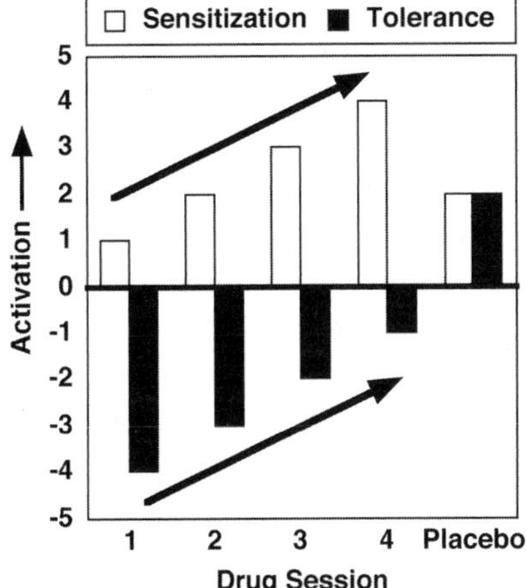

Figure 11.4
Idealized representation of the activating nature of both sensitization and tolerance phenomena. The effects of tolerance and sensitization are parallel over time: both processes lead to greater activation. Note that the conditioned response, apparent with placebo administration, is activating, whether the drug effect shows tolerance (for deactivating drug effects) or sensitization (for activating drug effects).

Table 11.4
Assumptions (Rules) of the SPFit Game

1. Each abused drug has multiple effects, both activating and deactivating.
2. The "host" (substance abuser) emits a counterresponse to the "parasite" (drug) that is activating or deactivating.
3. The interaction between the drug effect and the organismic counterresponse can be subadditive or synergistic.
4. The "host" seeks to optimize his or her payoff and to avoid strong negatives.

table 11.4. It is useful to classify drug effects as activating or deactivating. The prototypic activating response is locomotor activation, and the prototypic deactivating response is sedation or locomotor suppression. Other activating drug effects include hyperalgesia, hyperthermia, hypersexuality, and seizures; other deactivating effects include analgesia, hypothermia, respiratory depression, anticonvulsant, and psychomotor impairment effects. The value of this classification is that it is strongly related to patterns of drug conditioning: tolerance to deactivating drug

effects and sensitization to activating effects. It is also the case that conditioned responses to drug cues are almost invariably activating rather than deactivating.

The organismic counterresponses of the subject (abuser) represent the plays of the game. In the SPFit game, efforts to maximize the positive (activation, or increased SPFit in humans) and to minimize the negative (deactivation, or decreased SPFit in humans) lead to characteristic patterns of amplification and attenuation of the drug effect. The end result of the game is that the relentless effort to artificially inflate SPFit leads to drug-seeking behavior and addictive physiological states in some vulnerable individuals.

The primary point of the SPFit game is that the empirical data that are available in the Pavlovian drug conditioning literature agree well with those predicted by the game. For example, locomotor activating effects of cocaine are associated over repeated administrations with the development of activating counterresponses (apparent with exposure to cocaine-associated environmental cues). Moreover, the combination of the response to cocaine and the counterresponse is synergistic (superadditive), which produces strong sensitization. With deactivating drug effects, such as the analgesic effect of morphine, the activating counterresponse (hyperalgesia) combines in an inhibitory interaction (subadditive) with the effect of morphine. The result is strong conditioned tolerance. In both cases, the effects are adaptive in that they avoid strong negatives (overdose death) and produce strong positives (activation).

Other Applications

The SPFit game can be applied to various risk factors for substance abuse, as well as other experimental phenomena such as stress response dampening (Levenson, Sher, Grossman, Newman, & Newlin, 1980) and stress response enhancement (Sayette & Wilson, 1991) from drinking alcohol. There are constitutional differences between players such that they tend to counterrespond to abused drugs with an activating versus a deactivating physiological response. One might expect that an organism that emits a deactivating counterresponse to drugs would not be likely to become addicted; that is, the organism would be a poor "host" for abused drugs.

Sher (1991) described a behaviorally undercontrolled pathway to SUD, in which antisocial individuals develop substance abuse problems, and a negative affect pathway, in which anxious people escalate their use of drugs to problematic or addictive levels. SPFit theory argues that in

motor activation when the animal has a variety of discrete stimuli in the test chamber that may indicate the nature of the directed behavior. For example, would a male rat be more likely to mount a receptive female, or huddle with other rats in a cool environment, or attack a common prey after he received low doses of drugs of abuse?

The view that CMDA is a reward pathway is still current (Berridge & Robinson, 1998; Di Chiara, 1998; Grace, 1995; Koob & Nestler, 1997; Robbins & Everitt, 1999; Sutton & Beninger, 1999). However, some recent reviews of the literature have attempted to incorporate evidence of CMDA involvement in aversive motivation into new ideas about the functional roles of this system. For example, Salamone (1994) concluded that the CMDA is involved in both appetitive and aversive motivation, and Ikemoto and Panksepp (1999) suggested that nucleus accumbens dopamine is "an incentive property constructor" and has a role in both "invigoration" and incentive learning. These roles are not limited to reward learning, nor are they directly related to positive hedonic effects of drugs (Ikemoto & Panksepp, 1999). However, theorists, including Nesse and Berridge (1997), have not viewed the CMDA as a survival–reproductive motivation system.

Behavioral Control

Cabib and Puglisi-Allegra (1996) reported that dopamine outflow in the nucleus accumbens is increased by controllable/escapable stressful stimuli but decreased by uncontrollable/inescapable stressors. This finding is again consistent with the notion that the mesolimbic dopamine system controls survival and reproductive motivation rather than being a simple reward center. Escapable stress would be expected to enhance active motivational processes, while prolonged inescapable stress would be expected to reduce motivation profoundly. It would be difficult to argue that an escapable stressor is in some way rewarding, but it is clearly motivating. Escapable stress would be expected to strongly activate basic survival functioning; in contrast, inescapable stress, if sufficiently prolonged, would suppress these same functions as the animal "gives up" and adopts more primitive defenses, such as freezing.

Drive?

Therefore, we might amend our notion to suggest that the mesolimbic dopamine system is a physiological substrate for motivated behavior relevant to basic survival and reproductive functions. We might have concluded that it is a nonspecific GO center (as opposed to a NO-GO center) or drive (Ikemoto & Panksepp, 1999) system, but this fails to capture the highly directed nature of the behavior. This point is analogous to the argument made earlier that systemic injections of abused drugs produce general activation of locomotor behavior simply because there is no external stimulus to which the organism can orient and approach. SPFit theory predicts that animals that are administered drugs of abuse will exhibit specific, goal-oriented behavior directed toward biologically relevant (i.e., relevant to survival or reproductive fitness) stimuli if such stimuli are physically present in their environment. This prediction contrasts with the notion of CMDA as a nonspecific GO or drive system that is engaged by abused drugs.

Neuropsychological Hypothesis

We hypothesize that, in terms of the cerebral cortex, SPFit is more closely associated with right prefrontal function than with other cortical areas. First, SPFit is understood as an executive cognitive function because it is goal-oriented (i.e., survival and reproductive fitness), reflective (a self-assessment of Darwinian fitness), corrective (it seeks to maximize fitness and minimize threats to fitness), regulating (it exerts some degree of both inhibitory and excitatory control over mesolimbic functions), prioritizing (it places survival and then reproductive fitness above all other functions), and integrative (it coordinates a wide range of human functions, depending on their relevance to perceived survival and reproductive fitness). This executive aspect of SPFit leads us to infer prefrontal functions. Another reason for hypothesizing that SPFit has preferentially prefrontal cortical components is that aspects of working memory and autobiographical memory are inherent in the construct, and much work has illuminated prefrontal aspects of these functions. In addition, the research of Keenan and Platek and their colleagues (Keenan, Wheeler, Gallup, & Pascual-Leone, 2000; Lou et al., 2004; Platek, Keenan, Gallup, & Mohamed, 2004) has implicated right prefrontal and right parietal structures in different aspects of the self. While SPFit is not proposed as the self, it is one important component of self-representation that may share this degree of brain localization.

Having framed SPFit in terms of executive cognitive functions (at least in relation to its cortical components), we are immediately reminded of left prefrontal functioning. In contrast, we implicate right prefrontal functions for several reasons. First, the SPFit construct is not defined in verbal terms; in fact, we argue that this self-perception is typically nonverbal, particularly when the culture (such as most Western societies) does not explicitly endorse this thinking. For example, many societies place a high value on having large numbers of fertile offspring, and we would then expect verbal thinking to reflect that valuation. Second, SPFit is not construed as rigid and rule-based, characteristics often associated with left hemispheric specialization. Instead, we view SPFit as analogous to prosodic as opposed to lexical functions in expressive and receptive speech. Prosody is no less important, and often trumps lexical information in conveying meaning (such as in sarcasm). This also points to preferential right prefrontal functions. Third, the evidence is strong that the right hemisphere is functionally specialized to provide a sense of our bodies in three-dimensional space and for self-recognition. Finally, we appeal again to Keenan and Platek's (Keenan et al., 2000; Lou et al., 2004; Platek et al., 2004) work on self-representation and the right hemispheric specialization for at least some aspects of self-observations, self-recognition, and representation of one's own body.

Importantly, we hypothesize that right prefrontal cortical involvement in SPFit is preferential rather than exclusive. We would expect tasks that engage SPFit (i.e., that are perceived as directly relevant to survival or reproductive fitness) to activate many (but not all) cortical areas. Neuroimaging research has shown clearly that depending on the nature of the task and the social and other context in which the task is performed, a number of cortical areas are activated or deactivated, even when the task is critically dependent on only one or two areas. We argue that right prefrontal activation is likely to be common to a wide range of stimuli and tasks that engage SPFit, not that activation will be exclusively right prefrontal. A corollary of this is that we would not predict that stimuli or tasks designed to engage SPFit would lead to a right prefrontal deactivation. This right prefrontal hypothesis is testable with modern brain imaging techniques.

Modularity

The concept of modules of mind (Fodor, 1983) is a controversial theory that has been applied to comparative cognition and behavioral ecology. This concept was originally discussed in relation to cognitive-perceptual modules that preprocess sensory information into informational units such as phonetics, which are useful to higher levels of cognition (e.g., syntactic or grammatical processing). Fodor's (1983) argument was that perceptual modules are primarily hard-wired, domain-specific, informationally encapsulated, and impermeable.

Taking these concepts one at a time, one might understand hard-wiring in relation to traditional Mendelian genetic processes. The term *domain-specific* refers to characteristics of the module that selectively process information (e.g., phonemes) that is relevant to its specific function, and it ignores stimuli that are not. For example, inflective or affective components of speech might be processed by a different module of mind than those engaged by phonetic stimuli. The module is *informationally encapsulated* in the sense that informational processing and output from a specific module are not available to all other modules but only to those that depend on the information specific to that module. An example would be that word recognition receives information only from a phonetic module; however, the information from the phonetic module might not be accessible to verbal self-report. Finally, the concept of *impermeability* refers to the automatic or seemingly reflexive nature of the module's processing. For example, once learned, processing phonetics requires no conscious effort. Moreover, once learned, it is virtually impossible (absent brain lesions) to interfere with or to suspend this phonetic processing.

In Fodor's (1983) theory, modules may be classified as either horizontal or vertical. Examples of horizontal modules are attention, analogical reasoning, verbal reasoning, and other central processing functions that are general in their function. These horizontal modules can be activated by a broad range of stimuli (i.e., they are not domain-specific) and are accessible to many other modules (i.e., they are not informationally encapsulated). In contrast, vertical modules, such as phonetic and grammatical processing, are both domain-specific and informationally encapsulated. They become automatic and difficult to interfere with (i.e., impermeable).

Gallistel (1991) adapted Fodor's (1983) concept of modularity of mind to account for behaviors such as navigation in migratory birds, three-dimensional space localization in rats, and nonassociative (i.e., cognitive representational) interpretations of rodent behavior in Pavlovian experiments. Modularity may account for numerical counting behavior in rats (Boysen & Capaldi, 1993). In addition, Gardner's (1983) faculty psychology of human intellectual functions uses similar constructs to describe different types of intelligences, and Gazzaniga (1985) adopted a modular approach to brain organization.

SPFit theory proposes that mammals and humans have evolved a vertical module, the CMDA, that controls the motivation to survive and reproduce. This module is assumed to be partially hard-wired in the sense that the capacity for behavior directed toward survival and reproduction is transmitted across generations through traditional Mendelian processes. Moreover, there are individual differences in this motivational system, and these hard-wired aspects of the system are also transmitted genetically. Individual differences are central to an evolutionary process because genetic variation forms the basis for natural selection.

The SPFit motivational module is domain-specific inasmuch as it is engaged only by environmental stimuli that are perceived as relevant to survival and reproductive functions. This proposed module of mind is limited in the sense that it ignores many (irrelevant) day-to-day stimuli, such as when animals that are drug naive and nonusing humans are relatively unaffected by drug stimuli. Also related to domain specificity is the fact that the SPFit module is not a homunculus that controls all behavior and is removed by only one step from the self. This module is informationally encapsulated in the sense that it is only partially accessible to verbal self-report in humans. This limited accessibility depends on the extent to which the specific culture endorses concepts that are directly relevant to SPFit.

Finally, the concept of impermeability refers to the observation that by the time functions associated with vertical modules, such as SPFit, are learned (based on foundations of hard-wiring through Mendelian genetics), they become automatic and highly resistant to extinction. Impermeability may underlie the compulsive nature of addiction and the very high relapse rates following cessation of drug use in SUD.

Empirical Falsifiability

A large amount of existing data is consistent with SPFit theory, some of it uniquely fitting the theory. It was noted earlier that effects found with aversive and novel stimuli support the conclusion that the CMDA system with its many neurophysiological connections is a survival and reproductive motivation system. Evidence that there are situations in which people self-administer drugs with no measurable subjective effects (euphoria) are inconsistent with the view that the CMDA is a reward pathway, but are consistent with SPFit theory. The pioneering work of McClelland (1974) and Wilsnack (1974) on the effect of alcohol drinking on power motivation, sex differences, and self-esteem is perhaps uniquely explained by SPFit theory. The drug conditioning literature suggests patterns of activating and deactivating drug effects and conditioned responses that are modeled in the SPFit game. Autoshaping has been demonstrated with alcohol (Krank, 2003) and cocaine (Carroll & Lac, 1993, 1997, 1998) as the unconditioned stimulus.

In addition to existing data, SPFit theory makes a number of very specific predictions that would require the theory to be modified or abandoned if they were falsified. First, SPFit should be amenable to reliable and valid measurement, and it should not be redundant with self-efficacy or positive affect. Acute administration of drugs of abuse should increase SPFit, and drug withdrawal should decrease SPFit. For drugs with biphasic effects (whether as a function of the slope of the drug blood curve or of dose), this should be mirrored in biphasic effects on SPFit. These effects should depend on the risk status of the subjects, with high-risk individuals showing larger effects of acute administration and more pronounced biphasic effects, depending on the positive and negative affect pathways to SUD. The effect of drug cues should be to produce a transient increase in SPFit when the drug is available and a decrease in SPFit when it is not. In experimental users of drugs, the increase in SPFit should occur in response to the drug effect, but in established drug users, it should occur more in response to anticipatory drug cues rather than in response to the drug itself (see the SPFit analysis of Robinson and Berridge's [1993] theory, above). Placebos should reveal the organismic counterresponse to drugs in accordance with the predictions of the SPFit game, which should also vary according to the behaviorally undercontrolled versus negative affect pathway to SUD.

There is a specific set of predictions concerning drug craving and outcome expectancies. For example, craving should be associated with a distorted cognitive map or frame of reference for behaviors and situations that increase and decrease SPFit, compared with that seen in individuals who do not use or crave drugs. During experimental use of drugs, the distortion of these maps should be apparent in longitudinal studies, with progressively diminished salience to behaviors and situations for increasing SPFit that are culturally sanctioned, and new anchors for the limits of experience provided by the acute effects of abused drugs. In addition, craving and drug outcome expectancies should be correlated.

A final prediction is that SPFit should have both significant heritability and environmental effects. The latter could be examined by studying families in which one or both adoptive, foster, or stepparents have an SUD, compared to similar families without SUD (Newlin et al., 2006).

Conclusion

This chapter has focused on five concepts that make up SPFit theory: SPFit, reformulation of the functional role of the CMDA, the SPFit game, autoshaping/sign tracking/feature positivity, and modularity of mind. Leshner (1997) suggested the SUDs are "chronic, relapsing disorders of the brain" and that the addictive process is "like a switch being thrown in the brain." SPFit theory proposes that the motivation to artificially enhance SPFit is a factor that is common to experimental use, escalating abuse, addiction, and relapse to drug-taking behavior. This contrasts with most theoretical models of SUDs, where initiation and experimental use are thought to be related to peer influence and drug availability, escalating use to genetic vulnerability, addiction to unclear biological mechanisms, and relapse to conditioning or other psychological factors. In SPFit theory, the motivation is the same throughout the course of the disorder—to artificially boost SPFit.

The psychobiological mechanism by which the substance abuser learns that drugs will enhance SPFit (throwing the "brain switch") is similar to autoshaping as the user learns that drug use is an easy, reliable means to enhance SPFit. The neurophysiological substrate of this disorder of the brain is the CMDA and its modulating connections. The SPFit and the CMDA have been likened to a vertical module (Fodor, 1983) for survival and reproductive motivation that has formal characteristics that make it relatively impervious to interference. In humans, the acute effect of abused drugs is to artificially elevate SPFit, and craving and outcome expectancies about drugs are an outgrowth of sign-tracking/feature positivity effects. The SPFit game is an evolutionary game that models the physiological behavior of animals and people using empirical patterns of Pavlovian conditioning that describe sensitization and tolerance phenomena.

SUDs represent the hijacking of this motivational module by abused drugs. SPFit theory emphasizes puberty as a critical period in which SPFit achieves full expression, and consolidates and becomes less plastic. For this reason, young people are at particular risk for fixation/completion/concretization of SPFit by drug-taking behavior. Rather than SPFit consolidating around culturally sanctioned behaviors, SPFit can be hijacked by the highly reliable enhancement of SPFit by drugs of abuse.

Acknowledgments I am indebted to Jessica Luc, who provided editorial and clerical assistance, and to my many colleagues at NIDA–Intramural Research Program and RTI International, who have provided feedback on the theory presented in this chapter. I also thank James MacKillop, Bryan Castelda, and Shannon Coleman, who have been valuable collaborators in matters concerning SPFit and the SPFQ.

References

Ahmed, S. H., & Koob, G. F. (1997). Cocaine- but not food-seeking behavior is reinstated by stress after prolonged extinction. *Psychopharmacology*, *132*, 289–295.

Bandura, A. (1977). Self-efficacy: Toward a unifying theory of behavioral change. *Psychological Review*, *84*, 191–215.

Barkow, J. H., Cosmides, L., & Tooby, J. (1995). *The adapted mind: Evolutionary psychology and the generation of culture.* London: Oxford University Press.

Berridge, K. C., & Robinson, T. E. (1998). What is the role of dopamine in reward: Hedonic impact, reward learning, or incentive salience? *Brain Research Reviews*, *28*, 309–369.

Besson, A., Privat, A. M., Eschalier, A., & Fialip, J. (1998). Reversal of learned helplessness by morphine in rats: Involvement of a dopamine mediation. *Pharmacology, Biochemistry, and Behavior*, *60*, 519–525.

Bolles, R. C. (1972). Reinforcement, expectancy, and learning. *Psychological Review*, *79*, 394–409.

Boysen, S. T., & Capaldi, E. J. (1993). *The development of numerical competence: Animal and human models*. Hillsdale, NJ: Erlbaum.

Cabib, S., & Puglisi-Allegra, S. (1996). Different effects of repeated stressful experiences on mesocortical and mesolimbic dopamine metabolism. *Neuroscience*, *73*, 375–380.

Carroll, M., & Lac, S. T. (1993). Autoshaping i.v. cocaine self-administration in rats: Effects of nondrug alternative reinforcers on acquisition. *Psychopharmacology*, *110*, 5–12.

Carroll, M., & Lac, S. T. (1997). Acquisition of i.v. amphetamine and cocaine self-administration in rats as a function of dose. *Psychopharmacology*, *129*, 206–214.

Carroll, M., & Lac, S. T. (1998). Dietary additives and the acquisition of cocaine self-administration in rats. *Psychopharmacology*, *137*, 81–89.

Cloninger, C. R. (1987). Neurogenetic adaptive mechanisms in alcoholism. *Science*, *236*, 410–416.

Cloninger, C. R., Bohman, M., & Sigvardsson, S. (1981). Inheritance of alcohol abuse: Cross-fostering analysis of adopted men. *Archives of General Psychiatry*, *38*, 861–868.

Collins, G. P. (2000). Quantum game theory: Schrödinger's games. *Scientific American*, *282*, 28–29.

Cotton, N. (1987). The familial incidence of alcoholism. *Journal of Studies on Alcohol*, *40*, 89–116.

Cox, W. M., & Klinger, E. (1988). A motivation model of alcohol use. *Journal of Abnormal Psychology*, *97*, 168–180.

Davidson, R. J. (1992). Emotion and affective style: Hemispheric substrates. *Psychological Science*, *3*, 39–43.

Davidson, R. J., & Irwin, W. (1999). The functional neuroanatomy of emotion and affective style. *Trends in Cognitive Sciences*, *3*, 11–21.

Davison, G. C., Vogel, R. S., & Coffman, S. G. (1997). Think-aloud approaches to cognitive assessment and the articulated thoughts about simulated situations paradigm. *Journal of Consulting and Clinical Psychology, 65*, 950–958.

Di Chiara, G. (1998). A motivational learning hypothesis of the role of mesolimbic dopamine in compulsive drug use. *Journal of Psychopharmacology, 12*, 54–67.

Di Chiara, G., Acquas, E., Tanda, G., & Cadoni, C. (1993). Drugs of abuse: Biochemical surrogates of specific aspects of natural reward? *Biochemical Society Symposia, 59*, 65–81.

Di Chiara, G., & Imperato, A. (1988). Drugs abused by humans preferentially increase synaptic dopamine concentrations in the mesolimbic system of freely moving rats. *Proceedings of the National Academy of Sciences, U.S.A., 85*, 5274–5278.

Dudley, R. (2000). Evolutionary origins of human alcoholism in primate frugivory. *Quarterly Review of Biology, 75*, 3–15.

Dunn, M. E., & Goldman, M. S. (1998). Age- and drinking-related differences in memory organization of alcohol expectancies in 3rd, 6th, 9th, and 12th-grade children. *Journal of Consulting and Clinical Psychology, 66*, 579–585.

Dworkin, S. I., Mirkis, S., & Smith, J. E. (1995). Response-dependent versus response-independent presentation of cocaine: Differences in the lethal effects of the drug. *Psychopharmacology, 117*, 262–266.

Earleywine, M. (1994). Cognitive bias covaries with alcohol consumption. *Addictive Behavior, 19*, 539–544.

Erb, S., Shaham, Y., & Stewart, J. (1996). Stress reinstates cocaine-seeking behavior after prolonged extinction and a drug-free period. *Psychopharmacology, 128*, 408–412.

Fodor, J. A. (1983). *The modularity of mind: An essay on faculty psychology.* Cambridge, MA: MIT Press.

Gallistel, C. R. (1991). *Animal cognition.* Cambridge, MA: MIT Press.

Gardner, H. (1983). *Frames of mind: The theory of multiple intelligences*. New York: Basic Books.

Gazzaniga, M. S. (1985). *The social brain: Discovering the networks of the mind.* New York: Basic Books.

Goldman, M. S., Brown, S. A., Christiansen, B. A., & Smith, G. T. (1991). Alcoholism and memory: Broadening the scope of alcohol-expectancy research. *Psychological Bulletin, 110*, 137–146.

Goodwin, D. W., Schulsinger, F., Hermanson, L., Guze, S. B., & Winokur, G. (1973). Alcohol problems in adoptees raised apart from alcoholic biological parents. *Archives of General Psychiatry, 28*, 238–243.

Grace, A. A. (1995). The tonic/phasic model of dopamine system regulation: Its relevance for understanding how stimulant abuse can alter basal ganglia function. *Drug and Alcohol Dependence, 37*, 111–129.

Gratton, A., & Wise, R. A. (1994). Drug- and behavior-associated changes in dopamine-related electrochemical signals during intravenous cocaine self-administration in rats. *Journal of Neuroscience, 14*, 4130–4146.

Harmon-Jones, E., & Allen, J. J. B. (1998). Anger and frontal brain activity: EEG asymmetry consistent with approach motivation despite negative affective valence. *Journal of Personality and Social Psychology, 74,* 1310–1316.

Hemby, S. E., Co, C., Koves, T. R., Smith, J. E., & Dworkin, S. I. (1997). Differences in extracellular dopamine concentrations in the nucleus accumbens during response-dependent and response-independent cocaine administration in the rate. *Psychopharmacology, 133,* 7–16.

Hobfoll, S. E. (1989). Conservation of resources: A new attempt at conceptualizing stress. *American Psychologist, 44,* 513–524.

Ikemoto, S., & Panksepp, J. (1999). The role of nucleus accumbens dopamine in motivated behavior with special reference to reward-seeking. *Brain Research Reviews, 31,* 6–41.

Kalin, R., McClelland, D. C., & Kahn, M. (1965). The effects of male social drinking on fantasy. *Journal of Personality and Social Psychology, 95,* 441–452.

Kalivas, P. W., & Duffy, P. (1989). Similar effects of daily cocaine and stress on mesocorticolimbic dopamine neurotransmission in the rat. *Biological Psychiatry, 25,* 913–928.

Kalivas, P. W., & Samson, H. H. (1992). The neurobiology of drug and alcohol addiction. *Annals of the New York Academy of Sciences, 654,* 171–191.

Keenan, J. P., Wheeler, M. A., Gallup, G. G., Jr., & Pascual-Leone, A. (2000). Self-recognition and the right prefrontal cortex. *Trends in Cognitive Science, 9,* 338–344.

Konovsky, M., & Wilsnack, S. C. (1982). Social drinking and self-esteem in married couples. *Journal of Studies on Alcohol, 43,* 319–333.

Koob, G. F., & Nestler, E. J. (1997). The neurobiology of drug addiction. *Journal of Neuropsychiatry, 9,* 482–497.

Krank, M. D. (2003). Pavlovian conditioning with ethanol: Sign-tracking (autoshaping), conditioned incentive, and ethanol self-administration. *Alcohol: Clinical and Experimental Research, 27,* 1592–1598.

Lamb, R. J., Preston, K. L., Schindler, C. W., Meisch, R. A., Davis, F., & Katz, J. L., et al. (1991). The reinforcing and subjective effects of morphine in post addicts: A dose-response study. *Journal of Pharmacological Experimental Theory, 259,* 1165–1173.

Leigh, B. C. (1989). In search of the seven dwarves: Issues of measurement and meaning in alcohol expectancy research. *Psychological Bulletin, 110,* 137–146.

Leshner, A. (1997). Addiction is a brain disease, and it matters. *Science, 278,* 45–47.

Levenson, R. W., Sher, K. J., Grossman, L. M., Newman, J., & Newlin, D. B. (1980). Alcohol and stress response dampening: Pharmacological effects, expectancy, and tension reduction. *Journal of Abnormal Psychology, 89,* 528–538.

Long, J. C., et al. (1998). Evidence of genetic linkage to alcohol dependence on chromosome 4 and 11 from an autosome-wide scan of an American Indian population. *Americal Journal of Medical Genetics, 81,* 216–221.

Lou, H. C., Luber, B., Crupain, M., Keenan, J. P., Nowak, M., Kjaer, T. W., et al. (2004). Parietal cortex and representation of the mental self. *Proceedings of the National Academy of Sciences, U.S.A., 101*, 6827–6832.

Lyons, D., Whitlow, C. T., & Porrino, L. J. (1998). Multiphasic consequences of the acute administration of ethanol on cerebral glucose metabolism in the rat. *Pharmacology, Biochemistry, and Behavior, 61*, 201–206.

MacKillop, J., Castelda, B. A., & Newlin, D. B. (2004). Testing the SPFit model of addictive behavior: Development of the self-perceived Fitness Questionnaire (SPFQ). Poster session presented at the Human Behavior and Evolution Society Meetings, Berlin, Germany.

Maynard Smith, J. (1973). *Evolution and the theory of games*. Cambridge: Cambridge University Press.

McClelland, D. (1974). Proceedings: Drinking as a response to power needs in man. *Psychopharmacology Bulletin, 10*, 5–6.

McNamara, J. M., Gasson, C. E., & Houston, A. I. (1999). Incorporating rules for responding into evolutionary games. *Nature, 401*, 368–371.

Miller, G. (2001). *The mating mind: How sexual choice shaped the evolution of human nature*. New York: Random House.

Mucha, R. F., van der Kooy, D., O'Shaughnessy, M., & Bucenieks, P. (1982). Drug reinforcement studied by the use of place conditioning in rat. *Brain Research, 243*, 91–105.

Muntaner, C., Kumor, K. M., Nagoshi, C., & Jaffe, J. H. (1989). Intravenous cocaine infusions in humans: Dose responsitivity and correlations of cardiovascular vs. subjective effects. *Pharmacology, Biochemistry, and Behavior, 34*, 697–703.

Nesse, R. M., & Berridge, K. C. (1997). Psychoactive drug use in evolutionary perspective. *Science, 278*, 63–66.

Newlin, D. B. (1992). A comparison of conditioning and craving for alcohol and cocaine. In M. Galanter (Ed.), *Recent developments in alcoholism: Vol. 10. Alcohol and cocaine: Similarities and differences* (pp. 147–164). New York: Plenum Press.

Newlin, D. B. (1994). In the belly of the beast: Toward a motivational view of addiction. *Contemporary Psychology, 39*, 511–512.

Newlin, D. B. (1996). Craving for alcohol and cocaine: From inside the brain to the clinic. *Alcoholism: Clinical and Experimental Research, 20*, 45A–47A.

Newlin, D. B. (1999). Evolutionary game theory and multiple chemical sensitivity. *Toxicology and Industrial Health, 15*, 313–322.

Newlin, D. B. (2002). The self-perceived survival ability and reproductive fitness (SPFit) theory of substance use disorders. *Addiction, 97*, 427–446.

Newlin, D. B., Hotchkiss, B., Cox, W. M., Rauscher, F., & Li, T.-K. (1989). Autonomic and subjective responses to alcohol stimuli with appropriate control stimuli. *Addictive Behaviors, 14*, 625–630.

Newlin, D. B., Miles, D., Van Den Bree, M. B. M., Gupman, A., &. Pickens, R. W. (2000). Environmental and genetic transmission of substance use disorders in adoptive and step families. *Alcoholism: Clinical and Experimental Research, 24*, 1785–1794.

Newlin, D. B., & Thomson, J. B. (1990). Alcohol challenge with sons of alcoholics: A critical review and analysis. *Psychological Bulletin, 108*, 383–402.

Newman, J. P., Wolff, W. T., & Hearst, E. (1980). The feature-positive effect in adult human subjects. *Journal of Experimental Psychology: Human Learning and Memory, 6*, 630–650.

Nowak, M., & Sigmund, K. (1993). A strategy of win-stay, lose-shift that outperforms tit-for-tat in the Prisoner's Dilemma game. *Nature, 364*, 56–58.

Piazza, P. V., Deroche, V., Deminiere, J. M., Maccari, S., Le Moal, M., & Simon, H. (1993). Corticosterone in the range of stress-induced levels possesses reinforcing properties: Implications for sensation-seeking behaviors. *Proceedings of the National Academy of Sciences, U.S.A., 90*, 11738–11742.

Piazza, P. V., & Le Moal, M. (1998). The role of stress in drug self-administration. *Trends in Pharmacological Sciences, 19*, 67–74.

Pickens, R. W., Svikis, D. S., McGue, M., Likken, D. T., Heston, L. L., & Clayton, P. J. (1991). Heterogeneity in the inheritance of alcoholism: A study of male and female twins. *Archives of General Psychiatry, 48*, 19–28.

Pickens, R. W., & Thompson, T. (1968). Cocaine-reinforced behavior in rats: Effects of reinforcement magnitude and fixed-ratio size. *Journal of Pharmacology and Experimental Therapeutics, 161*, 122–129.

Platek, S. M., Keenan, J. P., Gallup, G. G., Jr., & Mohamed, F. B. (2004). Where am I? The neurological correlates of self and other. *Brain Research: Cognitive Brain Research, 19*, 114–122.

Pohorecky, L. (1977). Biphasic action of ethanol. *Biobehavioral Reviews, 1*, 231–244.

Prasad, B. M., Ulibarri, C., & Sorg, B. A. (1998). Stress-induced cross-sensitization to cocaine: Effect of adrenalectomy and corticosterone after short- and long-term withdrawal. *Psychopharmacology, 136*, 24–33.

Prescott, C. A., & Kendler, K. S. (1999). Genetic and environmental contributions to alcohol abuse and dependence in a population-based sample of male twins. *American Journal of Psychiatry, 156*, 34–40.

Ranaldi, R., Pocock, D., Zereik, R., & Wise, R. A. (1999). Dopamine fluctuations in the nucleus accumbens during maintenance, extinction, and reinstatement of intravenous D-amphetamine self-administration. *Journal of Neuroscience, 15*, 4102–4109.

Reboreda, J. C., & Kacelnik, A. (1993). The role of autoshaping in cooperative two-player games between starlings. *Journal of the Experimental Analysis of Behavior, 60*, 67–83.

Reich, T., Edenber, H. J., Goate, A., Williams, J. T., Rice, J. P., & Van Eerdewegh, P. (1998). Genome-wide search for genes affecting the risk for alcohol dependence. *Americal Journal of Medical Genetics, 81*, 207–215.

Robbins, T. W., & Everitt, B. J. (1999). Drug addiction: Bad habits add up. *Science, 398*, 567–570.

Robinson, T. E., & Berridge, K. C. (1993). The neural basis of drug craving: An incentive sensitization theory of addiction. *Brain Research Reviews, 18*, 247–291.

Salamone, J. D. (1994). The involvement of nucleus accumbens dopamine in appetitive and aversive motivation. *Behavioral Brain Research*, *61*, 117–133.

Sayette, M. A., & Wilson, G. T. (1991). Intoxication and exposure to stress: Effects of temporal patterning. *Journal of Abnormal Psychology*, *100*, 56–62.

Shaham, Y., & Stewart, J. (1995). Stress reinstates heroin-seeking in drug-free animals: An effect mimicking heroin, not withdrawal. *Psychopharmacology*, *119*, 334–341.

Sher, K. J. (1991). *Children of alcoholics: A critical appraisal of theory and research*. Chicago: University of Chicago Press.

Sherer, M. A., Kumor, K. M., & Jaffe, J. H. (1989). Effects of intravenous cocaine are partially attenuated by haloperidol. *Psychiatry Research*, *27*, 117–125.

Sigmund, K. (1995). *Games of life*. London: Penguin Books.

Sinervo, B., & Lively, C. M. (1996). The rock-paper-scissors game and the evolution of alternative male strategies. *Nature*, *380*, 240–243.

Skinner, B. F. (1938). *The behavior of organisms*. New York: Appleton-Century-Crofts.

Sorg, B. A. (1992). Mesocorticolimbic dopamine systems: Cross-sensitization between stress and cocaine. *Annals of the New York Academy of Sciences*, *28*(654), 136–144.

Sorg, B., & Kalivas, P. W. (1991). Effects of cocaine and footshock stress on extracellular dopamine levels in the ventral striatum. *Brain Research*, *559*, 29–36.

Sorg, B., & Kalivas, P. W. (1993). Effects of cocaine and footshock stress on extracellular dopamine levels in the medial prefrontal cortex. *Neuroscience*, *53*, 695–703.

Stewart, J., de Wit, H., & Eikelboom, R. (1984). Role of unconditioned and conditioned drug effects in the self-administration of opiates and stimulants. *Psychological Review*, *91*, 251–268.

Sutton, M. A., & Beninger, R. J. (1999). Psychopharmacology of conditioned reward: Evidence for a rewarding signal at D_1-like dopamine receptors. *Psychopharmacology*, *144*, 95–110.

Tiffany, S. T. (1990). A cognitive model of drug urges and drug-use behavior: Role of automatic and nonautomatic processes. *Psychological Review*, *97*, 147–168.

Tomie, A., Aguado, A. S., Pohorecky, L. A., & Benjamin, D. (1998). Ethanol induces impulsive-like responding in a delay of reward operant choice procedure: Impulsivity predicts autoshaping. *Psychopharmacology*, *139*, 376–382.

Turner, P. E., & Chao, L. (1999). Prisoner's Dilemma in an RNA virus. *Nature*, *398*, 441–443.

Weibull, J. W. (1996). *Evolutionary game theory*. Cambridge, MA: MIT Press.

Weingardt, K. R., Stacy, A. W., & Leigh, B. C. (1996). Automatic activation of alcohol concepts in response to positive outcomes of alcohol use. *Alcoholism: Clinical and Experimental Research*, *20*, 25–30.

Wilsnack, S. (1974). The effects of social drinking on women's fantasy. *Journal of Personality*, *42*, 43–61.

Wise, R. A. (2000). Addiction becomes a brain disease. *Neuron*, *26*, 27–33.

Wise, R. A., & Bozarth, M. A. (1987). A psychomotor stimulant theory of addiction. *Psychological Review*, *94*, 469–492.

Wise, R. A., Newton, P., Leeb, K., Burnette, B., Pocock, D., & Justice, J. B., Jr. (1995). Fluctuations in nucleus accumbens dopamine concentration during intravenous cocaine self-administration in rats. *Psychopharmacology*, *120*, 10–20.

Zinser, M. C., Fiore, M. C., Davidson, R. J., & Baker, T. B. (1999). Manipulating smoking motivation: Impact on an electrophysiological index of approach motivation. *Journal of Abnormal Psychology*, *108*, 240–254.

IV Spatial Cognition and Language

Does a brain develop to maximize the fitness of the individual that possesses that brain? If different reproductive strategies are employed by males and females, do these strategies have neural origins that lead to cognitive differences? Furthermore, why were certain cognitive abilities selected for and not others?

Spatial and language abilities are two cornerstones of research in cognitive psychology and cognitive neuroscience. There is a vast contribution of research from diverse disciplines in terms of these cognitive abilities. From Broca's initial findings to extensive research in voles to recent genetic manipulations and neuroimaging, uncovering the evolution of spatial and linguistic abilities is a tour through disparate disciplines. The research reported in this part demonstrates the diverse ways in which we can approach evolutionary cognitive neuroscience. Interestingly, we have found that our students rarely show any interest initially in learning about these cognitive abilities. However, once they see a Morris maze in person or feel the endocast of early *Homo* with a possible indentation in the left frontal region, there is generally an enthusiastic shift in their curiosity. For example, an irresistible study demonstrates that the hippocampus (anatomically) is dependent on navigational demands (Maguire et al., 2000). In London, taxicab drivers must pass an extensive test to get their cab license (referred to as "The Knowledge"). It was found that the hippocampal volume of these drivers, who needed to memorize all of the streets in London (as well as routes between landmarks), was significantly different from that of controls. In fact, regions of the right hippocampus reflected the experience of the driver, such that increased experience led to increased volume size.

Language occupies a unique place in human evolution. Questions about its evolution are key to understanding how and why we are such language masters. However, these questions are not simple to answer.

For example, the question. When did humans begin to speak? elicits debate and controversy to this day. Furthermore, the ultimate question of language—Why do we have verbal language abilities?—also remains unresolved. Yet, as our methods improve, our answers become clearer. In these instances, literally unearthing more data has proved extremely valuable. As we pinpoint language development on an evolutionary time scale, we can begin to understand its function. One method involves correlating language development with other abilities. For example, if language makes its first appearance 100,000 years ago, we know that *Homo* was using tools by that time. Clearly, one difficulty here will be providing significant evidence of language *not* being present in an individual or species. That is, future studies will be challenged to provide negative evidence of language.

But why speak in the first place? Is it the case that freeing our hands was the driving motivation for the formulation and development of spoken language? Is spoken language intimately tied to manual gesturing? To answer these questions, we again see the creativity of the evolutionary cognitive neuroscientist in action. For example, research is gathered from present-day behavior (examining the relation between verbal and nonverbal speech) as well as the physiology of archaic humans, not just in terms of the skull but in terms of the larynx and related structures as well.

It is clear that the early humans that developed these spatial and linguistic abilities had a reproductive advantage over others. However, as we examine the intermix of variables, the story unfolds in exciting new directions. For example, is it true that there is a relationship between language and the mirror neuron system? What is the influence of hormones on our spatial abilities, and are these differences due to seeking out mates?

It is in this part, as indicated, that both the methodologies and the topics are of significant interest to evolutionary cognitive neuroscience.

Reference

Maguire, E. A., Gadian, D. G., Johnsrude, I. S., Good, C. D., Ashburner, J., Frackowiak, R. S., & Frith, C. D. (2000). Navigation-related structural change in the hippocampi of taxi drivers. *Proceedings of the National Academy of Sciences, U.S.A.* 97, 4398–4403.

12 Sex Differences in Spatial Ability: Evolution, Hormones, and the Brain

David Andrew Puts, Steven J. C. Gaulin, and S. Marc Breedlove

In the study of sex differences in spatial ability, multiple levels of explanation—functional, phylogenetic, developmental, and proximate—have made complementary contributions to a more coherent view of a behavioral sex difference and its evolution, development, and neurobiology. In 1985, Wimer and Wimer commented that hippocampal function "has something to do with an adaptive difference in roles played by males and females of at least some species. Both our understanding of the operations performed by the hippocampus and of the nature of gender might benefit if a concerted attempt were made to understand what that something is" (p. 108).

The following year, Gaulin and Fitzgerald (1986) published the first of several papers that would begin to answer the question of why sex differences in the hippocampus have evolved in many species. Subsequent work by these authors and others would predict and find sex differences in the hippocampi of those species in which the sexes differ in the spatial problems confronted over their evolution.

But while evolutionary theory can predict the *presence* of neural sex differences, it cannot by itself predict what these differences will be or what will cause their development. This is because natural selection "sees" behaviors, not the underlying neural architecture. The proximate and ontogenetic causes of sex differences in spatial ability must be uncovered by careful anatomical, histological, cytological, molecular, and behavioral analysis. For example, Jacobs, Gaulin, Sherry, and Hoffman (1990) could predict sex differences in hippocampal volume across species only because previous work had shown that the hippocampus is related to spatial processing. Evolutionary theory could then inform hypotheses about the cross-species distribution of sex differences in the hippocampus. Likewise, behavioral neuroendocrinological research demonstrating the activational effects of sex hormones on spatial ability informs adaptive hypotheses about when, and in which species, these effects will be most pronounced. This chapter reviews the evolutionary, psychological, endocrinological, and neuroanatomical bases of sex differences in spatial cognition, in the hope of fostering such reciprocal contributions and a multilevel perspective.

Sex Differences

Homo Sapiens

With their influential book, *The Psychology of Sex Differences*, Maccoby and Jacklin (1974) made cognitive sex differences a topic of legitimate study and pointed to spatial ability as the most dramatic among these differences. They argued that, on average, males perform reliably better than females on a wide array of spatial tests. Subsequent meta-analyses (Linn & Petersen, 1985; Voyer, Voyer, & Bryden, 1995) confirmed this overall finding but also divided spatial skills more finely and estimated the magnitude of the sex difference in each of these areas. Using both psychometric (homogeneity of effect sizes) and cognitive (similarity of mental operations) criteria, this body of work has isolated three distinct types of spatial ability: spatial perception, mental rotation, and spatial visualization.

Spatial perception refers to the ability to recognize spatial relationships, for example, the horizontal, in spite of distracting or contradictory information. These tasks typically have a gravitational or kinesthetic component. Examples are the rod-and-frame test and the water-level task. Mental rotation is the ability to imagine two- or three-dimensional (2D, 3D) objects from a perspective other than the one depicted. The most widely used of these is the Vandenberg and Kuse (1978) mental rotation test. Spatial visualization tasks require the disembedding of a simple shape from a complex background. There is some question about whether spatial visualization can be reliably distinguished from what psychometricians call general fluid ability (the ability to form relationships among symbols), which is regarded as a nonspatial cognitive ability (Linn & Petersen, 1985). Examples of spatial visualization tasks include the embedded-figures test, the block design test, and the spatial relations subtest of the differential aptitude test.

Effect sizes (the difference between male and female means expressed in standard deviations) vary dramatically among these types of spatial ability (table 12.1.) The two most recent meta-analyses agree

Table 12.1
A Comparison of Effect Sizes for Three Types of Spatial Ability from Two Large Meta-analyses

	Weighted Effect Size	
Ability	Linn and Petersen, 1985	Voyer et al., 1995
Mental rotation	0.73*	0.56*
Spatial perception	0.44*	0.44*
Spatial visualization	0.13	0.19

*$P < 0.05$.
Note: Effect size is the difference between male and female performance on the same task, means expressed in standard deviations.

that mental rotation shows the largest sex difference and that the questionably spatial factor, spatial visualization, shows the smallest, often failing to reach statistical significance (Linn & Petersen, 1985; Voyer et al., 1995). Effect sizes within each of these three types of spatial ability are also heterogeneous and depend on task, presentation, and scoring details. For example, 2D mental rotation tasks show smaller effect sizes than 3D versions (Voyer et al., 1995). The Vandenberg and Kuse (1978) mental rotation test is a 3D test, but it can be scored one of two ways. Each of the 20 items has two correct and two incorrect answers. Each of the answers can be scored separately, which would yield a perfect score of 40, or an item can be scored correctly if and only if both choices are correct; this method yields a maximum score of 20. Effect sizes for the 40-point method lie between 0.50 and 0.75, whereas for the 20-point method they are larger, between 0.75 and 1.00 (Voyer et al., 1995). This scoring method with this test yields the largest reliable cognitive sex difference, unless, of course, one regards mating preferences as cognitive traits! Sex differences in mental rotations have been observed in African (Mayes & Jahoda, 1988; Owen & Lynn, 1993), East Indian (Owen & Lynn, 1993), and Asian (Mann, Sasanuma, Sakuma, & Masaki, 1990) populations, as well as in Western cultures.

In table 12.1, effect sizes for subjects of all ages are aggregated into a single group. In general, the larger the adult sex difference (as indicated by effect size) for a given type of spatial ability, the earlier during ontogeny that a reliable sex difference emerges. Thus, significant sex differences in mental rotation performance are regularly found even in prepubertal children. Significant sex differences in spatial perception generally arise during puberty. Although sex differences in spatial visualization are not significant when all ages are aggregated, they are significant among adults, with an effect size of 0.23 (Voyer et al., 1995).

Although they involve significant motor components, targeting and intercepting are sometimes discussed in the context of sexually dimorphic spatial abilities. Here again there is a significant male advantage, and it is measurable from childhood onward (Wickstrom, 1977). Although targeting and intercepting tasks seem to have obvious spatial components, for example, in trajectory prediction, their performance is not highly correlated with performance on the more conventional pencil-and-paper measures of spatial ability discussed above (Watson & Kimura, 1991). On the other hand, given the very large effect sizes observed on targeting tasks (1.0 to 1.5, Watson & Kimura, 1991) and their obvious ecological validity, these spatiomotor domains deserve further study. In particular, these tasks reveal primary abilities that might have been relatively direct targets of selection over human evolution.

Not all spatial tasks show an unambiguous male advantage. Recently, based on predictions from a particular evolutionary perspective, a female advantage on object-*location* memory has been demonstrated (McBurney, Gaulin, Devineni, & Adams, 1997; Silverman & Eals, 1992; Tottenham, Saucier, Elias, & Gutwin, 2003). Both pencil-and-paper and desktop versions of this task have been implemented; all require the ability to recall the location of items in arrays. These tasks tend to show a female advantage, but the effect size is not large (no meta-analysis is yet available), and the female advantage depends on details of the task and the presentation (see, e.g., Dabbs, Chang, Strong, & Milun, 1998; Montello, Lovelace, Golledge, & Self, 1999). For example, making the task explicit by telling participants that they will subsequently be asked about locations, or using abstract objects that are difficult to name, tends to eliminate or even reverse the female advantage (Choi & L'Hirondelle, 2005; Eals & Silverman, 1994). James and Kimura (1997) showed that when the positions of array objects are reciprocally exchanged there is a female advantage, but no sex difference is observed when objects are moved to new positions.

One possible explanation for these inconsistencies is that object-location memory tasks may require multiple cognitive processes, only some of which show a female advantage. Postma, Izendoorn, and De Hann (1998) attempted to decompose object-location memory, arguing that the task requires a spatial encoding of the occupied locations and a correct mapping of particular objects to particular locations. Unfortunately, they did not find a female advantage on any component of the task, so it is difficult to use their findings to explain the pattern of results seen in other studies of object-location memory.

From an evolutionary perspective, it seems appropriate to ask how and why these kinds of spatial skills evolved—what real-world challenges they were designed to address. Navigation is a plausible answer offered by numerous researchers (e.g., Gaulin & FitzGerald, 1986; Gray & Buffery, 1971; Halpern, 2000). There are surprisingly few real-world studies of navigation, probably because of the difficulty of implementing and scoring such tests, and fewer still have investigated the relationship between real-world wayfinding and performance on pencil-and-paper measures. Malinowski (2001) examined mental rotation ability and performance on a large-scale orienteering task among West Point cadets. Subjects were given the task of finding 10 waypoints distributed over an unfamiliar 6-km course, given only map coordinates and simple clues such as "in the valley." Performance on the orienteering task was positively correlated with mental rotation ability among men but not among women.

Montello et al. (1999) administered a large battery of spatial tests, some of them conventional pencil-and-paper tasks, some of them map-based tasks, and some of them involving real-world navigation. Using discriminant analysis, they discovered that performance on these various tasks could accurately assign 92% of their subjects to sex. An examination of those equations led the authors to support the emerging view that, with regard to real-world navigation, the sexes tend to exhibit different styles (e.g., Dabbs et al., 1998). Males exhibit better survey knowledge—they are better at understanding the relationships among locations that could be deduced from an aerial view or from a map. In the same contexts females exhibit better landmark knowledge—they are better at remembering particular locations, their contents, and their sequence along the route. Such a finding might accord well with the observation (above) that females exhibit superior object-location memory. Together these ideas suggest that, when environments are learned from maps, the sex difference in survey knowledge might be eliminated. This prediction agrees with the findings of Montello et al. (1999), but not with those of Malinowski (2001), whose participants were given maps but still exhibited a sex difference. A difference in scale might be responsible—Malinowski's course was an order of magnitude bigger than that of Montello et al. (1999)—but as yet no theory has explained why scale per se might affect male and female performance differently.

Virtual environments have also been used to study spatial problem solving in humans. Although some somatic cues (e.g., proprioceptive and vestibular input) are artificially absent in these studies, virtual environments provide an interesting bridge to the tools traditionally used to study spatial learning in rodents, that is, mazes. Moffat, Hampson, and Hatzipentalis (1998) administered a series of spatial and verbal tasks, along with computer-generated virtual mazes. Factor analysis was used to extract a spatial and a verbal factor from performance on the various nonmaze tests. When performance was measured either in terms of speed or of accuracy, males performed significantly better than females on the virtual mazes. In contrast to the findings of Montello et al. (1999), Moffat et al. (1998) found that virtual maze performance was correlated with their spatial factor in both sexes. On the other hand, they found that the verbal factor was also correlated with maze performance, but only among females. This finding, like the preferential use of landmarks by females, suggests that the sexes use different navigational strategies.

Astur, Ortiz, and Sutherland (1998) implemented a virtual version of the Morris water maze (MWM) task commonly used to study spatial learning in rodents. In the virtual task the subject uses a joystick to move about a "pool" in an attempt to find a hidden platform. The only available cues to the location of the platform are the landmarks and geometric features of the virtual "room" surrounding the pool. Three versions of the task gave progressively more helpful instruction, but all produced a significant male advantage, with an effect size between 0.50 and 1.00. In contrast, a control task in which the platform was visible produced no sex difference, suggesting that motivation, manual skill related to joystick use, and skill moving through virtual space were not causes of the observed sex difference.

In summary, the human data suggest most domains of spatial cognition show a significant male advantage, of at least moderate effect size, at least in adults. This finding holds across scales and presentations, from small-scale, pencil-and-paper and desktop tasks to walking-scale and real-world tasks, as well as for virtual instantiations of these tasks. Object-location memory may show a female advantage, but the effect size is typically small, and the precise task details that produce this sex difference have yet to be specified. In addition, scale may play a role, with larger scales accentuating the sex difference, but this idea requires further research.

Animal Models

A careful meta-analysis of the literature on sex differences in spatial ability among laboratory rodents (Jonasson, 2005) yields a clear but complex picture. This review concentrates on the two most frequently used paradigms, radial arm and water mazes, which isolate quite different components of spatial ability than the tasks used in the human literature. In particular, animal behaviorists have focused on a distinction between working and reference memory. For static objects, location is permanent and can be learned once and simply referred to; the ability to learn and recall such static information is called reference memory. In contrast, some objects are mobile or exhaustible, and thus their location must be frequently updated; the ability to update and retrieve this information is called working memory.

The contrast between working and reference memory can be illustrated in the radial arm maze (RAM) paradigm. A RAM has a central arena and a fixed number of arms, often eight, radiating from that arena. In one type of RAM experiment, a single reward is placed in each of the eight arms; the performance measure is the number of arms visited before all rewards are collected. If the animal remembers where it has collected rewards (working memory), it can attain a perfect score of 8. In a second type of experiment, some arms, perhaps four, never contain any reward. In these protocols, entering an arm where a reward has already been collected is a working-memory error, whereas entering an arm that has never contained a reward is a reference-memory error. In principle, either type of memory can be tested in either type of maze. For example, the MWM would be a reference-memory test if the location of the hidden platform were never changed, and a working-memory test if it were changed from one block of trials to the next.

This literature indicates that there is an overall male advantage, with an effect size of 0.60 in laboratory rodents. Reference-memory only experiments and MWM experiments yield effect sizes of about 0.50, whereas working-memory experiments and RAM experiments yield effect sizes approaching 0.70. Species differences, however, are large: the overall effect size for a sex difference in rats is 0.76 (and is somewhat variable among strains), but for mice it is only 0.18 and does not reach statistical significance (Jonasson, 2005).

The species difference suggests that selection may have been at work, and thus something about the history of these gene pools reflects the extent to which males and females differ on spatial performance. Unfortunately, it is probably impossible to reconstruct the founding populations and model the relevant selection pressures that might have operated in breeding colonies. A clear suggestion is that it would be useful to know something about male and female spatial performance in wild rodent populations. Some data are available and are reviewed in a theoretical context below. To foreshadow that discussion, polygynous species generally show spatially related sex differences, but monogamous species do not (Gaulin, 1992; Jacobs & Spencer, 1994).

The species difference is also relevant to questions regarding the neurobiology of sex differences in spatial ability. Most studies in this area have used rats as animal models, but several have also examined mice. In species, such as mice, in which consistent spatial sex differences have not been established, how should brain sexual dimorphisms be interpreted? At first it might appear that these brain dimorphisms must be unrelated to spatial ability. But consider that ancestors of laboratory mice may have differed by sex in spatial ability and underlying neurobiology. Many generations of artificial selection may have reduced spatial sex differences but spared some neural dimorphisms. A safer conclusion is that brain dimorphisms in species without spatial sex differences may be necessary but not sufficient for spatial sex differences. A final caveat is that particular brain sex dimorphisms may be sufficient to produce sex differences in spatial behavior in some species or strains but not in others.

There is intriguing evidence that rats may exhibit a strategic sex difference paralleling the apparent landmark-survey preferences seen in women and men, respectively. Solutions to mazes of any type must be based on some sort of reference. Those references could be relatively goal-specific, such as a landmark hung over the hidden platform in an MWM, or more distal, such as the shape of the room in which the maze sits. Experimental manipulation of these two types of cues suggests that female performance is more degraded when landmark cues are altered or withheld, whereas male performance is more adversely affected when global geometry is altered (by moving the maze in the room) or withheld (by curtaining it off) (Kolb & Cioe, 1996; Williams, Barnett, & Meck, 1990; Williams & Meck, 1991). Unfortunately, there have been no attempts to implement studies of object-location memory with rodents, either wild or domesticated.

Summarizing the rodent literature, it seems clear that sex differences in spatial performance are not restricted to humans. Laboratory rats and polygynous species of wild rodents show a distinct male advantage on various types of maze tasks. In laboratory mice and monogamous species of wild rodents, these differences are reduced or absent. Some sex-specific cue preferences also seem to be shared among humans and rodents. The cross-species distribution of these sex differences should constrain our hypothesizing about their proximate and ultimate causes.

Evolution

Theoretical Underpinnings

Because selection shapes organisms to match the demands of their environments and because, within most species, males and females tend to contact the environment in similar ways, the phenotypes of the two sexes tend to evolve in the same direction (Darwin, 1859). The mating context often provides an exception (Darwin, 1871). Particularly when the sexes have different maximal rates of reproduction, they will face different challenges in the mating arena (Clutton-Brock & Vincent, 1991; Trivers, 1972).

Consider the case where males can reproduce more rapidly than females because females invest more in each reproductive venture (e.g., via obligate gestation and lactation in mammals). In such a case, a male can return to the mating pool quite rapidly following copulation without compromising his fitness prospects. On the other hand, high levels of parental investment may remove females from the mating pool for extended periods of time. This means that the mating pool would typically include many more males than females. Such an imbalance produces disproportionate competition among males for mating opportunities. In contrast, females are not expected to compete for something in abundant supply. The result is that selection favors competitive traits in males more than in females, and thus their phenotypes diverge over evolutionary time precisely with respect to the traits that confer an advantage in mating competition. Of course, such traits are not limited to mere weaponry or sexual display structures. Cognitive and motivational systems are likely to be affected as well.

Not only sexual selection but also natural selection may occasionally produce sex differences. A classic case would be feeding niche differentiation in monogamous birds. Because of biparental care, the feeding success of each partner has an impact on the fitness of its mate. In these cases selection may favor adaptations that allow males and females to exploit different food resources, so as to reduce competition with their mates.

Both sexual selection and natural selection theories have generated hypotheses about the evolutionary basis for sex differences in spatial ability; Sherry and Hampson (1997) provide a review that integrates hormonal and evolutionary perspectives. Most of these hypotheses assume that the cognitive processes measured as "spatial ability" originally evolved in the service of real-world navigation. A further assumption is that relatively large ranges would have favored improvements in these abilities. The baseline observation that any such adaptive hypothesis must explain is that, in general, males perform better than females on most tests of spatial ability. This difference is not restricted to humans, being observed, for example, in laboratory rats and polygynous wild rodents. And as discussed earlier, a spatial domain in which human females outperform males, object-location memory, has also been discovered. A satisfactory evolutionary explanation would account for all three of these observations.

Sexual Selection and Spatial Ability

Sex differences in spatial ability could be explained by sexual selection if, for some reason, increments in spatial ability had a greater effect on the mating success of one sex than the other. Several such theories exist. Alexander (1979) has proposed that human warfare, which potentially eliminates male competitors and may involve the capture of wives, would have favored male range expansion, and hence put a premium on male spatial abilities. Hawkes (1990, 1991) views the hunting of animal prey as energetically inefficient compared to the gathering of plant foods. She thus explains hunting as a form of sexually selected male display; for some reason, females prefer males who are better hunters. The possibility that such males have better-fed offspring falls under a different, natural selection explanation (see below), and in any case is contradicted by Hawkes's data. She concludes that better hunters do have higher reproductive success, but only by virtue of their elevated sexual access to *other men's* wives (Hawkes, 1991). This hypothesis requires the plausible assumption that hunting requires a larger range than gathering.

The last sexual selection hypothesis appeals not to a uniquely human trait, such as warfare or sexual division of foraging labor, but to the mate-searching strategies that are precipitated by various kinds of

have confirmed that females more precisely recall the location of food items in a real-world environment (Gaulin, Krasnow, Truxaw, & New, 2005).

Thus, at present, the most plausible evolutionary explanation for the patterns of observed sex differences in human spatial cognition requires the conjunction of two models. Because both chimpanzees and humans are fundamentally polygynous with larger male ranges, we might plausibly assume that our common ancestor was as well. From this viewpoint, sexual selection arising out of male-male competition for access to mates favored an array of superior spatial skills in male protohumans for at least 7 million years. At some later point in human evolution the sexes began to concentrate on different ecological resources. This differential concentration in turn began to favor a distinctive spatial ability in females. This type of cumulative selection (Dawkins, 1986) is a hallmark of the evolutionary process.

Hormones and Development

Selection can create sex differences by favoring responses to sex-specific hormonal regimes. At their most basic level, these responses are molecular. Sex hormones bind to their receptors and modulate gene transcription. Because the sexes differ in the relative amounts of hormones secreted by their gonads, sex differences in gene expression result. Sex hormone–mediated gene transcription affects the growth, development and maintenance of the body, including the nervous system. In this section, we review evidence regarding the hormonal mediation of sex differences in spatial ability, and in the next we look at what neural substrates these hormones may be acting on to create spatial sex differences.

Organizational Hormonal Effects

Animal Models In rats, spatial ability is masculinized by testicular hormones during the perinatal period. Several studies have shown that neonatal castration impairs maze learning in males (Dawson, Cheung, & Lau, 1975; Isgor & Sengelaub, 2003; Joseph, Hess, & Birecree, 1978; Williams et al., 1990) and neonatal testosterone treatment improves maze performance in females (Dawson et al., 1975; Isgor & Sengelaub, 1998, 2003; Joseph et al., 1978; Roof, 1993b; Roof & Havens, 1992; Stewart, Skvarenina, & Pottier, 1975).

At present, however, it is unclear whether the effect of testosterone on spatial performance is mediated by the binding of testosterone to androgen receptors (ARs). This is because many androgens, including testosterone, may be converted into estrogens, such as estradiol, in the brain through a process called *aromatization* (after the enzyme aromatase), which may then masculinize behaviors by binding to estrogen receptors (ERs). Williams and colleagues (Williams et al., 1990; Williams & Meck, 1991) found that neonatal estradiol treatment masculinized spatial ability in female rats, suggesting that spatial sex differences may be ER-mediated. However, a subsequent study (Isgor & Sengelaub, 1998) found that prenatal estradiol treatment did not masculinize MWM performance in female Sprague-Dawley rats, whereas treatment with either testosterone or (the nonaromatizable androgen) dihydrotestosterone did. Of course, it is possible that both ARs and ERs are involved in masculinizing spatial ability in rats. The question of whether spatial ability in rats is masculinized via AR is likely to be answered in the near future by studies of rats with nonfunctional ARs.

Whether androgens masculinize spatial ability in rats directly or via aromatization, there appears to be an optimal level of early androgen exposure beyond which spatial ability actually declines. For example, early androgen treatment improves spatial ability in females but impairs it in gonadally intact males (Roof, 1993b; Roof & Havens, 1992).

Homo Sapiens As in experimental rodents, early androgens appear to masculinize spatial ability in humans, but pubertal androgens may be necessary for complete masculinization. The role of early androgens is supported by multiple lines of evidence. In one study, second-trimester testosterone levels in female fetuses positively predicted spatial abilities when these girls were 7 years old (Grimshaw, Sitarenios, & Finegan, 1995). In another study, girls with male twins exhibited better spatial ability, presumably because of in utero exposure to androgens produced by the male twin (Cole-Harding, Morstad, & Wilson, 1988). Further evidence for the role of early androgens comes from so-called natural experiments, developmental variations characterized by sex-atypical hormone signaling.

Turner syndrome Turner syndrome (TS) represents one such natural experiment. TS individuals have a 45,X karyotype and are phenotypically female, although they tend to be below average in stature and are infertile. Androgen and estrogen production are extremely low due to undifferentiated gonads (Hojbjerg Gravholt, Svenstrup, Bennett, & Sandahl Christiansen, 1999; Ross et al., 2002), and these hormonal abnormalities may be responsible for specific cognitive deficits in spatial ability (Nijhuis-van der Sanden, Eling, & Otten, 2003). Ross and colleagues (Ross et al., 2003) also found that 2 years of androgen treatment did not improve spatial ability in 26 adolescent (10–14 years) girls with TS. Because pubertal androgens probably improve spatial ability in males (see below), this lack of an effect of pubertal androgens in TS females indicates that early androgens may be necessary for later pubertal organizational effects.

Congenital adrenal hyperplasia Studies of congenital adrenal hyperplasia (CAH) provide further evidence for an organizational effect of androgens. In this condition, an enzyme deficiency causes precursors of cortisol to be shunted down the androgen pathway, leading to an overproduction of androgens from the adrenal glands. Although the hormonal abnormalities of CAH are treated shortly after birth, girls with CAH show signs of elevated prenatal androgen exposure (e.g., virilized genitalia) and tend to be masculinized along several behavioral dimensions (Berenbaum, 1999). Some studies have found CAH girls to exhibit masculinized spatial abilities (Hampson, Rovet, & Altmann, 1998; Hines et al., 2003; Perlman, 1973; Resnick, Berenbaum, Gottesman, & Bouchard, 1986), although others have not (Baker & Ehrhardt, 1974; Helleday, Bartfai, Ritzen, & Forsman, 1994; McGuire, Ryan, & Omenn, 1975). An early CAH study (Perlman, 1973) found that girls with CAH and boys outperformed control girls on one spatial test, but that girls with CAH performed worse than control girls on a spatial test in which no sex difference between controls was found. Because males normally outperform females on this latter test (Weschler Block Design test), this finding raises questions about the representativeness of the control samples. With the exception of this study, however, wherever significant differences between the spatial abilities of CAH and control females have been found, females with CAH have exhibited more masculine spatial abilities. Indeed, a recent meta-analysis (Puts, McDaniel, Jordan, & Breedlove, 2005) concluded that females with CAH have better spatial abilities than do control females across studies.

Complete androgen insensitivity syndrome Studies of females with complete androgen insensitivity syndrome (CAIS) further support the role of androgens in organizing spatial ability. CAIS individuals have a

46,XY karyotype and develop testes that remain undescended in the abdominal cavity. Despite producing normal to high male levels of testosterone, individuals with CAIS have nonfunctional ARs and so are phenotypically female (Imperato-McGinley et al., 1982). CAIS females thus have the potential to provide information about whether androgens masculinize spatial ability and whether they do so via the AR.

Imperato-McGinley and colleagues (Imperato-McGinley, Pichardo, Gautier, Voyer, & Bryden, 1991) found that females with CAIS performed significantly worse on spatial tasks than did their male relatives. On the surface, this finding seems to suggest that androgens masculinize spatial ability via ARs. However, it is also possible that females with CAIS exhibit less masculine spatial abilities because they were socialized in a manner concordant with their phenotypic gender. A more powerful comparison is that between CAIS females and their unaffected (46,XX) female relatives. If spatial ability is AR-mediated, then the spatial abilities of CAIS females should be even less masculine than those of their unaffected female relatives (who produce and receive some androgen message, if less than that of male relatives). In fact, this is precisely what Imperato-McGinley and her colleagues (1991) found. Even this comparison must be interpreted cautiously, however: it is possible that ovarian hormone production in unaffected females caused this difference with CAIS individuals.

Idiopathic hypogonadotropic hypogonadism Thus, CAIS studies indicate that androgens may masculinize spatial ability by acting directly on the AR, and CAH studies suggest that *prenatal* androgens are particularly important. However, evidence from individuals with idiopathic hypogonadotropic hypogonadism (IHH) indicates that pubertal androgenization may be necessary for complete masculinization of spatial ability. IHH males have a 46,XY karyotype but do not produce gonadotropin-releasing hormone (GnRH). GnRH stimulates the anterior pituitary to release luteinizing hormone, causing the testes to produce testosterone. Consequently, untreated IHH men have very low testosterone levels. IHH individuals have normal masculinization in utero, probably due to exposure to maternal luteinizing hormone, and their condition usually is not discovered until they fail to produce the testosterone surge required for puberty.

Hier and Crowley (1982) tested 19 such men on a battery of spatial and verbal tasks. Spatial (but not verbal) performance correlated positively with testicular size, indicating that androgen production affected spatial ability. The men with IHH were also compared with 19 eugonadal men and five men who had developed hypogonadism during or after an otherwise normal puberty. The spatial (but not verbal) scores of the men with IHH were significantly below those of the two control groups, which did not differ significantly from one another. Because both hypogonadal groups had plasma testosterone levels within the normal female range, but only the IHH men had below-normal levels during puberty, these results suggest that pubertal androgens have a positive effect on spatial ability that is undiminished if androgen levels subsequently decline (but see Cappa et al., 1988).

Activational Effects

Androgens appear to organize spatial ability, probably through the AR in humans, and possibly through aromatization in some other mammals. Sometimes gonadal hormones in adulthood also have activational effects on spatial ability, affecting the magnitude of sex differences. We should expect spatial behaviors to remain susceptible to hormonal fluctuations whenever maintaining plasticity in the neural systems underlying spatial ability has some net fitness benefit to the organism. This is likely to pertain when spatial demands change significantly and repeatedly (for example, seasonally). These conditions differ not only across species but also between the sexes.

Animal Models

Testosterone and polygyny In some species, such as meadow voles (*Microtus pennsylvanicus*) and deer mice (*Peromyscus maniculatus*), males expand their home ranges during the breeding season in order to increase access to mates (Galea, Kavaliers, & Ossenkopp, 1996; Galea, Kavaliers, Ossenkopp, & Hampson, 1995; Gaulin & FitzGerald, 1989). In both of these species, males outperform females on laboratory spatial tasks only during the breeding season (Galea et al., 1996; Gaulin & FitzGerald, 1989). These seasonal sex differences are probably due partly to testosterone levels, which are elevated in males during the breeding season (Galea & McEwen, 1999). On the other hand, in relatively nonseasonal species, such as rats, spatial ability appears to be comparatively unresponsive to testosterone after certain critical periods. We have known for decades, for example, that castration of male rats after the first 10 or so days of life has little effect on spatial ability (Commins, 1932).

Estrogen, fertility, and maternal care Although testosterone probably increases spatial ability and range size in males of seasonally breeding species, estrogens appear to have the opposite effect in intact females. For example, several studies have found impaired maze performance in female rats during days in the estrous cycle when estradiol levels are high (Diaz-Veliz, Soto, Dussaubat, & Mora, 1989; Frye, 1995; Warren & Juraska, 1997). Similarly, range size in the wild and maze performance in the laboratory decrease with elevated estradiol levels during the breeding season in female meadow voles (Galea et al., 1995), and female rats show impaired maze performance during the third trimester of pregnancy, when estradiol levels are highest (Galea et al., 2000). On the other hand, very low levels of estradiol also decrease spatial ability in females: maze performance is impaired by ovariectomy and restored by estradiol administration in female rats (Daniel, Fader, Spencer, & Dohanich, 1997; Luine, Richards, Wu, & Beck, 1998). Sherry and Hampson (1997) have suggested that responsiveness of spatial behavior to estradiol in these species constitutes a pregnancy-related adaptation. According to this hypothesis, the relatively low estradiol levels characteristic of early pregnancy increase spatial ability and ranging to aid females in foraging and locating suitable nest-building sites. Late in pregnancy, high estradiol levels decrease ranging behavior in preparation for nest building and parturition.

Homo Sapiens Although numerous studies purport to demonstrate activational effects of androgens on spatial ability in humans, a careful examination of the literature reveals that such effects are likely to be small or nonexistent. On the other hand, some evidence suggests that estrogens may have inhibitory activational effects on spatial ability in some groups.

Androgens Several studies have found significant relationships between current testosterone levels and spatial ability in between-subjects comparisons. Some of these studies have found simple linear relationships between testosterone levels and spatial ability in men (Silverman, Kastuk, Choi, & Phillips, 1999), pubertal boys (Hassler, 1992) and women (Hausmann, Slabbekoorn, Van Goozen, Cohen-Kettenis, & Gunturkun, 2000). Others have found evidence of a curvilinear relationship (Gouchie

& Kimura, 1991; Moffat & Hampson, 1996). In the latter studies, low and high testosterone levels are associated with poorer performance, and intermediate levels are associated with superior performance.

These studies suggest relationships between spatial abilities and testosterone levels, but the shape of the relationships (linear vs. curvilinear) remains unclear. Perhaps more important, between-subjects correlational studies leave questions about the temporal relationships between hormones and spatial ability. The problem with such tests is that circulating levels of hormones in adults may correlate with levels during some earlier life stage. For example, the gonads of some individuals may produce higher than normal androgen levels throughout life. If so, a correlation between adult hormone levels and spatial ability may simply reflect the effects of high androgen production during some earlier organizational period and a tendency for androgen production to continue to be relatively high later in life. Thus, between-subjects correlations often cannot address whether testosterone has activational or organizational effects on spatial ability.

Within-subjects correlational studies can better address whether testosterone activates spatial ability because these studies can show changes in spatial ability that might be caused by fluctuating hormone levels. For example, Moffat and Hampson (1996) found circadian changes in spatial ability that differed significantly by sex. Males tended to improve over the morning, whereas females exhibited the opposite trend. Because testosterone levels decrease over the morning in both sexes, and assuming that high testosterone levels augment female spatial ability but impair it in males, Moffat and Hampson suggested that the sex difference in performance change was the result of activating effects of testosterone. Although plausible, this hypothesis would be better supported by within-subjects correlations between changes in testosterone levels and changes in spatial performance. Without these data, we are left wondering whether the observed changes in spatial ability correlated with testosterone level changes in either sex, or whether another hormone or some other physiological change was responsible. Indeed, the only study to report these highly relevant correlations (Silverman et al., 1999) found no significant relationship between changes in men's testosterone levels and changes in their 3D mental rotation performance over a 12-hour period.

Of course, the best tests of potential causal relationships between current hormone levels and spatial ability involve hormone manipulations. Demonstrating that hormone treatment elicits a particular phenotypic change and that removal of treatment abolishes this effect constitutes strong evidence for the activating effects of the hormone on the phenotype. Although no studies of which we are aware have examined the effects of removing testosterone treatment, several have measured spatial performance before and after testosterone treatment.

Hier and Crowley (1982) found no difference in spatial ability after androgen therapy in a small sample of six androgen-deficient men. On the other hand, Van Goozen and colleagues (Van Goozen, Cohen-Kettenis, Gooren, Frijda, & Van de Poll, 1994) reported that 22 female-to-male transsexuals performed better at 2D mental rotation after 3 months of testosterone treatment than shortly before treatment was initiated. The authors interpreted this result as a clear demonstration that "the administration of androgens to females causes a shift in the direction of a masculine pattern of cognitive functioning" (p. 1155). However, no untreated controls were included in this study, so the improvement observed could have been due to practice rather than testosterone treatment.

Indeed, in a subsequent study of both female-to-male and male-to-female transsexuals, this time including male and female controls (Van Goozen, Cohen-Kettenis, Gooren, Frijda, & Van de Poll, 1995), subjects' spatial performance improved over time. The authors also reported that the changes in spatial performance differed significantly between these groups, but it appears that this interaction was driven by a slight decline in spatial performance in male-to-female transsexuals (treated with estrogen and antiandrogen) compared to improvement in the other three groups. In order to show that testosterone treatment improved spatial ability, it would have been necessary to show that testosterone-treated female-to-male transsexuals improved significantly more than did untreated females. Another study by these authors, this time without untreated controls, found similar results in hormone-treated individuals: improvement in testosterone-treated female-to-male transsexuals, and no improvement in estrogen- and antiandrogen-treated male-to-female transsexuals (Slabbekoorn, Van Goozen, Megens, Gooren, & Cohen-Kettenis, 1999). From these articles, it is impossible to determine whether testosterone treatment in adults causes an improvement in spatial ability or whether estrogen treatment inhibits it.

Several studies have performed the appropriate controlled comparisons to address whether testosterone treatment improves spatial learning in adults. Van Goozen and colleagues (Van Goozen, Slabbekoorn, Gooren, Sanders, & Cohen-Kettenis, 2002) again examined changes in spatial performance in hormone-treated transsexuals and untreated controls. Although scores improved on mental rotations tasks, there were no differences between groups in improvement on any of the tasks. Alexander and colleagues (Alexander et al., 1998) also found no improvement in visuospatial performance above that due to practice after 6 or more weeks of testosterone treatment in 10 eugonadal and 33 hypogonadal men. Likewise, Ross et al. (2003) observed no improvement in spatial abilities in 26 androgen-treated TS patients relative to placebo-treated TS controls, and Wolf and colleagues (2000) found no effect of a single testosterone injection relative to placebo in 30 elderly men.

O'Connor, Archer, Hair, and Wu (2001), in a well-designed, double-blind, placebo-controlled experiment, also found that testosterone treatment did not affect spatial ability in seven hypogonadal men relative to controls. On the other hand, these researchers observed a significant effect of testosterone treatment in eugonadal men. Whereas placebo group performance increased over three testing sessions, the performance of the testosterone-treated eugonadal group decreased on the second testing session and then showed normal improvement on the third. One interpretation of these results is that, within the normal female-male range, testosterone has little activational effect on spatial performance, but supraphysiological levels of circulating androgens (such as those in androgen-treated eugonadal men) impair spatial performance. However, given that another study (Alexander et al., 1998) failed to find an effect of testosterone treatment on eugonadal men of the same age group, this interpretation should be made cautiously.

Another placebo-controlled double-blind experiment found a significant effect of testosterone treatment on spatial performance in elderly men, but these results are peculiar as well. Janowsky, Oviatt, and Orwoll (1994) observed no significant difference in spatial performance between testosterone-treated and placebo-treated elderly men after 12 weeks of treatment. However, the testosterone-treated men improved slightly between tests, whereas the performance of the placebo-treated men decreased slightly, resulting in a significant interaction between treatment group and testing session. What seems most noteworthy is not the improvement in the testosterone-treated group but the lack of improvement in the placebo-treated group. Several studies (Alexander et al., 1998; O'Connor et al., 2001; Van Goozen et al., 1995, 2002; Wolf et al., 2000) have shown significant improvement with practice in untreated or placebo-treated controls on a variety of spatial tasks (including the block design task used by Janowsky et al.) over a range of between-test intervals subsuming that used by Janowsky et al. Thus, the significant "effect" of testosterone observed in this study may have been due to the absence of normal task learning in the control group.

In general, these findings—no within-subjects correlations between changes in testosterone levels and changes in spatial ability, and evidence against a testosterone treatment effect—suggest that, at least within the normal range of circulating levels, testosterone has no activational effect on spatial ability in humans. Perhaps this should not be surprising in a species with very low breeding seasonality.

Estrogens On the other hand, menstrual cycle variation in spatial performance (Hampson, 1990a, 1990b; Hampson & Kimura, 1988; Hausmann et al., 2000; Phillips & Silverman, 1997), between-subjects correlations (Hausmann et al., 2000) and the possible treatment effects of estrogens (Slabbekoorn et al., 1999; Van Goozen et al., 1995) suggest that estrogen may have inhibitory activating effects on spatial learning. Other studies have found no effect of estrogen treatment, however. Miles, Green, Sanders, and Hines (1998) and Van Goozen et al. (2002) found no effect of estrogen and antiandrogen treatment on mental rotation performance on male-to-female transsexuals. Moreover, in postmenopausal women, estrogen replacement *improved* performance on a prefrontal cortex/working memory-related spatial task (Duff & Hampson, 2000). Differences between studies in treatment groups (males vs. females, normally cycling vs. postmenopausal women), hormone treatments, and spatial tests may explain these discrepancies. In particular, estrogens may have an inverted U-shaped relationship to spatial ability in women, such that intermediate levels are associated with optimal spatial ability, as in some rodents.

The Brain

Sex-specific hormonal milieus appear to play a major role in causing sex differences in spatial ability, and they may do so by operating on brain regions such as the hippocampus, which is often larger in the sex with superior spatial ability. However, knowing, for example, that male meadow voles have larger hippocampi than females is not particularly informative about what precisely is causing spatial sex differences at the proximate level. Moreover, selection for superior spatial abilities in one sex may not always lead to sex differences in gross measures like hippocampal size. The neural substrate for spatial sex differences may be subtler, including differences in the sizes of smaller brains regions; differences in cell soma size, neuron density, or dendritic arborization; differences at the molecular level; or widely distributed but subtle differences in any of these measures, to name a few possibilities. The next sections review such finer-scale neural sex differences and their hormonal mediation in mammalian species that exhibit sex differences in spatial behavior.

The Hippocampal Complex

Animal Models The hippocampal complex is located in the medial temporal lobe and is associated with episodic memory and especially with spatial memory and navigation. In humans, the right hippocampus appears particularly important for spatial learning and recall (Maguire, Frackowiak, & Frith, 1996). The hippocampal complex comprises several regions, including the dentate gyrus (DG), the subiculum, and the hippocampus proper (cornu ammonis 1–3, CA1–CA3). Information enters the hippocampus via the DG, where it is transmitted to CA3, to CA1, and then to the subiculum. Males are apparently more reliant than females on the hippocampus for spatial processing in species in which males are advantaged at spatial tasks. This sex difference is illustrated by functional imaging studies in humans and lesion studies in laboratory animals. Lesions to the ventral hippocampus or the entorhinal cortex (the primary cortical input to the hippocampal complex) impair MWM performance in male but not in female Sprague-Dawley rats (Roof, Zhang, Glasier, & Stein, 1993; Silva-Gomez et al., 2003). Thus, the neural substrate for sex differences in spatial ability probably resides partly in the hippocampal complex. Sex differences have been found within several hippocampal subfields, including CA1, CA3, and the DG, as we will see.

Cornu ammonis 1

CA1 sex differences CA1 is one of the final cell fields in the processing and passage of information through the hippocampal complex before output to other brain regions. In species in which males exhibit superior spatial behavior, males tend to have a CA1 that contains larger pyramidal cells (large, multipolar neurons) and, at least in some regions, is larger in volume. For example, compared to females, male Sprague-Dawley rats have larger pyramidal cell bodies (Isgor & Sengelaub, 1998) and CA1 pyramidal cell field volumes (Isgor & Sengelaub, 1998; Madeira et al., 1992). Madeira and colleagues (1992) also estimated that male rats have more total CA1 pyramidal neurons than females, but Isgor and Sengelaub (1998) did not. Lavenex and colleagues (2000) found no sex difference in CA1 neuronal number among Eastern gray squirrels (in which males have larger home ranges than females), but found larger volumes in two CA1 cell layers (strata oriens and radiatum) in males than in females.

Finally, Cobb and Juraska (2004) found males of one mouse strain to have a larger-volume CA1 than females. But we recall here that across studies, there is no overall sex difference in spatial ability in mice. This suggests that, if a larger CA1 volume is necessary for male spatial superiority, it is not sufficient, or that some mouse strains may indeed display a sex difference on spatial tasks.

Hormonal mediation of CA1 sex differences In Sprague-Dawley rats, early exposure to sex steroids organizes at least two adult CA1 sexual dimorphisms. Isgor and Sengelaub (1998) treated pregnant dams with either flutamide (an antiandrogen), testosterone, estradiol, dihydrotestosterone, or no treatment, and their offspring were examined. Prenatally flutamide-treated males were castrated at birth, and males in another group that received no prenatal treatment were castrated as adults. CA1 pyramidal soma size and pyramidal cell field volume were subsequently measured in adult males and females of various treatment groups. Most notably, prenatal estradiol and testosterone masculinized females on these measures, but prenatal dihydrotestosterone did not. Because testosterone, but not dihydrotestosterone, is aromatizable into estradiol, these results indicate that androgens masculinize CA1 pyramidal soma size and pyramidal cell field volume via aromatization. In addition, adult castration did not feminize males on these measures, suggesting that the activational influences of testicular hormones are not required to masculinize these traits in adult rats.

A puzzling result concerns the prenatally flutamide-treated males. Flutamide blocks androgens by binding to the androgen receptor, so it might seem that flutamide treatment should not affect traits that are masculinized by androgens via aromatization. The finding that flutamide-treated males were not masculinized seemingly implicates AR mediation and contradicts the female data. Isgor and Sengelaub (1998) suggested that both prenatal testosterone and estradiol may be needed for the masculinization of these traits. However, neonatal castration, rather than flutamide treatment, may explain why males in this group were not masculinized. The critical period for masculinization of these CA1 traits may extend to postnatal day 1, when castration was performed on flutamide-treated males. If so, this group may have exhibited feminine CA1 morphology because castration removed their source of aromatizable testosterone neonatally, a possibility that accords well with the female data.

In contrast, Cobb and Juraska (2004) found no effect of ER-alpha knockout on CA1 volume in a mouse strain that is sexually dimorphic for this trait. One way to reconcile this finding with those of Isgor and Sengelaub (1998) in rats is that CA1 volume masculinization depends on the binding of estrogen to its other receptor (ER-beta). Alternatively, separate hormones may mediate CA1 dimorphisms in different species, or separate hormones may mediate different CA1 dimorphisms in the same species.

Some studies have also found androgen treatment effects on CA1 cell morphology (Leranth, Petnehazy, & MacLusky, 2003) and cytochemistry (Xiao & Jordan, 2002) in adult rats. However, given the lack of an effect of adult androgen manipulations on spatial ability in rats (see above), these neural treatment effects are probably not related to changes in spatial ability.

Cornu ammonis 3

CA3 sex differences CA3 is situated between the DG and CA2. As in CA1, CA3 pyramidal cell bodies and pyramidal cell field volumes are larger in male rats than in females (Isgor & Sengelaub, 1998). Moreover, rats exhibit sex differences in CA3 pyramidal cell dendritic branching (Isgor & Sengelaub, 2003; Juraska, Fitch, & Washburne, 1989) and length (Isgor & Sengelaub, 2003); thus, males' CA3 pyramidal cells have a greater volume of influence than do females' (Isgor & Sengelaub, 2003). Finally, although the number of synapses between mossy fibers (axons projecting from the DG) and the apical dendrites of CA3 pyramidal neurons is the same in male and female Sprague-Dawley rats, the density of such synapses is lower and the volume of the mossy fiber system is greater in males than in females (Madeira, Sousa, & Paula-Barbosa, 1991).

Hormonal mediation of CA3 sex differences Isgor and Sengelaub (2003) demonstrated that neonatal androgens masculinize several sexual dimorphisms in the rat CA3. Sprague-Dawley rats were divided into three low-androgen groups (ovariectomized females, sham-ovariectomized females, neonatally castrated males) and three high-androgen groups (sham-castrated males, neonatally castrated males treated with testosterone propionate from postnatal day 2, females treated with testosterone propionate on postnatal days 3 and 5). Relative to the low-androgen groups, the high-androgen groups were significantly masculinized in CA3 pyramidal cell length, dendritic branching, and volume of influence (volume of the gray matter from which a cell's dendrites can receive input) for nearly all two-group comparisons (Isgor & Sengelaub, 2003). It is not clear from this study whether the aromatization of testosterone into estradiol is involved in the development of any of these CA3 sex dimorphisms.

A previous study by these authors (Isgor & Sengelaub, 1998), however, neatly demonstrates that androgens directly masculinize two other CA3 sexual dimorphisms. In this study, females treated prenatally with testosterone or dihydrotestosterone were masculinized on pyramidal cell field volumes and soma sizes, whereas those treated with estradiol were not. Additionally, males with androgenic influences removed via prenatal flutamide treatment and neonatal castration were feminized on these traits. These results indicate that aromatization of androgen into estrogen is unnecessary for masculinization of CA3 pyramidal cell field volume and soma size. However, this study cannot rule out the possibility that sexual dimorphisms in these traits also depend on early *post*natal androgen action, because it was not demonstrated that similar postnatal treatments would not produce the same results.

Dentate gyrus

DG sex differences The dentate gyrus (so called because of its toothy appearance) consists of three cell layers, including the granule cell layer (DG-GCL). Rodents exhibit several sex differences in the DG-GCL, with males tending to have some combination of the following features: a more lateralized (right greater than left) and perhaps larger DG-GCL, with larger and perhaps more numerous and more densely packed granule cells.

In meadow voles (Galea, Perrot-Sinal, Kavaliers, & Ossenkopp, 1999) and juvenile rats (Roof, 1993a), the DG-GCL is wider in males than in females on the right side only. And in adult rats (Roof & Havens, 1992), the DG-GCL on both sides is wider in males than in females, but the right side is wider than the left side in males only. Interestingly, in both adult (Roof & Havens, 1992) and juvenile (Roof, 1993a) rats, MWM performance correlates with right DG-GCL width. The DG-GCL is also thicker in males than in females in adult (Roof & Havens, 1992) and juvenile (Roof, 1993a) rats. However, at least two studies (Isgor & Sengelaub, 1998; Madeira, Paula-Barbosa, Cadete-Leite, & Tavares, 1988) have found no sex differences in DG-GCL volume in rats. In some mouse strains, DG-GCL volume is also greater on the right than on the left in males only (Tabibnia, Cooke, & Breedlove, 1999).

DG granule cell nuclei tend to be larger in male mice (Wimer & Wimer, 1985) and in adult (Pfaff, 1966) but not juvenile (Roof, 1993a) rats, and male rats have more total DG granule cells than female rats (Madeira et al., 1988; but see Yanai, 1979). Finally, Wimer and Wimer (1985) found that males had higher DG granule cell densities than did females in each of six strains of house mice examined, but Yanai (1979) found no sex differences in this measure in Long-Evans or Wistar rats, suggesting that a sex difference in this measure may be unrelated to sex differences in spatial ability.

Hormonal mediation of DG sex differences Sex differences in the DG appear to be mediated by androgens: early postnatal testosterone treatment masculinizes DG morphology in female rats, and neonatal castration prevents DG masculinization in males. Pfaff (1966) found that neonatal castration prevents masculinization of the nuclear areas of DG neurons. Furthermore, testosterone treatment on postnatal days 3 and 5 masculinizes DG-GCL width in adult (Roof & Havens, 1992) and juvenile (Roof, 1993a) female rats. Early postnatal androgens also masculinize DG-GCL thickness in adult, but not juvenile, female rats (Roof, 1993a). Because juvenile (28-day-old) male and female rats differ in DG-GCL thickness regardless of neonatal testosterone treatment (Roof, 1993a), it is plausible that prenatal hormones contribute to juvenile sex differences in DG-GCL thickness and that early postnatal androgens contribute to maintaining these differences later in life. Roof and Havens (1992) also showed that testosterone treatment of female rats on postnatal days 3 and 5 lateralized DG-GCL width in adults (>90 days of age). This lateralization was also found in male, but not female, controls. A subsequent study (Roof, 1993a) confirmed that the effects of this early testosterone treatment on DG-GCL laterality were present in female rats by 28 days of age.

By themselves, these results cannot rule out the possibility that androgens contribute to sex differences in DG morphology by first being aromatized into estradiol. However, some evidence indicates that DG-GCL laterality is mediated directly by androgens. Tabibnia et al. (1999) found no laterality in DG-GCL volume in either sex of C57/BL6J mice with a defective structural gene for ARs, despite the fact that both sexes normally exhibit DG-GCL volume laterality in this mouse strain. This indicates that androgens act directly on some aspects of rodent DG morphology without first being aromatized into estradiol. In addition, knockout of ER-alpha in these mice did not affect DG volume, which is probably sexually dimorphic in this strain (Cobb & Juraska, 2004).

Some evidence also indicates that gonadal steroids may exert activational influences in DG morphology in some species. Spatial behavior changes gestationally in female meadow voles (Galea et al., 1995, 2000) and seasonally in males (Gaulin & FitzGerald, 1986), suggesting that the neural substrates for spatial behavior might be responsive to fluctuating sex steroid levels in this species. Indeed, Galea and colleagues (1999) found that DG width correlated with estradiol levels in adult female meadow voles and with testosterone levels in adult males.

Homo Sapiens Few studies have looked for sexual dimorphisms in the human hippocampus. Klekamp, Riedel, Harper, and Kretschmann (1991) reported significantly larger hippocampal volumes in males than in females in postmortem brain sections from adult Australian Aboriginals, but not in those from Caucasians. However, this study did not control for overall brain size, which is larger in males than in females. After controlling for cerebral volume in a quantitative MRI study, Giedd et al. (1996) found that the hippocampus was not significantly ($P = 0.25$) larger in 53 boys ages 4 to 18 than in 46 girls of the same age. However, the right hippocampus grew significantly faster in females (Giedd et al., 1996), and this differential growth may explain the MRI finding of Filipek, Richelme, Kennedy, and Caviness (1994) that young adult females had relatively larger hippocampi than did males.

Neither of these studies found significant sex differences in hippocampal volume laterality. (Giedd et al. found laterality in both sexes, Filipek et al. found laterality in neither.) On the other hand, Zaidel, Esiri, and Oxbury (1994) found greater densities of nucleolated cells on the left compared to the right hippocampi of males but not females in a sample of 52 unilateral hippocampi surgically removed from epileptic patients. In a voxel-based MRI study of 465 normal adults, men also had significantly more gray matter volume in the hippocampus and entorhinal cortex when white matter, CSF, and age were statistically controlled for (Good et al., 2001).

Finally, hippocampal activation patterns during spatial navigation appear to differ by sex. When navigating a virtual maze, the left hippocampus and the left parahippocampal gyrus were significantly more activated in men than in women relative to a control condition (Gron, Wunderlich, Spitzer, Tomczak, & Riepe, 2000). Indeed, relative to the control condition, these areas were significantly activated only in men. (Recall that the right hippocampus is most activated during spatial cognition in humans.) Using different spatial tests and a different control condition, Blanch, Brennan, Condon, Santosh, and Hadley (2004) found no sex differences in brain activation during spatial navigation. However, male and female performance differed significantly on only one of two spatial tasks used in this study, and the difference was small compared to that reported by Gron and colleagues. Moreover, unlike the control condition used in the Gron et al. study, which consisted of looking at a static screen image and pressing buttons as directed, the control condition used in the Blanch et al. study was itself a spatial task. Thus, it is unclear precisely what was measured in the Blanch et al. study when activation during the control spatial task was subtracted from activation during the experimental spatial task.

The Prefrontal and Parietal Cortices

Animal Models

Prefrontal and parietal cortical sex differences The prefrontal cortex (PFC) is the anterior region of the frontal cortex and is associated with attention to specific events in the environment and with behavioral planning. The PFC receives projections from the parietal cortex, which is associated with spatial perception and spatial working memory. Whereas males seem more reliant on the hippocampus for spatial problem solving, females appear more dependent on the prefrontal and possibly the parietal cortices. Like sex differential reliance on the hippocampus, differential reliance on the prefrontal and parietal cortices is suggested by functional imaging studies in humans and lesion studies in laboratory animals. In one lesion study, Long-Evans rats were PFC-lesioned and tested on MWM and RAM tasks (Kolb & Cioe, 1996). Females performed worse than nonlesioned controls, but males given identical lesions were unaffected on these tasks. However, males were not entirely unaffected by PFC lesions. On a test in which subjects were required to ignore extramaze cues and attend to a single cue on the maze wall, only lesioned males performed worse than controls (Kolb & Cioe, 1996).

Kolb and Cioe (1996) suggested that these results reflect the different strategies employed by males and females when solving spatial problems. Male rats apparently attend more to "configural" cues (distances and directions) when solving spatial problems and are more impaired in the absence of such cues, whereas females attend more to "specific" cues (landmarks) and are disrupted when landmarks are moved (Williams & Meck, 1991; Williams et al., 1990). This strategic sex difference closely parallels what has been observed in humans. Kolb and Cioe (1996) suggested that PFC lesions may interfere with subjects' ability to shift maze-solving strategies from dominant to less dominant strategies, and that this could explain the sexually dimorphic responses to lesions.

Alternatively, the PFC may aid more directly in tasks requiring landmark use. Because females tend to navigate using landmarks, this would explain why female navigation is more impaired generally by PFC lesions. This could also explain why PFC lesions disrupted landmark task acquisition in males but not in females (Kolb & Cioe, 1996); given females' reliance on landmarks, landmark tasks may be relatively easy for females, and the limited PFC lesions administered by Kolb and Cioe (1996) may have been insufficient to impair females' performance significantly on the single-cue landmark task.

If the PFC is involved in the processing of landmark cues, and if females are more reliant on both landmarks and the PFC for spatial navigation, one might expect some structural sex differences in this region. Indeed, Kolb and Stewart (1991) and Kavaliers, Ossenkopp, Galea, and Kolb (1998) found structural sex differences in both the prefrontal and the parietal cortical regions. Male Sprague-Dawley rats showed more pyramidal cell dendritic branching in parts of a medial PFC region called the anterior cingulate cortex (Kolb & Stewart, 1991). And in meadow voles, females had longer but fewer pyramidal cell dendrites in layer II/III of the prefrontal (cingulate) and parietal cortical regions (Kavaliers et al., 1998). This is a reversal of the pattern observed in the hippocampal complex. That is, given that males have larger cells, more dendritic branching, and so forth, in the hippocampus, on which they are more reliant for spatial processing, it might be expected that females would be greater on such measures in brain regions, such as the prefrontal and parietal cortices, on which they are more reliant than males. Differential reliance on the prefrontal and parietal cortices and the presence of sex differences in these regions suggest that the brain differences cause the differential reliance. However, it is also possible that these brain dimorphisms reflect sex differences in nonspatial functions, and these possibilities warrant further investigation.

Hormonal mediation of prefrontal and parietal cortical sex differences The developmental causes of sexual dimorphisms in the prefrontal and parietal cortices are poorly understood. However, one study implicates both organization by early androgens and activation by adult ovarian hormones. Stewart and Kolb (1994) found that adult ovariectomy in rats increased the dendritic arbor of layer II/III pyramidal neurons in the parietal cortex and moderately increased apical dendritic spine density, suggesting that ovarian hormones feminize dendritic morphology in the parietal cortex of adult female rats. In addition, intact neonatally testosterone propionate-treated females exhibited greater pyramidal neuronal dendritic arbor than did intact oil-treated females—a result indicating that early androgens masculinize parietal cortical dendritic morphology (Stewart & Kolb, 1994).

Homo Sapiens Little is known about sex differences in the human PFC and how these differences might translate into differential spatial abilities. In an fMRI study, the right superior and inferior parietal lobules and right PFC were significantly more activated during spatial navigation in women than in men (Gron et al., 2000). This finding suggests that there might be some sex differences in the human PFC. In a voxel-based MRI study of 465 normal adults, women had significantly increased gray matter concentration in the parietal cortical mantle compared to men, when white matter, CSF, and age were statistically controlled for (Good et al., 2001).

The Basal Forebrain

Animal Models

Sex differences in the basal forebrain The basal forebrain (BF) is a collection of structures located near the medial and ventral surfaces of the cerebral hemispheres. The BF has been implicated in attention, motivation, and memory. Cholinergic neurons (those using the neurotransmitter acetylcholine, ACh) in several BF structures, including the medial septal nucleus (MS), the vertical nucleus of the diagonal band of Broca (DBv), and the nucleus basalis magnocellularis (nBM), project to the hippocampus and frontal cortex and are important in memory (Bartus, Dean, Pontecorvo, & Flicker, 1985; Berger-Sweeney, 2003; Davies, 1985; Meck, Church, Wenk, & Olton, 1987). BF cholinergic neurotransmission appears to be involved specifically (but not exclusively) in spatial learning (Bachman, Berger-Sweeney, Coyle, & Hohmann, 1994; Meck, Smith, & Williams, 1988, 1989; Whishaw, 1985). A variety of evidence suggests that sex differences in BF cholinergic neurotransmission may underlie sex differences in spatial performance.

First, cholinergic neurotransmission is sexually dimorphic. Rats differ by sex in the expression of several cholinergic markers over development, including ACh levels (Hortnagl, Berger, Havelec, & Hornykiewicz, 1993), activities of acetylcholinesterase (the enzyme that breaks down ACh at the synapse) (Loy & Sheldon, 1987; Luine, Renner, Heady, & Jones, 1986; Smolen, Smolen, Han, & Collins, 1987) and choline acetyltransferase (the enzyme that synthesizes ACh) (Brown & Brooksbank, 1979; Luine et al., 1986), and uptake of high-affinity choline (a component of ACh) (Miller, 1983).

Second, the spatial performance of male and female rodents is differentially affected by cholinergic manipulations. Embryonic exposure to an inhibitor of acetylcholinesterase impaired female but not male rats on RAM and figure-8 mazes (Levin et al., 2002). On the other hand, dietary perinatal supplementation with choline had a more beneficial effect on RAM performance in male Sprague-Dawley rats compared to females (Williams et al., 1998). Moreover, treatment of adult mice with an ACh antagonist decreased spatial (noncued) MWM performance more in females than in males (Berger-Sweeney, Arnold, Gabeau, & Mills, 1995).

Finally, BF lesions affect spatial learning and associated cortical structure in a sexually dimorphic manner. Only male mice exhibited impaired adult MWM performance as a consequence of neonatal nBM lesions (Arters, Hohmann, Mills, Olaghere, & Berger-Sweeney, 1998). This impairment was greater on spatial than on cued MWM performance, and treatment affected neither activity levels nor learning or retention of nonspatial tasks (Arters et al., 1998). Neonatal nBM lesions also affected cortical layer II/III width differentially by sex, and lesion-related decreases in cortical layer IV and V widths correlated with spatial MWM performance in males only (Hohmann & Berger-Sweeney, 1998).

For these reasons, it is plausible that sex differences in BF cholinergic neurotransmission contribute to sex differences in spatial ability. However, the relationship between the BF and spatial sex differences is unclear. At least two explanations suggest themselves for the sexually dimorphic effects of neonatal nBM lesions (Arters et al., 1998) discussed earlier. One explanation is that BF afferents affect hippocampal and cortical development, and dimorphisms in these latter regions contribute directly to sex differences in spatial performance. This possibility receives some support from the finding that neonatal nBM lesions had sexually dimorphic effects on cortical structure in adult mice (Hohmann & Berger-Sweeney, 1998).

Another possibility is that the BF is involved in spatial problem solving, and that BF sex differences contribute directly to sex differences in spatial ability. This possibility is supported by the finding that MWM performance is significantly impaired in adult rats treated with an immunotoxin that destroys a type of cholinergic BF neurons (LeBlanc et al., 1999). Furthermore, some lesioned animals received cholinergic neuron grafts to the hippocampus. Grafted animals exhibited greater cholinergic innervation to the DG, and the level of cholinergic innervation to the DG correlated with MWM performance (LeBlanc et al., 1999).

Thus, sexual dimorphisms in the BF may contribute to sex differences in spatial ability by providing sexually dimorphic input to the cortex and hippocampus in adult animals, by playing a role in sexually dimorphic cortical and hippocampal development, or both.

Hormonal mediation of sex differences in the basal forebrain Although no studies of which we are aware have looked for possible organizing effects of sex hormones on cholinergic neurotransmission in the BF specifically, some studies have examined the effects of early sex hormone treatment on cholinergic markers in other brain regions. For example, Libertun, Timiras, and Kragt (1973) found that male and neonatally testosterone-treated female rats exhibited lower choline acetyltransferase activity than did control females in the preoptic-suprachiasmatic area of the hypothalamus, but not in the arcuate-mammillary area or the frontoparietal cortex. Brown and Brooksbank (1979) observed no significant effect of sex or neonatal testosterone treatment on choline acetyltransferase activity in several other rat brain regions. Thus, testosterone may have organizational effects on BF cholinergic neurotransmission in regions, such as the BF and the preoptic-suprachiasmatic area of the hypothalamus, that exhibit cholinergic sexual dimorphisms.

On the other hand, cholinergic markers in the adult female BF probably depend on the activational effects of estrogens (Gibbs, 1994, 1996, 1997; Gibbs & Aggarwal, 1998; Gibbs, Wu, Hersh, & Pfaff, 1994; Kompoliti et al., 2004; McMillan, Singer, & Dorsa, 1996; Singer, McMillan, Dobie, & Dorsa, 1998) and progesterone (Gibbs, 1996, 2000; Gibbs & Aggarwal, 1998). For example, ovariectomized adult Sprague-Dawley rats that received estrogen replacement exhibited increased cellular levels of choline acetyltransferase mRNA in the MS and nBM (Gibbs et al., 1994). Similar treatment of female rhesus monkeys elevated choline acetyltransferase in the DBv in both young and aged monkeys and decreased numbers of acetylcholinesterase-positive fibers in layer II of the frontal, insular, and cingulate cortices of aged monkeys (Kompoliti et al., 2004).

Environment

Hormonal differences cause sexual dimorphisms in spatial ability and its neural substrates. This is clear from experimental manipulations in animal models and from comparisons between members of the same chromosomal sex who differ in hormonal experience. Environmental differences also contribute to sex differences in spatial ability, and this is probably especially true in humans (e.g., Tracy, 1987). Although a consideration of environmental contributions to sex differences in spatial ability is beyond the scope of this chapter, we have already seen how performance in certain spatial tasks improves with practice.

Finally, it is important to consider the interaction between sex and environment. Sometimes an environmental change may increase or decrease a brain measure equally in both sexes. But often the effects of an environmental manipulation depend on the sex of the animal. Thus, a sex difference in one environment may be smaller, nonexistent, or even reversed in another. For example, the hippocampi of male and female laboratory rodents differ in their responses to stress and stress-related hormones. Adult Wistar rats exposed to restraint stress exhibited sexually dimorphic responses in mineralocorticoid and glucocorticoid (adrenal steroid hormones) receptor expression in several hippocampal areas (Kitraki, Kremmyda, Youlatos, Alexis, & Kittas, 2004). Moreover, treatment of pregnant guinea pigs with glucocorticoid (a stress-related hormone) resulted in sexually dimorphic responses in mineralocorticoid receptor expression in the hippocampi of their offspring (Liu, Li, & Matthews, 2001; Owen & Matthews, 2003).

These sexually dimorphic molecular responses to stress and stress-related hormones are associated with dimorphic behavioral responses. Restraint stress had divergent effects on spatial ability in Wistar rats, improving MWM performance in females while impairing it in males (Kitraki et al., 2004). Similarly, female Sprague-Dawley rats whose mothers were stressed during gestation exhibited improved RAM performance, while their male counterparts showed poorer performance (Bowman et al., 2004). Finally, females were more impaired than males on water maze performance after early postnatal treatment with a synthetic glucocorticoid (Vicedomini, Nonneman, DeKosky, & Scheff, 1986).

The hippocampal complexes of male and female rats also respond differently to social and sensory stimulation during maturation. Juraska and colleagues (Juraska, Fitch, Henderson, & Rivers, 1985) examined environmental effects on dendritic branching in the DG-GCL of hooded rats. In this study, littermates were randomly divided into environmentally enriched and isolated condition groups. Enriched condition rats were group-housed, given toys, and released daily into an open field with different toy arrangements. Isolated condition rats were individually housed and did not have access to toys or open field exploration. This environmental manipulation affected dendritic branching in the DG-GCL of females but not males. Within the isolated condition group, males showed more dendritic branching per neuron. However, the enriched condition increased dendritic branching in females, reversing the sex difference in dendritic branching in the DG-GCL. In contrast, another study by Juraska and colleagues (1989) found that dendritic branching in CA3 appeared to be more plastic in males in response to this environmental manipulation. Enriched condition males showed less branching in the proximal apical dendrites than did isolated condition males, leading to sex differences in dendritic arborization in CA3 pyramidal cells only in the enriched condition group.

Such sex differences in responsiveness to the environment highlight the degree to which male and female brains may differ across species, but they illustrate another important point: we will not necessarily observe adaptive sex differences in environments (like laboratories) that differ substantially from the environment in which the species evolved (Sherry, Forbes, Khurgel, & Ivy, 1993). Evolutionary theory specifies that

(1) there will be sex differences in spatial ability and related brain regions in species in which males and females have recurrently faced different spatial problems over their evolutionary histories, and (2) these sex differences will develop and persist in environments that are similar to those in which the species evolved. The more an organism's current environment differs from its ancestral one, the less confident we can be that the necessary environmental conditions will exist to allow the organism to develop adaptations to the ancestral environment.

Summary

The largest known cognitive sex differences in humans have been found in the arena of spatial ability. Males outperform females on tasks involving mental rotation and spatial perception, although recent research indicates a spatial domain (spatial-location memory) in which females outperform males. In laboratory rats and polygynous wild rodents, males exhibit superior maze learning, and recent work demonstrates a parallel human sex difference on virtual versions of rodent mazes. The spatial demands of relatively large ranges likely favored superior spatial abilities in males of polygynous mammal species. This sex difference is absent in monogamous rodents and reversed in brood-parasitic birds, in which females experience greater spatial demands. In humans, male superiority on some spatial tasks may have evolved as a result of a combination of polygynous ancestry with broader male ranging patterns and additional spatial demands imposed by hunting. Foraging for immobile resources may have selected for superior object-location memory in human females.

In both humans and rodents, early androgens appear to exert organizational masculinizing effects on spatial ability. It is likely that androgens masculinize rodent spatial ability both via ARs and by aromatization into estradiol before binding to ERs. In humans, spatial ability is probably AR-mediated. Both androgens and estrogens likely have activational effects on spatial ability in some rodents. Responsiveness to fluctuating androgen levels in adult male rodents may be an adaptation to breeding seasonality and accompanying changes in range size. Humans exhibit very low breeding seasonality, and despite assertions to the contrary, current evidence does not support androgens having activational effects on human spatial ability. On the other hand, reasonable data suggest that elevated estrogens in adult female rodents and humans may diminish spatial ability and behavior. These activational effects may represent an adaptation to changing spatial demands over pregnancy.

Androgens probably masculinize spatial ability by affecting multiple brain regions involved in spatial processing, including the hippocampal complex, the prefrontal and parietal cortices, and the basal forebrain. Masculinization in rodents is AR-mediated for some sexually dimorphic measures in these regions and ER-mediated for others, which accords with the idea that masculinization of rodent spatial ability occurs through steroid binding to both types of receptors.

Within the hippocampal complex, male rats have larger pyramidal cell soma and cell fields in CA1 and CA3 and have greater pyramidal cell dendritic branching and a more voluminous mossy fiber system in CA3 than do females. The DG-GCL is more lateralized and may be larger in some regions in male rats and meadow voles and in males of some mouse strains. DG granule cell nuclei may also be larger in male mice and rats. All of these traits are masculinized by prenatal or early postnatal androgens, but some may remain responsive to estradiol in adult females and to testosterone in adult males of seasonally breeding species. In humans, adult females may have relatively larger hippocampi, but males are apparently more lateralized on some cytological measures and have relatively more gray matter in the hippocampus and its primary cortical input, the entorhinal cortex. Men may also experience greater left hippocampal and parahippocampal activation during spatial processing.

By contrast, the parietal lobules and prefrontal cortex may be more activated during spatial navigation in women than in men, and women possess relatively more gray matter in the parietal cortical mantle than do men. In some strata of the prefrontal and parietal cortices of meadow voles, females have longer but fewer pyramidal cell dendrites. And in one area of the medial PFC, male rats appear to have more extensive pyramidal cell dendritic branching. In one study, dendritic arbor in the parietal cortex of female rats was masculinized by neonatal testosterone and by adult ovariectomy, suggesting that early androgens have masculinizing organizational effects and that estrogens have feminizing activational effects on dendritic morphology in these regions.

Some evidence also implicates sex differences in basal forebrain cholinergic neurotransmission in sex differences in spatial ability. This evidence includes sex differences in cholinergic neurotransmission and in the effects of cholinergic manipulations and neonatal BF lesions on spatial performance. The BF may affect spatial ability by direct involvement in spatial processing, by affecting the development of the cortex and hippocampus, or both.

In conclusion, the studies reviewed here in aggregate make it clear that there are widespread sex differences in spatial reasoning ability across mammalian species, including humans, such that males on average perform better than females on most tasks. Of course, there are some tasks on which females display better performance, including object-location memory in humans, which suggests that sex differences in spatial ability may be very specific for particular types of spatial reasoning tasks. The task specificity of these sex differences in human performance raises the question of whether selection has honed particular sexes to excel on particular tasks or whether cultural influences on the socialization of developing humans contribute to sex differences in performance. These are not mutually exclusive possibilities, but if cultural factors play an important role, then one could expect to see varying levels of sex differences in spatial ability across varying cultures or to see the magnitude of the sex difference in spatial ability change within the span of a few generations, which is sufficient time for culture to change but not for selection to alter the gene pool. There are some data to support both of these possibilities, so there may well be cultural factors mediating some of the sex differences in human spatial ability.

On the other hand, animal models suggest that selection has also contributed to the sex difference in spatial ability in mammals. For example, the several findings that there is a sex difference in spatial ability in one species but no sex difference in another, closely related species, and that the differing mating systems of the two species allow one to predict which will display a sex difference, is powerful evidence of sexual selection at work. Moreover, surveying sex differences in spatial ability across animals also suggests that selection can exaggerate, minimize, or reverse sex differences, indicating that it is an evolutionarily labile or malleable trait. If so, then there must be genes that augment spatial ability more in one sex than the other, which raises the question of what proximate mechanisms could provide such sex-selective augmentation.

Again, animal models inform the debate, as they indicate that steroid hormones, acting either early in development or in adulthood (or both), augment spatial reasoning in males more than in females. Several studies indicate a similar effect of steroid hormones in humans, which strengthens the notion that selection has contributed to sex differences in spatial reasoning in our own species. Those studies might tempt us to conclude that because hormones influence spatial reasoning, there is no role for experience to influence this behavior and therefore no opportunity for culture to exaggerate or minimize sex differences. Such a conclusion would be absurd, for several reasons. Just because steroids have some influence on human spatial reasoning does not in any way preclude other factors, including experience, from also affecting spatial reasoning. More interestingly, it is always possible that steroid hormones affect spatial reasoning by altering the individual's proclivities, leading the individual to seek out experiences that improve spatial reasoning. If so, then social factors could easily influence how fully an individual might indulge proclivities to sharpen her or his spatial reasoning abilities.

In the future, there will surely be additional comparisons of related species to further detail the evolutionary pressures that promote a sex difference in spatial reasoning. There will also be studies to flesh out the details of the proximate mechanisms underlying such sex differences: which steroid hormones are responsible, where do they act on the brain, what processes do they modulate there, and what are the consequences for brain development and adult behavior? These studies will be conducted in animal models and will serve to inform future inquiries about sex differences in human spatial reasoning.

The study of sex differences in spatial reasoning ability has already been a fruitful area of research for a deeper understanding of how evolutionary pressures can produce proximate mechanisms to alter the brain and thereby favor adaptive behaviors. We can feel fortunate that these same mechanisms also appear to apply, at least in part, to humans, so that we can look forward to a greater understanding of how evolution affects human behavior as this field of study continues to grow.

References

Alexander, G. M., Swerdloff, R. S., Wang, C., Davidson, T., McDonald, V., Steiner, B., et al. (1998). Androgen-behavior correlations in hypogonadal men and eugonadal men. II. Cognitive abilities. *Hormones and Behavior*, *33*(2), 85–94.

Alexander, R. D. (1979). *Darwinism and human affairs*. Seattle: University of Washington Press.

Arters, J., Hohmann, C. F., Mills, J., Olaghere, O., & Berger-Sweeney, J. (1998). Sexually dimorphic responses to neonatal basal forebrain lesions in mice. I. Behavior and neurochemistry. *Journal of Neurobiology*, *37*(4), 582–594.

Astur, R. S., Ortiz, M. L., & Sutherland, R. J. (1998). A characterization of performance by men and women in a virtual Morris water task: A large and reliable sex difference. *Behavioural Brain Research*, *93*(1–2), 185–190.

Bachman, E. S., Berger-Sweeney, J., Coyle, J. T., & Hohmann, C. F. (1994). Developmental regulation of adult cortical morphology and behavior: An animal model for mental retardation. *International Journal of Developmental Neuroscience*, *12*(4), 239–253.

Baker, S. W., & Ehrhardt, A. A. (1974). Prenatal androgen, intelligence, and cognitive sex differences. In R. C. Friedman, R. M. Richart, & R. L. Vande Wiele (Eds.), *Sex differences in behavior* (pp. 53–76). New York: John Wiley & Sons.

Bartus, R. T., Dean, R. L., Pontecorvo, M. J., & Flicker, C. (1985). The cholinergic hypothesis: A historical overview, current perspective, and future directions. *Annals of the New York Academy of Sciences*, *444*, 332–358.

Berenbaum, S. A. (1999). Effects of early androgens on sex-typed activities and interests in adolescents with congenital adrenal hyperplasia. *Hormones and Behavior*, *35*(1), 102–110.

Berger-Sweeney, J. (2003). The cholinergic basal forebrain system during development and its influence on cognitive processes: Important questions and potential answers. *Neuroscience and Biobehavioral Reviews*, *27*(4), 401–411.

Berger-Sweeney, J., Arnold, A., Gabeau, D., & Mills, J. (1995). Sex differences in learning and memory in mice: Effects of sequence of testing and cholinergic blockade. *Behavioral Neuroscience*, *109*(5), 859–873.

Blanch, R. J., Brennan, D., Condon, B., Santosh, C., & Hadley, D. (2004). Are there gender-specific neural substrates of route learning from different perspectives? *Cerebral Cortex*, *14*, 1207–1213.

Bowman, R. E., MacLusky, N. J., Sarmiento, Y., Frankfurt, M., Gordon, M., & Luine, V. N. (2004). Sexually dimorphic effects of prenatal stress on cognition, hormonal responses, and central neurotransmitters. *Endocrinology*, *145*(8), 3778–3787.

Brown, R., & Brooksbank, B. W. (1979). Developmental changes in choline acetyltransferase and glutamate decarboxylase activity in various regions of the brain of the male, female, and neonatally androgenized female rat. *Neurochemistry Research*, *4*(2), 127–136.

Cappa, S. F., Guariglia, C., Papagno, C., Pizzamiglio, L., Vallar, G., Zoccolotti, P., et al. (1988). Patterns of lateralization and performance levels for verbal and spatial tasks in congenital androgen deficiency. *Behavioural Brain Research*, *31*(2), 177–183.

Choi, J., & L'Hirondelle. (2005). Object location memory: A direct test of the verbal memory hypothesis. *Learning and Individual Differences*, *15*, 237–245.

Clutton-Brock, T. H., & Vincent, A. C. J. (1991). Sexual selection and the potential reproductive rates of males and females. *Nature*, *351*, 58–60.

Cobb, J. A., & Juraska, J. M. (2004). *No effect of estrogen receptor-alpha knockout on hippocampal volume in C57BL/6J mice.* Paper presented at the annual meeting of the Society for Neuroscience, San Diego.

Cole-Harding, S., Morstad, A. L., & Wilson, J. R. (1988). Spatial ability in members of opposite-sex twin pairs. *Behavioral Genetics*, *18*, 710.

Commins, W. D. (1932). The effect of castration at various ages upon learning ability of male albino rats. *Journal of Comparative Psychology*, *14*, 29–53.

Dabbs, J. M., Chang, E. L., Strong, R. A., & Milun, R. (1998). Spatial ability, navigation strategy, and geographic knowledge among men and women. *Evolution and Human Behavior*, *19*, 89–98.

Daniel, J. M., Fader, A. J., Spencer, A. L., & Dohanich, G. P. (1997). Estrogen enhances performance of female rats during acquisition of a radial arm maze. *Hormones and Behavior*, *32*(3), 217–225.

Darwin, C. (1859). *On the origin of species by means of natural selection*. London: Murray.

Darwin, C. (1871). *The descent of man and Selection in relation to sex*. London: Murray.

Davies, P. (1985). A critical review of the role of the cholinergic system in human memory and cognition. *Annals of the New York Academy of Sciences*, *444*, 212–217.

Dawkins, R. (1986). *The blind watchmaker*. New York: Norton.

Dawson, J. L., Cheung, Y. M., & Lau, R. T. (1975). Developmental effects of neonatal sex hormones on spatial and activity skills in the white rat. *Biological Psychology*, *3*(3), 213–229.

Dewsbury, D. A. (1981). An exercise in the prediction of monogamy in the field from laboratory data on 42 species of muroid rodents. *The Biologist*, *63*, 138–162.

Diaz-Veliz, G., Soto, V., Dussaubat, N., & Mora, S. (1989). Influence of the estrous cycle, ovariectomy and estradiol replacement upon the acquisition of conditioned avoidance responses in rats. *Physiology & Behavior*, *46*(3), 397–401.

Duff, S. J., & Hampson, E. (2000). A beneficial effect of estrogen on working memory in postmenopausal women taking hormone replacement therapy. *Hormones and Behavior*, *38*(4), 262–276.

Eals, M., & Silverman, I. (1994). The hunter-gatherer theory of spatial sex differences: Proximate factors mediating the female advantage in recall of object arrays. *Ethology and Sociobiology*, *15*, 95–105.

Filipek, P. A., Richelme, C., Kennedy, D. N., & Caviness, V. S., Jr. (1994). The young adult human brain: An MRI-based morphometric analysis. *Cerebral Cortex*, *4*(4), 344–360.

Frye, C. A. (1995). Estrus-associated decrements in a water maze task are limited to acquisition. *Physiology & Behavior*, *57*(1), 5–14.

Galea, L. A., Kavaliers, M., & Ossenkopp, K. P. (1996). Sexually dimorphic spatial learning in meadow voles *Microtus pennsylvanicus* and deer mice *Peromyscus maniculatus*. *Journal of Experimental Biology*, *199*(Pt. 1), 195–200.

Galea, L. A., Kavaliers, M., Ossenkopp, K. P., & Hampson, E. (1995). Gonadal hormone levels and spatial learning performance in the Morris water maze in male and female meadow voles, *Microtus pennsylvanicus*. *Hormones and Behavior*, *29*(1), 106–125.

Galea, L. A., & McEwen, B. S. (1999). Sex and seasonal differences in the rate of cell proliferation in the dentate gyrus of adult wild meadow voles. *Neuroscience*, *89*(3), 955–964.

Galea, L. A., Ormerod, B. K., Sampath, S., Kostaras, X., Wilkie, D. M., & Phelps, M. T. (2000). Spatial working memory and hippocampal size across pregnancy in rats. *Hormones and Behavior*, *37*(1), 86–95.

Galea, L. A., Perrot-Sinal, T. S., Kavaliers, M., & Ossenkopp, K. P. (1999). Relations of hippocampal volume and dentate gyrus width to gonadal hormone levels in male and female meadow voles. *Brain Research*, *821*(2), 383–391.

Gaulin, S. J. C. (1992). Evolution of sex differences in spatial ability. *Yearbook of Physical Anthropology*, *35*, 125–151.

Gaulin, S. J. C., & FitzGerald, R. W. (1986). Sex differences in spatial ability: An evolutionary hypothesis and test. *American Naturalist*, *127*, 74–88.

Gaulin, S. J. C., & FitzGerald, R. W. (1988). Home range size as a predictor of mating system in *Microtus*. *Journal of Mammology*, *69*, 311–319.

Gaulin, S. J. C., & FitzGerald, R. W. (1989). Sexual selection for spatial-learning ability. *Animal Behaviour*, *37*, 332–331.

Gaulin, S. J. C., FitzGerald, R. W., & Wartell, M. S. (1990). Sex differences in spatial ability and activity in two vole species (*Microtus ochrogaster* and *M. pennsylvanicus*). *Journal of Comparative Psychology*, *104*(1), 88–93.

Gaulin, S. J. C., & Hoffman, H. (1988). Evolution and development of sex differences in spatial ability. In L. L. Betzig, M. Borgerhoff-Mulder & P. W. Turke (Eds.), *Human reproductive behavior: A Darwinian perspective* (pp. 129–152). Cambridge: Cambridge University Press.

Gaulin, S. J. C., Krasnow, M., Truxaw, D., & New, J. (2005). *An ecologically valid foraging task yields a female spatial advantage and significant contect effects.* Paper presented at the annual meeting of the Human Behavior and Evolution Society, Austin, Tex.

Gaulin, S. J. C., & Wartell, M. S. (1990). Effects of experience and motivation on symmetrical-maze performance in the prairie vole (*Microtus ochrogaster*). *Journal of Comparative Psychology*, *104*(2), 183–189.

Gibbs, R. B. (1994). Estrogen and nerve growth factor-related systems in brain: Effects on basal forebrain cholinergic neurons and implications for learning and memory processes and aging. *Annals of the New York Academy of Sciences*, *743*, 165–196 [discussion 197–199].

Gibbs, R. B. (1996). Fluctuations in relative levels of choline acetyltransferase mRNA in different regions of the rat basal forebrain across the estrous cycle: Effects of estrogen and progesterone. *Journal of Neuroscience*, *16*(3), 1049–1055.

Gibbs, R. B. (1997). Effects of estrogen on basal forebrain cholinergic neurons vary as a function of dose and duration of treatment. *Brain Research*, *757*(1), 10–16.

Gibbs, R. B. (2000). Effects of gonadal hormone replacement on measures of basal forebrain cholinergic function. *Neuroscience*, *101*(4), 931–938.

Gibbs, R. B., & Aggarwal, P. (1998). Estrogen and basal forebrain cholinergic neurons: Implications for brain aging and Alzheimer's disease-related cognitive decline. *Hormones and Behavior*, *34*(2), 98–111.

Gibbs, R. B., Wu, D., Hersh, L. B., & Pfaff, D. W. (1994). Effects of estrogen replacement on the relative levels of choline acetyltransferase, trkA, and nerve growth factor messenger RNAs in the basal forebrain and hippocampal formation of adult rats. *Experimental Neurology*, *129*(1), 70–80.

Giedd, J. N., Vaituzis, A. C., Hamburger, S. D., Lange, N., Rajapakse, J. C., Kaysen, D., et al. (1996). Quantitative MRI of the temporal lobe, amygdala, and hippocampus in normal human development: Ages 4–18 years. *Journal of Comparative Neurology*, *366*(2), 223–230.

Good, C. D., Johnsrude, I., Ashburner, J., Henson, R. N., Friston, K. J., & Frackowiak, R. S. (2001). Cerebral asymmetry and the effects of sex and handedness on brain structure: A voxel-based morphometric analysis of 465 normal adult human brains. *NeuroImage*, *14*(3), 685–700.

Gouchie, C., & Kimura, D. (1991). The relationship between testosterone levels and cognitive ability patterns. *Psychoneuroendocrinology*, *16*(4), 323–334.

Gray, J. A., & Buffery, A. W. H. (1971). Sex differences in emotional and cognitive behavior in mammals, including man: Adaptive and neural bases. *Acta Psychologica*, *35*, 89–111.

Greenwood, P. J. (1980). Mating systems, philopatry, and dispersal in birds and mammals. *Animal Behaviour*, *28*, 1140–1162.

Grimshaw, G. M., Sitarenios, G., & Finegan, J. A. (1995). Mental rotation at 7 years: Relations with prenatal testosterone levels and spatial play experiences. *Brain and Cognition*, *29*(1), 85–100.

Gron, G., Wunderlich, A. P., Spitzer, M., Tomczak, R., & Riepe, M. W. (2000). Brain activation during human navigation: Gender-different neural networks as substrate of performance. *Nature Neuroscience*, *3*(4), 404–408.

Halpern, D. F. (2000). *Sex differences in cognitive abilities*. Mahwah, NJ: Erlbaum.

Hampson, E. (1990a). Estrogen-related variations in human spatial and articulatory-motor skills. *Psychoneuroendocrinology*, *15*(2), 97–111.

Hampson, E. (1990b). Variations in sex-related cognitive abilities across the menstrual cycle. *Brain and Cognition*, *14*(1), 26–43.

Hampson, E., & Kimura, D. (1988). Reciprocal effects of hormonal fluctuations on human motor and perceptual-spatial skills. *Behavioral Neuroscience*, *102*(3), 456–459.

Hampson, E., Rovet, J. F., & Altmann, D. (1998). Spatial reasoning in children with congenital adrenal hyperplasia due to 21-hydroxylase deficiency. *Developmental Neuropsychology*, *14*(2), 299–320.

Hassler, M. (1992). Creative musical behavior and sex hormones: Musical talent and spatial ability in the two sexes. *Psychoneuroendocrinology*, *17*(1), 55–70.

Hausmann, M., Slabbekoorn, D., Van Goozen, S. H., Cohen-Kettenis, P. T., & Gunturkun, O. (2000). Sex hormones affect spatial abilities during the menstrual cycle. *Behavioral Neuroscience*, *114*(6), 1245–1250.

Hawkes, K. (1990). Why do men hunt? Benefits for risky choices. In E. Cashdan (Ed.), *Risk and uncertainty in tribal and peasant economies* (pp. 146–166). Boulder, CO: Westview Press.

Hawkes, K. (1991). Showing off: Tests of an hypothesis about men's foraging goals. *Ethology and Sociobiology*, *12*, 29–54.

Helleday, J., Bartfai, A., Ritzen, E. M., & Forsman, M. (1994). General intelligence and cognitive profile in women with congenital adrenal hyperplasia (CAH). *Psychoneuroendocrinology*, *19*(4), 343–356.

Hier, D. B., & Crowley, W. F., Jr. (1982). Spatial ability in androgen-deficient men. *New England Journal of Medicine*, *306*(20), 1202–1205.

Hines, M., Fane, B. A., Pasterski, V. L., Mathews, G. A., Conway, G. S., & Brook, C. (2003). Spatial abilities following prenatal androgen abnormality: Targeting and mental rotations performance in individuals with congenital adrenal hyperplasia. *Psychoneuroendocrinology*, *28*(8), 1010–1026.

Hohmann, C. F., & Berger-Sweeney, J. (1998). Sexually dimorphic responses to neonatal basal forebrain lesions in mice. II. Cortical morphology. *Journal of Neurobiology*, *37*(4), 595–606.

Hojbjerg Gravholt, C., Svenstrup, B., Bennett, P., & Sandahl Christiansen, J. (1999). Reduced androgen levels in adult Turner syndrome: Influence of female sex steroids and growth hormone status. *Clinical Endocrinology*, *50*(6), 791–800.

Hortnagl, H., Berger, M. L., Havelec, L., & Hornykiewicz, O. (1993). Role of glucocorticoids in the cholinergic degeneration in rat hippocampus induced by ethylcholine aziridinium (AF64A). *Journal of Neuroscience*, *13*(7), 2939–2945.

Imperato-McGinley, J., Peterson, R. E., Gautier, T., Cooper, G., Danner, R., Arthur, A., et al. (1982). Hormonal evaluation of a large kindred with complete androgen insensitivity: Evidence for secondary 5 alpha-reductase deficiency. *Journal of Clinical Endocrinology and Metabolism*, *54*(5), 931–941.

Imperato-McGinley, J., Pichardo, M., Gautier, T., Voyer, D., & Bryden, M. P. (1991). Cognitive abilities in androgen-insensitive subjects: Comparison with control males and females from the same kindred. *Clinical Endocrinology*, *34*(5), 341–347.

Isgor, C., & Sengelaub, D. R. (1998). Prenatal gonadal steroids affect adult spatial behavior, CA1 and CA3 pyramidal cell morphology in rats. *Hormones and Behavior*, *34*(2), 183–198.

Isgor, C., & Sengelaub, D. R. (2003). Effects of neonatal gonadal steroids on adult CA3 pyramidal neuron dendritic morphology and spatial memory in rats. *Journal of Neurobiology*, *55*(2), 179–190.

Jacobs, L. F., Gaulin, S. J., Sherry, D. F., & Hoffman, G. E. (1990). Evolution of spatial cognition: Sex-specific patterns of spatial behavior predict hippocampal size. *Proceedings of the National Academy of Sciences, U.S.A.*, *87*(16), 6349–6352.

Jacobs, L. F., & Spencer, W. D. (1994). Natural space-use patterns and hippocampal size in kangaroo rats. *Brain, Behavior and Evolution*, *44*, 125–132.

James, T. W., & Kimura, D. (1997). Sex differences in remembering the locations of objects in an array: Location-shifts versus location-exchanges. *Evolution and Human Behavior*, *18*, 155–163.

Janowsky, J. S., Oviatt, S. K., & Orwoll, E. S. (1994). Testosterone influences spatial cognition in older men. *Behavioral Neuroscience*, *108*(2), 325–332.

Jonasson, Z. (2005). Meta-analysis of sex differences in rodent models of learning and memory: A review of behavioral and biological data. *Neuroscience and Biobehavioral Reviews*, *28*(8), 811–825.

Jones, C. M., Braithwaite, V. A., & Healy, S. D. (2003). The evolution of sex differences in spatial ability. *Behavioral Neuroscience*, *117*, 403–411.

Joseph, R., Hess, S., & Birecree, E. (1978). Effects of hormone manipulations and exploration on sex differences in maze learning. *Behavioral Biology*, *24*(3), 364–377.

Juraska, J. M., Fitch, J. M., Henderson, C., & Rivers, N. (1985). Sex differences in the dendritic branching of dentate granule cells following differential experience. *Brain Research*, *333*(1), 73–80.

Juraska, J. M., Fitch, J. M., & Washburne, D. L. (1989). The dendritic morphology of pyramidal neurons in the rat hippocampal CA3 area. II. Effects of gender and the environment. *Brain Research*, *479*(1), 115–119.

Kavaliers, M., Ossenkopp, K. P., Galea, L. A., & Kolb, B. (1998). Sex differences in spatial learning and prefrontal and parietal cortical dendritic morphology in the meadow vole, *Microtus pennsylvanicus*. *Brain Research*, *810*(1–2), 41–47.

Kitraki, E., Kremmyda, O., Youlatos, D., Alexis, M. N., & Kittas, C. (2004). Gender-dependent alterations in corticosteroid receptor status and spatial performance following 21 days of restraint stress. *Neuroscience*, *125*(1), 47–55.

Klekamp, J., Riedel, A., Harper, C., & Kretschmann, H. J. (1991). Morphometric study on the postnatal growth of the hippocampus in Australian Aborigines and Caucasians. *Brain Research*, *549*(1), 90–94.

Kolb, B., & Cioe, J. (1996). Sex-related differences in cortical function after medial frontal lesions in rats. *Behavioral Neuroscience*, *110*(6), 1271–1281.

Kolb, B., & Stewart, J. (1991). Sex-related differences in dendritic branching of cells in the prefrontal cortex of rats. *Journal of Neuroendocrinology*, *3*(1), 95–99.

Kompoliti, K., Chu, Y., Polish, A., Roberts, J., McKay, H., Mufson, E. J., et al. (2004). Effects of estrogen replacement therapy on cholinergic basal forebrain neurons and cortical cholinergic innervation in young and aged ovariectomized rhesus monkeys. *Journal of Comparative Neurology*, *472*(2), 193–207.

Lavenex, P., Steele, M. A., & Jacobs, L. F. (2000). Sex differences, but no seasonal variations in the hippocampus of food-caching squirrels: A stereological study. *Journal of Comparative Neurology*, *425*(1), 152–166.

LeBlanc, C. J., Deacon, T. W., Whatley, B. R., Dinsmore, J., Lin, L., & Isacson, O. (1999). Morris water maze analysis of 192-IgG-saporin-lesioned rats and porcine cholinergic transplants to the hippocampus. *Cell Transplant*, *8*(1), 131–142.

Leranth, C., Petnehazy, O., & MacLusky, N. J. (2003). Gonadal hormones affect spine synaptic density in the CA1 hippocampal subfield of male rats. *Journal of Neuroscience*, *23*(5), 1588–1592.

Levin, E. D., Addy, N., Baruah, A., Elias, A., Christopher, N. C., Seidler, F. J., et al. (2002). Prenatal chlorpyrifos exposure in rats causes persistent behavioral alterations. *Neurotoxicology and Teratology*, *24*(6), 733–741.

Libertun, C., Timiras, P. S., & Kragt, C. L. (1973). Sexual differences in the hypothalamic cholinergic system before and after puberty: Inductory effect of testosterone. *Neuroendocrinology*, *12*(2), 73–85.

Linn, M. C., & Petersen, A. C. (1985). Emergence and characterisation of gender differences in spatial abilities: A meta-analysis. *Child Development*, *56*, 1479–1498.

Liu, L., Li, A., & Matthews, S. G. (2001). Maternal glucocorticoid treatment programs HPA regulation in adult offspring: Sex-specific effects. *American Journal of Physiology and Endocrinology, and Metabolism*, *280*(5), E729–E739.

Lovejoy, C. O. (1981). The origin of man. *Science*, *211*, 341–350.

Loy, R., & Sheldon, R. A. (1987). Sexually dimorphic development of cholinergic enzymes in the rat septohippocampal system. *Brain Research*, *431*(1), 156–160.

Luine, V. N., Renner, K. J., Heady, S., & Jones, K. J. (1986). Age and sex-dependent decreases in ChAT in basal forebrain nuclei. *Neurobiology of Aging*, *7*(3), 193–198.

Luine, V. N., Richards, S. T., Wu, V. Y., & Beck, K. D. (1998). Estradiol enhances learning and memory in a spatial memory task and effects levels of monoaminergic neurotransmitters. *Hormones and Behavior*, *34*(2), 149–162.

Maccoby, E. E., & Jacklin, C. N. (1974). *The psychology of sex differences*. Stanford, CA: Stanford University Press.

Madeira, M. D., Paula-Barbosa, M., Cadete-Leite, A., & Tavares, M. A. (1988). Unbiased estimate of hippocampal granule cell numbers in hypothyroid and in sex-age-matched control rats. *Journal für Hirnforschung*, *29*(6), 643–650.

Madeira, M. D., Sousa, N., Lima-Andrade, M. T., Calheiros, F., Cadete-Leite, A., & Paula-Barbosa, M. M. (1992). Selective vulnerability of the hippocampal pyramidal neurons to hypothyroidism in male and female rats. *Journal of Comparative Neurology*, *322*(4), 501–518.

Madeira, M. D., Sousa, N., & Paula-Barbosa, M. M. (1991). Sexual dimorphism in the mossy fiber synapses of the rat hippocampus. *Experimental Brain Research*, *87*(3), 537–545.

Maguire, E. A., Frackowiak, R. S., & Frith, C. D. (1996). Learning to find your way: A role for the human hippocampal formation. *Proceedings of the Royal Society of London, B*, *263*, 1745–1750.

Malinowski, J. C. (2001). Mental rotation and real-world wayfinding. *Perceptual and Motor Skills*, *92*, 19–30.

Mann, V. A., Sasanuma, S., Sakuma, N., & Masaki, S. (1990). Sex differences in cognitive abilities: A cross-cultural perspective. *Neuropsychologia*, *28*, 1063–1077.

Mayes, J. T., & Jahoda, G. (1988). Patterns of visual-spatial performance and "spatial ability": Dissociation of ethnic and sex differences. *British Journal of Psychology*, *79*, 105–119.

McBurney, D. H., Gaulin, S. J. C., Devineni, T., & Adams, C. (1997). Superior spatial ability of women: Stronger evidence for the gathering hypothesis. *Evolution and Human Behavior*, *18*(3), 167–174.

McGuire, L. S., Ryan, K. O., & Omenn, G. S. (1975). Congenital adrenal hyperplasia. II. Cognitive and behavioral studies. *Behavioral Genetics*, *5*(2), 175–188.

McMillan, P. J., Singer, C. A., & Dorsa, D. M. (1996). The effects of ovariectomy and estrogen replacement on trkA and choline acetyltransferase mRNA expression in the basal forebrain of the adult female Sprague-Dawley rat. *Journal of Neuroscience*, *16*(5), 1860–1865.

Meck, W. H., Church, R. M., Wenk, G. L., & Olton, D. S. (1987). Nucleus basalis magnocellularis and medial septal area lesions differentially impair temporal memory. *Journal of Neuroscience*, *7*(11), 3505–3511.

Meck, W. H., Smith, R. A., & Williams, C. L. (1988). Pre- and postnatal choline supplementation produces long-term facilitation of spatial memory. *Developmental Psychobiology*, *21*(4), 339–353.

Meck, W. H., Smith, R. A., & Williams, C. L. (1989). Organizational changes in cholinergic activity and enhanced visuospatial memory as a function of choline administered prenatally or postnatally or both. *Behavioral Neuroscience*, *103*(6), 1234–1241.

Miles, C., Green, R., Sanders, G., & Hines, M. (1998). Estrogen and memory in a transsexual population. *Hormones and Behavior*, *34*(2), 199–208.

Miller, J. C. (1983). Sex differences in dopaminergic and cholinergic activity and function in the nigro-striatal system of the rat. *Psychoneuroendocrinology*, *8*(2), 225–236.

Moffat, S. D., & Hampson, E. (1996). A curvilinear relationship between testosterone and spatial cognition in humans: Possible influence of hand preference. *Psychoneuroendocrinology*, *21*(3), 323–337.

Moffat, S. D., Hampson, E., & Hatzipentalis, M. (1998). Navigation in a virtual maze: Sex differences and correlation with psychometric measures of spatial ability in human. *Evolution and Human Behavior*, *19*, 73–87.

Montello, D. R., Lovelace, K. L., Golledge, R. G., & Self, C. M. (1999). Sex-related differences and similarities in geographic and environmental spatial abilities. *Annals of the Association of American Geographers*, *89*, 515–534.

Murdock, G. P. (1967). *Ethnographic atlas*. Pittsburgh: University of Pittsburgh Press.

Nijhuis-van der Sanden, M. W., Eling, P. A., & Otten, B. J. (2003). A review of neuropsychological and motor studies in Turner syndrome. *Neuroscience and Biobehavioral Reviews*, *27*(4), 329–338.

O'Connor, D. B., Archer, J., Hair, W. M., & Wu, F. C. (2001). Activational effects of testosterone on cognitive function in men. *Neuropsychologia, 39*(13), 1385–1394.

Owen, D., & Matthews, S. G. (2003). Glucocorticoids and sex-dependent development of brain glucocorticoid and mineralocorticoid receptors. *Endocrinology, 144*(7), 2775–2784.

Owen, K., & Lynn, R. (1993). Sex differences in primary cognitive abilities among blacks, Indians and whites in South Africa. *Journal of Biosocial Science, 25*, 557–560.

Perlman, S. M. (1973). Cognitive abilities of children with hormone abnormalities: Screening by psychoeducational tests. *Journal of Learning Disabilities, 6*(1), 26–34.

Pfaff, D. W. (1966). Morphological changes in the brains of adult male rats after neonatal castration. *Journal of Endocrinology, 36*(4), 415–416.

Phillips, K., & Silverman, I. (1997). Differences in the relationship of menstrual cycle phase to spatial performance on two- and three-dimensional tasks. *Hormones and Behavior, 32*(3), 167–175.

Postma, A., Izendoorn, R., & De Hann, E. (1998). Sex differences in object location memory. *Brain and Cognition, 36*, 334–345.

Puts, D. A., McDaniel, M. A., Jordan, C. L., & Breedlove, S. M. (2005). Prenatal androgens and spatial ability in humans: Meta-analyses of CAH and 2D:4D studies. *Hormones and Behavior, 48*(1), 121.

Resnick, S. M., Berenbaum, S. A., Gottesman, I. I., & Bouchard, T. J. (1986). Early hormonal influences on cognitive functioning in congenital adrenal hyperplasia. *Developmental Psychology, 22*(2), 191–198.

Roof, R. L. (1993a). The dentate gyrus is sexually dimorphic in prepubescent rats: Testosterone plays a significant role. *Brain Research, 610*(1), 148–151.

Roof, R. L. (1993b). Neonatal exogenous testosterone modifies sex difference in radial arm and Morris water maze performance in prepubescent and adult rats. *Behavioural Brain Research, 53*(1–2), 1–10.

Roof, R. L., & Havens, M. D. (1992). Testosterone improves maze performance and induces development of a male hippocampus in females. *Brain Research, 572*(1–2), 310–313.

Roof, R. L., Zhang, Q., Glasier, M. M., & Stein, D. G. (1993). Gender-specific impairment on Morris water maze task after entorhinal cortex lesion. *Behavioural Brain Research, 57*(1), 47–51.

Ross, J. L., Roeltgen, D., Stefanatos, G. A., Feuillan, P., Kushner, H., Bondy, C., et al. (2003). Androgen-responsive aspects of cognition in girls with Turner syndrome. *Journal of Clinical Endocrinology and Metabolism, 88*(1), 292–296.

Ross, J. L., Stefanatos, G. A., Kushner, H., Zinn, A., Bondy, C., & Roeltgen, D. (2002). Persistent cognitive deficits in adult women with Turner syndrome. *Neurology, 58*(2), 218–225.

Sherry, D. F., Forbes, M. R., Khurgel, M., & Ivy, G. O. (1993). Females have a larger hippocampus than males in the brood-parasitic brown-headed cowbird. *Proceedings of the National Academy of Sciences, U.S.A., 90*(16), 7839–7843.

Sherry, D. F., & Hampson, E. (1997). Evolution and the hormonal control of sexually-dimorphic spatial abilities in humans. *Trends in Cognitive Sciences, 1*(2), 50–56.

Sherry, D. F., Jacobs, L. F., & Gaulin, S. J. (1992). Spatial memory and adaptive specialization of the hippocampus. *Trends in Neurosciences, 15*(8), 298–303.

Silva-Gomez, A. B., Bermudez, M., Quirion, R., Srivastava, L. K., Picazo, O., & Flores, G. (2003). Comparative behavioral changes between male and female postpubertal rats following neonatal excitotoxic lesions of the ventral hippocampus. *Brain Research, 973*(2), 285–292.

Silverman, I., & Eals, M. (1992). Sex differences in spatial abilities: Evolutionary theory and data. In J. Barkow, L. Cosmides & J. Tooby (Eds.), *The adapted mind: Evolutionary psychology and the generation of culture* (pp. 533–549). New York: Oxford University Press.

Silverman, I., Kastuk, D., Choi, J., & Phillips, K. (1999). Testosterone levels and spatial ability in men. *Psychoneuroendocrinology, 24*(8), 813–822.

Singer, C. A., McMillan, P. J., Dobie, D. J., & Dorsa, D. M. (1998). Effects of estrogen replacement on choline acetyltransferase and trkA mRNA expression in the basal forebrain of aged rats. *Brain Research, 789*(2), 343–346.

Slabbekoorn, D., Van Goozen, S. H., Megens, J., Gooren, L. J., & Cohen-Kettenis, P. T. (1999). Activating effects of cross-sex hormones on cognitive functioning: A study of short-term and long-term hormone effects in transsexuals. *Psychoneuroendocrinology, 24*(4), 423–447.

Smolen, A., Smolen, T. N., Han, P. C., & Collins, A. C. (1987). Sex differences in the recovery of brain acetylcholinesterase activity following a single exposure to DFP. *Pharmacology and Biochemistry of Behavior, 26*(4), 813–820.

Stewart, J., & Kolb, B. (1994). Dendritic branching in cortical pyramidal cells in response to ovariectomy in adult female rats: Suppression by neonatal exposure to testosterone. *Brain Research, 654*(1), 149–154.

Stewart, J., Skvarenina, A., & Pottier, J. (1975). Effects of neonatal androgens on open-field behavior and maze learning in the prepubescent and adult rat. *Physiology and Behavior, 14*(3), 291–295.

Tabibnia, G., Cooke, B. M., & Breedlove, S. M. (1999). Sex difference and laterality in the volume of mouse dentate gyrus granule cell layer. *Brain Research, 827*(1–2), 41–45.

Tottenham, L. S., Saucier, D., Elias, L., & Gutwin, C. (2003). Female advantage for spatial location memory in both static and dynamic environments. *Brain and Cognition, 53*, 381–383.

Tracy, D. M. (1987). Toys, spatial ability, and science and mathematics achievement: Are they related? *Sex Roles, 17*, 115–138.

Trivers, R. R. (1972). Parental investment and sexual selection. In B. Cambell (Ed.), *Sexual selection and the descent of man, 1871–1971* (pp. 136–179). London: Heinemann.

Van Goozen, S. H., Cohen-Kettenis, P. T., Gooren, L. J., Frijda, N. H., & Van de Poll, N. E. (1994). Activating effects of androgens on cognitive performance: Causal evidence in a group of female-to-male transsexuals. *Neuropsychologia*, *32*(10), 1153–1157.

Van Goozen, S. H., Cohen-Kettenis, P. T., Gooren, L. J., Frijda, N. H., & Van de Poll, N. E. (1995). Gender differences in behaviour: Activating effects of cross-sex hormones. *Psychoneuroendocrinology*, *20*(4), 343–363.

Van Goozen, S. H., Slabbekoorn, D., Gooren, L. J., Sanders, G., & Cohen-Kettenis, P. T. (2002). Organizing and activating effects of sex hormones in homosexual transsexuals. *Behavioral Neuroscience*, *116*(6), 982–988.

Vandenberg, S. G., & Kuse, A. R. (1978). Mental rotations: A group test of three-dimensional spatial visualization. *Perceptual and Motor Skills*, *47*, 599–604.

Vicedomini, J. P., Nonneman, A. J., DeKosky, S. T., & Scheff, S. W. (1986). Perinatal glucocorticoids disrupt learning: A sexually dimorphic response. *Physiology and Behavior*, *36*(1), 145–149.

Voyer, D., Voyer, S., & Bryden, M. P. (1995). Magnitude of sex differences in spatial abilities: A meta-analysis and consideration of critical variables. *Psychological Bulletin*, *117*(2), 250–270.

Warren, S. G., & Juraska, J. M. (1997). Spatial and nonspatial learning across the rat estrous cycle. *Behavioral Neuroscience*, *111*(2), 259–266.

Watson, N. V., & Kimura, D. (1991). Nontrivial sex differences in throwing and intercepting: Relation to psychometrically-defined spatial functions. *Personality and Individual Differences*, *12*, 375–385.

Whishaw, I. Q. (1985). Cholinergic receptor blockade in the rat impairs locale but not taxon strategies for place navigation in a swimming pool. *Behavioral Neuroscience*, *99*(5), 979–1005.

Wickstrom, R. L. (1977). *Fundament motor patterns*. Philadelphia: Lea and Febiger.

Williams, C. L., Barnett, A. M., & Meck, W. H. (1990). Organizational effects of early gonadal secretions on sexual differentiation in spatial memory. *Behavioral Neuroscience*, *104*(1), 84–97.

Williams, C. L., & Meck, W. H. (1991). The organizational effects of gonadal steroids on sexually dimorphic spatial ability. *Psychoneuroendocrinology*, *16*(1–3), 155–176.

Williams, C. L., Meck, W. H., Heyer, D. D., & Loy, R. (1998). Hypertrophy of basal forebrain neurons and enhanced visuospatial memory in perinatally choline-supplemented rats. *Brain Research*, *794*(2), 225–238.

Wimer, R. E., & Wimer, C. (1985). Three sex dimorphisms in the granule cell layer of the hippocampus in house mice. *Brain Research*, *328*(1), 105–109.

Wolf, O. T., Preut, R., Hellhammer, D. H., Kudielka, B. M., Schurmeyer, T. H., & Kirschbaum, C. (2000). Testosterone and cognition in elderly men: A single testosterone injection blocks the practice effect in verbal fluency, but has no effect on spatial or verbal memory. *Biological Psychiatry*, 47(7), 650–654.

Xiao, L., & Jordan, C. L. (2002). Sex differences, laterality, and hormonal regulation of androgen receptor immunoreactivity in rat hippocampus. *Hormones and Behavior*, 42(3), 327–336.

Yanai, J. (1979). Strain and sex differences in the rat brain. *Acta Anatomica (Basel)*, *103*(2), 150–158.

Zaidel, D. W., Esiri, M. M., & Oxbury, J. M. (1994). Sex-related asymmetries in the morphology of the left and right hippocampi? A follow-up study on epileptic patients. *Journal of Neurology*, *241*(10), 620–623.

13 The Evolution of Sex Differences in Spatial Cognition

Ruben C. Gur, Sarah L. Levin, Farzin Irani, Ivan S. Panyavin, Jaime W. Thomson, and Steven M. Platek

Sex differences in spatial cognition have been well documented. Behavioral, imaging and comparative studies show that males tend to perform better on spatial tasks involving mental rotation, while females excel in object location and spatial working memory tasks. Numerous evolutionary explanations have been offered for this phenomenon including theories of foraging, life history, and male range size. Evolutionary explanations of these spatial abilities are best viewed as an amalgamation of theories. We suggest a *why* and *how* approach to these theories, such that foraging theories (i.e., hunter-gatherer hypothesis) and male range hypotheses explain *why* sex differences in spatial abilities developed; while the *how* is explained through male-range size and female choice hypotheses. We argue that evolutionary theories provide an umbrella under which behavioral, biological, and comparative findings of sex differences can be understood. Biological hypotheses have implicated hormonal, genetic, and neural factors.

Human and animal studies have provided well-documented differences between males and females across various cognitive domains. Males and females differ on spatial abilities, as well as on other cognitive skills such as verbal and quantitative abilities (Maccoby & Jacklin, 1974). Currently, it is generally accepted that males perform better on certain spatial tasks, whereas females outperform males on some verbal and memory tasks (Caplan, MacPherson, & Tobin, 1985; Collins & Kimura, 1997; Delgado & Prieto, 1996; Maccoby & Jacklin, 1974; McGivern et al., 1997). Among the sex differences in cognition, those involving spatial abilities are the most robust. It is possible that the dramatic difference in responsibilities of the sexes (i.e., hunters vs. gatherers) created a sexual dimorphism in the processing of spatial tasks and in performance on certain types of spatial tasks. These sex differences have important evolutionary origins and continue to play an important role in our day-to-day interactions with our environment. Evolutionary theories provide an umbrella under which behavioral, biological, and comparative findings of sex differences can be understood.

This chapter begins with a review of various evolutionary hypotheses proposed to explain the current observations of sexual dimorphism in spatial abilities. Insights obtained from studies in voles and cowbirds are discussed, followed by a review of evidence obtained from behavioral studies in humans. The biological underpinnings of the sexually selected sex differences in spatial skills in humans are also reviewed, with a focus on the current state of evolved hormonal, genetic, and neural mechanisms. Finally, the current ecological validity of these sexually dimorphic traits is discussed in the context of the important role these traits play in our day-to-day interactions with our environment.

Sex Differences: Evolutionary Hypotheses

Numerous evolutionary hypotheses have been proposed to explain the observed sex differences in spatial ability in humans (Jones, Braithwaite, & Healy, 2003; Sherry & Hampson, 1997). These hypotheses can be roughly divided into three categories: foraging theories, life history theories, and theories based on sexual selection. We suggest that sex differences in spatial abilities developed through the acclimation of naturally and sexually selected traits developed to further advance species existence.

Foraging theories postulate that male foraging involved hunting, which forced our ancestors to cover larger terrain and venture farther from home (i.e., in pursuit of an animal, food, and so on). Over generations, this male foraging pattern resulted in the emergence of a male-dominated spatial ability involving mental manipulation of the environment. This is supported by robust findings of a male advantage on tasks involving three-dimensional (3D) mental rotations of object, which are discussed later (Geary, Gilger, & Elliott-Miller, 1992; Levin, Mohamed, & Platek, 2005; McBurney, Gaulin, Devineni, & Adams, 1997; Moffat, Hampson, & Hatzipantelis, 1998; O'Laughlin & Brubaker, 1998; Silverman, Kastuk, Choi, & Phillips, 1999). Conversely, female foraging involved gathering of food from stationary sources and resulted in the evolution of favored domains of spatial abilities such as object location and object memory (Silverman & Eals, 1992). This female advantage in object location memory or spatial working memory is also well documented and is discussed later in the chapter (Duff & Hampson, 2001; James & Kimura, 1997; McBurney et al., 1997; Tottenham, Saucier, Elias, & Gutwin, 2003).

Life history theories include the dispersal hypothesis, which predicts better spatial ability in the sex that travels farther from its natal area. The pattern of dispersal is predicted by the type of mating system employed in a given species: in systems where males have to defend their mates, males show more dispersal, whereas in systems in which males defend their resources, females show more postjuvenile dispersal (Greenwood, 1980). Perhaps surprisingly, in the majority of nonindustrialized human populations, females between birth and adulthood do in fact disperse farther than men (Koenig, 1989). However, it could be argued that the observation of such a pattern of behavior in the present does not necessarily mean that the same pattern was taking place during the environment of evolutionary adaptedness (EEA). Perhaps for this reason, the predicted advantage in spatial abilities one would expect to be associated with greater dispersal (i.e., spatial navigation) is not evident in females.

Another version of the life history hypothesis, the fertility and parental care hypothesis, has been proposed by Sherry and Hampson (1997). They advanced the notion that reduced mobility in females during the environment of evolutionary adaptedness could have served to increase females' reproductive success by redirecting energy resources from ranging to maintaining fertility and menstrual cyclicity, to increase the success of conception and successful pregnancy and reduce the risk of predation, rape, or accident for mother (and offspring). In support of this idea they offered evidence of reduced mobility during pregnancy from studies of both animals and humans. This hypothesis would predict that males possess better navigational spatial ability or, more accurately, that females' ability to navigate their environment is not as well developed as the males' ability.

The last set of theories postulates sex differences in spatial ability as sexually selected traits. Proposed by Gaulin and FitzGerald (1986, 1989), the male range size hypothesis asserts that in species where males have larger home ranges than females, males cover a larger area in order to compete for reproductive access to as many females as possible (Gaulin, 1992; Gaulin & FitzGerald, 1989). *Home range* refers to the physical space or territory that a member of a species travels during the day. As such, over time, the evolution of superior spatial abilities is sexually selected for in subsequent generations.

A similar hypothesis, the male warfare hypothesis (see Geary, 1995, for a review), states that male superiority in spatial ability is also due to ancestral males' need to cover a large territory. However, this need was not for hunting, as assumed by the male foraging hypothesis. Instead, the increased territory coverage may have been to wage small-scale warfare and to ambush other males, presumably to kidnap their females, capture their resources, and protect their own resources from being seized (Jones et al., 2003). Yet another sexual selection hypothesis deals with the idea that the ability to hunt successfully is a reproductively desirable trait, and as such guides the female's choice of mates. For females, the ostensible benefits of pair bonding with a successful hunter are thought to include not only a reliable supply of resources and provisions but also a share of genetic material to endow her offspring with a greater likelihood of surviving and mating.

The evolutionary perspectives presented above have received varying degrees of empirical support from human- and animal-based research. There is research supporting various aspects of these hypotheses. We suggest that it might be beneficial to amalgamate the perspectives of these theories into formulations of *why* and *how* sex differences evolved. For instance, foraging theories (i.e., the hunter-gatherer hypothesis) and male range hypotheses could explain *why* sex differences in spatial abilities developed, while male range size and female choice hypotheses could explain *how* these sexually dimorphic traits developed. Before examining the applicability of these theories, however, we need to establish that there are in fact sex differences in spatial cognition. Behavioral support for these sex differences in humans and animals is presented in the next section.

Insights from Voles and Cowbirds

Research into the evolutionary basis of spatial abilities can be traced back to the observations of Gaulin and colleagues in different species of *Microtus* voles (Gaulin & FitzGerald, 1986). Although most species in the genus *Microtus* are polygamous (e.g., the meadow vole, *Microtus pennsylvanicus*), a few pursue the monogamous reproductive strategy (e.g., the pine vole, *Microtus pinetorum*, and the prairie vole, *Microtus ochrogaster*). There are no notable exceptions between the polygamous and monogamous species except for the time when mating season arrives. Although male and female species of pine and prairie voles share the same home range year round, males of the meadow vole increase their home range significantly—by as much as five times—during the breeding season, compared with females of the species (such behavior does not occur in immature meadow voles) (Jacobs, Gaulin, Sherry, & Hoffman, 1990). This increase in the amount of terrain covered gives the polygamous males more access to a greater number of reproductively viable females in the area, thus increasing their chances of maximizing their fitness. In the monogamous species, however, such an increase would prove inconsequential, since any female a male is likely to encounter will have already pair bonded, providing no reproductive consequences. Instead, it is likely that pine and prairie voles pursue *K*-selection strategies of maximizing their fitness by investing more care into their young.

This distinction between males and females of the polygamous meadow vole made them a great specimen for the study of sex differences in spatial ability. The existence of differences in reproductive behavior between males and females of meadow voles speaks of larger environmental and evolutionary pressures that have exerted an influence throughout evolution. Much as in the rest of the animal world, where monogamy is an exception rather than the rule (Daly & Wilson, 1983), female meadow voles bear all the consequences of carrying and caring for their offspring, which makes them much more selective in their choice of breeding mate. A potential mate must possess certain characteristics that the female deems necessary to ensure the viability of the offspring, so that the offspring in turn may have offspring of their own (that is not to say that the females of monogamous species do not utilize the same strategy). Males, on the other hand, mate opportunistically: they bear no responsibility for rearing offspring, and thus experience no cost of copulating with as many females as they can during the mating period.

Male meadow voles increasing their home range is a way to increase their access to reproductively viable females with which they otherwise would not come in contact. This adaptation, though by no means unique to the voles, introduces a selective pressure for spatial ability in males that is absent in the females; males with lower ability to navigate their environment would have left fewer offspring, and their genes would have been selected against. Conversely, males with higher spatial ability (i.e., the ability to cover more terrain during the breeding period) left more descendants, and their genes, some containing the code for spatial ability, perpetuated. In other words, males compete with other males for reproductive access—precisely the condition for sexual selection which Darwin outlined in *The Descent of Man* (1874). Sexual selection, unlike natural selection, which is thought to contribute to survival of the species, is concerned with the evolution of traits that provide an advantage in competing for mates:

> Sexual selection depends on the success of certain individuals over others of the same sex, in relation to the propagation of the species; while natural selection depends on the success of both sexes, at all ages, in relation to the general conditions of life. The sexual struggle is of two kinds: in the one it is between the individuals of the same sex, generally the males, in order to drive away or kill their rivals, the females remaining passive; while in the other, the struggle is likewise between the individuals of the same sex, in order to excite or charm those of the opposite sex, generally the females, which no longer remain passive, but select the more agreeable partners. (Darwin, 1874, p. 639)

Darwin went on to outline further properties necessary for a sexually selected trait: it will be more developed in one sex than the other, it appears at and not before sexual maturity, it may be utilized only during some part of the year (i.e., the breeding season), and it is demonstrated to rivals or potential mates. Granted, in the case of meadow voles the sexual struggle does not reach the level of "killing their rivals," nor is there a demonstration of superior spatial ability between them, as one might observe in birds of paradise showing off or rams butting heads—the competition between male meadow voles is not direct.

The sex difference in spatial ability between male and female meadow voles was further investigated in a series of laboratory experiments that involved a variety of maze tasks (Gaulin & FitzGerald, 1986), on which males outperformed females. No such difference exists in monogamous species of voles. Sex differences in any given trait occur only if there is selective pressure for that trait in one sex only; there is no such pressure in the monogamous voles. In polygamous voles, however, the selection for range size affected only the males; therefore, we can conclude that sex difference in spatial ability is indeed a trait that is sexually selected for, though it is by no means exclusive to the males.

Another example in the animal world can be drawn from a species of birds known as brown-headed cowbird (*Molothrus ater*). This bird is a brood parasite; it creates no nest of its own, instead laying up to 40 eggs per year in the nests of other birds (Scott & Ankney, 1980). Not only are female cowbirds responsible for laying their eggs in the nests of potential "parents" unaided by the males of their species, they are also the ones that locate such nests, some time prior to parasitizing them. In other words, we can assume that the demand to navigate and recall 3-D environments in female cowbirds is greater than that of males. Jacobs and colleagues (1990) theorized that species that exhibit adaptive sex differences in spatial ability also likely exhibit differences in neural structure. This idea was applied as an a priori hypothesis to brown-headed cowbirds, where the hippocampus was identified as the neural structure of interest due to its role in spatial learning. Sherry, Farbes, Khurgel, and Ivy (1993) compared the size of the hippocampal complex in six male and six female cowbirds, as well as in two species of birds closely related to cowbirds but that are not brood parasites (the red-winged blackbird, *Agelaius phoeniceu*, and the common grackle, *Quiscalus quiscula*). As was predicted, female cowbirds were found to have a larger hippocampal complex than the males, while no such difference was noted in the other species examined. Similar investigation in voles produced the same results: in the polygamous meadow vole, males had a greater hippocampal volume than females, and there was no difference in monogamous pine and prairie voles. It seems, then, that in any given species it is not sex in and of itself, but rather the space and how it is utilized given specific reproductive pressures that is the best predictor of sexually dimorphic spatial ability and its underlying neural substrate. The results of this evolved sexual dimorphism in spatial abilities are observed differences in behavioral, biological, and functional studies in humans, as discussed in the next section.

Sex Differences in Humans: Behavioral Evidence

Historically, along with other sex differences in cognition, differences in spatial abilities were contentious (Caplan et al., 1985). These contentions often arose in part because of the lack of consensus on the concept of spatial ability, as well as the inclusion of different aspects of "spatial abilities" in studies of spatial cognition (Caplan et al., 1985; Voyer, Voyer, & Bryden, 1995). Before we describe the current status of behavioral evidence for human sex differences in spatial cognition, we consider the various spatial tasks that have been used to investigate this ability in humans.

Spatial cognition tasks have typically measured spatial perception, spatial visualization, mental rotation, and targeting. Spatial perception involves the ability to position stimuli such as lines despite distracting information such as tilted frames (e.g., rod-and-frame task). Making judgments on the orientations of an array of lines is another type of spatial perception task (e.g., the judgment of line orientation task, or JOLO). Drawing the water level on a picture of a tilted glass half-full of water is another spatial perception task. Spatial visualization tasks have involved the ability to use analytic strategies to manipulate spatial information (e.g., embedded-figures test). Mental rotation tasks measure the ability to mentally rotate figures rapidly and accurately in 2D or 3D space. Targeting performance tasks assess the ability to aim projectiles accurately at a specified point in space. These tasks involve spatiotemporal judgments about dynamic or moving objects. Other tasks include the generation and maintenance of a spatial image that is later used as part of a cognitive task.

Early investigations into the magnitude of sex differences on these various tasks led to conflicting findings about the variability in spatial abilities that could be accounted for by sex (Hyde, 1981; Linn & Petersen, 1985; Voyer et al., 1995). Although some argued that sex differences accounted for only a small percentage of the variance (Fairweather, 1976; Hyde, 1981; Kimball, 1981), others found that only sex differences in spatial perception and mental rotation were robust (Linn & Petersen, 1985; Voyer et al., 1995). There has also been debate about the stability of these sex differences across time. Some have suggested that sex differences have decreased over time (Baenninger & Newcombe, 1995; Crawford, Chaffin, & Fitton, 1995; Stumpf, 1995), while others have highlighted the stability of sex differences on tasks such as mental rotation (Casey, 1996a; Masters & Sanders, 1993; Stumpf, 1995).

There has been an abundance of research in the last few decades investigating various aspects of spatial abilities using a variety of tasks. A recent meta-analytic review found that sex differences favoring males were clearly established in some areas of spatial abilities tapped by tasks such as mental rotation, Cards Rotation Test, the Spatial Relations subtest of the Primary Mental Abilities (PMA) test, Paper Form Board, rod-and-frame test, and the Block Design subtest of the Weschler Adult Intelligence Scale (WAIS) (Voyer et al., 1995). Repeatedly, the literature points to mental rotation as the most robust sex difference in cognition (Casey, 1996a; Linn & Petersen, 1985; Masters & Sanders, 1993; Voyer et al., 1995) even though some have downplayed the size of this effect (Crawford et al., 1995). Boys as young as 10 years typically surpass females on tasks dealing with mental rotation (Johnson & Meade, 1987). However, females typically outperform males on tasks dealing with object location (Eals & Silverman, 1994; Silverman & Eals, 1992) and spatial working memory (Duff & Hampson, 2001; McBurney et al., 1997).

From a cognitive and proximate perspective, the observed differences between the sexes on these tasks could be due to different cognitive strategies used by the sexes to solve spatial tasks (Pezaris & Casey, 1991). For instance, males may excel on spatial tasks requiring maintenance and manipulation of information in working memory, while females may excel on tasks that require rapid access to and retrieval of information from stored memory (Halpern & Wright, 1996). This could be one explanation for why males perform well on tasks that benefit from combining new strategies such as mental rotation ability (Stumpf, 1995). But how and why did these differences in strategies evolve?

An evolutionary perspective would argue that spatial skills and strategies evolved due to factors such as home range size and reproductive behavior that would have resulted in survival and reproductive advantages during the EEA. In other words, sex differences in spatial cognition are the result of selection for cognitive programs that have recurrently allowed our ancestors to solve adaptive problems. Reproductive behavior is a set of actions designed to maximize one's fitness. Because of the fundamental asymmetry in most mammalian reproduction (i.e., females bear all the consequences of pregnancy and rearing for the offspring), females have evolved to be highly selective in choosing their mates. As a result, in the EEA, females chose to mate with males who had larger home ranges and had become good hunters as a result of enhanced spatial abilities. Males, however, evolved to be opportunistic in their mating practices, attempting to maximize their fitness with the highest number of females possible. These fundamental differences in reproductive behavior may then have given rise to sexual dimorphisms and differences in specific domains of functioning. These male-range-size and female-choice hypotheses could explain the evolutionary basis of the observed sex differences in abilities and strategies on the various spatial tasks. Furthermore, hunting tactics could have also played an important role. In order to implement and advance hunting strategies, ancestral males may have developed unique ways of viewing their environments. While females needed to focus their attention on the location and identity of objects in order to remember where the best places for food collection were, males needed to attend to the location of themselves in space. This might have involved active strategies, such as knowing where their weapon was in relation to their prey. This could have involved the use of compensatory strategies to maximize their likelihood of finding and hunting their food. Since active efforts at problem solving result in more efficient solutions, males might have adopted strategies involving rotating objects in space relative to themselves. This might have furthered their chances of locating and hunting their prey. In return, females chose males with dominant spatial manipulation skills (hence better hunters), and males mated with females who were in the location of their prey. Sexual selection therefore could have played a role in the evolution of sexual dimorphism in these spatial abilities.

Biological Underpinnings of Sex Differences

In addition to sexual differences on behavioral tasks, there has been a concurrent evolution of biological changes in males and females to maintain these sex differences. Sex differences in spatial abilities have been linked to differences in hormones, genes, and brain morphology (Casey, 1996a; Casey, Nuttall, & Pezaris, 1999), all of which are discussed in this section.

Hormones

Numerous studies have examined the role of hormones in proximate explanations of sexual dimorphism in spatial abilities. The difference in neural (Gur et al., 2000; Jordan et al., 2002; Thomsen et al., 2000) and behavioral (Geary et al., 1992; McBurney et al., 1997; Moffat et al., 1998; O'Laughlin & Brubaker, 1998; Silverman, Kastuk, Choi, & Phillips, 1999) evidence correlates with hormonal variation both prenatally (Witelson, 1991) and during adulthood (Christiansen & Knussmann, 1987; Galea, Kavaliers, Ossenkopp, & Hampson, 1995; Van Goozen, Cohen-Kettenis, Gooren, Frijda, & Van De Poll, 1995), and as a function of menstrual cycle phases (Frye, 1995; Hausmann, Slabbekoorn, Van Goozen, Cohen-Kettenis, & Gunturkun, 2000; McCormick & Teillon, 2001) in animals and humans. Studies of hormone variation in the female menstrual cycle have shown improvement in performance on spatial tasks during the menstrual phase and decreased performance during the midluteal phase (Hampson, 1990; Hampson & Kimura, 1988; Hausmann et al., 2000; McCormick & Teillon, 2001; Phillips & Silverman, 1997). It seems that either increased levels of testosterone or decreased levels of estradiol (or both) correlate with improved performance on spatial tasks (Berenbaum, Korman, & Leveroni, 1995).

Studies of humans with congenital disorders such as congenital adrenal hyperplasia (CAH) and idiopathic hypogonadotropic hypogonadism (IHH) offer further insights into hormonal influence on spatial abilities. CAH is a disorder characterized by high levels of prenatal androgens; however, these levels tend to normalize with postnatal treatment. The underlying problem is deficiency in an enzyme, usually 21-hydroxylase (21-OH), needed to produce cortisol (New & Levine, 1984). Males with IHH lack the hormones necessary to undergo puberty and develop the structures needed for lower levels of testosterone. Hines et al. (2003) found a correlation between males affected with CAH (i.e., they had increased levels of testosterone) and decreased or impaired performance on spatial tasks. Females with CAH have shown increased spatial abilities compared with their unaffected siblings (Berenbaum, 1992). Males with IHH (i.e., they had decreased levels of testosterone) showed decreased spatial abilities (Berenbaum et al., 1995). Silverman et al. (1999) found a positive relationship between testosterone levels and performance on mental rotation tasks. These contradictory findings can be attributed to an inverted U-shaped or curvilinear relationship between testosterone level and performance on spatial tasks. Specifically, as testosterone levels increase, performance on spatial tasks increases until an optimal level of testosterone is achieved, and then an inverse relationship emerges. The findings of Hines et al. (2003) suggest that surpassing this peak or optimal testosterone level in males adversely affects performance on spatial tasks. In females, it maybe that an "optimal" level of testosterone is not reached, since baseline levels of testosterone are lower overall.

Findings from comparative studies also suggest that increased androgen levels in neonatal environments of rats are correlated with increased spatial abilities in adulthood (Sherry & Hampson, 1997). Conversely, estradiol has been negatively correlated with spatial abilities in female meadow voles on a spatially oriented water maze (Galea et al., 1995), suggesting that estradiol may hinder performance on spatial tasks. A review of the findings from rodent studies suggested that females exposed to elevated levels of androgen or its metabolites during early development show enhanced performance on spatial tasks, whereas males exposed to reduced levels of androgens show impaired performance (Williams & Meck, 1991). However, in one study of sex-reversed female mice, mice with the Y chromosome performed significantly better on a Morris water maze task than did mice with two X chromosomes (Stavnezer et al., 2000). This suggests that, among other things, there may be both hormonal *and* genetic contributions to the sexual dimorphism in spatial abilities.

Genes

Casey (1996b) reviewed the various conceptualizations of biological-environmental influences of spatial dimorphism. *Biological-environmental interactions* are those in which combinations of biological and environmental relationships result in consequences not predicted by the two factors individually. For example, animals with one type of genetic inheritance may benefit more when exposed to spatial experiences than those with another type of genetic makeup. *Passive gene-environmental correlations* are those in which individuals with a particular genetic makeup within a family provide an environment for their child that is strongly influenced by their own biological makeup. For instance, a mother who inherits good spatial skills may then pass on her skills to her daughters through her genes, but her daily choices also affect her daughters' environment. *Reactive gene-environment correlations* are those in which individuals in the environment react to biological characteristics of the person. For example, a mother who has a daughter with a genetic predisposition toward spatial activities may give in and buy Legos because of her daughter's insistence. In this way, the mother is reacting to the child's biological predispositions and fostering spatial skills as a result. *Active gene-environment correlations* are those in which people actively seek out experiences related to their traits and abilities. For example, owing to her genetic predisposition a child may actively seek out spatially oriented toys. Casey (1996b) introduced these theories, as well as a "bent twig" hypothesis, to explain individual differences in females in spatial ability, particularly in women who proved exceptions to the male advantage in mental rotation ability. Based on her work with women who excelled on spatial tasks, she hypothesized that these women can be expected to have a combination of inherited genetic potential and prior experiences that allows them to succeed on such tasks. However, these hypotheses are insufficient to explain the inherited mechanism for the observed sexual dimorphism in spatial skills in the majority of cases. Furthermore, these hypotheses are unable to explain why there is a sex difference in the first place. Is there a sex-linked chromosomal mechanism of inheritance? Does the presence of the Y chromosome, which is sexually dimorphic and present only in males, also lead to sex differences in behavior?

A recent study investigated this in an animal model by seeking evidence for genetic influences on some sexually dimorphic morphological structures and influences of the Y chromosome on behavior (Stavnezer et al., 2000). Strains of mice were used that had sex-reversed XY females, which were then compared with their XX female siblings. This allowed direct assessment of the influence of the Y chromosome in a female phenotype. They found that XY females performed significantly better than XX females on the Morris hidden platform spatial maze. These findings suggest that males may have both a genetic and a hormonal mechanism for a spatial advantage over females. Evolution may have then passed on this sex-linked trait through the generations.

Brain Morphology

Another evolved biological change has been brain changes in hemispheric organization. Several hypotheses have linked sex differences in cognitive function to hemispheric organization, since the right hemisphere is involved in spatial abilities and the left is involved in verbal tasks (Hiscock et al., 1995; Witelson, 1976). Improvements in functional neuroimaging technology have allowed direct exploration of the neural basis for sex differences in spatial cognition.

Studies examining lateralized activation in regions implicated in spatial processing have found that men show greater right hemisphere activation in these regions, and this right-lateralized activation is associated with better performance (Gur et al., 1994; Howard, Fenwick, Brown, & Norton, 1992; Wendt & Risberg, 1994; see also Levin et al., 2005). An fMRI study by Gur et al. (2000) found that hypothesized region-of-interest (ROI)–based analysis revealed a right-lateralized increase in activation for a judgment of line orientation spatial task in men. An image-based analysis revealed a distributed network of cortical regions, with more right lateralization for the spatial task in both sexes but with some unique left activation in men. Increased task difficulty produced more circumscribed activation for the spatial task, and the results suggest that failure to activate the appropriate hemisphere in regions directly involved in task performance may explain these sex differences in performance. Bilateral activation in a distributed cognitive system underlies sex differences in spatial task performance.

On mental rotation tasks, where there is the most robust support for sexual dimorphism in spatial skills, the data from imaging studies investigating hemispheric differences are inconsistent (Dietrich et al., 2001; Jordan, Wustenberg, Heinze, Peters, & Jancke, 2002; Tagaris et al., 1996, 1998; Thomsen et al., 2000; Unterrainer, Wranek, Staffen, Gruber, & Ladurner, 2000; Weiss et al., 2003). It is unclear whether differences in brain activation patterns between men and women are due to task performance or sex-related hemispheric organization (Weiss et al., 2003). Some have found different cortical activation patterns depending on performance level rather than sex (Tagaris et al., 1998; Unterrainer et al., 2000), others have reported gender-specific activation patterns without sex differences in performance (Jordan et al., 2002; Thomsen et al., 2000), and still others have found both a behavioral and an activation pattern difference (Levin et al., 2005). The discrepancies are surprising, since behavioral data are most consistent in showing robust sex differences on spatial tasks involving mental rotation. Performance levels may be a potential confound in all of these experiments.

Therefore, in order to eliminate the confounding influences of overall performance levels, Jordan et al. (2002) investigated cortical activation patterns in males and females who did not differ in overall level of performance on three mental rotation tasks. They found that women exhibited significant bilateral activation in the intraparietal sulcus and the superior and inferior parietal lobule, as well as in the inferior temporal gyrus and the premotor areas. Men showed significant activation in the right parieto-occipital sulcus, the left intraparietal sulcus, and the left superior parietal lobule. They also found that both men and women showed activation of the premotor areas but men also had additional significant activation of the left motor cortex. No significant activation was found in the inferior temporal gyrus. These results suggested that there are genuine sex differences in brain activation patterns during mental rotation activities even when performances are similar.

Levin et al. (2005) further investigated the neural basis of tasks involving mental rotation and object location. Previous findings of a male advantage on the mental rotation task, with faster responses and fewer errors, were confirmed. This was further corroborated with fMRI results showing a strong sex difference, with males showing significantly more overall cortical activation during the mental rotation task than women. Interestingly, performance appeared to be modulated by activation in the parahippocampal gyrus, implicating this region as one of importance for males when solving spatial tasks.

On a task involving object location memory to recall placement and similarity of objects, Levin et al. (2005) did not find a performance difference. However, their neuroimaging findings indicated a sex difference, with greater activation on the object location task for females. Interestingly, males and females seemed to use similar cortical areas when performing these tasks. Specifically, each sex revealed significantly more activation in the medial frontal lobes for tasks at which they excelled: mental rotation for males and object location for females. Overall, these fMRI studies provide further support for evolved changes in the brain, possibly due to males' and females' specialized responsibilities (i.e., hunting and gathering), which can account for sex differences in performance on spatial tasks.

Sex Differences Outside the Laboratory

Many of the sex differences in spatial abilities and strategies that have been found in the laboratory are also evident in today's real-world environment. Research suggests that males are more likely to use Euclidean strategies involving distances and direction and to be more accurate with these strategies, while females are more likely to use topographic strategies using landmarks. In an early study by McGuiness and Sparks (1983), college students were asked to draw maps of the campus that could help someone unfamiliar with it. While both sexes included most major buildings on their maps, women included few connectors (roads, bridges, and paths) between the landmarks. The authors also found that males showed a greater topographic sense of the campus, placing buildings more accurately with respect to spatial coordinates and showing more routes and connectors. Women, however, showed a more accurate sense of distance and were more accurate in their placement of buildings with respect to absolute distance. In a second experiment they examined whether the omission of connectors by women was due to memory difficulties or to a lack of relevance of these items. They found that women knew some of the connectors and included them when specifically requested to do so. However, even when instructed to include connectors, the female students were consistently less accurate than the male students. These results suggest that connectors, especially roads, are both less relevant and less memorable for women than for men. This dimorphism fits with our ancestral male's ability to navigate in the EEA in order to hunt for food, and the female's ability to sense and remember distances during food gathering.

In another study, Ward, Newcombe, and Overton (1986) examined how male and female undergraduates gave directions from perceptually available maps and memorized maps. They coded six different aspects of direction giving: the use of landmarks, use of relational terms, use of cardinal directions, use of mileage estimates, and frequency of omission and commission errors. They found that males used more mileage estimates and cardinal directions and made fewer errors than females, even though the use of cardinal directions and mileage estimates was rarer in relation to the opportunities to use them compared with the use of landmarks and relational terms. Galea and Kimura (1993) controlled for possible effects of extramap superior visual memory in females on their memory for landmarks and investigated the relationship between accuracy of performance and geometric or landmark knowledge. They asked participants to learn a route to criterion through a novel map. They found that males made fewer errors and took fewer trials to reach criterion, but females remembered more landmarks both on and off the route than males. They also found that superior memory for landmarks was not accounted for by a superior visual item memory. Males also seemed to outperform females in knowledge of the Euclidean properties of the map. However, despite the pronounced sex differences in knowledge retained from the maps, both males' and females' performance was related to spatial ability rather than to landmark recall. Overall, these findings suggest that our evolved ability to spatially navigate ourselves based on sex is currently a real phenomenon that extends beyond both the EEA and the laboratory.

There is also evidence of differences between the sexes with respect to geographic knowledge. Studies have shown that males know more than females about far-off regions of the world. Eve, Price, and Counts (1994) investigated what variables might be related to geographic literacy by giving a large sample of college students a survey assessing aspects of geographic knowledge such as general geography, map reading, intercultural literacy, and icon recognition. They found that gender was a strong predictor of performance on this task, with males outperforming females. In another study, the ability to locate cities on outline maps of the United States and of the local region of residence differed based on sex throughout the life span, with males outperforming females (Beatty, 1989). Even after age 70 years, while accuracy declined slightly for women, it remained stable for men. This sex difference in geographic knowledge could not be explained by education, region of residence, or travel experience (Beatty & Troester, 1987). Again, the implications of this work extend to our understandings of the current ecological validity for these sexually dimorphic traits and the important role that these traits play in our day-to-day interactions with our environment. A hunter-gatherer hypothesis adapted through sexual selection for males as they expanded their home range and for females as they selected the highest-quality mates.

Conclusion

Evolutionary theories provide an umbrella under which behavioral, biological, and comparative research can be understood. Just as the construction of a house is supported by its framework, the evidence discussed in this chapter supports evolutionary concepts and hypotheses. Having a theoretical framework for conducting research also enables us to further expand our knowledge of cognition, and of sex differences in cognition.

In this chapter we have approached evolutionary explanations of spatial abilities as an amalgamation of approaches under the evolutionary meta-theoretical framework. We have proposed a *why* and *how* approach to integrate the current theories from an evolutionary perspective. The foraging theories (i.e., hunter-gatherer hypothesis) and male range hypotheses explain *why* sex differences in spatial abilities developed. Males, being the hunters, needed to mentally manipulate their environment in order to navigate through it and obtain an advantage in hunting. Females, however, maintained superiority in object location memory or spatial working memory in order to effectively gather food for their family. The *how* is explained through male range size: a larger home range enables greater access to females and therefore increased opportunity for males to procreate and pass down their genes. The female choice hypothesis further elaborates on the *how* by explaining the evolutionary principle of sexual selection. These theories are supported by behavioral, biological, and comparative evidence. Further research is needed to clarify the critical impact of these sex differences on our present, past, and future.

References

Baenninger, M., & Newcombe, N. (1995). Environmental input to the development of sex-related differences in spatial and mathematical ability. *Learning and Individual Differences*, *7*(4), 363–379.

Beatty, W. W. (1989). Geographical knowledge throughout the lifespan. *Bulletin of the Psychonomic Society*, *27*(4), 379–381.

Beatty, W. W., & Troester, A. I. (1987). Gender differences in geographical knowledge. *Sex Roles*, *16*(11–12), 565–590.

Berenbaum, S. A., Korman, K., & Leveroni, C. (1995). Early hormones and sex differences in cognitive abilities. *Learning and Individual Differences*, *7*(4), 303–321.

Caplan, P. J., MacPherson, G. M., & Tobin, P. (1985). Do sex-related differences in spatial abilities exist? A multilevel critique with new data. *American Psychologist*, *40*(7), 786–799.

Casey, M. B. (1996a). Gender, sex, and cognition: Considering the interrelationship between biological and environmental factors. *Learning and Individual Differences*, *8*(1), 39–53.

Casey, M. B. (1996b). Understanding individual differences in spatial ability within females: A nature/nurture interactionist framework. *Developmental Review*, *16*(3), 241–260.

Casey, M. B., Nuttall, R. L., & Pezaris, E. (1999). Evidence in support of a model that predicts how biological and environmental factors interact to influence spatial skills. *Developmental Psychology*, *35*(5), 1237–1247.

Christiansen, K., & Knussmann, R. (1987). Sex hormones and cognitive functioning in men. *Neuropsychobiology*, *18*(1), 27–36.

Collins, D. W., & Kimura, D. (1997). A large sex difference on a two-dimensional mental rotation task. *Behavioral Neuroscience*, *111*(4), 845–849.

Crawford, M., Chaffin, R., & Fitton, L. (1995). Cognition in social context. *Learning and Individual Differences*, *7*(4), 341–362.

Daly, M., & Wilson, M. (1983). *Sex, evolution, and behavior* (2nd ed.). Belmont, CA: Wadsworth.

Darwin, C. (1874). *The descent of man and Selection in relation to sex* (2nd ed.). London: John Murray.

Delgado, A. R., & Prieto, G. (1996). Sex differences in visuospatial ability: Do performance factors play such an important role? *Memory & Cognition*, *24*(4), 504–510.

Dietrich, T., Krings, T., Neulen, J., Willmes, K., Erberich, S., Thron, A., et al. (2001). Effects of blood estrogen level on cortical activation patterns during cognitive activation as measured by functional MRI. *NeuroImage*, *13*(3), 425–432.

Duff, S. J., & Hampson, E. (2001). A sex difference on a novel spatial working memory task in humans. *Brain and Cognition*, *47*(3), 470–493.

Eals, M., & Silverman, I. (1994). The hunter-gatherer theory of spatial sex differences: Proximate factors mediating the female advantage in recall of object arrays. *Ethnology and Sociobiology*, *15*, 95–105.

Eve, R. A., Price, B., & Counts, M. (1994). Geographic illiteracy among college students. *Youth & Society*, *25*(3), 408–427.

Fairweather, H. (1976). Sex differences in cognition. *Cognition*, *4*, 231–280.

Frye, C. A. (1995). Estrus-associated decrements in a water maze task are limited to acquisition. *Physiology & Behavior*, *57*(1), 5–14.

Galea, L. A. M., Kavaliers, M., Ossenkopp, K.-P., & Hampson, E. (1995). Gonadal hormone levels and spatial learning performance in the Morris water maze in male and female meadow voles. *Microtus pennsylvanicus. Hormones and Behavior*, *29*(1), 106–125.

Galea, L. A. M., & Kimura, D. (1993). Sex differences in route-learning. *Personality and Individual Differences*, *14*(1), 53–65.

Gaulin, S. J. C. (1992). Evolution of sex differences in spatial ability. *Yearbook of Physical Anthropology*, *35*, 125–151.

Gaulin, S. J. C., & FitzGerald, R. W. (1986). Sex differences in spatial ability: An evolutionary hypothesis and test. *American Naturalist*, *127*, 74.

Gaulin, S. J. C., & FitzGerald, R. W. (1989). Sexual selection for spatial-learning ability. *Animal Behavior*, *37*, 322–331.

Geary, D. C. (1995). Sexual selection and sex differences in spatial cognition. *Learning and Individual Differences*, *7*(4), 289–301.

Geary, D. C., Gilger, J. W., & Elliott-Miller, B. (1992). Gender differences in three-dimensional mental rotation: A replication. *Journal of Genetic Psychology*, *153*(1), 115–117.

Greenwood, P. J. (1980). Mating systems, philopatry and dispersal in birds and mammals. *Animal Behaviour*, *28*, 1140–1162.

Gur R. C., Ragland J. D., Resnick S. M., Skolnick B. E., Jaggi J., Muenz L., et al. (1994). Lateralized increases in cerebral blood flow during performance of verbal and spatial tasks: Relationship with performance level. *Brain and Cognition*, *24*(2), 244–258.

Gur, R. C., Alsop, D., Glahn, D., Petty, R., Swanson, C. L., Maldjian, J. A., et al. (2000). An fMRI study of sex differences in regional activation to a verbal and a spatial task. *Brain and Language*, *74*(2), 157–170.

Halpern, D. F., & Wright, T. M. (1996). A process-oriented model of cognitive sex differences. *Learning and Individual Differences*, *8*(1), 3–24.

Hampson, E. (1990). Estrogen-related variations in human spatial and articulatory-motor skills. *Psychoneuroendocrinology*, *15*(2), 97–111.

Hampson, E., & Kimura, D. (1988). Reciprocal effects of hormonal fluctuations on human motor and perceptual-spatial skills. *Behavioral Neuroscience*, *102*(3), 456–459.

Hausmann, M., Slabbekoorn, D., Van Goozen, S. H. M., Cohen-Kettenis, P. T., & Gunturkun, O. (2000). Sex hormones affect spatial abilities during the menstrual cycle. *Behavioral Neuroscience*, *114*(6), 1245–1250.

Hines, M., Fane, B. A., Pasterski, V. L., Mathews, G. A., Conway, G. S., & Brook, C. (2003). Spatial abilities following prenatal androgen abnormality: Targeting and mental rotations performance in individuals with congenital adrenal hyperplasia. *Psychoneuroendocrinology*, *28*(8), 1010–1026.

Hiscock, M., Israelian, M., Inch, R., Jacek, C., et al. (1995). Is there a sex difference in human laterality? II. An exhaustive survey of visual laterality studies from six neuropsychology journals. *Journal of Clinical and Experimental Neuropsychology*, *17*(4), 590–610.

Howard, R., Fenwick, P., Brown, D., & Norton, R. (1992). Relationship between CNV asymmetries and individual differences in cognitive performance, personality and gender. *International Journal of Psychophysiology*, *13*(3), 191–197.

Hyde, J. S. (1981). How large are cognitive gender differences? A meta-analysis using $!w^2$ and d. *American Psychologist*, *36*(8), 892–901.

Imperato-McGinley, J., Pichardo, M., Gautier, T., Voyer, D., & Bryden, M. P. (1991). Cognitive abilities in androgen-insensitive subjects: Comparison with control males and females from the same kindred. *Clinical Endocrinology*, *34*(5), 341–347.

Jacobs, L. F., Gaulin, S. J., Sherry, D. F., & Hoffman, G. E. (1990). Evolution of spatial cognition: Sex-specific patterns of spatial behavior predict hippocampal size. *Proceedings of the National Academy of Sciences, U.S.A.*, *87*(16), 6349–6352.

James, T. W., & Kimura, D. (1997). Sex differences in remembering the locations of objects in an array: Location-shifts versus location-exchanges. *Evolution and Human Behavior*, *18*(3), 155–163.

Johnson, E., & Meade, A. (1987). Developmental patterns of spatial ability: An early sex difference. *Child Development*, *58*(3), 725–740.

Jones, C. M., Braithwaite, V. A., & Healy, S. D. (2003). The evolution of sex differences in spatial ability. *Behavioral Neuroscience*, *117*(3), 403–411.

Jordan, K., Wustenberg, T., Heinze, H.-J., Peters, M., & Jancke, L. (2002). Women and men exhibit different cortical activation patterns during mental rotation tasks. *Neuropsychologia*, *40*(13), 2397–2408.

Kimball, M. M. (1981). Women and science: A critique of biological theories. *International Journal of Women's Studies*, *4*, 318–338.

Koenig, W. D. (1989). Sex-biased dispersal in the contemporary United States. *Ethology and Sociobiology*, *10*(4), 263–277.

Levin, S. L., Mohamed, F. B., & Platek, S. M. (2005) Common ground for spatial cognition? A behavioral and fMRI study of sex differences in mental rotation and spatial working memory. *Evolutionary Psychology*, *3*, 227–254.

Linn, M. C., & Petersen, A. C. (1985). Emergence and characterization of sex differences in spatial ability: A meta-analysis. *Child Development*, *56*(6), 1479–1498.

Maccoby, E. E., & Jacklin, C. N. (1974). *The psychology of sex differences*. Stanford, CA: Stanford University Press.

Masters, M. S., & Sanders, B. (1993). Is the gender difference in mental rotation disappearing? *Behavioral Genetics*, *23*(4), 337–341.

McBurney, D. H., Gaulin, S. J. C., Devineni, T., & Adams, C. (1997). Superior spatial memory of women: Stronger evidence for the gathering hypothesis. *Evolution and Human Behavior*, *18*(3), 165–174.

McCormick, C. M., & Teillon, S. M. (2001). Menstrual cycle variation in spatial ability: Relation to salivary cortisol levels. *Hormones and Behavior*, *39*(1), 29–38.

McGee, M. G. (1979). Human spatial abilities: Psychometric studies and environmental, genetic, hormonal, and neurological influences. *Psychological Bulletin*, *86*(5), 889–918.

McGivern, R. F., Huston, J. P., Byrd, D., King, T., Siegle, G. J., & Reilly, J. (1997). Sex differences in visual recognition memory: Support for a sex-related difference in attention in adults and children. *Brain and Cognition*, *34*(3), 323–336.

McGuiness, D., & Sparks, J. (1983). Cognitive style and cognitive maps: Sex differences in representations of a familiar terrain. *Journal of Mental Imagery*, 7(2), 91–100.

Moffat, S. D., Hampson, E., & Hatzipantelis, M. (1998). Navigation in a "virtual" maze: Sex differences and correlation with psychometric measures of spatial ability in humans. *Evolution and Human Behavior*, *19*(2), 73–87.

New, M. I., & Levine, L. S. (1984). Recent advances in 21-hydroxylase deficiency. *Annual Review Of Medicine*, *35*, 649–663.

O'Laughlin, E. M., & Brubaker, B. S. (1998). Use of landmarks in cognitive mapping: Gender differences in self report versus performance. *Personality and Individual Differences*, *24*(5), 595–601.

Pezaris, E., & Casey, M. B. (1991). Girls who use "masculine" problem-solving strategies on a spatial task: Proposed genetic and environmental factors. *Brain and Cognition*, *17*(1), 1–22.

Phillips, K., & Silverman, I. (1997). Differences in the relationship of menstrual cycle phase to spatial performance on two- and three-dimensional tasks. *Hormones and Behavior*, *32*(3), 167–175.

Scott, D., & Ankney, C. (1980). Fecundity of the brown-headed cowbird in southern Ontario. *Auk 97*, 677–683.

Sherry, D., Forbes, M., Khurgel, M., & Ivy, G. (1993). Females have a larger hippocampus than males in the brood-parasitic brown-headed cowbird. *Proceedings of the National Academy of Sciences, U.S.A.*, *90*, 7839–7843.

Sherry, D. F., & Hampson, E. (1997). Evolution and the hormonal control of sexually-dimorphic spatial abilities in humans. *Trends in Cognitive Sciences*, *1*(2), 50–56.

Signorella, M. L., & Jamison, W. (1986). Masculinity, femininity, androgyny, and cognitive performance: A meta-analysis. *Psychological Bulletin*, *100*(2), 207–228.

Silverman, I., & Eals, M. (1992). Sex differences in spatial abilities: Evolutionary theory and data. In J. H. Barkow, L. Cosmides, & J. Tooby (Eds.), *The adapted mind* (pp. 533–549). New York: Oxford University Press.

Silverman, I., Kastuk, D., Choi, J., & Phillips, K. (1999). Testosterone levels and spatial ability in men. *Psychoneuroendocrinology*, *24*(8), 813–822.

Stavnezer, A. J., McDowell, C. S., Hyde, L. A., Bimonte, H. A., Balogh, S. A., Hoplight, B. J., et al. (2000). Spatial ability of XY sex-reversed female mice. *Behavioural Brain Research*, *112*(1–2), 135–143.

Stumpf, H. (1995). Gender differences in performance on tests of cognitive abilities: Experimental design issues and empirical results. *Learning and Individual Differences*, *7*(4), 275–287.

Tagaris, G. A., Kim, S.-G., Strupp, J. P., Andersen, P., Ugurbil, K., & Gerogopoulos, A. P. (1996). Quantitative relations between parietal activation and performance in mental rotation. *Neuroreport*, *7*(3), 773–776.

Tagaris, G. A., Richter, W., Kim, S.-G., Pellizzer, G., Andersen, P., Uurbil, K., et al. (1998). Functional magnetic resonance imaging of mental rotation and memory scanning: A multidimensional scaling analysis of brain activation patterns. *Brain Research Reviews*, *26*(2–3), 106–112.

Thomsen, T., Hugdahl, K., Ersland, L., Barndon, R., Lundervold, A., Smievol, A. I., et al. (2000). Functional magnetic resonance imaging (fMRI) study of sex differences in a mental rotation task. *Medical Science Monitor*, *6*(6), 1186–1196.

Tottenham, L. S., Saucier, D., Elias, L., & Gutwin, C. (2003). Female advantage for spatial location memory in both static and dynamic environments. *Brain and Cognition*, *53*(2), 381–383.

Unterrainer, J., Wranek, U., Staffen, W., Gruber, T., & Ladurner, G. (2000). Lateralized cognitive visuospatial processing: Is it primarily gender-related or due to quality of performance? A HMPAO-SPECT study. *Neuropsychobiology*, *41*(2), 95–101.

Van Goozen, S. H. M., Cohen-Kettenis, P. T., Gooren, L. J. G., Frijda, N. H., & Van De Poll, N. E. (1995). Gender differences in behaviour: Activating effects of cross-sex hormones. *Psychoneuroendocrinology*, *20*(4), 343–363.

Voyer, D., Voyer, S., & Bryden, M. P. (1995). Magnitude of sex differences in spatial abilities: A meta-analysis and consideration of critical variables. *Psychological Bulletin*, *117*(2), 250–270.

Waber, D. P. (1976). Sex differences in cognition: A function of maturation rate? *Science*, *192*(4239), 572–574.

Ward, S. L., Newcombe, N., & Overton, W. F. (1986). Turn left at the church, or three miles north: A study of direction giving and sex differences. *Environment & Behavior*, *18*(2), 192–213.

Weiss, E., Siedentopf, C. M., Hofer, A., Deisenhammer, E. A., Hoptman, M. J., Kremser, C., et al. (2003). Sex differences in brain activation pattern during a visuospatial cognitive task: A functional magnetic resonance imaging study in healthy volunteers. *Neuroscience Letters*, *344*(3), 169–172.

Wendt, P. E., & Risberg, J. (1994). Cortical activation during visual spatial processing: Relation between hemispheric asymmetry of blood flow and performance. *Brain and Cognition*, *24*(1), 87–103.

Williams, C. L., & Meck, W. H. (1991). The organizational effects of gonadal steroids on sexually dimorphic spatial ability. *Psychoneuroendocrinology*, *16*(1–3), 155–176.

Witelson, S. F. (1976). Sex and the single hemisphere: Specialization of the right hemisphere for spatial processing. *Science*, *193*(4251), 425–427.

Witelson, S. F. (1991). Neural sexual mosaicism: Sexual differentiation of the human temporo-parietal region for functional asymmetry. *Psychoneuroendocrinology*, *16*(1–3), 131–153.

14 The Evolution of Language: From Hand to Mouth

Michael C. Corballis

In 1866, seven years after the publication of Darwin's *Origin of Species*, the Linguistic Society of Paris famously banned all discussion of the evolution of language. The main difficulty, it seems, was the widespread belief that language was uniquely human, so that there was no evidence to be gained from the study of nonhuman animals. This meant that language must have evolved some time after humans split off from the great apes. Because there was little evidence to be gained from the fossil evidence, any theory as to how language evolved was largely a matter of speculation—and no doubt argument. Of course, evolution was itself a contentious issue, and was attacked vigorously by the Church. In the case of language, the conflict between science and religion would have been exacerbated by the long-standing view that language was gifted by God.

In more recent if not more enlightened times, the ban seems to have been lifted, but the contention remains. Although there are probably few who would argue that language is a gift from God, there are still those who maintain that language evolved in a single step, in all-or-none fashion. This is sometimes called the "big bang" theory of language evolution and has been most clearly articulated by the linguist Derek Bickerton (1995), but it is also implicit in much of the writing of Noam Chomsky. It smacks a little of the miraculous, although the appeal is more likely to be to a fortuitous genetic mutation (e.g., Crow, 2002) or to some emergent physical property rather than to God. Against this is the view that language evolved in incremental fashion, through natural selection, as maintained by Pinker and Bloom (1990) and elaborated by Jackendoff (2002). This view is clearly aligned with the Darwinian view of human evolution.

A related issue is whether the origins of language are to be found in the behavior or communication systems of animals or, more specifically, of our primate forebears. In 1966, Chomsky wrote as follows:

> The unboundedness of human speech, as an expression of limitless thought, is an entirely different matter [from animal communication], because of the freedom from stimulus control; and the appropriateness to new situations. . . . Modern studies of animal communication so far offer no counterevidence to the Cartesian assumption that human language is based on an entirely different principle. Each known animal communication system either consists of a fixed number of signals, each associated with a specific range of eliciting systems or internal states, or a fixed number of "linguistic dimensions," each associated with a non-linguistic dimension. (pp. 77–78)

Although Chomsky has also argued that language did not evolve through natural selection, the idea that language is uniquely human has not precluded the notion that it is a product of natural selection. For example, although Pinker and Bloom (1990; see also Pinker, 1994) argue that language evolved in the hominid lineage through natural selection, they agree with Chomsky that nothing resembling language has been demonstrated in any nonhuman species.

Chomsky appears to have moderated his conclusion in a recent coauthored article (Hauser, Fitch, & Chomsky, 2002), in which it is argued that there is a distinction between what the authors call the faculty of language in the broad sense (FLB) and the faculty of language in the narrow sense (FLN). It is clear that most animals, including primates, are able to make communicative sounds and perform intentional acts. It is also clear that primates can use sounds or actions in symbolic fashion. For example, vervet monkeys give different warning cries to distinguish between a number of different threats, such as snakes, hawks, eagles, or leopards. When a monkey makes one of these cries, the troop members act appropriately, clambering up trees in response to a leopard call or running into the bushes in response to an eagle call (Cheney & Seyfarth, 1990). These cries bear no obvious relation to the sounds emitted by the predators they stand for, and in that sense they are symbolic. More compellingly, perhaps, the bonobo Kanzi is able to use visual symbols on a keyboard to refer to objects and actions; these symbols are again abstract in the sense that they were deliberately chosen by the human keepers so as not to resemble what they stand for (Savage-Rumbaugh, Shanker, & Taylor, 1998).

Recursion

Based on such arguments, Hauser et al. (2002) argue that FLB is shared by other species, including birds and other mammals, although they also point out that the use of symbols in these examples does not mean that the symbols have all of the properties of *words*. The critical ingredient that is missing from FLB, and that characterizes FLN, is *recursion*. Recursion lies at the heart of grammar, and enables us to create a potential infinity of sentences that convey an infinity of meanings. Recursive language is well understood even by quite young children, as illustrated by the well-known children's story:

This is the house that Jack built.
This is the malt that lay in the house that Jack built.
This is the rat that ate the malt that lay in the house that Jack built.
This is the cat that worried the rat that ate the malt that lay in the house that Jack built.

Young children quickly understand that the sentence can be extended ad infinitum. The recursive rules of grammar also allow phrases to be moved around instead of simply being tacked on to the beginning. For example, if one wanted to highlight *malt* in the story, one could embed phrases as follows:

The malt that the rat that the cat killed ate lay in the house that Jack built.

It seems clear that this highly flexible, recursive property is absent from communication among nonhuman species. Although many birds emit long sequences of sounds, these are essentially repetitive, born of insistence, perhaps, rather than an attempt to convey new information. The same is true of primates, and it is evident even at the acoustic level that there is a variation and novelty in human vocal output that simply does not exist in the vocalizations of primates, even great apes (Arcadi, 2000). A visitor from Mars would soon discern that there is something special about human vocal output, even if she had no understanding of what was being said.

Protolanguage

Compared to the recursive sophistication of human language, animal communication systems are at best only weakly combinatorial. For the past half-century or so, there have been strenuous attempts to teach language to the great apes, and especially to our closest relatives, the chimpanzee and bonobo. It soon emerged that chimpanzees are essentially unable to speak; in one famous example, a baby chimpanzee reared in a human family proved able to articulate only three or four words, and was soon outstripped by the human children in the family (Hayes, 1952). It was then realized that the failure to speak may have resulted from deficiencies of the vocal apparatus, and perhaps of cortical control of vocal output, and subsequent attempts have been based on manual action and visual representations. For example, the chimpanzee Washoe was taught over 100 manual signs, based loosely on American Sign Language, and was able to combine signs into two- or three-"word" sequences to make simple requests (Gardner & Gardner, 1969). The bonobo Kanzi has mastered the use of a keyboard containing 256 symbols representing objects and actions, and can construct meaningful sequences by pointing to the symbols. He supplements this vocabulary with gestures of his own invention. Although he makes full use of this vocabulary, his manual utterances appear to be limited to only two or three "words." Surprisingly, though, he has shown an impressive ability to follow instructions conveyed in spoken sentences with as many as seven or eight words (Savage-Rumbaugh et al., 1998).

There seems to be a general consensus, though, that these exploits are not *language*. As Pinker (1994, p. 340) put it, the great apes "just don't 'get it.'" Kanzi's ability to understand spoken sentences, although seemingly impressive, was shown to be roughly equivalent to that of a $2^1/_2$-year-old girl (Savage-Rumbaugh et al., 1998) and is probably based on the extraction of two or three key words rather than on a full decoding of the syntax of the sentences. His ability to produce symbol sequences is also at about the level of the average 2-year-old human. In human children, grammar typically emerges between the ages of 2 and 4 years, so that the linguistic capabilities of Kanzi and other great apes is generally taken as equivalent to that of children in whom grammar has not yet emerged. Bickerton (1995, p. 339), who wrote that "[t]he chimps' abilities at anything one would want to call grammar were next to nil," has labeled this pregrammatical level of linguistic performance "protolanguage."

Bickerton has further suggested, however, that protolanguage may be the precursor of true language, not only in development but also in evolution, an idea adopted by Jackendoff (2002) in a recent influential book. Yet protolanguage has been taught to such diverse creatures as the great apes, dolphins, a sea lion, and an African Gray parrot, implying parallel evolution. Further, it has never been observed in the wild. An alternative view, then, is that it is not a precursor to language but rather is indicative of a general problem-solving ability. For example, chimpanzees have been observed to solve mechanical problems by combining implements, such as joining two sticks together to rake in food that would not be reachable using either stick alone (Kohler, 1925; Tomasello, 1996). The combining of symbols to achieve some end, such as food, may in principle be no different.

Nevertheless, protolanguage may also be said to characterize the communication skills of the 2-year-old child, as well as persons with expressive aphasia following damage to Broca's area—or perhaps, as more recent research suggests, damage to the left precentral area of the insula, a cortical structure underlying the frontal and temporal lobes (Dronkers, 1996). Given that protolanguage seems to underlie language both in development and in terms of neural representation, it is perhaps reasonable to suppose that language did in fact grow out of protolanguage, whether conceived as a linguistic ability or simply as a capacity for problem solving, in human evolution. The question then is, when did this happen?

From Protolanguage to Language

Because great apes have not acquired any communicative skill beyond protolanguage, despite strenuous efforts to teach them, we can be reasonably sure that the language capacity evolved some time after the split between the hominid line and the line leading to modern chimpanzees and bonobos. The earliest fossil skull tentatively identified as a bipedal hominid is *Sahelanthropus tchadensis*, discovered in Chad, and dated between 6 and 7 million years ago (mya) (Brunet et al., 2002). This date is probably very close to the time of the chimpanzee-hominid split, estimated at between 6.3 and 7.7 mya by a DNA-DNA hybridization technique (Sibley & Ahlquist, 1984). Another early fossil, *Orrorin tugenensis*, is perhaps more securely identified as bipedal, and is dated from between 5.2 and 5.8 mya (Galik et al., 2004). The early hominids were distinguished from the great apes by facultative bipedalism, but with respect to brain size and what little is known of their cognitive capacities, they probably were little different from present-day chimpanzees. There is little reason to suppose that grammatical language emerged within the 4 or 5 million years of early hominid evolution.

Dramatic changes began to occur beginning some 2 mya, with the emergence of the genus *Homo*. Stone tool industries have been dated from about 2.5 mya in Ethiopia (Semaw et al., 1997) and have been tentatively identified with *H. rudolfensis*. However, these tools, which belong to the Oldowan industry, are primitive, and some have suggested that *H. rudolfensis* and *H. habilis*, the hominid traditionally associated with the Oldowan, should really be considered australopithecines (e.g., Wood, 2002). The true climb to humanity and to language probably began with the emergence of the larger-brained *H. erectus* around 1.8 mya, and the somewhat more sophisticated Acheulian tool industry dating from around 1.5 mya (Ambrose, 2001). But even tool manufacture may not be an especially good guide to the advance of cognition, since the Acheulian industry remained fairly static for over a million years and even persisted into the culture of early *H. sapiens* some 125,000 years ago (Walter et al., 2000).

Other changes associated with *H. erectus* may give a better guide to the emergence of more sophisticated cognition. *Erectus* marked the progression from facultative to obligate bipedalism and the full striding gait characteristic of modern humans. From about 1.6 mya, some members of this species strode out of Africa and into Asia, and *erectus* fossils in Java have been dated to as recently as 30,000 years ago (Swisher et al., 1994). But perhaps the surest signs of intellectual advance have to do with the size and development of the brain.

Bigger Brains

According to estimates based on fossil skulls, brain size increased from 457 cc in *Australopithecus africanus*, to 552 cc in *H. habilis*, to 854 cc in early *H. erectus* (also known as *H. ergaster*), to 1,016 cc in later *H. erectus*, to 1,552 in *H. neanderthalensis*, and back to 1,355 cc in *H. sapiens* (Wood & Collard, 1999). These values depend partly on body size, which probably explains why *H. neanderthalensis*, being slightly larger than modern humans, also had a slightly larger brain, but the picture is clearly one of a progressive increase, first clearly evident in early *Homo*.

Indeed, Chomsky (1975, p. 59) has suggested that language may have arisen simply as a consequence of possessing an enlarged brain, without the assistance of natural selection:

> We know very little about what happens when 10^{10} neurons are crammed into something the size of a basketball, with further conditions imposed by the

> specific manner in which this system developed over time. It would be a serious error to suppose that all properties, or the interesting structures that evolved, can be "explained" in terms of natural selection.

Nevertheless, the increase in brain size itself may have depended on natural selection, and recent research has brought to light two genetic mutations that may have a bearing on this. It is of some interest that both mutations resulted in the inactivation of genes, suggesting that we may owe our humanity at least in part to the *loss* of genes rather than to the incorporation of new ones.

One of these mutations has to do with a gene on chromosome 7 that encodes the enzyme CMP-*N*-acetylneuraminic acid (CMP-Neu5Ac) hydroxylase (CMAH). An inactivating mutation of this gene has resulted in a deficiency in humans of the mammalian sialic acid *N*-glycolylneuraminic acid (Neu5Gc). This acid appears to be absent in Neanderthal fossils as well as in humans, but it is present in present-day primates. It also seems to have been downregulated in the chimpanzee brain, and through mammalian evolution, leading to speculation that inactivation of the CMAH gene may have removed a constraint on brain growth in human ancestry (Chou et al., 2002). Chou et al. applied molecular clock analysis to the CMAH genes in chimpanzees and other great apes, as well as to the pseudogene in humans, which indicated that the mutation occurred some 2.1 mya, leading up to the expansion in brain size.

The other inactivating mutation that also may have contributed to the increase in brain size has to do with a gene on chromosome 7 that encodes for the myosene heavy chain MYH16, responsible for the heavy masticatory muscles in most primates, including chimpanzees and gorillas, as well as the early hominids. Molecular clock analysis suggests that this gene was inactivated around 2.4 mya, leading to speculation that the diminution of jaw muscles and their supporting bone structure removed a further constraint on brain growth (Stedman et al., 2004). It is a matter of further speculation as to why this seemingly deleterious mutation became fixed in the ancestral human population. It may have had to do with the change from a predominantly vegetable diet to a meat-eating one, or it may have had to do with the increasing use of the hands rather than the jaws to prepare food (Currie, 2004).

More generally, however, these mutations, and the resulting increase in brain size, may have been selected because of a change in environment. With the global shift to a cooler climate after 2.5 mya, much of southern and eastern Africa probably became more open and sparsely wooded (Foley, 1987). This left the hominids not only more exposed to attack from dangerous predators, such as saber-tooth cats, lions, and hyenas, but also obliged to compete with them as carnivores. The solution was not to compete on the same terms but to establish what Whiten (1999, p. 175) has termed a cognitive niche, relying on social cooperation and intelligent planning for survival. It seems reasonable to suppose that language evolved in this context. Language is expensive in terms of neural circuitry, as it takes up a good deal of the left hemisphere (and some of the right) in modern humans (Dick et al., 2001), so selection for more sophisticated language may itself have been one of the drivers of increasing brain size.

Although Bickerton (1995) and others have argued that language evolved in a single step from protolanguage, it is more likely that it developed gradually, through progressive refinements based on natural selection. One scenario has been proposed by Jackendoff (2002) and is summarized in figure 14.1. Although recursion is not explicitly mentioned in Jackendoff's scheme, it seems likely that phrase structure may have evolved before the flexible, recursive properties of grammar emerged, allowing phrases to be sequenced and embedded according to flexible rules.

Postnatal Growth

There is reason to suppose that recursive grammar depends not on brain size per se but rather on postnatal growth of the brain. Human development is characterized by "secondary altriciality"; that is, the human brain undergoes most of its growth after birth. In macaque newborns, the brain at birth has a volume of about 70% of adult size; in chimpanzees the figure is about 40%; in humans, it is only 25% (Coqueugniot, Hublin, Veillon, Houet, & Jacob, 2004). The brain is at its most plastic during growth, which may explain the so-called critical period for the development of language. The optimal period is probably between 2 and 4 years, although there is evidence that grammar can be acquired later in childhood (Vargha-Khadem et al., 1997), but probably not after puberty.

The idea that growth may be critical to the emergence of recursive grammar also gains some support from work with artificial networks. Elman (1993) tried to determine whether a network with recurrent loops could acquire the rules underlying sequences of symbols by testing whether the network could learn to predict the next symbol in the sequence. The network was able to learn simple grammars but was at

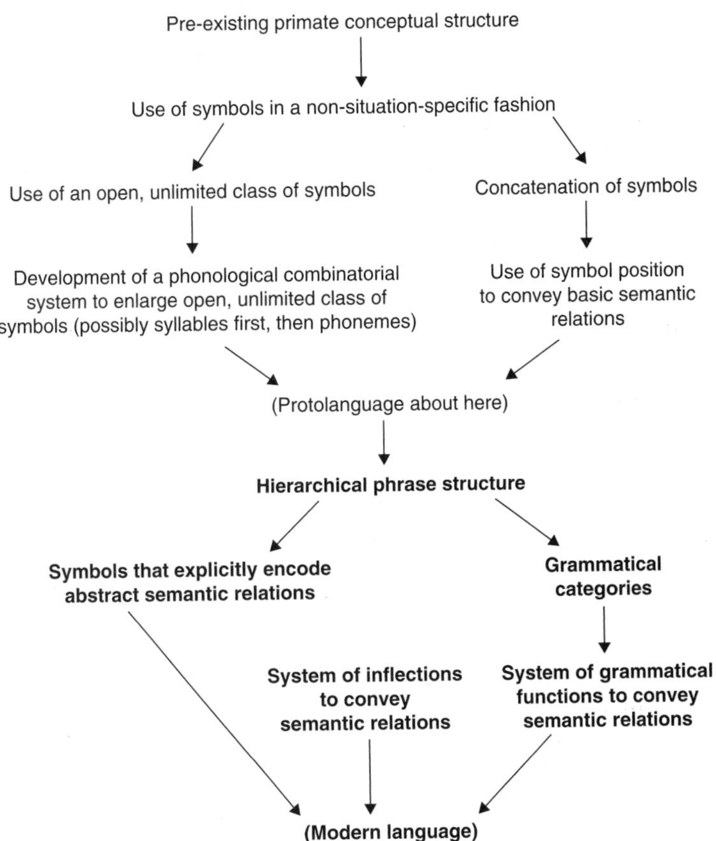

Figure 14.1
Hypothesized evolutionary steps in the evolution of language. Sequential steps are ordered top to bottom; parallel, independent steps are shown side by side. Steps unique to humans are shown in bold type (after Jackendoff, 2002).

first unable to deal with recursive grammars in which phrases were embedded in other phrases. This problem was at least partially surmounted when Elman introduced a "growth" factor, which he simulated by degrading the system early on so that only global aspects of the input were processed, and then gradually decreasing the "noise" in the system so that it was able to process more and more detail. When this was done, the system was able to pick up some of the recursive quality of grammar, and so begin to approximate the processing of true language.

One might expect secondary altriciality to have evolved along with the increasing brain size, since restrictions on the size of the birth canal would presumably have forced earlier birth; indeed, it has been estimated that if it were to conform to the general primate pattern, human birth should occur after 18 months of gestation, not 9 months (Krogman, 1972). Yet there is recent evidence that early *H. erectus* showed an apelike pattern rather than a humanlike pattern of postnatal brain growth, suggesting that secondary altriciality may not have emerged until fairly late in the genus *Homo* (Coqueugniot et al., 2004; Dean et al., 2001). Evidence based on dental enamel suggests that this feature may not have been present even in *H. neanderthalensis*, but had at least begun to emerge in *H. antecessor* and *H. heidelbergensis*, the immediate forerunners of *H. sapiens* (Ramirez-Rossi & Bermudez de Castro, 2004). The difference between *H. sapiens* and earlier *Homo* may lie not in the relative size of the brain at birth but in the speed of brain growth. Komarova and Nowak (2001) suggest that language is more accurate the longer the period of acquisition, but this must be balanced against the high cost of learning. The pressure may have been toward longer periods of growth as language skills became more critical to biological fitness.

In summary, the expansion of brain size from some 2 mya may well have signaled the beginnings of more sophisticated language and the emergence of phrase structure and grammatical categories. But the emergence of fully recursive language may not have come about until later in the evolution of *Homo*, when postnatal growth was characterized not only by secondary altriciality but also by a slowing of postnatal growth. This scenario may well provide an evolutionary framework for the sequential processes of language evolution outlined by Jackendoff (2002).

Did Language Evolve from Manual Gesture?

Language is often equated with speech. Yet it has become increasingly evident that the signed languages invented by deaf communities down the ages have all of the grammatical and semantic sophistication of speech (Armstrong, Stokoe, & Wilcox, 1995; Emmorey, 2002; Neidle, McLaughlin, Bahau, & Lee, 2000). Chomsky (2000, pp. 121–122) puts it succinctly:

> Though highly specialized, the language faculty is not tied to specific sense modalities, contrary to what was assumed not long ago. Thus, the sign language of the deaf is structurally very similar to spoken language, and the course of acquisition is very similar. Large-scale sensory deficit seems to have limited effect

on language acquisition. . . . The analytic mechanisms of the language faculty seem to be triggered in much the same ways whether the input is auditory, visual, even tactual, and seems to be localized in the same brain areas, somewhat surprisingly.

Further, language is seldom wholly one or the other. Both manual and facial gestures normally accompany speech and are closely synchronized with it, implying a common source, and the gestures carry part of the meaning (Goldin-Meadow & McNeill, 1999). The visible movements of the face can also influence the perception of speech, as in the McGurk effect, in which dubbing sounds onto a mouth that is saying something different alters what the hearer actually hears (McGurk & MacDonald, 1976). Although we can communicate without having to see the person we are talking to, as on radio or cell phone, speech in the natural world is rendered more eloquent and meaningful with the addition of bodily movements.

These considerations raise the possibility that language itself might have evolved from manual gestures rather than from animal calls, although it may always have been a mixture of both—perhaps with gestures punctuated by grunts gradually giving way to vocalizations embellished by gestures. The idea that language may have its roots in manual gesture goes back at least to the eighteenth-century philosopher Condillac (1971/1746), but has been advocated many times since, often independently (e.g., Armstrong, 1999; Armstrong et al., 1995; Corballis, 2002; Givòn, 1995; Hewes, 1973; Rizzolatti & Arbib, 1998). From an evolutionary point of view, the idea makes some sense, because nonhuman primates have little if any cortical control over vocalization but excellent cortical control over the hands and arms. Human speech required extensive anatomical modifications, including changes to the vocal tract and to innervation of the tongue, and the development of cortical control over voicing via the pyramidal tract. The vocalizations of other primates are probably largely emotional, controlled by the limbic system rather than the cortex (Ploog, 2002). The human equivalents are laughing, crying, shrieking, and the like. The modifications necessary for articulate speech arrived late in hominid evolution, and may not have been complete until the emergence of *H. sapiens*—or even later. We have also seen that efforts to teach great apes anything resembling speech have proven futile, but there has been reasonable success in teaching them to communicate through gestures or by pointing to visual symbols.

The Late Emergence of Speech

Fossil evidence suggests that articulate speech emerged late in hominid evolution, which gives further grounds for supposing that it may have been preceded by a gestural system. One piece of evidence has to do with the hypoglossal canal at the base of the tongue. The hypoglossal nerve, which passes through this canal and innervates the tongue, is much larger in humans than in great apes, probably because of the important role of the tongue in speech. Fossil evidence suggests that the size of the hypoglossal canal in early australopithecines, and perhaps in *H. habilis*, was within the range of that in modern great apes, while that of the Neanderthal and early *H. sapiens* skulls was well within the modern human range (Kay, Cartmill, & Barlow, 1998), although this has been disputed (DeGusta, Gilbert, & Turner, 1999). A further clue comes from the finding that the thoracic region of the spinal cord is relatively larger in humans than in nonhuman primates, probably because breathing during speech involves extra muscles of the thorax and abdomen. Fossil evidence indicates that this enlargement was not present in the early hominids or even in *H. ergaster*, dating from about 1.6 mya, but was present in several Neanderthal fossils (MacLarnon & Hewitt, 1999).

The production of articulate speech in humans depends on the lowering of the larynx. According to P. Lieberman (1998; P. Lieberman, Crelin, & Klatt, 1972) this adaptation was incomplete even in the Neanderthals of 30,000 years ago, and their resultant poor articulation would have been sufficient to keep them separate from *H. sapiens*, leading to their eventual extinction. This work remains controversial (e.g., Gibson & Jessee, 1999), but there is other evidence that the cranial structure underwent changes subsequent to the split between anatomically modern and earlier, archaic *Homo*, such as the Neanderthals, *H. heidelbergensis*, and *H. rhodesiensis*. One such change is the shortening of the sphenoid, the central bone of the cranial base from which the face grows forward, resulting in a flattened face (D. E. Lieberman, 1998). D. E. Lieberman speculates that this is an adaptation for speech, contributing to the unique proportions of the human vocal tract, in which the horizontal and vertical components are roughly equal in length. This configuration, he argues, improves the ability to produce acoustically distinct speech sounds, such as the vowel [i] (D. E. Lieberman, McBratney, & Krovitz, 2002). It is not seen in Neanderthal skeletal structure (see also Vleck, 1970), suggesting that it emerged in our own species within the past 500,000 years. Another adaptation unique to *H. sapiens* is neurocranial globularity, defined as the roundness of the cranial vault in the sagittal, coronal, and transverse planes, which is likely to have increased the relative size of the temporal and/or frontal lobes relative to other parts of the brain (D. E. Lieberman et al., 2002). These changes may reflect more refined control of articulation and also, perhaps, more accurate perceptual discrimination of articulated sounds.

These various findings suggest that the tinkering of the brain and cranial configuration for the refinement of speech continued into *H. sapiens* after the split from the Neanderthals and other archaic species of *Homo*, and perhaps even into the past 100,000 years of human evolution, as I shall suggest later. Nevertheless, grammatical language may well have arisen considerably earlier, and may have been conveyed initially by means of manual and facial gestures, increasingly augmented and eventually replaced by vocalization.

Speech as Gesture

The idea that language may have evolved from manual gestures receives some support from evidence that speech itself is better considered a gestural system than an acoustic one. Traditionally, speech has been regarded as made up of discrete elements of sound, called phonemes. It has been known for some time, though, that phonemes do not exist as discrete units in the acoustic signal (Joos, 1948) and are not discretely discernible in mechanical recordings of sound, such as a sound spectrograph (Liberman, Cooper, Shankweiler, & Studdert-Kennedy, 1967). One reason for this is that the acoustic signal corresponding to individual phonemes varies widely, depending on the context in which it is embedded. This has led to the view that phonemes exist only in the minds of speakers and hearers, and the acoustic signal must undergo complex transformation for individual phonemes to be perceived as such. Yet we can perceive speech at remarkably high rates, up to at least 10–15 phonemes per second, which seems at odds with the idea that some complex, context-dependent transformation is necessary.

These problems have led to the alternative view, known as articulatory phonology (Browman & Goldstein, 1995), that speech is better understood as comprised of articulatory gestures. Six articulatory organs—the lips, the velum, the larynx, and the blade, body, and root of the tongue—produce these gestures. Each is controlled separately, so that individual speech units are comprised of different combinations of movements. The distribution of action over these articulators means that the elements overlap in time, which makes possible the high rates of production and perception. Unlike phonemes, speech gestures can be discerned by mechanical means, through radiography, magnetic resonance imaging (MRI), and palatography (Studdert-Kennedy, 1998).

This still raises the question, though, of how these gestures are perceived; our ability to understand speech over the radio or telephone is incontrovertible evidence that speech can be understood from the acoustic stream alone. The short (but still incomplete) answer is that speech is understood in terms of the articulatory gestures that produce it, rather than in terms of elementary sound units. This is the so-called motor theory of speech perception (Liberman et al., 1967). Although we can understand the radio announcer, there is abundant evidence that watching people speak can aid understanding of what they are saying. Nevertheless, the process by which articulatory information is extracted from the acoustic signal is not fully understood, although some insight has come from the recent discovery of what has been termed the mirror system in the brain.

The Mirror System

Neurons in the region of F5 in the ventral premotor cortex of the monkey typically fire when the animal makes grasping movements with the hand or mouth. A subset of those cells, dubbed mirror neurons, also fire when the animal observes another individual making the same movements. In the monkey, these responses require the presence of a target and do not respond to actions that merely mimic an action in the absence of a target, nor do they respond to a target alone (Gallese, Fadiga, Fogassi, & Rizzolatti, 1996; Rizzolatti, Fadiga, Fogassi, & Gallese, 1996). This direct mapping of perceived action onto the production of action seems to provide a platform for the evolution of language, and to support, albeit indirectly, the motor theory of speech perception. Furthermore, the area of the human brain that corresponds most closely to area F5 in the monkey includes Broca's area, which is one of the main cortical areas underlying the production of speech. This suggests that speech may have arisen from cortical structures that initially had to do with manual action rather than with vocalization (Rizzolatti & Arbib, 1998).

It has also become apparent that mirror neurons are part of a more general mirror system that involves other regions of the brain as well. The superior temporal sulcus (STS) also contains cells that respond to observed biological actions, including grasping actions (Perrett et al.,

1989), although few if any respond when the animal itself performs an action. F5 and STS are connected to area PF in the inferior parietal lobule, where there are also neurons that respond both to the execution and the perception of actions. These neurons are now known as PF mirror neurons (Rizzolatti, Fogassi, & Gallese, 2001). Other areas, such as the amygdala and orbitofrontal cortex, may also be part of the mirror system.

A similar system has been inferred in humans, based on evidence from electroencephalography (Muthukumaraswamy et al., 2004), magnetoencephalography (Hari et al., 1998), transcranial magnetic stimulation (Fadiga, Fogassi, Pavesi, & Rizzolatti, 1995), and functional magnetic resonance imaging (fMRI) (Iacoboni et al., 1999). Unlike the mirror system in monkeys, the human mirror system appears to be activated by movements that need not be directed toward an object (Rizzolatti et al., 2001), although there is evidence that it is activated more by actions that are object-directed than by those that are not object-directed (Muthukumaraswamy et al., 2004). Activation by non-object-directed action may reflect adaptation of the system for more abstract signaling, as in signed languages. The mirror system in humans appears to involve areas in the frontal, temporal, and parietal lobes that are homologous to those in the monkey, although there is some evidence that they tend to be lateralized to the left hemisphere in humans, especially in the frontal lobes (Iacoboni et al., 1999, Nishitani & Hari, 2000). It is well established that manual apraxia, especially for actions involving fine motor control, is associated with left hemisphere damage (Heilman, Meador, & Loring, 2000). It is possible that the incorporation of vocalization into the mirror system, perhaps unique to *H. sapiens*, resulted in lateralization of the manual as well as of the vocal system (Corballis, 2003).

The mirror system leads to what has been termed the direct-matching hypothesis, according to which we understand actions by mapping the visual representations of observed actions onto the motor representations of the same actions (Rizzolatti et al., 2001). This system is tuned to the perception of actions that have a "personal" reference. Evidence from an fMRI study shows, for example, that it is activated when people watch mouth actions, such as biting, lip smacking, and oral movements involved in vocalization (e.g., speech reading, barking), performed by people, but not when they watch such actions performed by a monkey or a dog. Actions belonging to the observer's own motor repertoire are mapped onto the observer's motor system, while those that do not belong are not; instead, they are perceived in terms of their visual properties

(Buccino et al., 2004). Watching speech movements, and even stills of a mouth making a speech sound, activates the mirror system, including Broca's area (Calvert & Campbell, 2003). This is consistent with the idea that language may have evolved from visual display that included movements of the face.

Although most of the evidence on the mirror system has to do with visual input, area F5 of the monkey also contains what might be termed acoustic mirror neurons. These respond to the sounds of actions, such as tearing paper or breaking a peanut, as well as to the performance of those actions. That is, even in the monkey, the direct-matching hypothesis is not restricted to visual input (Kohler et al., 2002). There is no evidence for mirror neurons in the monkey that fire to both the production and perception of vocalization. It is likely, though, that vocalization was incorporated into the mirror system in humans, and probably only in humans or our hominid forebears (Ploog, 2002), providing the mechanism for the motor theory of speech perception.

The next question is when vocalization was added to the system. A clue to this comes from the *FOXP2* gene.

The *FOXP2* Gene

About half of the members of three generations of an extended family in England, known as the KE family, are affected by a disorder of speech and language. The disorder is evident from the affected child's first attempts to speak and persists into adulthood (Vargha-Khadem, Watkins, Alcock, Fletcher, & Passingham, 1995). The disorder is now known to be due to a point mutation on the *FOXP2* gene (forkhead box P2) on chromosome 7 (Fisher, Vargha-Khadem, Watkins, Monaco, & Pembrey, 1998; Lai, Fisher, Hurst, Vargha-Khadem, & Monaco, 2001). For normal speech to be acquired, two functional copies of this gene seem to be necessary.

The nature of the deficit in the affected members of the KE family, and therefore the role of the *FOXP2* gene, have been debated. Some have argued that *FOXP2* is involved in the development of morphosyntax (Gopnik, 1990), and it has even been identified more broadly as "the grammar gene" (Pinker, 1994). Subsequent investigation suggests, however, that the core deficit is one of articulation, with grammatical impairment a secondary outcome (Watkins, Dronkers, & Vargha-Khadem, 2002). It may therefore play a role in the incorporation of vocal articulation into the mirror system.

This is supported by a study in which fMRI was used to record brain activity in both affected and unaffected members of the KE family while they covertly generated verbs in response to nouns (Liégeois et al., 2003). Whereas unaffected members showed the expected activity concentrated in Broca's area in the left hemisphere, affected members showed relative *under*activation in both Broca's area and its right hemisphere homologue, as well as in other cortical language areas. They also showed *over*activation bilaterally in regions not associated with language. However, there was bilateral activation in the posterior superior temporal gyrus; the left side of this area overlaps Wernicke's area, important in the comprehension of language. This suggests that affected members may have generated words in terms of their sounds rather than in terms of articulatory patterns. Their deficits were not attributable to any difficulty with verb generation itself, since affected and unaffected members did not differ in their ability to generate verbs overtly, and the patterns of brain activity were similar to those recorded during covert verb generation. Another study based on structural MRI showed morphological abnormalities in the same areas (Watkins, Vargha-Khadem, et al., 2002).

The *FOXP2* gene is highly conserved in mammals, and in humans it differs in only three places from that in the mouse. Nevertheless, two of the three changes occurred in the human lineage after the split from the common ancestor with the chimpanzee and bonobo. A recent estimate of the date of the more recent of these mutations suggests that it occurred "since the onset of human population growth, some 10,000 to 100,000 years ago" (Enard et al., 2002, p. 871). If this is so, then it might be argued that the final incorporation of vocalization into the mirror system was critical to the emergence of modern human behavior, often dated to the Upper Paleolithic (Corballis, 2004).

It is unlikely, though, that the *FOXP2* mutation was the only event in the transition to speech, which undoubtedly went through several steps and involved other genes (Marcus & Fisher, 2003). Moreover, the *FOXP2* gene is expressed in the embryonic development of structures other than the brain, including the gut, heart, and lung (Shu, Yang, Zhang, Lu, & Morrisey, 2001). It may have even played a role in the modification of breath control for speech (MacLarnon & Hewitt, 1999). A mutation of the *FOXP2* gene may nevertheless have been the most recent event in the incorporation of vocalization into the mirror system, and thus the refinement of vocal control to the point that it could carry the primary burden of language.

The idea that the critical mutation of the *FOXP2* gene occurred less than 100,000 years ago is indirectly supported by recent evidence from African click languages. Two of the many groups that make extensive use of click sounds are the Hadzabe and San, who are separated geographically by some 2,000 kilometers, and genetic evidence suggests that the most recent common ancestor of these groups goes back to the root of present-day mitochondrial DNA lineages, perhaps as early as 100,000 years ago (Knight et al., 2003). This could mean that clicks were a prevocal way of adding sound to facial gestures, prior to the *FOXP2* mutation. Evidence from mitochondrial DNA suggests that modern humans outside of Africa date from groups who migrated from Africa from around 52,000 years ago (Ingman, Kaessmann, Pääbo, & Gyllensten, 2000), and these groups may have already developed autonomous speech, leaving behind African speakers who retained click sounds. The only known non-African click language is Damin, an extinct Australian aboriginal language, and migrations to Australia may have been earlier than the presumed migrations of 52,000 years ago (Thorne et al., 1999). This is not to say that the early Australians and Africans did not have full vocal control of speech; rather, click languages may be simply a vestige of earlier languages in which vocalization was not yet part of the mirror system giving rise to autonomous speech.

Why Speech?

According to the account presented here, the transition from manual to vocal language was not abrupt. This raises the question, though, of why the transition took place at all. The signed languages of the deaf clearly show that manual languages can be as sophisticated as vocal ones. Further, the transition to speech involved the lowering of the larynx, which greatly increased the risk of choking to death. Clearly, the evolutionary pressure toward speech must have been strong. But why?

There are a number of possible answers. First, a switch to autonomous vocalization would have freed the hands from necessary involvement in communication, allowing increased use of the hands for manufacture and tool use. Indeed, vocal language allows people to speak and use tools at the same time, leading perhaps to pedagogy (Corballis, 2002). It may explain the so-called "human revolution" (Mellars & Stringer, 1989), manifest in the dramatic appearance of more sophisticated tools, bodily ornamentation, art, and perhaps music, dating from some 40,000 years ago in Europe, and maybe earlier in Africa

(McBrearty & Brooks, 2000). This may well have come about because of the switch to autonomously vocal language, made possible by the *FOXP2* mutation (Corballis, 2004).

Although manual and vocal language can be considered linguistically equivalent, there are other advantages to vocalization. Speech is less attentionally demanding than signed language; one can attend to speech with one's eyes shut, or when watching something else. Speech also allows communication over longer distances, as well as communication at night or when the speaker is not visible to the listener. The San, a modern hunter-gatherer society, are known to talk late at night, sometimes all through the night, to resolve conflict and to share knowledge (Konner, 1982). Boutla, Supalla, Newport, and Bavelier (2004) have shown that the span of short-term memory is shorter for American Sign Language than for speech, suggesting that voicing may have permitted longer and more complex sentences to be transmitted, although the authors claim that the shorter memory span has no impact on the linguistic skill of signers.

A possible scenario for the switch is that there was selective pressure for the face to become more extensively involved in gestural communication as the hands were increasingly engaged in other activities. Our species had been habitually bipedal from some 6 or 7 mya, and from some 2 mya was developing tools, which would have increasingly involved the hands. The face had long played a role in visual communication, and it plays an important role in present-day signed languages (e.g., Neidle et al., 2000). Consequently, there may have been pressure for intentional communication to move to the face, including the mouth and tongue. Gesturing may then have retreated into the mouth, so there may have been pressure to add voicing in order to render movements of the tongue more accessible—through sound rather than sight. In this scenario, speech is simply gesture half swallowed, with voicing added. Even so, lip-reading can be a moderately effective way to recover the speech gestures, and, as mentioned earlier, the McGurk effect illustrates that speech is in part a visual medium. Adding voicing to the signal could have had the extra benefit of allowing a distinction between voiced and unvoiced phonemes, increasing the range of speech elements.

Changes in the mode of communication can have a dramatic influence on human culture, as illustrated by the invention of writing and more recently by e-mail and the Internet. These changes were relatively sudden, and cultural rather than biological. The change from manual to vocal communication, in contrast, would have been slow, driven by natural selection and involving biological adaptations, but it may have had no less an impact on human culture—and therefore, perhaps, on human fitness.

Conclusion

Fully grammatical language appears to be a uniquely human accomplishment. Other animals are capable of understanding symbolic representations, and perhaps even of segmenting speech, at least to the point of isolating words. Besides the bonobo (Savage-Rumbaugh et al., 1998), these animals may include the African Gray parrot (Pepperberg, 2002) and the domestic dog (Kaminsky, Call, & Fischer, 2004). But there is no evidence that nonhuman animals can decode or generate grammar and so create and understand a potentially infinite variety of sentences. At best, they are at the level of the 2-year-old human child, with a level of communication lacking the generative, recursive property of fully developed language. They have protolanguage.

The emergence of language from protolanguage may have occurred late in hominid evolution, though not so late as to represent an evolutionary "big bang." The steps toward grammar may have begun some 2 mya, with the emergence of larger-brained hominids, and continued over the next 1.5 million years or thereabouts. The final step may have been full recursion, depending perhaps on secondary altriciality and the slowing of postnatal brain growth. This process may not have been complete even in the Neanderthals, who survived until some 30,000 years ago, and may not have been fully developed in the line leading to *H. sapiens* until the emergence of that species around 170,000 years ago.

There is also evidence that fully articulate speech evolved late and may not have been complete until less than 100,000 years ago, with the mutation of the *FOXP2* gene allowing vocalization to be incorporated into the mirror system. Evidence from mtDNA suggests that modern humans migrated out of Africa some 52,000 years ago (Ingman et al., 2000), eventually replacing all other hominids, including the Neanderthals in Europe, *H. erectus* in Asia, and even groups of *H. sapiens* who had migrated earlier. What was it that led to the dominance of these late migrants? I have suggested that it may have been the consequences of the emergence of fully articulate speech, resulting in improved technology, perhaps including more lethal weaponry, and a more coherent culture (Corballis, 2004). An alternative, perhaps, is that fully recursive language itself did not evolve until within the past 100,000 years. Rather than talking our forebears out of existence, we may have lost them in a recursive loop. Or maybe the invading hordes out of Africa simply brought diseases that the indigenous populations were not resistant to.

One might have thought that an understanding of how language evolved would have been beyond the reach of science. That, presumably, was the view in 1866, when the Linguistic Society of Paris banned all discussion of the topic. Nevertheless, the past decade in particular has produced an extraordinarily rich accumulation of evidence from multiple sources, all of which appear to be converging on common themes, if not yet on an agreed scenario. In 1866, very little was known about the transitions from ape to human, but modern archaeology has given us a remarkably detailed account of what our hominid forebears must have been like. From skeptical talk of a "missing link" we now have evidence of over 20 hominid species separating us from our common ancestry with the chimpanzee and bonobo (Wood, 2002). Detailed inspection of hominid fossils has provided evidence of brain size and growth characteristics, and modern biochemistry has elucidated the timing of critical events, such as the ape-hominid split, and the late migration out of Africa. There are techniques for dating genetic mutations, and this chapter has identified three that may be of significance to the understanding of the evolution of language—two dating from just over 2 mya and one from something under 100,000 years ago.

We also now understand much better what language is actually like, how it differs from other forms of communication, and how it develops. It has only recently become clear that the signed languages of the deaf are true grammatical languages and not impoverished signaling systems. With the advance of brain imaging, the neurophysiology of language is increasingly understood, and work on the so-called mirror system has led to important insights into how language might be better understood as part of a more general system for understanding biological motion, instead of a rather abstract coding system beyond any affinity with our animal heritage.

The scenario sketched in this chapter may be wrong, but we can be sure that evidence will continue to accumulate. The trick will be to integrate the diverse sources, and so gain a better appreciation of how we became such compulsive chatterboxes. This chapter, I hope, has been a start.

References

Ambrose, S. H. (2001). Paleolithic technology and human evolution. *Science, 291*, 1748–1752.

Arcadi, A. C. (2000). Vocal responsiveness in male wild chimpanzees: Implications for the evolution of language. *Journal of Human Evolution, 39*, 205–223.

Armstrong, D. F. (1999). *Original signs: Gesture, sign, and the source of language*. Washington, DC: Gallaudet University Press.

Armstrong, D. F., Stokoe, W. C., & Wilcox, S. E. (1995). *Gesture and the nature of language*. Cambridge: Cambridge University Press.

Bickerton, D. (1995). *Language and human behavior*. Seattle: University of Washington Press.

Boutla, M., Supalla, T., Newport, E. L., & Bavelier, D. (2004). Short-term memory span: Insights from sign language. *Nature Neuroscience, 7*, 997–1002.

Browman, C. P., & Goldstein, L. F. (1995). Dynamics and articulatory phonology. In T. van Gelder & R. F. Port (Eds.), *Mind as motion* (pp. 175–193). Cambridge, MA: MIT Press.

Brunet, M., Guy, F., Pilbeam, D., Mackaye, H. T., Likius, A., Ahounta, D., et al. (2002). A new hominid from the Upper Miocene of Chad, Central Africa. *Nature*, 418, 145–151.

Buccino, G., Lui, F., Canessa, N., Patteri, I., Lagravinese, G., Benuzzi, F., et al. (2004). Neural circuits involved in the recognition of actions performed by nonconspecifics: An fMRI study. *Journal of Cognitive Neuroscience, 16*, 114–126.

Calvert, G. A., & Campbell, R. (2003). Reading speech from still and moving faces: The neural substrates of visible speech. *Journal of Cognitive Neuroscience, 15*, 57–70.

Cheney, D. L., & Seyfarth, R. M. (1990). *How monkeys see the world*. Chicago: University of Chicago Press.

Chomsky, N. (1966). *Cartesian linguistics: A chapter in the history of rationalist thought*. New York: Harper & Row.

Chomsky, N. (1975). *Reflections on language*. New York: Pantheon.

Chomsky, N. (2000). Language as a natural object. In *New horizons in the study of language and mind* (pp. 106–133). Cambridge: Cambridge University Press.

Chou, H.-H., Hakayama, T., Diaz, S., Krings, M., Indriati, E., Leakey, M., et al. (2002). Inactivation of CMP-*N*-acetylneuraminic acid hydroxylase occurred prior to brain expansion during human evolution. *Proceedings of the National Academy of Sciences, U.S.A., 99*, 11736–11741.

Condillac, E. B. de (1971). *An essay on the origin of human knowledge* (T. Nugent, trans.). Gainesville, FL: Scholars Facsimiles and Reprints. (Original work published 1746)

Coqueugniot, H., Hublin, J.-J., Veillon, F., Houet, F., & Jacob, T. (2004). Early brain growth in *Homo erectus* and implications for cognitive ability. *Science, 431*, 299–302.

Corballis, M. C. (2002). *From hand to mouth: The origins of language*. Princeton, NJ: Princeton University Press.

Corballis, M. C. (2003). From mouth to hand: Gesture, speech, and the evolution of right-handedness. *Behavioral and Brain Sciences*, *26*, 199–260.

Corballis, M. C. (2004). The origins of modernity: Was autonomous speech the critical factor? *Psychological Review*, *111*, 543–552.

Crow, T. J. (2002). Sexual selection, timing, and an X-Y homologous gene: Did *Homo sapiens* speciate on the Y chromosome? In T. J. Crow (Ed.), *The speciation of modern* Homo sapiens (pp. 197–216). Oxford: Oxford University Press.

Currie, P. (2004). Muscling in on hominid evolution. *Nature*, *428*, 373–374.

Dean, C., Leakey, M. G., Reid, D., Shrenk, F., Schwartz, G. T., Stringer, C., et al. (2001). Growth processes in teeth distinguish modern humans from *Homo erectus* and earlier hominins. *Nature*, *414*, 628–631.

DeGusta, D., Gilbert, W. H., & Turner, S. P. (1999). Hypoglossal canal size and hominid speech. *Proceedings of the National Academy of Sciences, U.S.A.*, *96*, 1800–1804.

Dick, F., Bates, E., Wulfeck, B., Utman, J. A., Dronkers, N. F., & Gernsbacher, M. A. (2001). Language deficits, localization, and grammar: Evidence for a distributed model of language breakdown in aphasic patients and neurologically intact individuals. *Psychological Review*, *108*, 759–788.

Dronkers, N. F. (1996). A new brain region for coordinating speech articulation. *Nature*, *384*, 159–161.

Elman, J. (1993). Learning and development in neural networks: The importance of starting small. *Cognition*, *48*, 71–99.

Emmorey, K. (2002). *Language, cognition, and brain: Insights from sign language research*. Hillsdale, NJ: Erlbaum.

Enard, W., Przeworski, M., Fisher, S. E., Lai, C. S. L., Wiebe, V., Kitano, T., et al. (2002). Molecular evolution of *FOXP2*, a gene involved in speech and language. *Nature*, *418*, 869–871.

Fadiga, L., Fogassi, L., Pavesi, G., & Rizzolatti, G. (1995). Motor facilitation during action observation: A magnetic stimulation study. *Journal of Neurophysiology*, *73*, 2608–2611.

Fisher, S. E., Vargha-Khadem, F., Watkins, K. E., Monaco, A. P., & Pembrey, M. E. (1998). Localisation of a gene implicated in a severe speech and language disorder. *Nature Genetics*, *18*, 168–170.

Foley, R. (1987). *Another unique species: Patterns in human evolutionary ecology*. Harlow, U.K.: Longman Scientific and Technical.

Galik, K., Senut, B., Pickford, M., Gommery, D., Treil, J., Kuperavage, A. J., & Eckhardt, R. B. (2004). External and internal morphology of the BAR 1002′00 *Orrorin tugenensis* femur. *Science*, *305*, 1450–1453.

Gallese, V., Fadiga, L., Fogassi, L., & Rizzolatti, G. (1996). Action recognition in the premotor cortex. *Brain*, *119*, 593–609.

Gardner, R. A., & Gardner, B. T. (1969). Teaching sign language to a chimpanzee. *Science*, *165*, 664–672.

Gibson, K. R., & Jessee, S. (1999). Language evolution and expansions of multiple neurological processing areas. In B. J. King (Ed.), *The origins of language: What nonhuman primates can tell us*. Santa Fe, NM: School of American Research Press.

Givòn, T. (1995). *Functionalism and grammar*. Philadelphia: John Benjamins.

Goldin-Meadow, S., & McNeill D. (1999). The role of gesture and mimetic representation in making language the province of speech. In M. C. Corballis & S. E. G. Lea (Eds.), *The descent of mind* (pp. 155–172). Oxford: Oxford University Press.

Gopnik, M. (1990). Feature-blind grammar and dysphasia. *Nature, 344*, 715.

Hari, R., Forss, N., Avikainen, S., Kirveskari, E., Salenius, S., & Rizzolatti, G. (1998). Activation of human primary motor cortex during action observation: A neuromagnetic study. *Proceedings of the National Academy of Sciences, U.S.A., 95*, 15061–15065.

Hauser, M. D., Fitch, W. T., & Chomsky, N. (2002). The faculty of language: What is it, who has it, and how did it evolve? *Science, 298*, 1569–1579.

Hayes, C. (1952). *The ape in our house*. London: Gollancz.

Heilman, K. M., Meador, K. J., & Loring, D. W. (2000). Hemispheric asymmetries of limb-kinetic apraxia: A loss of deftness. *Neurology, 55*, 523–526.

Hewes, G. W. (1973). Primate communication and the gestural origins of language. *Current Anthropology, 14*, 5–24.

Iacoboni, M., Woods, R. P,, Brass, M., Bekkering, H., Mazziotta, J. C., & Rizzolatti, G. (1999). Cortical mechanisms of human imitation. *Science, 286*, 2526–2528.

Ingman, M., Kaessmann, H., Pääbo, S., & Gyllensten, U. (2000). Mitochondrial genome variation and the origin of modern humans. *Nature, 408*, 708–713.

Jackendoff, R. (2002). *Foundations of language: Brain, meaning, grammar, evolution*. Oxford: Oxford University Press.

Joos, M. (1948). *Acoustic phonetics* (Language Monograph No. 23). Baltimore: Linguistic Society of America.

Kaminsky, J., Call, J., & Fischer, J. (2004). Word learning in a domestic dog: Evidence for "fast mapping." *Science, 304*, 1682–1683.

Kay, R. F., Cartmill, M., & Barlow, M. (1998). The hypoglossal canal and the origin of human vocal behavior. *Proceedings of the National Academy of Sciences, U.S.A., 95*, 5417–5419.

Knight, A., Underhill, P. A., Mortensen, H. M., Zhivotovsky, L. A., Lin, A. A., Henn, B. M., et al. (2003). African Y chromosome and mtDNA divergence provides insight into the history of click languages. *Current Biology, 13*, 464–473.

Kohler, E., Keysers, C., Umilta, M. A., Fogassi, L., Gallese, V., & Rizzolatti, G. (2002). Hearing sounds, understanding actions: Action representation in mirror neurons. *Science, 297*, 846–848.

Kohler, W. (1925). *The mentality of apes*. New York: Routledge & Kegan Paul.

Komarova, N. L., & Nowak, M. A. (2001). Natural selection of the critical period for language acquisition. *Proceedings of the Royal Society of London Series, B, Biological Sciences, 268*, 1189–1196.

Konner, M. (1982). *The tangled wing: Biological constraints on the human spirit*. New York: Harper.

Krogman, W. M. (1972). *Child growth.* Ann Arbor: University of Michigan Press.

Lai, C. S., Fisher, S. E., Hurst, J. A., Vargha-Khadem, F., & Monaco, A. P. (2001). A novel forkhead-domain gene is mutated in a severe speech and language disorder. *Nature, 413*, 519–523.

Liberman, A. M., Cooper F. S., Shankweiler, D. P., & Studdert-Kennedy, M. (1967). Perception of the speech code. *Psychological Review, 74*, 431–461.

Lieberman, D. E. (1998). Sphenoid shortening and the evolution of modern cranial shape. *Nature*, 393, 158–162.

Lieberman, D. E., McBratney, B. M., & Krovitz, G. (2002). The evolution and development of cranial form in *Homo sapiens. Proceedings of the National Academy of Sciences, U.S.A., 99*, 1134–1139.

Lieberman, P. (1998). *Eve spoke: Human language and human evolution.* New York: Norton.

Lieberman, P. (2002). On the nature and evolution of the neural bases of human language. *Yearbook of Physical Anthropology, 45*, 36–62.

Lieberman, P., Crelin, E. S., & Klatt, D. H. (1972). Phonetic ability and related anatomy of the new-born, adult human, Neanderthal man, and the chimpanzee. *American Anthropologist, 74*, 287–307.

Liégeois, F., Baldeweg, T., Connelly, A., Gadian, D. G., Mishkin, M., & Vargha-Khadem, F. (2003). Language fMRI abnormalities associated with *FOXP2* gene mutation. *Nature Neuroscience, 6*, 1230–1237.

MacLarnon, A., & Hewitt, G. (1999). The evolution of human speech: The role of enhanced breathing control. *American Journal of Physical Anthropology, 109*, 341–363.

Marcus, G. F., Fisher, S. E. (2003). *FOXP2* in focus: What can genes tell us about speech and language? *Trends in Cognitive Science, 7*, 257–262.

McBrearty, S., & Brooks, A. S. (2000). The revolution that wasn't: A new interpretation of the origin of modern human behavior. *Journal of Human Evolution, 39*, 453–563.

Mellars, P. A., & Stringer, C. B. (Eds.) (1989). *The human revolution: Behavioural and biological perspectives on the origins of modern humans.* Edinburgh: Edinburgh University Press.

McGurk, H., & MacDonald, J. (1976). Hearing lips and seeing voices. *Nature, 264*, 746–748.

Muthukumaraswamy, S. D., Johnson, B. W., & McNair, N. A. (2004). Mu rhythm modulation during observation of an object-directed grasp. *Brain Research: Cognitive Brain Research, 19*, 195–201.

Neidle, C., Kegl, J., MacLaughlin, D., Bahan, B., & Lee, R. G. (2000). *The syntax of American Sign Language.* Cambridge, MA: MIT Press.

Nishitani, N., & Hari R. (2000). Temporal dynamics of cortical representation for action. *Proceedings of the National Academy of Sciences, U.S.A., 97*, 913–918.

Pepperberg, I. M. (2002). In search of King Solomon's ring: Cognitive and communicative studies of Gray parrots (*Psittacus erithacus*). *Brain, Behavior, and Evolution, 59*, 54–67.

Perrett, D. I., Harries, M. H., Bevan, R., Thomas, S., Benson, P. J., Mistlin, A. J., et al. (1989). Frameworks of analysis for the neural representation of animate objects and actions. *Journal of Experimental Biology*, *146*, 87–113.

Pinker, S. (1994). *The language instinct*. New York: Morrow.

Pinker, S., & Bloom, P. (1990). Natural language and natural selection. *Behavioral and Brain Sciences*, *13*, 707–784.

Ploog, D. (2002). Is the neural basis of vocalisation different in non-human primates and *Homo sapiens*? In T. J. Crow (Ed.), *The speciation of modern* Homo Sapiens (pp. 121–135). Oxford: Oxford University Press.

Ramirez-Rossi, F. V., & Bermudez de Castro, J. M. (2004). Surprisingly rapid growth in Neanderthals. *Nature*, *428*, 936–939.

Rizzolatti, G., & Arbib, M. A. (1998). Language within our grasp. *Trends in Cognitive Sciences*, *21*, 188–194.

Rizzolatti, G., Fadiga, L., Fogassi, L., & Gallese V. (1996). Premotor cortex and the recognition of motor actions. *Brain Research: Cognitive Brain Research*, *3*, 131–141.

Rizzolatti, G., Fogassi, L., & Gallese V. (2001). Neurophysiological mechanisms underlying the understanding and imitation of action. *Nature Reviews*, *2*, 661–670.

Savage-Rumbaugh, S., Shanker, S. G., & Taylor, T. J. (1998). *Apes, language, and the human mind*. New York: Oxford University Press.

Semaw, S. P., Renne, P., Harris, J. W. K., Feibel, C. S., Bernor, R. L., Fessweha, N., et al. (1997). 2.5-million-year-old stone tools from Gona, Ethiopia. *Nature*, *385*, 333–336.

Shu, W. G., Yang, H. H., Zhang, L. L., Lu, M. M., & Morrisey, E. E. (2001). Characterization of a new subfamily of winged-helix/forkhead (Fox) genes that are expressed in the lung and act as transcriptional repressors. *Journal of Biological Chemistry*, *276*, 27488–27497.

Sibley, C. G., & Ahlquist, J. E. (1984). The phylogeny of hominoid primates, as indicated by DNA-DNA hybridisation. *Journal of Molecular Evolution*, *20*, 2–15.

Stedman, H. H., Kozyak, B. W., Nelson, A., Thesier, D. M., Su, L. T., Low, D. W., et al. (2004). Myosin gene mutation correlates with anatomical changes in the human lineage. *Nature*, *428*, 415–418.

Studdert-Kennedy, M. (1998). The particulate origins of language generativity: From syllable to gesture. In J. R Hurford, M. Studdert-Kennedy, & C. Knight (Eds.), *Approaches to the evolution of language* (pp. 169–176). Cambridge: Cambridge University Press.

Swisher, C. C., III, Curtis, G. H., Jacob, A. C., Getty, A. G., Suprojo, A., & Widiasmoro. (1994). Age of the earliest known hominids in Java, Indonesia. *Science*, *263*, 1118–1121.

Thorne, A., Grün, R., Mortimer, G., Spooner, N. A., Simpson, J. J., McCulloch, M., et al. (1999). Australia's oldest human remains: Age of the Lake Mungo human skeleton. *Journal of Human Evolution*, *36*, 591–612.

Tomasello, M. (1996). Do apes ape? In J. Galef & C. Heyes (Eds.), *Social learning in animals: The roots of culture* (pp. 319–346). New York: Academic Press.

Vargha-Khadem, F., Carr, L. J., Isaacs, E., Brett, E., Adams, C., & Mishkin, M. (1997). Onset of speech after left hemispherectomy in a nine-year-old boy. *Brain*, *120*, 159–182.

Vargha-Khadem, F., Watkins, K. E., Alcock, K. J., Fletcher, P., & Passingham, R. (1995). Praxic and nonverbal cognitive deficits in a large family with a genetically transmitted speech and language disorder. *Proceedings of the National Academy of Sciences, U.S.A.*, 92, 930–933.

Vleck, E. (1970). Etude comparative onto-phylogénétique de l'enfant du Pech-de-L'Azé par rapport à d'autres enfants néanderthaliens. In D. Feremback (Ed.), *L'enfant Pech-de-L'Azé* (pp. 149–186). Paris: Masson.

Walter, R. C., Buffler, R. T., Bruggemann, J. H., Guillaume, M. M. M., Berhe, S. M., Negassi, B., et al. (2000). Early human occupation of the Red Sea coast of Eritrea during the last interglacial. *Nature*, *405*, 65–69.

Watkins, K. E., Dronkers, N. F., & Vargha-Khadem, F. (2002). Behavioural analysis of an inherited speech and language disorder: Comparison with acquired aphasia. *Brain*, *125*, 452–464.

Watkins, K. E., Vargha-Khadem, F., Ashburner, J., Passingham, R. E., Connelly, A., Friston, K. J., et al. (2002). MRI analysis of an inherited speech and language disorder: Structural brain abnormalities. *Brain*, *125*, 465–478.

Whiten, A. (1999). The evolution of deep social mind in humans. In M. C. Corballis & S. E. G. Lea (Eds.), *The descent of mind* (pp. 172–193). Oxford: Oxford University Press.

Wood, B. (2002). Hominid revelations from Chad. *Nature*, *418*, 134–135.

Wood, B., & Collard, M. (1999). The human genus. *Science*, *284*, 65–71.

V Self-Awareness and Social Cognition

Humans, though not unique in their capacities, are certainly masters of higher-order cognition. We routinely employ language, abstract reasoning, and self-directed awareness in everyday life, beginning at an early age. These abilities are seen in all cultures and have been the foundation of civilization.

Yet many species without these abilities have been reproducing successfully for many more generations than humans. These abilities are certainly costly, requiring, for example, calories and oxygen to drive the brain that supports them. The self you possess, or the language that you use, must pay reproductive benefits, else these traits ought to have been selected against.

Certainly, these abilities provide benefits at the social level. A social system predicated on higher-order awareness can arise rapidly (as in humans), adjust quickly, and demonstrate flexibility. Whereas some other animals have larger social systems than humans, the complexity and richness of human social interaction is unparalleled. Social neuroscience, the field that examines the relations between individuals and themselves at the neural level, is a research domain that examines these variables (Decety and Keenan, 2006). Within this field, one can examine the adaptive nature of having a brain that excels in social interactions but also enables higher-order abilities in nonsocial domains, such as solitary food gathering and, in part, some of the spatial abilities previously reviewed.

Why have a sense of self? Why have such complex abilities? Where and how does the brain create such cognitive flexibility, and why would it be adaptive? Why endure the cost of such a brain? These questions are presented in Part V, as is the idea of theory of mind, the ability to "mindread" or to infer the thoughts of others. Theory of mind and self-awareness form the basis for complex interactions with both self and others. Watching a good detective movie or American soap opera will readily reveal these phenomena in action ("Joe knew that Jen thought that Bill was thinking the child was his, which made Bill jealous because he thought Jen loved him").

The chapters in this part present an overview of primate behavior, the neuroimaging of self and other, the pervasive developmental disorders (in which we see deficits in higher-order cognition), and a number of applications of these ideas, such as the social prosthetic system, which is a unique way of considering these variables.

Is it possible that sex differences exist in these variables, as is suggested by work on autism and Asperger's syndrome? If so, why might such differences exist? Is there an adaptive reason for sex differences in higher-order cognition? Perhaps, as suggested, the abilities of the parents must be considered in terms of the child's development of these abilities, and perhaps with a large genetic component. Furthermore, the adaptiveness of such a transmission must be considered.

Without a self, psychoanalytic psychologists would be hard-pressed to make a living. Resentment and trauma, for example, involve reliving some event with negative affective components. Furthermore, the self gives rise to emotions that don't always feel comfortable, such as embarrassment and shame. Also, the self appears directly tied to deception, at least in some of its definitions. This side of the self is examined in some detail, as both the self and deception are intimately tied to our current cognitive repertoire. Taken together, the flexibility of metacognition is likely key to the development of our brain and our survival.

Reference

Decety, J. E., & Keenan, J. P. (2006). Social neuroscience: A new journal. *Social Neuroscience*, *1*, 1–4.

15 The Evolution of Human Mindreading: How Nonhuman Primates Can Inform Social Cognitive Neuroscience

Laurie R. Santos, Jonathan I. Flombaum, and Webb Phillips

The Secret of Our Evolutionary Success

For social species like our own, evolutionary success requires more than the basics of finding some food and a mate. To survive and reproduce, humans must successfully navigate a rather complicated social world. Each day, we are required to interact with countless other humans, all behaving in ways that we must predict, interpret, and in some cases manipulate. It's a daunting task, even for large-brained primates like us. Luckily for each of our genetic legacies, we're pretty good at it.

We owe this social sophistication in large part to a remarkable cognitive shortcut—the capacity to think about behaviors in terms of mental states such as intentions, desires, thoughts, and beliefs, an ability commonly subsumed under theory of mind. The term *theory of mind*, originally coined by Premack and Woodruff (1978), serves as a somewhat literal portrayal of why this cognitive shortcut is considered so remarkable. As Premack and Woodruff originally noted, our theory of mind is theory-like in that it is a set of inferences about things that are not directly observable—namely, states of the mind—which are then used to make predictions about an individual's future behaviors. That humans have developed a theory about something we cannot observe directly is a fascinating computational feat.

In the decades since Premack and Woodruff's landmark paper, theory of mind (hereafter ToM) has become a hot topic of research in many different fields: cognitive development, social psychology, comparative psychology, abnormal psychology, and philosophy of mind, to name only a few. Not surprisingly, researchers in cognitive neuroscience have followed suit with investigations into the mechanisms underlying human ToM abilities. Such work has revealed that a constellation of neural areas seem to take part in our processing of other minds.

Researchers have identified areas as diverse as parts of visual cortex, the amygdala, the anterior cingulate cortex, the medial prefrontal cortex, and the superior temporal sulcus. At present, a vast array of experimental work seeks to understand the unique role played by each of these centers in our ability to represent the mental states of others. This work has been elegantly summarized in a number of recent reviews (e.g., Allison, Puce, & McCarthy, 2000; Gallagher & Frith, 2003; Saxe, Carey, & Kanwisher, 2004), so we will not focus on it here.

Instead, we highlight here a potential methodological challenge that neuroscientific investigations of ToM have now begun to face. To date, explorations into the neural substrates of ToM have focused primarily on the subjects to whom we can confidently attribute these abilities—ourselves. Consequently, much of the research exploring the neural basis of ToM has utilized noninvasive imaging techniques, such as functional magnetic resonance imaging (fMRI). Although noninvasive techniques have already provided some marvelous insights into the global systems involved in human ToM processing, these techniques are limited for studying some aspects of ToM. The first limitation has to do with the problem of dissociating the smaller components of ToM processing at the neural level. ToM is composed of a host of elemental cognitive abilities, some of which are likely to activate very close neural populations. Present functional imaging techniques are not well suited to examining adjacent neural populations in this way, and thus may miss critical differences between the smaller component processes that give rise to ToM (see discussion in Saxe et al., 2004). Likewise, anatomically separate brain regions may become jointly activated by a number of distinct cognitive processes, making it difficult to decipher the role of each of these areas in different aspects of our ToM abilities. Finally, functional imaging techniques are not well suited for one of the major goals of a neuroscientific inquiry into ToM abilities: the development of neuropharmacological treatments for clinical disorders of ToM reasoning, such as autism (e.g., Baron-Cohen, 1995; Schultz, Romanski, & Tsatsanis, 2000; Siegal & Varley, 2002). The exclusive use of human subjects ethically precludes the pharmacological experimentation essential for eventually treating such disorders (Amaral, 2002; Machado & Bachevalier, 2003). In light of these limiting aspects of noninvasive techniques, the field of cognitive neuroscience would greatly benefit from the development of a primate model of human ToM abilities. Using a primate model would afford the opportunity to engage the substrates of ToM with more sensitive measures of neural activity and the potential to develop new neurochemical assays.

The need for a primate model of ToM raises some tough questions for primate researchers. Do any primates other than humans actually possess ToM? Or is ToM a uniquely human cognitive specialization? Thankfully for neuroscientists, researchers in comparative cognition have gained considerable insight into these controversial questions in just the past few years. Here we review recent empirical and philosophical advances in the study of ToM abilities in nonhuman primates (hereafter primates), presenting some of our own work investigating these abilities in macaque monkeys.

Before we turn to this review, however, we note an important caveat. The challenge of investigating representations of unobservable entities such as mental states is amplified when one hopes to study these representations in an organism that cannot speak. As a consequence, the comparative study of ToM has enjoyed lively (and often productive) debates as to which (if any) experimental paradigms successfully demonstrate ToM abilities in primates. Although others may disagree (e.g., Povinelli, 2004), we will make the case that the available data demonstrate that some primates do in fact theorize about the mental states of others in some of the same ways as adult humans do. However, we hope that differences in the specific interpretation of recent experimental findings will not obscure what we believe are the most vital conclusions drawn from these experiments, conclusions that we hope even methodological skeptics can agree with: (1) Some primates succeed in competitive versions of ToM tasks, even though they fail on nearly identical cooperative versions. (2) Primates in these experiments succeed in interpreting exactly the kinds of stimuli that individuals with clinical deficits, such as autism, appear unable to interpret. (3) There remains the possibility that the cells in primate cortex known to encode the location of another individual's eye gaze are involved in sophisticated aspects of social reasoning, such as stealing food from a competitor.

The rest of this chapter is divided into three sections. The first section reviews recent findings in primate ToM, with an emphasis on how primates appear to succeed on social reasoning tasks that are extremely similar to ones on which they have famously tended to fail. This section focuses more on historical changes in the study of primate social reasoning and less on the kinds of ToM representations that these experiments reflect. The second section revisits this suite of experimental work, presenting both skeptical and generous views of what these findings might mean, with an eye toward the more theoretical aspects of this work (i.e., are primate representations of other minds different from human ones?). Finally, we turn to the implications that recent behavioral work might have for neuroscientists studying the underlying circuitry of human ToM and the etiology of diseases such as autism.

A History of Primate Theory of Mind: From Cooperative to Competitive Paradigms

Anyone who has observed primates knows that they spontaneously exhibit a number of natural behaviors that seem consistent with a rich understanding of the mental states of others. Primates demonstrate countless examples of functional deception both in the wild and in captivity. For instance, they refrain from announcing the presence of food if more dominant individuals are nearby, and they conceal objects by hiding them from other individuals' view (see Byrne & Whiten, 1988; Mitchell, 1999; Whiten & Byrne, 1997). Similarly, a number of primate species naturally recognize where other individuals are looking (see Tomasello & Call, 1997, for a review). Chimpanzees (*Pan troglodytes*), for example, spontaneously follow the gaze of both human experimenters (Itakura, 1996; Povinelli & Eddy, 1996b, 1997; Tomasello, Hare, & Fogelman, 2001) and conspecifics (Tomasello, Call, & Hare, 1998), and can track an individual's line of regard past barriers and distracters to a target object (Povinelli & Eddy, 1996b, 1997; Tomasello, Hare, & Agnetta, 1999).

Despite this impressive range of seemingly insightful social reasoning, however, until very recently primates have typically failed on experimental tasks that attempted to expose whether these animals represent the contents of others' mental states, as opposed to merely responding, albeit adaptively, to others' behaviors (for reviews, see Heyes, 1998; Povinelli, 2000; Tomasello & Call, 1997). That is, although animals may respond to cues that reflect the knowledge of others—for example, they may turn their heads to follow the gaze of others—they may not represent the mental states of others in any meaningful way, or be able to use the information that they represent for solving more sophisticated social problems. Rhesus monkeys (*Macaca mulatta*), for example, fail to recognize the difference between guessing and knowing on a hidden food task (Povinelli, Parks, & Novak, 1991) and overlook the knowledge state of their offspring when alerting them to the presence of predators (Cheney & Seyfarth, 1990). Primates have even failed on one of the most simplified ToM tasks possible, a game of pick-the-one-the-experimenter-is-looking-at, also known as the object choice task. Here, a human experimenter (or sometimes a trained conspecific) attempts to communicate the location of a hidden food by either looking directly at it or pointing and gesturing toward it. Monkeys and apes in a number of different laboratories have failed on this task, choosing randomly between indicated and ignored locations (Anderson, Montant, & Schmitt, 1996; Anderson, Sallaberry, & Barbier, 1995; Call, Agnetta, & Tomasello, 2000; Call, Hare, & Tomasello, 1998; Itakura, Agnetta, Hare, & Tomasello, 1999; Itakura & Anderson, 1996; Peignot & Anderson, 1999; Povinelli, Bierschwale, & Cech, 1999; Povinelli & Eddy, 1996c). Similarly, chimpanzees with extensive experience interacting socially with human caretakers fail to take into account what different experimenters can see when choosing whom to ask for food (Povinelli & Eddy, 1996a). Without training, chimpanzees failed to distinguish between an experimenter with a bucket on her head and one holding a bucket in her arms (though these chimpanzees also had extensive experience playing with buckets and wearing them on their own heads), and an experimenter whose entire head orientation was directed at them and one who was directed away from them. Even after training, the apes in these experiments continued to beg for food from experimenters unable to see them, leading the authors to conclude that chimpanzees possess little understanding of the nature of visual attention, let alone mental states such as beliefs, desires, and intentions.

The apparent discrepancy between naturalistic observations of primate social behavior and negative experimental findings in laboratory tasks recently prompted some researchers to seek a new approach to examining what primates know about the minds of other individuals. This new approach employs the general insight that some kinds of cognitive abilities (particularly those that are domain-specific; see Cosmides & Tooby, 1994) may only become engaged in experimental settings that mimic the ecologically relevant context for which these abilities evolved. Thus, as a number of researchers have proposed (Byrne & Whiten, 1988; Hare, 2001; Humphrey, 1976), if primates actually understand the mental states of other individuals, then the most likely domain in which they would use this information is when vying with other individuals in ecologically relevant situations.

In a series of pioneering experiments, Hare, Call, Tomasello and their colleagues (Hare, Call, Agnetta, & Tomasello, 2000; Hare, Call, & Tomasello, 2001; for a review, see Tomasello, Call, & Hare, 2003) noticed that previous empirical tests on primate ToM differed in at least one important way from the situations in which primates in the wild appear to reason about mental states. Specifically, primates tend to appear the most socially adept when competing with one another (e.g., Hare, 2001), for example, when attempting to conceal the presence of contested food or gain sexual access to contested females. Unfortunately, most previous experiments on primate ToM did not incorporate this feature of competition; most past tests instead required cooperation among individuals. In the object choice paradigm, for example, the importance of where a human is looking can only be understood if the subject first appreciates the experimenter's cooperative intent to communicate the location of the hidden food. Similarly, in the context of Povinelli and Eddy's (1996a) begging experiments with chimpanzees, the visual awareness of human caretakers is only relevant if the subject appreciates that a human's response to the begging gesture will be the voluntary decision to give the beggar food. It remains to be seen if this is indeed how chimpanzees and other primates perceive these inherently cooperative interactions.

Hare and colleagues (2000) attempted, therefore, to capitalize on what appears to be primates' natural inclination to reason about mental states when competing with others for scarce food resources. To do so, they set up a situation in which chimpanzees could compete over access to hidden pieces of food. In this competitive foraging paradigm, subordinate and dominant chimpanzees were positioned on alternate sides of a middle cage that contained two pieces of food (figure 15.1). In some cases the food was placed such that the dominant individual could see only one of the two pieces. Hare and colleagues hypothesized that if the subordinate individual was sensitive to what the dominant individual could and could not see, then he should selectively attempt to retrieve only those foods that the dominant individual could not see. Subordinate chimpanzees did just this, successfully retrieving the food that the dominant individual could not see. Similarly, subjects were more likely to approach food that was hidden from the dominant's view by a visual barrier (Hare et al., 2000) and food that the dominant individual did not know was placed in a particular location (Hare et al., 2001). In a more recent series of studies, Hare, Call, and Tomasello (2006) presented chimpanzees with the opportunity to approach a competitive human experimenter in an attempt to obtain contested food. Again, chimpanzees were more likely to try to approach food that the experimenter was not watching. Taken together, the competitive studies of Hare and colleagues suggest that chimpanzees do have a rich notion of what other individuals can and cannot see, at least in certain competitive situations.

But chimpanzees are not the only species to succeed in ToM experiments involving competition. Recently, our laboratory has extended

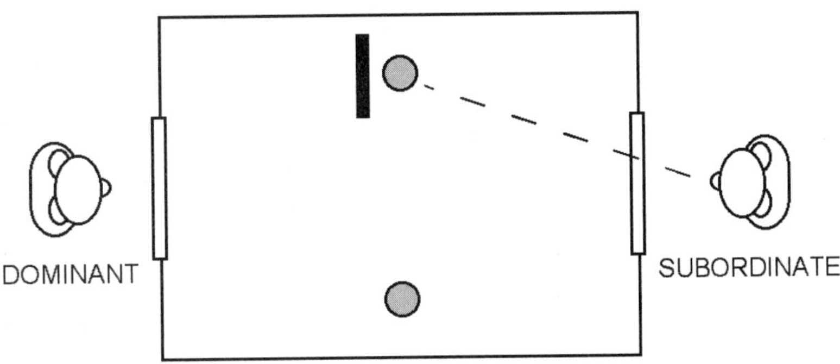

Figure 15.1
An aerial view of the chimpanzee competition paradigm (adapted from Hare et al., 2000). Subordinate chimpanzees selectively approached the piece of food that the dominant individual could not see.

the findings of Hare and colleagues to another primate species, the rhesus macaque. To do so, we attempted to develop a competitive paradigm that was similar to classic ToM experiments in the hope that such a methodology would help to illuminate just why competition is such a fertile test bed for ToM-like reasoning. To this end, we (Flombaum & Santos, 2005) developed a competitive foraging task for a population of free-ranging macaques living at the Cayo Santiago field station in Puerto Rico.

The Cayo Santiago macaques spend much of their day foraging for monkey chow provisioned to them by human experimenters. However, because of the long history of research on this island, the monkeys in this population have also developed a compelling interest in the foods that they see human experimenters eating. (Human foods tend to be sweeter and more exciting than the monkey chow they normally eat.) Because of their curiosity about human foods (see Santos, Hauser, & Spelke, 2001), the macaques often approach experimenters and attempt to steal their food, so much so that we and our colleagues typically use this approach behavior as a dependent measure in empirical studies with this population (see Flombaum, Kundey, Santos, & Scholl, 2004; Hauser, 2001; Hauser, Carey, & Hauser, 2000; Phillips & Santos, 2006; Santos, 2004; Santos et al., 2001, 2002; Sulkowski & Hauser, 2001). In these studies, subjects always appeared somewhat apprehensive of getting close to the experimenters involved in testing, suggesting the macaques may view humans as potentially dangerous competitors. As such, we reasoned that these monkeys should be motivated to take human food only when they can do so without being detected. Subjects chose the competitor with his eyes covered.

Using this logic, we investigated whether the monkeys spontaneously took into account the direction of a human experimenter's gaze when attempting to steal food (Flombaum & Santos, 2005). In each experiment, two male experimenters—a.k.a. the competitors—approached a lone monkey. Both competitors then placed a platform holding a grape on the ground and then turned in a particular way relative to the grape. In the first study, one competitor turned to face the grape, while the other turned his back to the grape (figure 15.2A). Both experimenters then froze for one minute and allowed the subject to attempt to steal the grape. We predicted that subjects should steal the grape from the competitor who could not see the grape, namely, the one with his back turned. Our rhesus monkey subjects did just this, choosing to approach the experimenter with his back to the grape. We then explored whether the monkeys could use more subtle cues to what competitors can see. As in the first study, the macaques selectively retrieved the grape from a competitor whose head and eyes were oriented away (Experiment 3), whose eyes alone were oriented away (Experiment 4, figure 15.2B), or whose gaze was blocked by a small barrier (Experiments 5 and 6, figure 15.2C). Taken together, these results demonstrate that macaques, like chimpanzees (Hare et al., 2006), spontaneously use information about what a human competitor can see when determining which contested grape to steal. Interestingly, our subjects selectively chose between two humans whose postures were nearly identical to those used in Povinelli and Eddy's (1996a) famous begging experiments with chimpanzees (see figure 15.2); unlike our subjects, the chimpanzees in Povinelli and Eddy's studies chose randomly between the two experimenters. The only substantive difference between our experiments and those with chimpanzees is that success in the chimpanzee studies required approaching and then cooperating with the experimenter who could see the object food, whereas success in our experiments required avoiding and competing with the experimenter who could not see.

In a second series of studies (Flombaum & Santos, 2004), we examined whether rhesus monkeys further understand that seeing leads to knowing. Do they know that a competitor who has not *seen* where a piece of food is doesn't *know* where that food is? And do they exploit such false knowledge when competing for food? To examine these questions, we used a slightly different setup (figure 15.3): a single human competitor crouched behind a ramp apparatus with two horizontal platforms, one at its top and one at its bottom. Above each platform stood a canopy that prevented the experimenter from seeing what was on the

Figure 15.2
Stimuli used in Flombaum and Santos (2005) to test rhesus monkeys' use of information about what a competitor can and cannot see in stealing food (a grape). (A) Experiment 1. Subjects chose to steal a grape placed on the ground by the competitor (human) now with his back turned to the grape. (B) Experiment 4. Subjects chose to steal a grape from the competitor who had averted his eyes to the side. (C) Experiment 6. Subjects chose to steal a grape from the competitor whose gaze was blocked by a small barrier. These results indicate that macaques, like chimpanzees, use information about what a competitor can see when determining which contested food to take—that visible or invisible to the competitor.

platform. The platforms were connected by a slightly inclined horizontal ramp and were positioned such that a food object placed on the upper platform could potentially roll down the ramp and come to rest on the lower platform.

During testing, the apparatus was positioned such that the subject, who was seated in front of the apparatus, could see a piece of food when placed on either of the two platforms, but the competitor, who was seated behind the apparatus, could not. This positioning set up a situation in which the subject knew exactly where the food was at all times, but the competitor, who lacked direct visual access, had to rely on his knowledge of where the food was originally placed. In one condition, the competitor began by placing two grapes at the top of the ramp. One of these grapes then appeared to secretly roll down the apparatus (in fact, its rolling was triggered covertly by the experimenter). Because the experimenter was unable to see the grape roll to its new position, he had no knowledge that it had moved. If subjects recognized that the competitor no longer *knew* where the grape that rolled was, then they should selectively try to steal that grape rather than the one that was still in its original location, a location that the competitor knew about. When the competitor was behind an opaque barrier, subjects did just this; without training, macaques knew to approach the grape that the experimenter didn't know had rolled. In contrast, when no visual barrier was in place (and thus the experimenter had knowledge of the grape's rolling; see figure 15.3 for this setup), subjects approached the two grapes randomly,

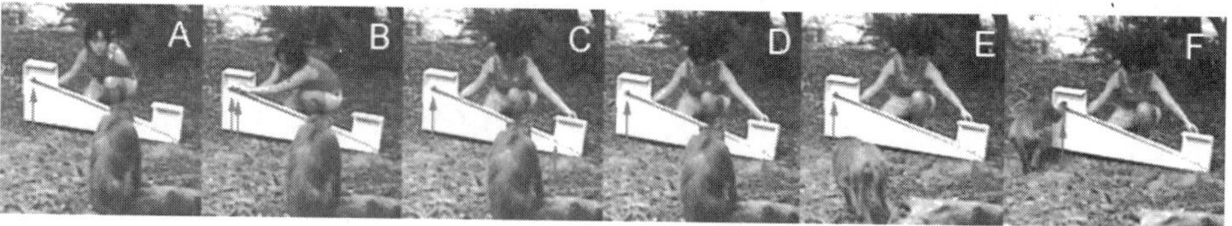

Figure 15.3
The competition event used in Flombaum and Santos (2004). (A, B) The experimenter (food competitor) placed two grapes at the top of the platform (grapes indicated by arrows). (C, D) The experimenter then sat back, released a hidden switch, and allowed one of the two grapes to roll down the ramp as the subject watched. (E, F) The subjects then had one full minute to approach one of the grapes. Here the subject chose the top grape. When the experimenter was behind a screen and could not confirm the position of a grape released by a lever to roll to a lower position, the macaques chose the lower grape. When no visual barrier was in place, the macaques' approach was random.

no longer showing a preference for the rolled grape. Again, like chimpanzees (Hare et al., 2001), our macaque subjects seemed to take into account what competitors did and did not know about a food's location.

Overall, we believe that our recent studies, together with the experiments of Hare and colleagues, build a strong case for ToM reasoning in apes and monkeys. In these studies, primates behaved as though they appreciated the mental states of others and could take specific account of how the experimenter's eye orientation and the presence of visual barriers constrained the contents of mental states. We consider whether this is the best way to interpret these results in the next section. As we have emphasized previously, however, what seems to us the most valuable contribution of these experiments is the demonstration that animals who have previously performed abysmally on tasks requiring a certain set of skills—in this case, reasoning about where individuals are looking—suddenly perform excellently when the same skills are required on a different kind of task. As others have highlighted (see Hare, 2001; Tomasello et al., 2003), the most critical difference between past and present ToM tasks seems to involve this aspect of competition for scarce resources. Therefore, it appears that competition may come to play a central role in our understanding of the nature of ToM reasoning more generally, in both humans and nonhuman primates. Additionally, as we discuss more in the final section, understanding why primates succeed on competitive tasks may shed some light on how the networks involved in human ToM become engaged during our normal social interactions, and the reasons why these networks do not appear to function in clinical disorders such as autism (Baron-Cohen, 1995).

Primate Theory of Mind: An Open-and-Shut Case?

Thus far we have tried to describe the state of the art in primate ToM experiments in as neutral representational terms as possible, but it will come as no surprise that we favor a rather generous view of the kinds of representations that these experiments endow on our closest living relatives. In particular, we believe that these competitive experiments establish that chimpanzees and rhesus monkeys are able to attribute perceptions, sometimes known as perspective taking, as well as knowledge to others. Thus, they represent what others *see* as well as what others *know*.

Over the past several years, however, a compelling deflationary alternative to this account has been put forward by Povinelli and colleagues (Povinelli, 2000, 2004; Povinelli & Bering, 2002; Povinelli & Vonk, 2003, 2004). According to their view, which they call the behavioral-abstraction hypothesis, primates may appear to be insightful social thinkers in competitive experiments, but they do not actually represent *mental states*. Instead, primates represent the *behaviors of others* and the correlations among sets of behaviors, postures, and actions. Take, for example, rhesus monkeys' performance in one of our studies, successfully stealing food from the competitor facing away from the food versus one facing toward the food (Flombaum & Santos, 2005, Experiment 1). The behavioral-abstraction hypothesis would claim that monkeys succeed by representing "his back is facing me, therefore he probably will not respond to my approach aggressively" as opposed to "his back is facing me, therefore he cannot *see* me." Primates, under this view, do not represent the mental states of others but rather the likely future behaviors of others given their current behaviors and postures. To be sure, the behavioral-abstraction hypothesis is not behaviorism; Povinelli and colleagues do credit nonhuman primates with the capacity for mental representations, but such representations only involve others' behaviors, not their mental states. Under their view, nonhuman primates are not naive psychologists but instead naive behaviorists.

At the core of our disagreement is that, as Povinelli and colleagues correctly point out, almost any experiment can be described with both a mentalisitic account, such as our own, or their behavioral-abstraction account. Essentially, in the context of any experiment with a behavior-dependent measure, one can always replace "therefore, he cannot *see* me" or "therefore, he does not *know*" with a phrase such as "therefore, he probably will not respond aggressively." But before we look more closely at this issue, it should be clear that more is at stake in this debate than merely the semantically most appropriate way to describe some experiments that appear justly described either way. What is at stake is the extent to which primates represent the events of these experiments in the same way as humans do. The goal of this research enterprise is not only to discover whether primates possess a ToM, but to learn whether primates possess a ToM so that we can develop a neuroscientific animal model of human abilities. According to the behavioral-abstraction hypothesis, this project is dead in the water. Our own introspections and natural language make it clear that we humans can reason about unobservable mental states, not just about behavioral correlations. Primates, on the other hand, do not reason about mental states, and moreover, if they did, we couldn't really know it in the first place. With respect to ToM, then, comparing our brains to theirs would be like comparing apples and oranges.

We propose that, in fact, primates do reason about unobservable mental states, and that they do so with the same basic cognitive systems that humans use to reason about mental states. The problem with the behavioral-abstraction hypothesis is that the better primates perform in increasingly difficult and controlled experiments, the more the behavioral-abstraction hypothesis becomes indistinguishable from a mentalistic account, and the more it becomes implausible. One can always substitute a mentalisitic term such as "he desires *X*" with a set of potential behaviors such as "he will probably approach *X*." This is because the purpose of representing another individual's mental states is to accurately predict their future behaviors. Thus, the chain of events in any assessment of another individual's mental states includes the prediction of behavior—something like "an occluder blocks his view, therefore he cannot see, therefore does not know, and therefore he will not approach."

This is not to say that an animal could not succeed with a set of representations that skipped mental states altogether and correlated behaviors directly to environmental cues, including other behaviors. For example, an animal that only represents "an occluder blocks his view, therefore he will not approach" is not representing mental states, though that animal would obviously succeed in some experiments. But it becomes increasingly unlikely that this is *all* animals are doing when the set of environmental cues that they can accurately correlate with certain behaviors appears to be highly sophisticated and specific. That is, correlations only work when one knows what features to use as units in the correlation. Most of the time, when an animal's eyes point forward, their mouth points forward as well. Therefore, we might expect that a primate interested in knowing when someone else is going to approach them will believe that approach is likely when either their eyes or their mouth point forward. Indeed, in Povinelli and Eddy's (1996a) begging experiments, chimpanzees appeared to fail for just this reason: without direct training, chimpanzees exhibited no begging preference between someone with a barrier covering her eyes compared to someone with a barrier covering her mouth, nor between someone whose entire face, including their eyes, was oriented forward and someone whose eyes alone were oriented away. A "simple" system of correlations is well suited for explaining errors of overgeneralization like the ones demonstrated by chimpanzees in these earlier experiments.

But the animals tested in competitive paradigms more recently appeared *not* to make errors of overgeneralization (see Flombaum & Santos, 2005; Hare et al., 2000, 2001, 2006). Instead, they appeared finely tuned to exactly the variables, postures, and behaviors in their environments that are good predictors of future behavior (surely because these features are factors that constrain mental states such as seeing and knowing). This suggests that primates possess a specialized system that identifies and analyzes only those factors that are relevant to problems of social reasoning—what in essence boils down to a ToM system. In our experiments (Flombaum & Santos, 2004), monkeys correctly avoided an experimenter with an occluder over his eyes but not his mouth (Experiment 6), an experimenter whose eyes alone were averted to the side (Experiment 4), and even an experimenter whose profile faced the monkey but whose eyes faced the contested food (compared to an experimenter with his profile to the monkey as well, but with his back to the contested food; Experiment 2). If the monkey's goal is to predict which of the human competitors will not foil their plan to steal the contested food, then it must do so by correlating all of the human competitors' past patterns of approach with where the competitor's eyes are pointing relative to (1) the monkey's current position, (2) the monkey's position at all stages along its approach to the food, and (3) a variety of things in the environment that could cover the competitor's eyes. This alone is an astounding feat, but even more astounding is the fact the monkeys successfully ignore many other correlated but irrelevant variables, such as where the competitor's mouth or nose is pointing; these features may have correlated reliably with the competitor's approach in past encounters, but they are not causally related to what the experimenter can see, and therefore they do not actually constrain where he will approach. Identifying and then analyzing exactly the variables that will best predict future behavior (because they constrain mental states) is, functionally, what a system that is specialized for ToM does, and it is a system that, in our opinion, we appear to share with monkeys and apes.

No doubt, monkeys and apes, as well as humans, also possess the ability to compute massive correlations between current behaviors and future behaviors. In fact, we believe that it is the difference between a general system for computing statistical probabilities and a finely tuned ToM system that helps to explain a pattern of data that would otherwise appear inconsistent. Namely, how can we account for the fact that primates tend to fail on cooperative ToM tasks but succeed on nearly identical competitive ones? It is insufficient to explain these results merely by saying that the cooperative experiments do not, for whatever reason, engage the system that allows for success on the competitive tasks. Such an explanation neglects to specify the system that accounts for behavior on failed cooperative tasks. Thus, if the behavioral-abstraction hypothesis accounts for the success of primates in competitive experiments, as Povinelli and colleagues would have it, then what accounts for their failure on cooperative ones? Primates may behave incorrectly, even foolishly, in cooperative experiments, but not always randomly: chimpanzees know to beg from a human facing forward and not one facing backward, they just don't know to distinguish between one with his eyes averted and one with his eyes forward (among other similarly subtle differences). This type of behavior, as we described previously, is precisely the symptom that one would expect of a system that computes nothing more than correlations. Indeed, after further investigation, Povinelli and Eddy (1996a) found that chimpanzees tended to beg for food from the experimenter who showed more "frontal aspect" (more surface area on the front of the body). The chimpanzees had, therefore, tried to correlate posture and behavior, but they failed on the begging task simply because they correlated behavior to the wrong kind of posture. It is important to note, however, that if a system for correlating postures and behaviors successfully accounts for the stereotypical failures exhibited in cooperative experiments, then it cannot also account for the systematic successes exhibited in nearly identical competitive experiments. Essentially, the pattern of failures in cooperative experiments and successes in competitive experiments necessitates two sets of explanations, and therefore the use of two different cognitive systems. We agree that behavioral-abstraction (or whatever system computes massive correlations between behaviors) likely accounts for the patterns of failures observed in primate cooperative experiments. It seems to us, however, that a different system—namely, a ToM system—accounts for successes in competitive experiments.

Finally, we acknowledge that the kinds of arguments we have put forward here in support of primates possessing a ToM rely on revisiting a large suite of experiments that included primate failures as well as successes. Therefore, our arguments do not by any means constitute an open-and-shut empirical demonstration that animals, under unambiguous terms, possess the kind of ToM that humans do. These issues aside, there is another important sense in which it can be said that primates possess ToM. That is, they possess ToM in the practical sense that they possess abilities that are clearly relevant to studying human ToM neurally. This is the major focus of the next section of the chapter. Our working hypothesis is that whenever primates reason about the minds of others—whether by abstracting from their behaviors or with the use of a ToM system—they appear to engage the same brain areas that humans appear to use in similar situations (Allison et al., 2000). In addition, they appear to reason about others in exactly the ways that autistic children seem unable to (see Baron-Cohen, 1995, for a review). Thus, even if we cannot be sure of how monkeys are representing the actions of others, we can be sure that we are studying systems in monkeys that are highly relevant to understanding human ToM.

Brains That Think About Minds and Brains That Don't

We now turn to a somewhat more practical and pressing issue in the study of primate ToM. As reviewed in the previous section, recent competitive studies have provided some of the first behavioral evidence that primates use the direction of an individual's eye gaze to make predictions about that individual's knowledge and perceptions. What implications do these findings have for our understanding of ToM neurally? Do competitive paradigms bring us any closer to a better primate model of ToM abilities? In short, can these behavioral experiments somehow help us to understand how brains come to think about minds?

As a first pass, it would seem that these new findings may help clarify data that are already available on social processing in the primate brain, data that in some cases are now more than a decade old. Very elegant single-unit work by Perrett and colleagues (Jellema et al., 2000, 2002; Perrett et al., 1985, 1992; Perrett & Mistlin, 1990), for example, has shown that the primate brain possesses some areas that are dedicated for the processing of head and eye gaze. This work suggests that cells in the rhesus monkey superior temporal sulcus (STS) are sensitive to the position of another individual's eyes, often independent of that individual's head and body orientation. Perrett and colleagues originally interpreted the function of these areas as encoding where another individual's eyes were pointed. Functional imaging studies in humans (reviewed by Saxe et al., 2004), however, have demonstrated that the human STS is sensitive to more than just eye position; in contrast to the single-unit data, the human STS seems to represent goal-directed actions and perceptions—what other individuals *want* and *see*. Using fMRI techniques, Pelphrey and colleagues, for example, have observed that the STS is modulated both by the context of where another person is looking (Pelphrey, Singerman, Allison, & McCarthy, 2003) and that person's perceived goal (Pelphrey, Viola, & McCarthy, 2004). Observing such an STS function in humans makes sense in light of the macaque single-unit findings: because the STS is able to disambiguate head direction and eye orientation (see Perrett et al., 1990), it seems ideally suited for the further process of computing the content of what another individual sees. But could this neural area be encoding something similarly complex in macaques as well? Previously, few researchers would have thought so. Nonetheless, our behavioral experiments with macaques, the very animals studied in this physiology work, make it clear that this species is capable of representing more than just where others are looking; they, too, seem to represent what others see. For this reason, it is possible that the macaque STS is also capable of encoding what others see, and therefore is necessary to mediate the decisions that macaques successfully make in our competition experiments, decisions about which competitor to avoid. Thus, the macaque STS may do more than simply record where eyes are pointed. This region may generate representations of *seeing* and *not seeing* that can be used in task-relevant ways by the organism. As such, though it remains uncertain whether the monkey STS actually "does ToM," it is likely that this system provides an important foundation for sophisticated social reasoning, and therefore can serve as a site for understanding the neural basis of such reasoning, even in a primate model.

Auspiciously, the capacity to detect what others can and cannot see is a capacity ripe for cognitive neuroscientific study; it is one of a suite of ToM abilities lacking in autism and therefore one for which an animal model would be most beneficial. In addition to their well-known problems representing the mental states of others (Baron-Cohen, Leslie, & Frith, 1985; Perner, Frith, Leslie, & Leekham, 1989), autistics also have a well-documented deficit in attending to and interpreting eye gaze (see Baron-Cohen, 1995, for a review), even at a young age (Charman et al., 1997). Similarly, autistic children rarely use proto-declarative gestures intended to change another individual's direction of gaze (Baron-Cohen, 1989). Beyond just attending to gaze, however, autistics seem to have problems interpreting the meaning of one's gaze—in particular, that it affects a person's mental state. For example, Baron-Cohen, Campbell, Karmiloff-Smith, and Grant (1995) found that autistic children do not use the direction of an individual's gaze to determine what that individual wants or knows, even though they succeed in following gaze and can accurately determine where an individual is looking. Rhesus macaques, on the other hand, seem to do both. In our experiments, for example, macaques relate the orientation of a competitor's gaze to his future disposition with respect to the grape, and more than autistics seem to do. For these reasons, a primate model of STS function may provide important insights into the reasons for cognitive deficits in autism. Such a model may even be necessary. The subtle cognitive impairments observed in autism likely reflect deficits at the neural level that are similarly subtle; some autopsy studies even suggest these deficits may involve impairments in a select population of neurons or even specific varieties of cells (Courchesne, 1997). Neural investigations at such a detailed level require a physiological model.

Not only do rhesus monkeys, which seem to reason about the perceptions of others, suffice as such a neurophysiological model of autism, but their idiosyncratic patterns of performance in ToM experiments might make them an especially interesting model to study. This is because their failure in cooperative ToM experiments juxtaposed with their success in nearly identical competitive ones affords the chance to understand not only what makes their social reasoning skills work, but also what makes them not work. At the heart of the logic of cognitive neuroscience is the importance of dissociations, and in rhesus monkeys we seem to have a natural dissociation between reasoning about eye gaze in competitive experiments and tracking it in cooperative ones. Such a dissociation behaviorally may provide some clues to understanding why autistic children seem similarly able to track gaze, but not to reason about it.

Moreover, this dissociation in monkeys between reasoning and tracking in different contexts might naturally map onto a dissociation between different neural systems. One outstanding question in the study of human ToM abilities, for instance, concerns the role that the amygdala plays in reasoning about eye gaze (see Baron-Cohen & Ring, 1994). It has been established both in monkeys (Leonard, Rolls, Wilson, & Baylis, 1985) and humans (Kawashima et al., 1999) that the amygdala shows sensitivity to eye gaze orientation. Moreover, amygdala damage results both in a generalized impairment of reasoning about mental states and in reasoning about gaze in particular (Fine, Lumsden, & Blair, 2001). Likewise, removal of the amygdala results in an impairment in the perception of gaze direction (Young et al., 1995). Nevertheless, several brain areas, including the STS, are sensitive to eye gaze prior to the amygdala in the processing of visual information (see Brothers, 1990). Because of this, neuroscientists are still unclear about the unique role that the amygdala plays in processing eye gaze information. One possibility is that the amygdala serves to amplify STS responses to eye gaze (Allison et al., 2000; Leonard et al., 1985). Such a view fits with some known facts about the amygdala, including its function in binding emotional content with sensory information (Adolphs, Russell, & Tranel, 1999). Thus the amygdala may prove the critical link in reasoning not only about where another individual is looking but what they *see*, as well (see Baron-Cohen & Ring, 1994).

Primates may fail in cooperative versions of ToM because these kinds of contexts fail to engage the amygdala emotionally, although competitive interactions obviously do. The amygdala is known to become engaged when monkeys participate in competitive situations. Moreover, monkeys with amygdalectomies show little understanding of their own place in the social dominance hierarchy (Kling, Lancaster, & Benitone, 1970), suggesting again the importance of this structure in negotiating competitive interactions. For these reasons, a monkey viewing stimuli in a competitive experiment should evince a highly active amygdala interacting with a highly active STS. A monkey viewing the same stimuli in cooperative experiments might show no such activation in the amygdala. Perhaps this lack of amygdala activity in cooperative studies is at the root of poor performance. In fact, we would hypothesize that monkeys with amygdala lesions should demonstrate an ability to follow eye gaze but be unable to use this information correctly in competitive paradigms like the one that we have developed. Consequently, they should fail to favor a competitor who cannot see them when selecting which contested food item to approach.

Dissociating the functions of the amygdala and STS is only one among many neural dissociations to be made if we are to understand how the human ToM network works. Similar experiments could afford a better understanding of how face perception and reasoning about eye gaze may rely on subtly different cognitive and neural resources (Allison et al., 2000; Campbell, Heywood, Cowey, Regard, & Landis, 1990), as well as the role of executive functions and regions of frontal cortex in ToM (Fine et al., 2001; Stuss, Gallup, & Alexander, 2001). Overall, the availability of a sensitive behavioral paradigm with a laboratory primate allows for investigations into the neural substrates of ToM with a degree of sensitivity unavailable from imaging methods with humans.

Conclusion

In 1978, Premack and Woodruff posed the question of whether the chimpanzee had a theory of mind. Several decades later, their question has sparked not only rich comparative work on social cognition in primates but also tremendous interest and empirical progress in other areas of cognitive neuroscience as well. In this chapter, we have argued that we can now go beyond their original question. The evidence we have reviewed, incorporating both experimental successes and failures, supports the view that we share a number of ToM capacities with at least some of our primate relatives. With this evidence in place, cognitive neuroscientists, together with comparative psychologists, are poised to begin asking more detailed questions about the mechanisms underlying social cognition, and potentially treating disorders involving these mechanisms. It is clear to us that only a broad interdisciplinary perspective—like the one advocated in this volume—can provide a richer picture of how brains of all kinds succeed in representing the minds of others.

Acknowledgments We thank Drs. Brian Scholl and Jeremy Gray for their helpful comments on the manuscript. L.R.S. was supported by Yale University and J.I.F. was supported by an NSF Predoctoral Fellowship. We also thank the staff of the Cayo Santiago Field Station for their support of our empirical work on rhesus monkey theory of mind abilities.

References

Adolphs, R., Russell, J. A., & Tranel, D. (1999). A role for the human amygdala in recognizing emotional arousal from unpleasant stimuli. *Psychological Science*, *10*, 167–171.

Allison, T., Puce, A., & McCarthy, G. (2000). Social perception from visual cues: Role of the STS region. *Trends in Cognitive Sciences, 4*, 267–278.

Amaral, D. G. (2002). The primate amygdala and the neurobiology of social behavior: Implications for understanding social anxiety. *Biological Psychiatry*, *51*, 11–17.

Anderson, J. R., Montant, M., & Schmitt, D. (1996). Rhesus monkeys fail to use gaze direction as an experimenter-given cue in an object-choice task. *Behavioural Processes*, *37*, 47–55.

Anderson, J. R., Sallaberry, P., & Barbier, H. (1995). Use of experimenter-given cues during object-choice tasks by capuchin monkeys. *Animal Behaviour*, *49*, 201–208.

Baron-Cohen, S. (1989). Perceptual role-taking and proto-declarative pointing in autism. *British Journal of Developmental Psychology*, *7*, 113–127.

Baron-Cohen, S. (1995). *Mindblindness: Essay on autism and the theory of mind*. Cambridge, MA: MIT Press.

Baron-Cohen, S., Campbell, R., Karmiloff-Smith, A., & Grant, J. (1995). Are children with autism blind to the mentalistic significance of the eyes? *British Journal of Developmental Psychology*, *13*, 379–398.

Baron-Cohen, S., Leslie, A. M., & Frith, U. (1985). Does the autistic child have a "theory of mind"? *Cognition*, *21*, 37–46.

Baron-Cohen, S., & Ring, H. (1994). A model of the mindreading system: Neuropsychological and neurobiological perspectives. In C. Lewis & P. Mitchell (Eds.), *Children's early understanding of mind: Origins and development* (pp. 183-207). Hillsdale, NJ: Erlbaum.

Brothers, L. (1990). The neural basis of primate social communication. *Motivation & Emotion*, *14*, 81–91.

Byrne, R. W., & Whiten, A. (1988). *Machiavellian intelligence: Social expertise and the evolution of intellect in monkeys, apes, and humans*. Oxford: Oxford University Press.

Call, J., Agnetta, B., & Tomasello, M. (2000). Cues the chimpanzees do and do not use to find hidden objects. *Animal Cognition*, *3*, 23–34.

Call, J., Hare, B. A., & Tomasello, M. (1998). Chimpanzee gaze following in an object-choice task. *Animal Cognition*, *1*, 89–99.

Campbell, R. Heywood, C. A., Cowey, A., Regard, M., & Landis, T. (1990). Sensitivity to eye gaze in prosopagnosic patients and monkeys with superior temporal sulcus ablation. *Neuropsychologia*, *28*, 1123–1142.

Charman, T., Swettenham, J., Baron-Cohen, S., Cox, A., Baird, G., & Drew, A. (1997). Infants with autism: An investigation of empathy, pretend play, joint attention, and imitation. *Developmental Psychology*, *33*, 781–789.

Cheney, D. L., & Seyfarth, R. M. (1990). Attending to behaviour versus attending to knowledge: Examining monkeys' attribution of mental states. *Animal Behaviour*, *40*, 742–753.

Cosmides, C., & Tooby, J. (1994). Origins of domain specificity: The evolution of functional organization. In L. A. Hirschfeld & S. A. Gelman (Eds.), *Mapping the mind: Domain specificity in cognition and culture* (pp. 85–116). New York: Cambridge University Press.

Courchesne, E. (1997). Brainstem, cerebellar and limbic neuroanotomical abnormalities in autism. *Current Opinion in Neurobiology*, *7*, 269–278.

Fine, C., Lumsden, J., & Blair, J. R. (2001). Dissociation between "theory of mind" and executive functions in a patient with early left amygdala damage. *Brain, 124*, 287–298.

Flombaum, J. I., Kundey, S. M., Santos, L. R., & Scholl, B. J. (2004). Dynamic object individuation in rhesus macaques: A study of the tunnel effect. *Psychological Science, 15*, 795–800.

Flombaum, J. I., & Santos, L. R. (April, 2004). *What rhesus monkeys* (Macaca mulatta) *know about what others can and cannot see.* Paper presented at the 14th Biennial Meeting of the International Society on Infant Studies, Chicago.

Flombaum, J. I., & Santos, L. R. (2005). Rhesus monkeys attribute perceptions to others. *Current Biology, 15*, 447–452.

Gallagher, H. L., & Frith, C. D. (2003). Functional imaging of "theory of mind." *Trends in Cognitive Sciences, 7*, 77–83.

Hare, B. (2001). Can competitive paradigms increase the validity of experiments on primate social cognition? *Animal Cognition, 4*, 269–280.

Hare, B., Call, J., Agnetta, B., & Tomasello, M. (2000). Chimpanzees know what conspecifics do and do not see. *Animal Behaviour, 59*, 771–785.

Hare, B., Call, J., & Tomasello, M. (2001). Do chimpanzees know what conspecifics know? *Animal Behaviour, 61*, 139–151.

Hare, B., Call, J., & Tomasello, M. (2006, January 10). Chimpanzees deceive a human competitor by hiding. *Cognition.* [Epub ahead of print]

Hauser, M. D. (2001). Searching for food in the wild: A nonhuman primate's expectations about invisible displacement. *Developmental Science, 4*, 84–93.

Hauser, M. D., Carey, S., & Hauser, L. B. (2000). Spontaneous number representation in semi-free-ranging rhesus monkeys. *Proceedings of the Royal Society of London, B, Biological Sciences, 267*, 829–833.

Heyes, C. M. (1998). Theory of mind in nonhuman primates. *Behavioral and Brain Sciences, 21*, 101–134.

Humphrey, N. K. (1976). The social function of intellect. In P. B. G. Bateson & R. A. Hinde (Eds.), *Growing points in ethology* (pp. 303–317). Cambridge: Cambridge University Press.

Itakura, S. (1996). An exploratory study of gaze-monitoring in nonhuman primates. *Japanese Psychological Research, 38*, 174–180.

Itakura, S., Agnetta, B., Hare, B., & Tomasello, M. (1999). Chimpanzee use of human and conspecific social cues to locate hidden food. *Developmental Science, 2*, 448–456.

Itakura, S., & Anderson, J. R. (1996). Learning to use experimenter-given cues during an object-choice task by a capuchin monkey. *Cahiers de Psychologie Cognitive, 15*, 103–112.

Jellema, T., Baker, C. I., Wicker, B., & Perrett, D. I. (2000). Neural representation for the perception of the intentionality of actions. *Brain and Cognition, 44*, 280–302.

Jellema, T., Baker, C. I., Oram, M. W., & Perrett, D. I. (2002). Cell populations in the superior temporal sulcus of the macaque and imitation. In A. N. Meltzoff and W. Prinz (Eds.), *The imitative mind: Development, evolution, and brain bases* (pp. 267–290). Cambridge: Cambridge University Press.

Kawashima, R., Sugiura, M., Kato, T., Nakamura, A., Hatano, K., Ito, K., et al. (1999). The human amygdala plays an important role in gaze monitoring: A PET study. *Brain, 122*, 779–783.

Kling, A., Lancaster, J., & Benitone, J. (1970). Amygdalectomy in the free-ranging vervet. *Journal of Psychiatric Research, 7*, 191–199.

Leonard, C. M., Rolls, E. T., Wilson, F. A. W., & Baylis, G. C. (1985). Neurons in the amygdala of the monkey with responses selective for faces. *Behavioural Brain Research, 15*, 159–176.

Machado, C. J., & Bachevalier, J. (2003). Non-human primate models of childhood psychopathology: The promise and the limitations. *Journal of Child Psychology & Psychiatry & Allied Disciplines, 44*, 64–87.

Mitchell, R. W. (1999). Deception and concealment as strategic script violation in great apes and humans. In S. T. Parker & R. W. Mitchell (Eds.), *The mentalities of gorillas and orangutans: Comparative perspectives* (pp. 295–315). Cambridge: Cambridge University Press.

Peignot, P., & Anderson, J. R. (1999). Use of experimenter-given manual and facial cues by gorillas (*Gorilla gorilla*) in an object-choice task. *Journal of Comparative Psychology, 113*, 253–260.

Pelphrey, K. A., Singerman, J. D., Allison, T., & McCarthy, G. (2003). Brain activation evoked by perception of gaze shifts: The influence of context. *Neuropsychologia, 41*, 156–170.

Pelphrey, K. A., Viola, R. J., & McCarthy, G. (2004). When strangers pass: Processing of mutual and averted social gaze in the superior temporal sulcus. *Psychological Science, 15*, 598–603.

Perner, J., Frith, U., Leslie, A. M., & Leekham, S. (1989). Exploration of the autistic child's theory of mind: Knowledge, belief, and communication. *Child Development, 60*, 689–700.

Perrett, D. I., Smith, P. A. J., Potter, D. D., Mistlin, A. J., Head, A. S., Milner, A. D., & Jeeves, M. A. (1985). Visual cells in the temporal cortex sensitive to face view and gaze direction. *Proceedings of the Royal Society of London, B, Biological Sciences, 223*, 293–317.

Perrett, D. I., Hietanen, J. K., Oram, M. W., & Benson, P. J. (1992). Organization and functions of cells responsive to faces in the temporal cortex. *Philosophical Transactions of the Royal Society of London, B, 335*, 23–30.

Perrett, D. I., & Mistlin, A. J. (1990). Perception of facial characteristics by monkeys. In W. C. Stebbins & M. A. Berkley (Eds.), *Comparative perception, Vol. 2: Complex signals* (pp. 187–215). Oxford: John Wiley & Sons.

Phillips, W., & Santos, L. R. (2006). *How rhesus monkeys* (Macaca mulatta) *reason about the insides of objects.* Manuscript submitted for publication.

Povinelli, D. J. (2000). *Folk physics for apes.* Oxford: Oxford University Press.

Povinelli, D. J. (2004). Behind the ape's appearance: Escaping anthropomorhism in the study of other minds. *Daedalus: Journal of the American Academy of Arts and Sciences*, Winter, 29–41.

Povinelli, D. J., & Bering, J. M. (2002). The mentality of apes revisited. *Current Directions in Psychological Science*, *11*, 115–119.

Povinelli, D. J., Bierschwale, D. T., & Cech, C. G. (1999). Comprehension of seeing as a referential act in young children, but not juvenile chimpanzees. *British Journal of Developmental Psychology*, *17*, 37–60.

Povinelli, D. J., & Eddy, T. J. (1996a). What young chimpanzees know about seeing. *Monographs of the Society for Research in Child Development*, *61*, 1–152.

Povinelli, D. J., & Eddy, T. J. (1996b). Chimpanzees: Joint visual attention. *Psychological Science*, *7*, 129–135.

Povinelli, D. J., & Eddy, T. J. (1996c). Factors influencing young chimpanzees' (*Pan troglodytes*) recognition of attention. *Journal of Comparative Psychology*, *110*, 336–345.

Povinelli, D. J., & Eddy, T. J. (1997). Specificity of gaze-following in young chimpanzees. *British Journal of Developmental Psychology*, *15*, 213–222.

Povinelli, D. J., Parks, K. A., & Novak, M. A. (1991). Do rhesus monkeys (*Macaca mulatta*) attribute knowledge and ignorance to others? *Journal of Comparative Psychology*, *105*, 318–325.

Povinelli, D. J., & Vonk, J. (2003). Chimpanzee minds: Suspiciously human? *Trends in Cognitive Sciences*, *7*, 157–160.

Povinelli, D. J., & Vonk, J. (2004). We don't need a microscope to explore the chimpanzee mind. *Mind and Language*, *19*, 1–28.

Premack, D., & Woodruff, G. (1978). Does the chimpanzee have a theory of mind? *Behavioral and Brain Sciences*, *1*, 515–526.

Santos, L. R. (2004). "Core knowledges": A dissociation between spatiotemporal knowledge and contact mechanics in a non-human primate? *Developmental Science*, *7*, 167–174.

Santos, L. R., Sulkowski, G. M., Spaepen, G. M., & Hauser, M. D. (2002). Object individuation using property/kind information in rhesus macaques (*Macaca mulatta*). *Cognition*, *83*, 241–264.

Santos, L. R., Hauser, M. D., & Spelke, E. S. (2001). Representations of food kinds in the rhesus macaques (*Macaca mulatta*): An unexplored domain of knowledge. *Cognition*, *82*, 127–155.

Saxe, R., Carey, S., & Kanwisher, N. (2004). Understanding other minds: Linking developmental psychology and functional neuroimaging. *Annual Review of Psychology*, *55*, 87–124.

Schultz, R. T., Romanski, L. M., & Tsatsanis, K. D. (2000). Neurofunctional models of autistic disorder and Asperger syndrome: Clues from neuroimaging. In A. Klin & F. R. Volkmar (Eds.), *Asperger syndrome* (pp. 172–209). New York: Guilford Press.

Siegal, M., & Varley, R. (2002). Neural systems involved in theory of mind. *Nature Reviews: Neuroscience*, *3*, 462–471.

Stuss, D. T., Gallup, G. G., Jr., & Alexander, M. P. (2001). The frontal lobes are necessary for "theory of mind." *Brain*, *124*, 279–286.

Sulkowski, G. M., & Hauser, M. D. (2001). Can rhesus monkeys spontaneously subtract? *Cognition*, *79*, 239–262.

Tomasello, M., & Call, J. (1997). *Primate cognition.* Oxford: Oxford University Press.

Tomasello, M., Call, J., & Hare, B. (1998). Five primate species follow the visual gaze of conspecifics. *Animal Behaviour, 55,* 1063–1069.

Tomasello, M., Call, J., & Hare, B. (2003). Chimpanzees understand psychological states: The question is which ones and to what extent. *Trends in Cognitive Sciences, 7,* 153–156.

Tomasello, M., Hare, B., & Agnetta, B. (1999). Chimpanzees follow gaze geometrically. *Animal Behaviour, 58,* 769–777.

Tomasello, M., Hare, B., & Fogleman, T. (2001). The ontogeny of gaze following in chimpanzees and rhesus macaques. *Animal Behaviour, 61,* 335–343.

Whiten, A., & Byrne, R. W. (1997). *Machiavellian intelligence II: Extensions and evaluations.* Cambridge: Cambridge University Press.

Young, A. W., Aggleton, J. P., Hellawell, D. J., Johnson, M., Broks, P., & Hanley, J. R. (1995). Face processing impairments after amygdalotomy. *Brain, 118,* 15–24.

16 Social Cognition and the Evolution of Self-Awareness

Farah Focquaert and Steven M. Platek

The hypothesis that human self-awareness[1] is adaptive because of the benefits it brings for understanding others' behavior, in terms of both cooperation and competition, is well-known (Gallup, 1982; Humphrey, 1980, 1982, 1986; Platek, Keenan, Gallup, & Mohamed, 2004). Understanding our own behavior allows us to relate to and understand the behavior of others, which opens up possibilities for more sophisticated social behavior, both in competitive and in cooperative terms. This is what Nicholas Humphrey pointed to when he wrote that the possession of an "inner eye" allowed our ancestors to "raise social life to a new level," and likely ultimately allowed for the evolution of human culture (Humphrey, 1986, p. 11). The origin of human self-awareness reflects our ancestors' need *to understand*, *respond to*, and *manipulate* each other's behavior. Hunter-gatherer societies were based on "an unprecedented degree of interdependency, reciprocity and trust," which also led to "unprecedented opportunities for an individual to manoeuvre and outmanoeuvre others in the group." Leaving the forest behind, our hunter-gatherer ancestors found themselves in a "community of familiar souls" that made more complex cooperative and competitive behavior very rewarding strategies to pursue (Humphrey, 1982, p. 476 passim). Possessing an explanatory framework for one's own and others' behavior that allows the prediction of one's own and others' behavior within the social group provides a possible adaptive solution to more demanding social life. Gallup (1997) similarly stresses the importance of both competition and cooperation for the evolution of human self-awareness.

The hypothesis that human self-awareness evolved because of the benefits it brings in respect to understanding others encompasses two

1 Self-awareness is considered in this chapter to be introspective awareness, or knowing/understanding one's own mental states.

distinct topics that are further explored in this chapter. First, it implies that self-awareness is related to other-awareness to the extent that self-awareness actually evolved because of the benefits to the individual from understanding others. Second, it implies that the presence of both competition and cooperation was pivotal to the evolution of human self-awareness. Besides being beneficial for understanding others, self-awareness probably also evolved because of the benefits that accompany the ability to understand one's own behavior, emotions, thoughts, intentions, and the like. Humphrey (1982) alluded to the importance of consciousness, specifically self-consciousness, to one's own behavior. It provides us with a unique explanatory model of our own behavior. Split-brain research has led to the discovery of a "left-hemipshere interpreter" that somehow brings together all the different pieces—self-processing encompasses a variety of brain systems distributed across both the right and left hemispheres—and seems to provide an explanatory framework for our own behavior (Platek et al., 2006; Roser & Gazzaniga, in press). Understanding one's own behavior makes sense, especially with reference to pursuing meaningful, goal-directed behavior. Our capacity for mental time travel (Suddendorf & Corballis, 1997) allows us to compare current situations with (distant) past and future ones. Having an "I" to project into the future allows us to plan our behavior with reference to our wants, desires, intentions, beliefs, and emotions, which in turn confers on us a motivation to pursue very diverse, future-directed, long-term goals. By comparing past experiences with future ones, we are able to choose the best future strategy to pursue while not making the same mistakes over and over again. Mental trial and error is much less costly than actual trial and error, eliminating at least some future mistakes that could potentially remove us from the mating environment. The ability to foresee how others will react to one's own behavior (i.e., mental simulation) and to make this part of the decision-making process reduces the likelihood of failure even more. In the end, human goal-directed behavior, especially cooperative and competitive, future-directed behavior, becomes a flexible and diverse enterprise.

Possibly, self-awareness and self-related emotions (Lewis, 1992, 1995, 2000, 2003) are required to formulate and achieve future-directed long-term goals, both for the individual and for society.

This chapter considers the following questions: Which selection pressures allowed human self-awareness to evolve? How are these ultimate explanations related to proximate explanations of human self-awareness? We show that current theories on the possible adaptive benefits of self-awareness (self- and other-related processing) are tightly linked at the proximate level, further validating current ultimate explanations. The simulation theory of mindreading and mirror neuron research in monkeys and humans shows that self-understanding and other-understanding are implemented, at least in part, by similar brain substrates and possibly information-processing architecture (Macrae, Heatherton, & Kelley, 2004; Platek, Keenan, et al., 2004; Platek et al., 2006). Proximate explanations of self-awareness teach us that the neurological mechanisms that underlie self-awareness at various levels are relevant both for understanding and guiding our own behavior and for understanding and taking into account others' behavior. Across different species, such as monkeys, nonhuman great apes, and humans, different levels of self-awareness (pre-mirror self-recognition [pre-MSR], MSR, reflective or introspective awareness) seem to go hand in hand with different ontogenetic levels of mental state attribution (theory of mind [ToM], empathy). Combining both ultimate and proximate explanations of psychological traits allows us to deepen our understanding of these traits and can guide us in resolving outstanding questions. This is the goal of the adaptationist program Cosmides and Tooby (2000b) have set forth, and captures the spirit of the recently developed field of evolutionary cognitive neuroscience developed in this book.

Other-Understanding

Are Self-Awareness and Mental State Attribution Related Traits?

Decades ago, Humphrey (1980) proposed that we understand others' behavior by virtue of understanding our own behavior. Having an introspectionist's explanatory model of our own behavior allowed the modeling of others' behavior by analogy to our own. Gallup (1982) claimed that the emergence of self-awareness may be equivalent to the emergence of mind, mindreading being a byproduct of self-awareness. Gallup hypothesized that only those species capable of self-recognition would be able to engage in a number of introspectively based social strategies, such as empathy, sympathy, and intentional deception–mental state attribution (MSA). In accordance with Humphrey's theory, Gallup's model suggests that in order to infer the mental experiences of others, individuals must have a sense of their own experiences (i.e., species that are self-aware could use their experiences to model or infer the mental states of others). Both theories place a premium on simulation or mental modeling when conferring the benefits of self-awareness.

Is there any evidence that suggests that human self-awareness is important for MSA (ToM, empathy, other), which would imply that as self-awareness skills increase, similar effects should be attributable to MSA in general? And if so, does simulation play an important role in this process?[2]

Self-Awareness and Autism Spectrum Conditions: Asperger Syndrome

Theory of Mind "Why is that person behaving that way?" For most of us, questions like these are relatively easy to answer. In general, human social behavior is immediately and instantaneously grasped. For others, especially individuals with autism spectrum conditions (autism, Asperger syndrome [AS], high-functioning autism), a different picture emerges. These individuals are said to have impaired or absent MSA (ToM empathy) skills, and have profound difficulties in navigating human social life. At the same time, their self-awareness and perspective-taking abilities also seem to be impaired. Their ToM deficit is often explained in representational terms, defending the idea that certain metarepresentational abilities (Leslie, 1987; Perner, 1991) responsible for representing mental states in normal adults are lacking or impaired in these individuals. Frith and Happé (1999) have pointed to the role Leslie's account of metarepresentations might play. Leslie's view[3] on metarepresentation basically refers to a kind of "attitude marking" of mental states, allowing us to define mental states as involving wants, desires, intentions, beliefs, and so on.[4] This could be described as reflecting a person's introspective skills. I believe both Perner's account of metarepresentation and Leslie's account are crucial for human mindreading abilities. Before going into more detail, I will briefly review Perner's account.

According to Perner, the key to understanding the mind lies in the child's general development of mental representations. The child

---

2 Possibly allowing for both theory theory-mediated and simulation theory-mediated processes in third-person mindreading. See Carruthers and Smith (1996) for a review.

3 According to Goldman and Sripada (2005), Leslie (1994) defends a "theory theory" view of MSA, which is not in line with our proposal. We believe that the process of attitude marking Leslie describes is induced by simulation.

4 Why this requires metarepresentation is not entirely clear. See Jarrold et al. (1994).

progresses from having an innate sensitivity to behavioral expressions of mental states to having a "situation theory of behavior," which relates mental states to situations, and finally to acquiring a "representational theory of mind," which relates mental states to internal representations (Perner, 1991, p. 283). This development is accompanied by a progression from primary presentations (a single model of the world) to secondary representations (multiple models of the world) to, finally, metarepresentational abilities (modeling models).

> This ability [metarepresentation] is necessary if a child is to compare a model of another's mental model with their own knowledge of the world (as is required in a false belief task for example). (Jarrold, Carruthers, Smith, & Boucher, 1994, p. 453)

Metarepresentational abilities allow the child to relate mental states to internal representations of situations, which in turn allows both for misrepresentation and for a more thorough understanding of social behavior. Metarepresentation, or understanding mental states *as* mental states, entails knowing that mental states represent situations in a certain way, and do not necessarily represent reality as such. In the famous Sally/Anne task, Sally looks for the marble where she left it, behaving not in relation to the situation as such (the marble is moved to the box by Anne) but in relation to her internal representation of the situation, which locates the marble not in the box but in the basket, where she left it. Perner's view can be described as attributing a higher-level ability for perspective taking.[5] For example, although a child without metarepresentational skills knows what another child can and cannot see (= reality), she does not know how another child looking at the same object from a different angle sees that object (this would amount to a representation of a representational relationship, or modeling models).[6] Because children with autism fail standard ToM tasks like the Sally/Anne task, it seems that they are, probably in a similar way, unable to "place themselves in someone else's shoes" and infer what that other person is thinking. But how do we determine that their ToM difficulties are not solely attributable to a metarepresentational deficit in Perner's sense, or

---

5 This idea was brought forward by Robert Mitchell, personal communication, 2004.

6 It appears that chimpanzees are also able to perform this kind of first-level perspective taking and would probably fail at second-level perspective-taking tasks requiring metarepresentation in Perner's sense (Hare et al., 2000, 2001, 2003; Tomasello & Call, 1997).

what I have referred to as the absence of more elaborate perspective-taking skills?

Individuals with AS and high-functioning individuals with autism show us that metarepresentation in Perner's sense is not sufficient to explain their MSA or "mindreading" difficulties. Whereas individuals with AS often succeed at standard ToM tasks, which at least endows them with the necessary perspective-taking skills required for third-person mindreading, in comparison with normal adults they are still impaired in their general MSA skills. I follow Frith and Happé (1999) in claiming that their impairment is probably due, at least in part, to a metarepresentational deficit in Leslie's sense. Hulbert, Happé, and Frith (1994) have tested the introspective skills of three individuals with AS and have shown their inner world to be very much unlike that of normal adults. Reports obtained from normal participants would typically include descriptions of verbal inner experience, visual images, feelings (located in the body), and unsymbolized thinking (thoughts without words or pictures associated with them), singly or in combination. The individuals with AS, however, described visual inner experience exclusively; no verbal or unsymbolized thinking was reported. Also—and this is very important—their introspective skills were tightly linked to their ToM abilities. The greater the difficulties noted in self-awareness, the greater the problems with ToM tasks. According to Happé (2003), it is as if individuals with autism do not understand their own thoughts and emotions *as* thoughts and emotions. Temple Grandin (1995), a famous veterinarian with AS, describes her inner world as a form of "thinking in pictures" (also the title of one of her books). She describes herself as understanding only simple emotions, such as fear, anger, happiness, and sadness. She does not know what complex emotions in human relationships are; they are beyond her comprehension. Her emotions are more like those of a child, she says, not like those of adults. According to Oliver Sacks (Grandin, 1995), she does not exhibit certain more complex emotions such as shame and embarrassment. These are social self-conscious emotions (Lewis, 2000, 2003) that require a sense of self/other that presumably is lacking or impaired in individuals with autism spectrum conditions (ASCs).

Self-conscious emotions such as exposure embarrassment first emerge between ages $2^{1}/_{2}$ and 3 years, after the emergence of self-recognition. These emotions come about through self-reflection (Lewis, 2003). Evaluative embarrassment, pride, shame, and guilt are described as later-emerging self-conscious evaluative emotions (Sullivan, Bennett,

& Lewis, 2003). Impaired introspection might prevent individuals with AS from acknowledging and attributing certain self-conscious emotions, especially the later-emerging evaluative emotions. Also, the pleasure Oliver Sacks gets from watching an astonishing nature scene is something Temple Grandin does not know about. She is denied those kinds of feelings, she says. Although she knows Sacks must be experiencing an intense feeling of beauty, she does not know what this must *really* be like. Not having had similar experiences herself, how could she fully grasp what Sacks is experiencing? It seems that the route to simulation is somehow blocked by having a totally different inner world than normal individuals, one that does not draw on metarepresentational skills (in Leslie's sense) to the same extent as it does in normal individuals. Whereas individuals with AS often pass standard ToM tasks, they do not seem to possess the immediate, spontaneous grasp of other people's behavior found in normal individuals. It seems that not only is perspective taking crucial, having an understanding of one's own experiences, thoughts, and emotions is vital to developing a humanlike ToM system. So, self-awareness allows you to be good at understanding others.[7] And as the research of Hulbert et al. (1994) implies, the better you are at understanding your own behavior, the better you are at understanding the behavior of others. Looking at the various difficulties individuals with ASCs are likely to encounter during social interaction, and taking into consideration how their introspective and ToM skills go hand in hand, it appears that human self-awareness is extremely beneficial for navigating a socially complex world such as our own. Human self-awareness opens up new opportunities for cooperation, deception, and the like, all of which have proven to be crucial for survival in complex social groups (Keenan, Gallup, & Falk, 2003). The important lesson to draw here is that both introspective skills and perspective-taking skills are required to place yourself in someone else's shoes and be able to infer the mental states of others (from standing in those shoes). Whereas individuals with AS probably have the basic perspective-taking skills, they are unable to spontaneously and adequately describe the inner world of others. Their introspective abilities do not give a way to understand the

7 Basically, our view is that (1) metarepresentation in Perner's sense allows us to take someone else's viewpoint, (2) simulation is crucial for determining the specific content of others' mental states, and (3) "metarepresentations" in Leslie's sense (knowledge of one's own mental states) are used as input when trying to do so. Of course, in probably many cases, this process can be shortened by resorting to some kind of rules of behavior that have been interiorized along the way.

mental worlds of others. Their mental state attribution is more mechanical (D. Legoff, personal communication, 2005). The better their introspective skills, however, the better they are at understanding others (Frith & Happé, 1999). Moreover, Johnson et al. (2004, 2005) found a significant correlation between levels of self-awareness, as assessed by the private part of the Fenigstein, Scheier, & Buss (1975) Self-Consciousness Scale, and levels of deception detection in normal females.

Empathy Human empathy, sometimes called cognitive empathy, is tightly linked to human ToM skills. Baron-Cohen and Wheelwright (2004, p. 168) define empathy as "the drive or ability to attribute mental states to another person/animal, [one that] entails an appropriate affective response in the observer to the other person's mental state," thus construing both an affective and a cognitive component. Again, it is important to recall that ASCs have been described as empathy disorders. Individuals with AS apparently have a specific cognitive style that combines a high level of systemizing skills with a low level of empathizing skills. According to Baron-Cohen (2002), ASCs constitute an extreme version of a pattern of cognitive difference that extends across the entire population. Also, Platek et al. (2006) have found that adolescents with AS show deficits on both self-recognition tasks and the "Reading the Mind in the Eyes" test developed by Baron-Cohen et al. (2001).

In accordance with Baron-Cohen's view that human empathy entails both a cognitive and an affective component, recent work by Völlm et al. (2006) has shown that ToM and empathy draw on common, distinct areas of brain activation. Shared areas include the medial prefrontal cortex, temporoparietal junction, and temporal poles. (These areas refer to the cognitive component.) The empathy condition, in comparison to the ToM condition, revealed enhanced activation of the paracingulate, anterior and posterior cingulate, and amygdala. From these results, the recruitment of networks involved in emotional processing appears to be crucial for empathic processing. It is important to note that amygdala (and other limbic) dysfunction is associated with ASCs (Baron-Cohen, 2004). Whether the impaired MSA skills found in individuals with ASCs result from an affective disorder (which might result in a cognitive disorder) or from separable affective and cognitive disorders, both contributing to the problem, remains to be determined. A recent study by Dalton et al. (2005) seems to indicate that "over-arousal" of the amygdala might lie at the root of face processing impairments in individuals with ASCs and as such either causes or contributes to further dysfunctional processing at more cognitive levels. This means that the difficulties these individuals experience when processing emotional faces might not be caused by impairments in the face processing areas (fusiform gyrus, STS) per se, as was previously assumed, but instead may be fueled by abnormal amygdala activation. The idea is that emotional face stimuli cause this overarousal by being too salient, and so motivate these individuals to look away, preventing processing altogether. Being able to determine which brain networks contribute to the impairment of MSA skills in these individuals would teach us a lot about how MSA comes about in normal individuals.

Comparative Data

Mirror Self-Recognition and Theory of Mind When observations in the great apes are taken into consideration, the data seem to add up. All the great ape species have a basic ability for MSR, although gorillas have been considered an exception to the rule (Barth, Povinelli, & Kant, in press; Gallup, 1994, 1997; Povinelli, 1994; Povinelli & Cant, 1995), implicating the existence of a basic level of self-awareness or self-concept (Gallup, 1982). According to current theorizing, all anthropoid apes (i.e., chimpanzees, bonobos, orangutans, and gorillas) are capable of passing the mark test, and thus show evidence of MSR. However, in comparison with humans, not every individual ape does (Anderson, 1994; de Waal, Dindo, Freeman, & Hall, 2005). About 43% of chimpanzees, 50% of orangutans, and 31% of gorillas involved in MSR experiments showed signs of MSR.

According to Gallup (1982), MSR reflects a sense of self that equally enables "mind." In Gallup's view, self-awareness subsumes consciousness and mind. *Mind* in his definition is the ability to monitor one's own mental states, in the sense of being able to differentiate among feelings of hunger, anger, fear, and the like. Evidence of mind, according to Gallup, is evidence of introspection. Organisms that lack the capacity to become the object of their own attention should be unable to introspect. In turn, the ability of an organism to impute mental states to others is related to whether or not the organism can introspect, because it presupposes the capacity to monitor such states in the organism itself. According to Gallup's theory, animals that fail to show MSR should also fail to show any signs of sympathy, empathy, intentional deception, sorrow, and the like.

Gallup (1982) concluded that whereas monkeys appear to be mindless, showing no signs of MSR, chimpanzees, on the other hand, entered a cognitive domain that set them apart from most other primates, showing behaviors that are consistent with a capacity to attribute mental states to others based on introspection. It is indeed true that the great apes, which show signs of MSR, also exhibit some forms of mental state attribution (Hare, Call, Agnetta, & Tomasello, 2000; Hare, Call, & Tomasello, 2001). However, whether this expression is based on introspective abilities remains to be determined. Human introspective awareness still appears to be one step up, endowing humans with a unique sense of self-awareness, which constitutes the putative human condition. Moreover, some form of MSA has recently been shown in monkeys (Flombaum & Santos, 2005; see also Santos et al., Chapter 15, this volume), and new discoveries have come up with respect to MSR and MSA in monkeys. The picture is no longer as clear-cut as was previously thought.

In the following discussion, we look at the existence and entanglement of MSR and MSA in nonhuman great apes (chimpanzees, orangutans) and monkeys (capuchins, rhesus macaques). We defend the idea that MSR might indeed be a precursor to mind or a marker of mind (de Waal et al., 2005; Gallup, 1982; Platek et al., 2006) and that this entanglement displays a more or less gradual process from no signs of MSR and MSA to humanlike introspective awareness and MSA skills. We propose that there is a gradual connection between MSR and MSA within and across primate species. In the case of humans, it appears that this connection is a direct connection in which the level of self-awareness (i.e., introspective awareness) goes hand in hand with the level of MSA skills. Whether or not this is the case in any nonhuman primate species remains to be determined. Most likely, however, some form of self-awareness (basic ability to distinguish self from other) is required for ToM-like or empathy-like behaviors. In addition, we discuss the presence or absence of competitive and cooperative social life and behavior as related to different levels of self-awareness. In reference to the importance of competitive social settings for MSA to be observable in chimpanzee societies, we argue that the explosion of cooperative behavior strategies during the human environment of evolutionary adaptedness (EEA) allowed for the next step and led to a kind of introspective awareness unique to humans. According to recent studies by Reno, Meindl, McCollum, and Lovejoy (2003, 2005), the growth pattern in early hominids (*A. afarensis*), when compared with the growth pattern of modern humans, chimpanzees, and gorillas, is most similar to that of modern humans. This specific growth pattern in early hominids implies that their social system and ecology must have been different from that of chimpanzees (Larsen, 2003; Reno et al., 2003, 2005). Although chimpanzees are more cooperative than baboons and orangutans, for example, early hominids' social life was still very different:

> Therefore, the social system and ecology of human ancestors, who evolved a characteristic growth pattern, must have been different from that of chimpanzees, perhaps because of their new habitat of open land. (Hamada & Udono, 2002, p. 283)

As Humphrey mentioned, our hunter-gatherer ancestors' new habitat of open land might have led to new opportunities in terms of cooperative and competitive social strategies, and we hypothesize this could have increased the fitness benefits associated with human self-awareness. What is important here is that "unlike any other species, humans cooperate with non-kin in large groups" (Boyd, Gintis, Bowles, & Richerson, 2003, p. 1), and "apes cannot, it seems, engage in future-directed cooperation" (Brinck & Gärdenfors, 2003, p. 486), whereas humans do. In fact, this is exactly what allows for more flexible cooperative behavior (Brinck & Gärdenfors, 2003). We argue that human-like self-awareness (which encompasses episodic memory and the ability to travel far back or forward in mental time) allows for a more sophisticated form of MSA, which in turn allows for more complex and flexible social behavior (in terms of competition *and* cooperation).

Chimpanzees (*Pan troglodytes*) can be said to possess a basic ability for MSR. In agreement with the idea that some form of self-awareness is required for other-understanding, chimpanzees show a wide range of what can be considered intelligent social behavior, such as tactical deception (Whiten & Byrne, 1988) and certain other skills related to understanding seeing and knowing states. They seem to understand what their conspecifics can and cannot see[8] (Hare et al., 2000, 2001), and they also seem to understand intentions in others (Call, Hare, Carpenter, & Tomasello, 2004). There is evidence that the great apes recognize certain mental states in others, and show appreciation of both attention and intention in others (Suddendorf & Whiten, 2001), although there are no signs that they understand beliefs in others. The great apes probably understand some psychological states, such as knowing and intending,

---

8 For an alternative view, see Karin-D'Arcy and Povinelli (2002).

but they may fail other tasks that require an understanding of (1) visual aspectuality, (2) alternative interpretations in conspecifics, and (3) misrepresentations or false beliefs (Hare et al., 2001). It is clear that they do not possess a full-blown humanlike ToM (Tomasello, Call, & Hare 2003). When it comes to setting aside their own vantage point and imagining how things look from a conspecific's perspective, they would probably fail. Following the third-way hypothesis of Hare et al. (2001), they succeed at Level 1 perspective-taking tasks but probably not at Level 2 perspective-taking tasks, the latter being the same as possessing metarepresentational skills in Perner's (1991) sense. The great apes must have some understanding of intentions and knowing states, but they somehow lack the ability to place themselves in others' shoes and analyze behavior accordingly. They might have what Perner (1991) calls an indirect grasp of mental states, relating them to situations as such and not to internal representations of situations. This same line of reasoning can be found in Gómez's (2004) work on nonhuman ToM skills. Gómez considers the possibility that nonhuman primates can represent overt mental states such as attention and intention (which presumably come from motor cognitive knowledge), whereas covert mental states such as knowing and believing remain out of reach. They might have some understanding of knowing states (Hare et al., 2001), but not in the same flexible way as humans do. Knowing states in nonhuman primates probably relate to their ability to understand the permanence of objects in the world:

> At the very least, nonhuman primates must be able to do something similar with overt mental states like attention and intentions—to conserve them through superficial transformations, that is, to understand that if a chimpanzee tried to reach a banana a minute ago, he might still have this intention (perhaps inside him), even if it is not visible now. (Goméz, 2004, p. 234)

According to Gómez, this kind of understanding of overt mental states, as in understanding intention, attention, and possibly some forms of knowledge, is also present in humans. Humans probably have evolved new ways of understanding behavior, as related to covert mental states, but these didn't necessarily replace more basic mindreading skills in terms of overt mental states. So, humans, chimpanzees, bonobos, orangutans, and possibly gorillas all share this similar ability, whereas only humans evolved a complete ToM system in terms of covert mental states. Possibly, MSA in terms of overt mental states was evolution's solution to competitive social situations, whereas MSA in terms of covert mental states is highly beneficial when confronted with a social EEA that is highly dominated by both competition and cooperation. As Hare et al. (2001) and Hare and Tomasello (2004) mention, competition is a much more natural mode of interaction for chimpanzees than cooperation and communication. Hare et al. (2001) discuss the possibility that ToM skills in chimpanzees evolved as specific adaptations for competition (Gómez, 2004), whereas human MSA skills probably evolved, as Humphrey and Gallup mention, because of the recurrent need for competition *and* cooperation during human evolutionary history. Gómez (1998) conducted experiments with a female orangutan, Dona, which clearly displayed the importance of competitive settings for MSA skills in nonhuman great apes. The noncompetitive/cooperative framework involved the following situation: Dona would sit in her cage, in front of which were two boxes locked with padlocks. The keys to the padlocks were kept in a different container. A "baiter" would enter the room, take the keys, open one of the padlocks, and place food in the box. A few seconds later the "giver" would enter the room and *ask* Dona where the food was (or wait for her to make a request). When Dona pointed to one box, the giver would collect the keys, open the padlock, give the food to Dona, and put the keys back.

This scenario was repeated several times. In the experimental trial, the baiter, after baiting the box, would hide the keys in a hiding place in the room and leave. If Dona understood the mental state of "ignorance," she would have been expected to point not only to the food but also to the hiding place of the keys when the giver entered the room. The control condition involved hiding the keys in the presence of the giver, or the giver himself hiding the keys, which would not be expected to evoke the same response of pointing both to the food and to the hiding place. Dona failed to perform as expected on six experimental trials and thus showed no sign of MSA at all. However, when a competitive element was introduced into the experimental setup, Dona successfully passed the test, thus showing signs of MSA. When the keys were hidden by a "stranger" (unfamiliar to Dona) who entered the room after the giver had left, she correctly pointed at both the food and the hiding place of the key in all seven experimental trials. It is clear from this and other research that competitive settings play a crucial role when it comes to MSA in nonhuman great apes.

Whereas understanding overt mental states might be enough for survival in competitive social problem situations, the mix of competition and cooperation in human evolution might have favored the evolution of more elaborate mindreading skills. As Humphrey mentioned, covert mental states allowed for more sophisticated behavior in terms of both competition and cooperation. Surely, both competition and cooperation are extremely benefited by the ability to understand others' behavior in terms of mental states, as in desires, emotions, beliefs, and the like. We need only consider modern humans, especially those with AS. Autobiographies like those of Temple Grandin give us an idea of how difficult human social life becomes when an individual is unable to relate to the thoughts and feelings of others. Understanding someone else's behavior in much the same way as you understand your own appears extremely important for human social life. Individuals with AS often became isolated because they have a great deal of difficulty building genuine, long-lasting, meaningful social relationships. For the most part, friends are work-related, and individuals with AS often stay single or marry another individual with AS. Often these individuals have problems finding a work environment that allows them to be themselves and does not stigmatise them for their odd social conduct. Especially the more subtle cues of human social life remain a mystery. Frith and Happé (1999) mention that individuals with AS often mistake jokes for lies and have difficulty separating sarcasm from outright deception. Also, Temple Grandin (1995) says that her understanding of deception involves the calculating of situational cues, as she is unable to grasp someone else's deceptive intentions in a more direct manner, for instance, by reading facial expressions. It might be the case that human social behavior as we know it could only have flourished because of our ancestors' ability to, through mental simulation, really relate to one another, being able to resort to their first-person experience in trying to decipher others' mental states.

Gallup, Anderson, and Shillito (2002) claim that monkeys, which fail to show any signs of MSR, do not show any signs of responding to mental states at all. In an attempt to show MSR in monkeys, Gallup (1997) gave a crab-eating macaque 5 months (more than 2,400 hours) of mirror exposure. After 5 months the monkey still did not show any signs of self-recognition and apparently continued to respond to her mirror image as if it involved a conspecific. Suarez and Gallup (1986) gave a pair of rhesus monkeys continuous mirror exposure for 17 years, and failed to find evidence for MSR in these animals. Recent work by Paukner, Anderson, and Fujita (2004) has again confirmed that no signs of MSR can be found in monkeys. However, Flombaum and Santos (2005) did find evidence of some forms of MSA in monkeys—comparable to chimpanzee MSA—and de Waal et al. (2005) have shown that the statement that monkeys treat their mirror image merely as a conspecific is questionable. These studies are in line with Gallup's original hypothesis that MSR is a marker of mind, but they might require us to broaden our perspective and incorporate monkeys (or at least capuchins) within this framework. Possibly, capuchin monkeys represent an intermediate step between species that show no signs of MSR and MSA and nonhuman great apes that do show some MSR and MSA skills (although not in the same way humans do). More and more data provide evidence in favor of a link between the level of self-awareness and the level of MSA within and across primate species.

It is commonly believed that monkeys responded to their mirror image as if it were a conspecific, displaying social behavior patterns commonly observed when confronted with an actual conspecific. However, de Waal et al. (2005) looked more closely at the precise manner in which capuchin monkeys (large-brained New World primates) respond to their mirror image by comparing their behavior patterns in front of the mirror with their behavior patterns when confronted with a conspecific. Male and female monkeys (although it was more pronounced in the female monkeys) reacted differently on both occasions, indicating that it is more likely that these monkeys do not treat their reflection as a mere conspecific but more as "something special." From the very onset, it seems that they do not mistake their mirror image for an actual conspecific. Whereas it did not seem to the observers that the capuchin monkeys actually recognized themselves in the mirror (in comparison to an encounter with an actual conspecific), females did treat their reflection much more positively, engaging in lots of eye contact and lip-smacking. It almost seemed as if they were flirting with their reflection, appearing to be much less anxious than in the presence of a female stranger. When confronted with a stranger, they avoided eye contact and kept their offspring closer to their body. Male capuchin monkeys, although manifesting different behavior patterns, reacted more ambiguously, displaying both friendly reactions and signs of distress, such as squealing and curling up, which they did not show in front of strangers. Strangers were threatened much more, though, bringing the balance between negative and positive reactions toward negative when confronted with a stranger and toward positive when looking in the mirror. "Capuchin monkeys notice immediately that their mirror image is not a regular stranger, or perhaps no stranger at all," de Waal et al. said. We have to emphasize, however, that monkeys clearly do not reach the same level of understanding as humans and apes. Possibly, their self-mirror-image understanding is comparable to pre-MSR human infants, who show exploratory behavior in front of mirrors and apparently understand the impossibility of reciprocal interaction with the mirror (de Waal et al., 2005). However it is also defendable that the capuchin monkeys' reactions to the mirror image were fueled by the monkeys' confusion about "the other monkey" acting strangely. The reaction of a capuchin monkey to its mirror image obviously differs from, say, that of a bird or a mouse looking in a mirror. Whether this is to be explained by either the presence of some form of pre-MSR or the possession of certain social processing skills that allow the capuchin monkey to detect the "odd behavior" that is being displayed remains to be determined. One potential interpretation is that the capuchins recognized visual cues of kinship; that is, they saw the "other" monkey as a kin member, and thus acted positively toward the conspecific. A direct comparison between strangers, kin, and mirror self-reflections has not been undertaken, but misperception of a mirror self-reflection as potential kin would be in line with the recent hypothesis of Platek et al. (2005, 2006; see also Chapter 9, this volume) that kin recognition activates a positive valence approach mechanism in the brain.

A similar picture seems to emerge concerning their ToM skills. Whereas chimpanzees have been shown to have some ToM skills, such as understanding seeing versus nonseeing, previous research on monkeys did not show similar results (Anderson, Montant, & Schmitt, 1996; Hare, Addessi, Call, Tomasello, & Visalberghi, 2003). Although rhesus macaques spontaneously follow the gaze of others, they seem unable to use this information to reason about what others see and know (Anderson et al., 1996). However, recent work by Flombaum and Santos (2005) showed that rhesus macaques do possess some ToM skills related to seeing versus nonseeing. The experimental setup involved an ecologically valid competitive framework: stealing grapes. During six experimental trials, the monkeys were given the opportunity to take grapes from one of two human experimenters ("competitors"). They had a choice of stealing grapes from, on the one hand, (1) an experimenter whose gaze was turned away or who was otherwise prevented from seeing the grapes, or, on the other hand, (2) an experimenter whose gaze was directed toward the grapes and who would be able to detect the stealing. (This series of experiments is described in Chapter 15.) The idea was that the monkeys would be motivated to steal only when they could do so without being detected. It was expected that they would show a preference for the first scenario, which would establish that macaques do more than just follow others' gaze. It would mean that they correctly infer information from eye gaze and adjust their behavior accordingly, which amounts to possessing at least one aspect of humanlike ToM skills. During the six different trials, the monkeys selectively retrieved grapes from the experimenter who was unable to see the grape rather than from the experimenter who was visually aware of the scene. For example, the monkeys were able to selectively retrieve grapes from an experimenter whose eyes alone were oriented 45° away. Unlike in noncompetitive paradigms (Anderson et al., 1996), the macaques were able to adjust their behavior based on the proper cue: eye gaze. Because the task involved representing the gaze of both experimenters (which one could see the grape and which one could not) and making a choice on the basis of this knowledge (to approach the experimenter who could not see the grape), it shows that rhesus monkeys not only follow another's gaze but also seem to "spontaneously reason about another individual's visual perception" (Flombaum & Santos, 2005). Applying the competitive paradigm hypothesis to MSA skills in monkeys not only provided additional evidence in favor of this hypothesis but also is in line with a "gradualist connection of self-awareness and MSA" in monkeys, nonhuman great apes, and humans, endowing monkeys with a minimum of ToM skills also found in chimpanzees. Although monkeys are said to lack MSR, something special seems to be going on in the case of capuchin monkeys (de Waal et al., 2005). Future research should investigate the hierchical nature of representing others and one's own mental states as recursively evolving mechanisms, that is, a coevolutionary process that might occur differently in different species.

Empathy Because empathy in humans is part of our broader MSA skills, a similar linkage between self-awareness and empathy should be found in monkeys and nonhuman great apes if we wish to defend a gradualist connection between self-awareness and MSA. De Waal (1996) previously argued that humans, who possess MSR, show more complex manifestations of empathy than monkeys, which do not. Expressions of emotional connectedness in monkeys most likely do not go beyond instances of emotional contagion, also observed in animals such as rats and pigeons. Emotional contagion refers to the emotional, physiological linkage between animals and can be described as the "spreading of all forms of emotion from one individual to another" (Preston & de Waal, 2002, p. 286). Emotional contagion is a self-focused process that does not involve a distinction between self- and other-generated emotions (it appears to occur automatically, without conscious reflection) as is the case in humanlike or cognitive empathy. Aside from emotional contagion, humans additionally experience more cognitive forms of empathy that involve an awareness of others' emotional state as the cause of one's own mental state. Preston and de Waal (2002, p. 287) regard empathy as a "general class of behavior that exists across species, to different degrees of complexity." They claim it is beneficial for any kind of cooperative, social species to be emotionally affected by the distress of their conspecifics. It is especially adaptive when one considers the role it plays in group living, the mother-infant bond, and interactions between conspecifics in general (Preston & de Waal, 2002).

The important findings for our current hypothesis of a gradualist connection of self-awareness and MSA within and across primate species is that capuchin monkeys, which show signs of pre-MSR and ToM, equally appear to be a special case with regard to empathic-like behaviors. Although capuchin monkeys, like other monkeys, do not initiate consoling behaviors, they do provide reassurance to distressed conspecifics seeking contact (de Waal et al., 2005). Unlike the postconflict behaviors observed in macaque monkeys, uninvolved third-party capuchin monkeys allowed the proximity and contact of capuchin monkeys that were the recipients of aggression during non-food-related fights, and often reciprocated this behavior through grooming, play, or the exchange of friendly signals. A similar though nonsignificant pattern was found for food-related conflicts (Verbeek & de Waal, 1997). Also, aside from humans and chimpanzees, capuchin monkeys are the only other primate species known to engage in cooperative hunting. Primate research has shown that capuchins share food spontaneously and will do so even more after cooperative food-obtaining efforts (de Waal, 2000).

As far as nonhuman great ape species are concerned, considerable anecdotal data on chimpanzees and bonobos show remarkable instances of aiding behavior and consoling behavior that resemble cognitive forms of empathy found in humans (de Waal, 1997; Goodall, 1990). Parr's research (2001) revealed that chimpanzees possess a basic kind of emotional awareness that humans and chimpanzees can be said to have in common. Her research with chimpanzees showed that the observation of specific facial expressions in conspecifics may lead to a more general understanding of others' emotional state, not just understanding as related to a specific situation or social context. On a matching-to-meaning task, chimpanzees were able to correctly relate short emotional video scenes portraying either negative or positive emotional valence—

for example, the scene "inject" = chimpanzees being injected with darts and needles—to pictures portraying either negative or positive emotional facial expressions—for example, scream faces and bared-teeth displays. She denotes that this type of emotional awareness can be considered a likely precursor to more cognitive forms of empathy thus far found only in humans. Nonetheless, chimpanzee MSA skills still remain very much situation driven.[9] If one defends the hypothesis that mental time travel—as in projecting oneself into the distant past and future—is uniquely human, and if one adopts the view that nonhuman great apes are, in comparison to humans, very much limited in their ability to travel back and forward in time (Suddendorf & Busby, 2003; Suddendorf & Corballis, 1997), it appears that humans are endowed with a kind of introspective awareness that allows them to take their mental states (thoughts, beliefs, desires, emotions) off-line (Gómez, 2004), providing them with an explanatory framework of their behavior that allows them to assess others' behavior in much the same way. Humans are capable of understanding and manipulating their own and others' behavior in a metarepresentational way that allows for a flexibility that goes beyond any direct or even remotely direct linkage to specific situations or social events. Empathy is skill that is very much related and beneficial to cooperative social settings and that has been shown to exist in a range of different species to greater or lesser degrees. According to Tomasello et al. (in press), selection for good collaborators involves selection for individuals that (1) are good at intention reading and (2) have a strong motivation to share psychological states with others. We defend the idea that selection must also have been on MSA skills in general. Even if there is a motivation to share psychological states with others, without a clear grasp of one's own psychological states, how can one share these with others? The fitness benefits that could be gained from acquiring more flexible and sophisticated cooperative behavior strategies during the human EEA are unknown to any species other than our hunter-gatherer ancestors. Having the ability to correctly assess others' goals and drives by virtue of understanding our own goals and drives enables this kind of flexible and complex social behavior.

---

9 In a similar vein, Vonk and Povinelli (in press) hypothesize that chimpanzees, unlike humans, cannot represent *unobservables*. They claim that Parr's (2001) meaning-to-matching task is still linked to observables and does not require any "generalized understanding." The chimpanzees' reactions could be the result of learning based on behavioral abstractions that are related to specific situations, without the need for any conceptual understanding or recognition of the underlying emotions.

Preliminary Verdict?

What exactly does our hypothesis entail? We do not wish to say that the level of self-awareness in a species determines the level of MSA, although we do state that as self-awareness gradually gets more sophisticated or complex, apparently so too does MSA. Whether there is a direct relationship between the two remains an open and empirical, as well as hotly debated, question. In humans, it certainly appears to be the case that one's level of self-awareness more or less defines one's level of MSA. In this section we look at cognitive neuroscientific evidence in favor of the hypothesis that human MSA can be defined as a byproduct of human self-awareness. The debate remains open as to whether or not this is the case in general. Although some form of self-awareness might be necessary for any kind of MSA to evolve, it might not be the case that the ToM skills found in chimpanzees are a byproduct of their specific level of self-awareness.[10] This issue remains to be dealt with and additional data is required. With regard to humans and monkeys, the promising new field of evolutionary cognitive neuroscience provides us with some preliminary answers.

Cognitive Neuroscience Data: Proximate Theorizing

The hypothesis that (human) self-awareness evolved because of the fitness benefits it brings in terms of other-understanding implies that simulation is the basis of other-understanding. It implies that we use our own experiences to understand and predict those of others. This claim lies at the basis of simulation theories of mindreading.[11] Mirror neuron research in monkeys and humans provides neurocognitive evidence in favor of this theory. Mirror neurons, or F5 neurons in the ventral premotor cortex of the monkey, fire both when the monkey performs a particular action and when the monkey observes the same action in a conspecific (di Pelligrino, Fadiga, Fogassi, Gallese, & Rizzolatti, 1992; Gallese, Fadiga, Fogassi, & Rizzolatti, 1996; Rizzolatti, Fadiga, Gallese, & Fogassi, 1996). In humans, a similar mirror neuron system (MNS) has

10 Perhaps their MSA skills are more along the lines of the calculative and learned MSA skills found in individuals with AS.

11 The opposing theory is the "theory theory" view of mindreading, which claims that MSA is mediated by some kind of theorizing, either innate or learned, and does not involve the simulation of others' experiences. For a summary and discussion, see Carruthers and Smith (1996).

been found that comprises at least Broca's region, the primary motor cortex, the superior temporal sulcus (STS) area, and the parietal cortex (Fadiga, Fogassi, Pavesi, & Rizzolatti, 1995; Gallese, Fogassi, Fadiga, & Rizzolatti, 2002; Gallese & Goldman, 1998; Hari et al., 1998; Iacoboni et al., 1999; Nishitani & Hari, 2000; Rizzolatti & Craighero, 2004; Rizzolatti et al., 1996). Avikainen, Forss, and Hari (2002) looked at area SI and SII as related to the human MNS and found similar activation patterns for action execution and observation in SI and SII. They advocate that SI and SII be considered part of the human MNS, or at least as brain structures contributing to mirror neuron activity. Their results are not at all surprising since the motor cortex and somatosensory cortices are densely reciprocally connected.

According to Gallese and Goldman (1998), mirror neuron research in both monkeys and humans provides evidence in favor of the simulation theory of mindreading. They describe the MNS in monkeys as a "primitive version, or possibly a precursor in phylogeny, of a simulation heuristic that might underlie mind-reading" (Gallese & Goldman, 1998, p. 498). This hypothesis makes sense, since the activation of mirror neurons in monkeys when they observe actions does not generate any action execution on the part of the monkey itself. It creates a state in the observer that matches the state of the executor without generating similar behavioral correlates in the observer. They do not claim that mirror neurons alone, especially in humans, lead to action understanding, although they probably constitute the most basic step in this process (Gallese & Goldman, 1998). Mirror neuron activity creates an action execution plan in the observer that appears to be taken off-line in such a way that it does not generate motor execution but remains available for processing. Our motor system seems to simulate observed actions in others *as if* we were executing the very same actions ourselves. According to this view, the generation of a similar state in the observer creates a simulation that allows for third-person action understanding from a first-person perspective:

> According to this hypothesis, "understanding" is achieved by modelling a behaviour as an action with the help of a motor equivalence between what the others do and what the observer does. (Gallese, 2001, p. 39)

Action understanding is viewed as arising out of "the 'penetration' of visual information into the experiential ('first person') motor knowledge of the observer" (Gallese, Keysers, & Rizzolatti, 2004, p. 396). Evidence for this view comes from two monkey studies (Kohler et al., 2002;

Umiltà et al., 2001) that showed that mirror neurons also fire when the actual visual features of the action are hidden, implying that mirror neuron activation represents the meaning of actions and thus underlies action understanding. In one of the studies the monkeys either saw an entire action being performed or were prevented from seeing the final and crucial part of that action. Whereas in the one condition they would see a hand actually grasping an object, in the second condition a screen would prevent them from seeing the actual hand-object interaction. So, the action would still be alluded to—the monkey "knew" the object was behind the screen—but the visual features of this process, which typically elicit mirror neuron activation, would be hidden. Apparently, more than half of the recorded neurons also responded in the hidden condition (Gallese et al., 2004; Umiltà et al., 2001). A recent fMRI study by Iacoboni et al. (2005) appears to show that mirror neuron activity not only relates to action recognition but additionally codes for intention. An object-grasping movement shown within a context revealed increased activity in the right inferior frontal cortex, which has mirror neuron properties, in contrast to observing either the same object-grasping movement without context or the same context without object-grasping movement. Moreover, comparisons of the same object-related grasping movement in a "drinking" situation in one case and a "cleaning" situation in the other yielded a much stronger response for the drinking situation, whereas comparing both contexts without the grasping movement did not produce reliable differences (Iacoboni et al., 2005). Moreover, Uddin, Kaplan, Molnar-Szakacs, Zaidel, and Iacoboni (2005) recently found that the MNS in humans might also be involved in "maintaining representations of self and others" as related to self- and other-recognition. Self-face recognition appears to involve a right hemisphere MNS that matches the face stimulus (one's own face) to an internal representation of the self (Uddin et al., 2005).

Gallese et al. (2004) proposed that emotional understanding is equally driven by an "action mirror matching mechanism," but this time pertaining to visceromotor areas. Basically, in their view, both first-person and third-person understanding of social interaction depend on the activation of cortical motor or visceromotor systems, which, by simulating observed actions or emotions, allows for a direct, experiential first-person understanding of third-person behavior. The mere observation of an action or emotion triggers the activation of the neural substrate, or at least part of it, involved when performing or experiencing that action or emotion oneself. Goldman and Sripada (2005) refer to this view as the "unmediated resonance model" of simulation. Crucial to the model of Gallese et al. (2004) is the fact that simulation is acquired by motor or visceromotor activation. A different approach was suggested by Adolphs, Damasio, Tranel, Cooper, & Damasio (2000), which Goldman and Sripada (2005, p. 206) refer to as "reverse simulation with [an] 'as if' loop." Adolphs et al. (2000) speculate on the possibility of a visual-somatosensory pathway directly generating emotional simulation in the observer. This model is explained as a special instance of the general "reverse simulation" model that postulates a causal role for facial mimicry in emotional recognition and decoding. According to the latter model, an individual observing an emotionally expressive face will start by mimicking the facial expression of the target, which in turn leads to the production of that very same emotion, or at least traces of it, in the observer. This would enable the observer to classify the emotion based on her own emotional state and thus lead to emotion recognition. Crucial to this model is the fact that facial movements occur prior to the emotional experience. This view is compatible with recent mirror neuron data if one assumes that the facial movements are part of a more general mirroring system. It is also in line with emotional contagion studies (Parr, 2001; Platek et al., 2003, 2005; Preston & de Waal, 2002) and recent work by Carr, Iacoboni, Dubeau, Mazziotta, and Lenzi (2003) that showed robust activation in premotor areas in normal individuals during the observation of expressive emotional faces, in line with the mirror neuron hypothesis that action representation mediates emotion recognition. Although Adolphs et al. (2000) recognize that this might be what is happening, they speculate on the possibility that a direct visual-somatosensory connection might surpass the role of facial mimicry in emotion recognition and directly produce the emotional state in the observer (as in directly producing what it would feel like to have the required facial expression). Basically, the reverse simulation with "as if" loop model proposes a similar mechanism as the unmediated resonance model, the only difference being that the latter model places the burden of simulation on the motor areas and allows for modulation of the representations associated with viewing the expressive face further down the line, and/or acknowledges the possibility of a separate mechanism for more cognitive forms of emotion recognition. The reverse simulation with "as if" loop model places the burden of emotional simulation on somatosensory face areas. Observing emotions in others triggers a somatosensory map of the expressed emotion "as if" the observer was experiencing the emotion. A neural map of the body state that is associated with the observed emotion would be recreated in the somatosensory-related cortices and would allow emotion recognition (retrieval of knowledge about the emotion) (Adolphs, Tranel, & Damasio, 2003).

So, on the one hand, we have the reverse simulation with "as if" loop model, which holds that simulation works at the level of the somatosensory cortex for emotion recognition by creating the associated body state in the observer of the expressive face in the target, and on the other hand, we have the unmediated resonance model, which holds that simulation works at the motor level, in which mirror neuron activity provides the observer with direct experiential information of the target's emotion by creating a motor schema of the observed action/emotion resembling one's own experience of similar actions/emotions. So, somatosensory areas are important only inasmuch as they are also motor structures or motor-related structures.

Singer et al. (2004) conducted a study on pain processing in self and other that revealed common activation in the anterior cingulate cortex (ACC) and the anterior insula (AI) when receiving a pain stimulus oneself and perceiving a similar pain stimulus in others (not directly observing, but anticipating). Somatosensory areas were activated only for the self condition. The ACC and AI are said to code for the subjective and affective dimension of the pain stimulus. According to Singer et al. (2004), the ACC and AI are central for empathic experiences related to pain. When subjects experience pain, it is assumed that an image of the body's internal state is mapped onto the sensorimotor cortex as well as the mid- and posterior dorsal insula. That sensory representation of the body's internal state in the posterior insula is initially re-represented in the AI on the same side of the brain and then remapped to the other side of the brain in the right AI by way of callosal transmission. At the same time, there are projections to the ACC. Activation of both areas results in the generation of both a feeling and an affective state. The ACC and AI appear to be crucial for the representation of internal bodily states as well as emotional awareness. The work of Singer et al. (2004) suggests that empathizing with someone else's pain involves the activation of these second-order representations, providing individuals with a subjective feel of someone else's pain, and does not require a detailed sensory representation of the pain stimulus, as is the case in one's own pain experiences.

Morrison, Lloyd, and Roberts (2004) found similar results when comparing one's own pain experience with the visual stimuli of someone else receiving a similar pain stimulus. SI did not show significant activation for the visual stimuli. Both studies provide evidence in favor of a simulation theory of other-understanding, although only part of the pain matrix involved in experiencing pain in oneself provides for a matching state in the observer. According to Gallese et al. (2004), Singer's study is in line with their visceromotor mirror matching theory of emotional understanding in others, since AI and ACC are motor-related areas. Carr et al. (2003) found largely similar patterns of activation for both observation and imitation of emotional face expressions, comprising the premotor face area, the dorsal sector of the pars opercularis of the inferior frontal gyrus, the STS, the insula, and the amygdala. Assuming that the left amygdala is associated with explicit representational content of observed emotions (Morris, Öhman, & Dolan, 1998), the right lateralized activation of the amygdala during imitation of facial expressions in the study of Carr et al. suggests that empathic resonance does not require explicit representational content, implying that the empathic experience relies on an experiential mechanism as proposed for the action mirror matching system by Gallese et al. (2004). However, it is important to note that this mechanism extends to the amygdala, a non-motor-related brain area. Avikainen et al. (2002) showed that observations of hand actions resulted in activation patterns in SI and SII similar to the mirror neuron activation traditionally found in Broca's region, STS, the ventral premotor cortex, and the posterior parietal cortex (see Gallese, 2001; Gallese et al., 2002). They propose to incorporate SI and SII as part of the human MNS, or at least as brain structures contributing to the MNS function (Avikainen et al., 2002). Also, on the basis of 108 patients with focal brain lesions, Adolphs et al. (2000) claim that integrity of the right somatosensory cortices (SI and SII) is required for emotional face recognition. Moreover, a recent study by Blakemore, Bristow, Bird, Frith, and Ward (2005) showed activity of SI and SII for the mere observation of touch (head or neck being touched).

So, there are mirror matching or simulation systems in the human brain for action (see above), emotion (Adolphs et al., 1994, 1999, 2003; Calder, Keane, Manes, Antoun, & Young, 2000; Lawrence, Calder, McGowan, & Grasby, 2002; Sprengelmeyer et al., 1999; Wicker et al., 2003), pain (Morrison et al., 2004; Singer et al., 2004), and possibly also for touch (Blakemore et al., 2005). Interestingly, self- and other-related MSA/ToM tasks both draw on the medial prefrontal cortex (MPFC) (Frith & Frith, 1999, 2003), which implies that simulation might work on every level of social processing. Kelley et al. (2002) found that the

MPFC is involved in self-referential processing and contributes to the formation of self-relevant memories. Gusnard, Akbudak, Shulman, and Raichle (2001) found MPFC activity for introspective judgments. Similarly, Johnson et al. (2002) found that the MPFC is involved in self-reflection. Self-reflective thought activated the anterior regions of the MPFC. Reflecting on current emotions also activates this region (Gusnard et al., 2001; Lane, Reiman, Ahern, Schwartz, & Davidson, 1997). Gallagher and Frith (2003) found that ToM tasks predominantly activate the anterior region of the MPFC, the STS, and the temporal poles bilaterally. They defend the hypothesis that the MPFC is involved in processing "decoupled" or off-line representations. Interestingly, McCabe, Houser, Ryan, Smith, and Trouard (2001) found activation in the anterior region of the MPFC associated with cooperation. Participants took place in a two-person trustworthiness paradigm involving reciprocal exchange. Individuals with the highest cooperation scores show significant increases in the activation of the MPFC during human-human interactions as opposed to human-computer interactions.

The hypothesis that self-awareness and MSA share, at least in part, a neurocognitive suite of processing is brought forward by Platek, Keenan, et al. (2004), Macrae et al. (2004), Keenan, Gallup, and Falk (2003), Happé (2003), Vogeley et al. (2001), Vogeley and Fink (2003), and Frith and Happé (1999).

Based on current evidence, it is compelling to believe that mirror neuron–like properties can be found outside of areas currently considered to be the core of the motoric MNS in humans. Possibly, more cognitive forms of social interaction require both the visceromotor mirror neuron area and more cognitive brain areas, which appear to function by a similar mechanism of simulation, to work together in order to generate other-understanding. That cognitive brain processes are devoid of any experiential processing is unlikely, since one cannot fully dissociate reason from emotion in normal humans.[12] Goldman and Sripada (2005) claim that emotional states, especially in reference to the six basic emotions—happiness, fear, anger, disgust, surprise, and sadness—most likely differ in cognitive style from other subdomains of MSA such as propositional mental states (beliefs and desires). However, it is arguable that different kinds of MSA depend, at least in part, on shared mechanisms. For example, self-conscious emotions (shame, guilt, and pride; Lewis, 2000) can be seen as an intermediate phenomenon between basic emo-

12 This is possibly the case to greater or lesser degrees in ASCs.

tions and propositional mental states, sharing neurocognitive processing with both. Whereas primary or basic emotions require a self to experience the emotion, self-conscious emotions require a self (the cognitive idea of "me") both to produce and to experience these emotions (Lewis, 2003). Also, Gallese (2001) proposed that our ability to ascribe mental states to others (ToM), the way in which we mirror ourselves in others' behavior and entertain social relationships, might all have one common root, empathy. Empathy builds a bridge between oneself and others. Gallese (2001) proposes to extend the concept of empathy to account for all aspects of social behavior that involve creating a link between oneself and others. His view is in line with the idea that no meaningful human interaction is devoid of affective states. This is exactly what we believe, and why we defend the hypothesis that simulation is doing its job on different levels, not only at the motor level. Gallese (2001), when formulating his "shared manifold hypothesis," mentioned the possibility of a neural matching mechanism in a variety of non-motor-related brain areas, although Gallese et al. (2004), as mentioned, stress the importance of a motor-related mirror matching system for action and emotion understanding. The question remains open. As a final remark related to mirror neuron research, we would like to mention that mirror neuron impairments have been related to ASCs (Gallese, 2003; Villalobos, Mizuno, Dahl, Kemmotsu, & Müller, 2005; Williams, Whiten, & Singh, 2004; Williams, Whiten, Suddendorf, & Perrett, 2001). According to Williams et al. (2001), mirror neuron dysfunction may account for the MSA difficulties typically found in ASCs. They propose that the ToM deficit found in autism and ASCs might be due to a simulation deficit resulting from mirror neuron dysfunction. Whereas in normal individuals, mirror neuron function allows for the action, intention, and emotion understanding of others' behavior (Gallese et al., 2004; Uddin et al., 2005), mirror neuron dysfunction in ASCs presumably obstructs this process. The findings of Villalobos et al. (2005) suggest abnormalities in the frontal components of the dorsal stream in autistic individuals (abnormal functional connectivity between the primary visual cortex and inferior frontal lobe), which is consistent with the mirror neuron dysfunction hypothesis. Research on face processing in ASCs has shown abnormal processing in the STS as compared to normal individuals (Pelphrey, Morris, & McCarthy, 2005), and another recent study hypothesized that the abnormalities related to FG and amygdala activation when individuals with ASCs process faces might be related to their gaze fixation patterns (STS) and concurrent overarousal of the amygdala when fixating on the eye region (Dalton et al., 2005). In normal individuals the STS is indirectly connected to the ventral premotor cortex (vPM) by virtue of the inferior parietal lobule, which are all mirror neuron–relevant areas. One of the major inputs to the vPM comes from the inferior parietal lobule, which is in turn reciprocally connected to the STS region (Gallese, 2001). If ASCs can be related to mirror neuron impairments, this would imply a disruption or lack of experiential processing that may result in overall impairments in other-understanding. Such a deficit would probably allow for a more cognitive understanding of social behavior but would disrupt a more direct, spontaneous grasp of others' behavior. This would be in line with our proposal that first-person experiential access to one's own experiences (reflective self-awareness) is pivotal for human MSA skills. The data of Singer et al. (2004) also provide evidence for our theory by showing that only part of the pain matrix involved in pain processing is activated when attributing pain to others. It is indeed adaptive to experience one's own pain on a sensory level, whereas it makes more sense to understand others' pain only on a subjective, affective level. The take-home message here is that this subjective, affective understanding is based on one's own sensory experiences. Either direct or indirect, it seems that other-understanding is fueled by self-understanding. In a similar vein, Gallese (2001) speculates that the MNS may have originally developed (or evolved) because of the benefits it brings in terms of better controlling one's own action performance. Its later generalization to the actions of others may then be used for different purposes, namely, understanding others' actions. Thus, self-awareness does more than just allow us to understand others.

Self-Understanding: Monitoring and Guiding One's Own Behavior

In human evolution, selection was for understanding one's own behavior better and, we hypothesize, not only for the purpose of understanding others better but also to understand, guide, and control one's own behavior better, to learn from mistakes, and to inhibit certain future behaviors when past situations already provided information that certain strategies were not the most rewarding to pursue. As we have outlined so far, similar brain processes underlie both the understanding of one's own behavior and the understanding of others' behavior. Moreover, action understanding inherently captures intention understanding, and the same mechanism that allows an individual to grasp his or her own actions is implemented when understanding others' actions (Blakemore

& Decety, 2001; Iacoboni et al., 2005; Jackson & Decety, 2004). It has been claimed that we distinguish self-generated and externally generated movements by way of an internal predictor, or forward model, that uses this intentional information to determine whether an action is one's own (Blakemore, 2003; Blakemore & Decety, 2001; Miall & Wolpert, 1996). Patients with delusions of control, such as individuals with schizophrenia, appear to confuse self-produced actions with externally produced actions and sensations. These patients describe their thoughts, speech, and actions as if they were initiated by some kind of alien control acting upon them. Overactivity of the parietal cortex and cerebellum appears to be responsible for the feeling that one's own movements are generated by alien control. Proprioception (awareness of one's body in physical space), the ability to locate one's body parts in space without having visual confirmation, is said to reside in the parietal lobe. The loss of proprioception involves an inability to properly control one's own body movement and may result in a feeling of disembodiment. Sacks (1985) describes the case of Christina, the "disembodied lady," who, after losing her proprioceptive skills, no longer recognized her body as her own. It was as if her body were dead, not real. Patients who experience such a loss of control over their own body parts experience a loss of agency that can severely disrupt their sense of self, even to the point that they no longer have a unitary sense of self, as if their self had disintegrated. Similar losses of self-identity have been reported in phantom limb cases (Ramachandran, 1998; Ramachandran & Hirstein, 1998).

One's sense of agency and ownership of one's body is crucial for the coherence and unitary feel of one's self. Body awareness constitutes only one part of an individual's self-awareness, but it appears to be vital for one's sense of agency and self-control. According to Gallese and Umiltà (2002), the self can be understood as an adaptive tool that gives coherence to different levels of representation. Moreover, they claim that the human level of self-awareness may be traced back to more ancient mechanisms, identifying the human version as a sophisticated homologue of "lower" levels of self-monitoring.

The notion that human self-awareness (by virtue of being able to project an "I" into the distant past and future) enables complex, flexible, and future-oriented behavior relates to the existence of a "left hemisphere interpreter" (Roser & Gazzaniga, 2004) as being responsible for our feeling of unity by explaining the world around us and our behavior in it. According to Roser and Gazzaniga (in press), "interpreting one's conscious experience in a coherent manner underlies a conception of oneself as a mental entity, with continuity through time, and with control over one's actions." Self-consciousness endows us with an explanatory framework of our behavior that allows for coherence and human agency, both in the present and in the future.

> Generation of explanations about our perceptions, memories, and actions, and the relationships among them, leads to the construction of a personal narrative that ties together elements of our conscious experience into a coherent whole. (Roser & Gazzaniga, 2004, p. 58)

Having an "I" to project into the future allows for mental trial and error (in self goal-directed behaviors, as well as in social situations) that is much less costly than actual trial and error.[13] Anticipating future events and working out the best possible strategies to pursue in light of past experiences leaves us better prepared for the future (Suddendorf & Busby, 2003). Possibly, our personal narrative provides us with the motivation for long-term goal-directed behavior. Together with our ability for mental time travel (Suddendorf & Whiten, 2001), it enables us to project ourselves into the future in terms of our wants, desires, intentions, and goals, and to take into account past mistakes and try to eliminate at least some future ones when planning future behavioral strategies. Taking into account how others might react to your own future behavior, and being able to change your behavioral strategy accordingly, allows for even greater chances at successful long-term problem-solving behavior. So, self-awareness in terms of covert mental states—having a personal narrative—might have evolved because of the benefits gained by the ability to monitor and control one's own behavior not only in the near future but also, and even more so, in the distant future. This kind of control lends itself to the kind of flexibility and complexity that can only be found in human (social) behavior. Introspective awareness allows for the assessment[14] of one's own and others' goals, desires, and wants, which in turn prepares the way for more sophisticated cooperative and competitive social strategies. According to recent theorizing (see Beer, Heerey, Keltner, Scabini, & Knight, 2003), self-conscious evaluative emotions regulate complex social behavior, in terms of regulating approach and inhibition tendencies that could threaten social relations. These emotions are said to regulate social behavior in ways that promote social harmony.

---

13 Actual trial and error is more likely to confer greater risks on its effectors in terms of survival and reproduction, such as death, not finding resources on time (food, shelter), not having access to mates, and so on.

14 According to McCabe et al. (2001), this kind of assessment is required for humanlike cooperative behavior strategies.

They tend to discourage inappropriate social behavior, reinforce appropriate social behavior, repair social relationships, and regulate one's overall social behavior.

Does this mean that our self is confined to or predominantly processed by the left hemisphere? Not quite. The right hemisphere is very important for processing self-related information, and patients with right hemisphere damage frequently experience apparent distortions in their self-image/unity of self. These conditions include asomatognosia, hemispatial neglect, anosognosia, Capgras' syndrome, Frégoli syndrome, and others (Feinberg, 2001; Feinberg & Keenan, 2005). Also, Platek, Keenan, et al. (2004) refer to the following evidence: from self-face identification tasks, there is growing evidence in humans that information about the self involves the right prefrontal cortex (e.g., Keenan, Freund, Hamilton, Ganis, & Pascual-Leone, 2000; Platek et al., 2006). This is true not only for self-face recognition but also for autobiographical memory (Levine et al., 1998), self-descriptive adjectives (Craik et al., 1999), MSA (Platek et al., 2005), and knowledge of one's own body parts (see Gallup & Platek, 2002).

Moreover, a recent study by one of the authors (Platek et al., 2005) defends a bilateral, distributed model of self-awareness that may reconcile previously adhered right hemisphere or left hemisphere dominant models. In the past, several studies appeared to defend a right hemisphere dominant model of self-awareness (Decety & Chaminade, 2003; Gallup, Anderson, & Platek, 2003; Jackson & Decety, 2004; Keenan, Gallup, & Falk, 2003; Keenan, McCutcheon, & Pascual-Leone, 2001; Keenan, Nelson, & Pascual-Leone, 2001; Keenan, Wheeler, & Ewers, 2003; Lou et al., 2004; Platek, Thomson, et al., 2004; Sugiura et al., 2000). Platek et al. (2005) found right superior frontal gyrus and inferior parietal lobe activation specific to self-face processing that is consistent with this right hemisphere/frontal-parietal model of self-awareness. However, they additionally found activation to self-face in the left middle temporal gyrus and right medial frontal lobe, indicating that the network responsible for self-face recognition is more extensive than previously thought. Their data support a more complex bilateral network (Kircher et al., 2001), similar to those implicated in social cognition and mirror neuron research, for both perceptual and executive aspects of self-face processing that cannot be reduced to a simplistic hemispheric dominance model. It also implies, as mentioned in Platek et al. (2005), the existence of a complex bilateral model of human self-awareness that encompasses right frontal and parietal regions, as well as left temporal and medial frontal lobes. This model is proposed to work across different levels of processing, which allows us to reconcile existing findings of both right and left hemisphere involvement in self-processing.

Conclusion

We defend the hypothesis that self-awareness and MSA are gradually linked both across and within primate species. Primatological, neurocognitive, and neuropathological data suggest a close linkage between self-awareness and MSA (e.g., ToM and empathizing) in humans and nonhuman primates. Moreover, it appears that in humans there is a direct relationship between self-awareness and MSA. Whether or not this is the case in other primate species needs to be determined.

Neurocognitive and neuropathological research provides proximate explanations for ultimate theories about the evolution of self-awareness, further suggesting that self-awareness evolved because of benefits entailed in self-understanding and other-understanding. Both ultimate and proximate theories on the functions of our psychological adaptations should be considered if one wishes to understand these phenomena and determine how they are implemented in the brain. Both evolutionary and cognitive neuroscience approaches to self-awareness provide evidence in favor of a simulation heuristic for human mindreading. Whereas ultimate explanations of human self-awareness have mostly focused on the benefits it brings in terms of understanding others, there are adaptive, neuropathological, and neurocognitive reasons to stress the benefits involved in having a unified sense of self from a first-person perspective. Moreover, neurocognitive research on self-face recognition, ToM, empathy, and mirror neurons shows that both phenomena, self-awareness and MSA, are tightly linked in terms of their functional neuroanatomy. However, they also show that the self cannot be reduced to MSA in general, and vice versa. More research on the resemblances and differences between these psychological phenomena is needed to achieve a more thorough proximate understanding of human and nonhuman self-awareness. This process would be most successful when undertaken in terms of ultimate explanations that can guide our understanding at the proximate level, which can then further extend our knowledge of their evolutionary history.

Acknowledgment Farah Focquaert is supported by the Scientific Fund for Research–Flanders (Belgium).

References

Adolphs, R., Damasio, H., Tranel, D., Cooper, G., & Damasio, A. R. (2000). A role for somatosensory cortices in the visual recognition of emotion as revealed by three-dimensional lesion mapping. *Journal of Neuroscience*, *20*(7), 2683–2690.

Adolphs, R., Tranel, D., & Damasio, A. R. (2003). Dissociable neural systems for recognizing emotions. *Brain and Cognition*, *52*, 61–69.

Adolphs, R., Tranel, D., Damasio, H., & Damasio, A. R. (1994). Impaired recognition of emotion in facial expressions following bilateral damage to the human amygdala. *Nature*, *372*, 669–672.

Adolphs, R., Tranel, D., Hamann, S., Young, A. W., Calder, A. J., Phelps, E. A., et al. (1999). Recognition of facial emotion in nine individuals with bilateral amygdala damage. *Neuropsychologia*, *37*, 1111–1117.

Anderson, J. R., Montant, M., & Schmitt, D. (1996). Rhesus monkeys fail to use gaze direction as an experimenter-given cue in an object-choice task. *Behavioural Processes*, *37*, 47–55.

Anderson, J. R. (1994). The monkey in the mirror: A strange conspecific. In S. T. Parker, R. W. Mitchell & M. L. Boccia (Eds.), *Self-awareness in animals and humans: Developmental perspectives* (pp. 315–329). New York: Cambridge University Press.

Anderson, J. R., Myowa-Yamakoshi, M., & Matsuzawa, T. (2004). Contagious yawning in chimpanzees. *Proceedings of the Royal Society of London: Biology Letters*, *271*(Suppl. 6), S468–S470.

Avikainen, S., Forss, N., & Hari, R. (2002). Modulated activation of the human SI and SII cortices during observation of hand actions. *NeuroImage*, *15*, 640–646.

Baron-Cohen, S. (2001). *The essential difference: The truth about the male and female brain.* New York: Perseus.

Baron-Cohen, S. (2004). The cognitive neuroscience of autism. *Journal of Neurology, Neurosurgery, and Psychiatry 75*, 945–948.

Baron-Cohen, S., & Wheelwright, S. (2004). The Empathy Quotient: An investigation of adults with Asperger syndrome or high functioning autism, and normal sex differences. *Journal of Autism and Developmental Disorders*, *34*(2), 164–175.

Baron-Cohen, S., Wheelwright, S., Lawson, J., Griffin, R., & Hill, J. (2002). The exact mind: Empathising and systemising in autism spectrum conditions. In U. Goswami (Ed.), *Handbook of cognitive development.* Oxford: Blackwell.

Baron-Cohen, S., Wheelwright, S., Skinner, R., Martin, J., & Clubley, E. (2001). The autism-spectrum quotient (AQ): Evidence from Asperger syndrome/high-functioning autism, males and females, scientists and mathematicians. *Journal of Autism and Developmental Disorders*, *31*, 5–17.

Barth, J., Povinelli, D. J., & Cant, J. G. H. (in press). Bodily origins of self. In D. Beike, J. Lampinen, & D. Behrend (Eds.), *The self and memory.* New York: Psychology Press.

Beer, J. S., Heerey, E. A., Keltner, D., Scabini, D., & Knight, R. T. (2003). The regulatory function of self-conscious emotion: Insights from patients with orbitofrontal damage. *Journal of Personality and Social Psychology*, *85*(4), 594–604.

Blakemore, S.-J. (2003). Deluding the motor system. *Consciousness and Cognition*, *12*, 647–655.

Blakemore, S.-J., & Decety, J. (2001). From the perception of action to the understanding of intention. *Nature Reviews: Neuroscience*, *2*, 561–567.

Blakemore, S.-J., Bristow, D., Bird, G., Frith, C., & Ward, J. (2005). Somatosensory activations during the observation of touch and a case of vision-touch synaesthesia. *Brain*, *128*, 1571–1583.

Boyd, R., Gintis, H., Bowles, S., & Richerson, P. J. (2003). The evolution of altruistic punishment. *Proceedings of the National Academy of Sciences, U.S.A.*, *100*(6), 3531–3535.

Brinck, I., & Gärdenfors, P. (2003). Co-operation and communication in apes and humans. *Mind and Language*, *18*(5), 484–501.

Byrne, R. W., & Whiten, A. (1992). Cognitive evolution in primates: The 1990 database. *Primate Report*, *27*, 1–101.

Calder, A. J., Keane, J., Manes, F., Antoun, N., & Young, A. W. (2000). Impaired recognition and experience of disgust following brain injury. *Nature Neuroscience*, *3*(11), 1077–1078.

Call, J., Hare, B., Carpenter, M., & Tomasello, M. (2004). "Unwilling" versus "unable": Chimpanzees' understanding of human intentional action. *Developmental Science*, *7*(4), 488–498.

Carr, L., Iacoboni, M., Dubeau, M. C., Mazziotta, J. C., & Lenzi, G. L. (2003). Neural mechanisms of empathy in humans: A relay from neural systems for imitation to limbic areas. *Proceedings of the National Academy of Sciences, U.S.A.*, *100*(9), 5497–5502.

Carruthers, P., & Smith, P. K. (Eds.). (1996). *Theories of theories of mind*. Cambridge: Cambridge University Press.

Cosmides, L., & Tooby, J. (2000a). Introduction. In M. Gazzaniga (Ed.), *The new cognitive neurosciences*. Cambridge, MA: MIT Press.

Cosmides, L., & Tooby, J. (2000b). Toward mapping the evolved functional organization of mind and brain. In M. Gazzaniga (Ed.), *The new cognitive neurosciences*. Cambridge, MA: MIT Press.

Craik, F. I. M., Moroz, T. M., Moscovitch, M., Stuss, D. T., Winocur, G., Tulving, E., et al. (1999). In search of the self: A positron emission tomography study. *Psychological Science*, *10*(1), 26–34.

Dalton, K. M., Nacewicz, B. M., Johnstone, T., Schaefer, H. S., Gernsbacher, M. A., Goldsmith, H. H., et al. (2005). Gaze fixation and the neural circuitry of face processing in autism. *Nature Neuroscience*, *8*, 519–526.

de Waal, F. B. M. (1996). *Good natured: The origins of right and wrong in humans and other animals*. Cambridge, MA: Harvard University Press.

de Waal, F. B. M. (1997). *Bonobo: The forgotten ape*. Berkeley and Los Angeles: University of California Press.

de Waal, F. B. M. (2000). Payment for labour in monkeys. *Nature*, *404*, 563.

de Waal, F. B. M., Dindo, M., Freeman, C. A., & Hall, M. J. (2005). The monkey in the mirror: Hardly a stranger. *Proceedings of the National Academy of Sciences, U.S.A.*, *102*(32), 11140–11147.

Decety, J., & Chaminade, T. (2003). When the self represents the other: A new cognitive neuroscience view on psychological identification. *Consciousness and Cognition*, *12*, 577–596.

di Pellegrino, G., Fadiga, L., Fogassi, L., Gallese, V., & Rizzolatti, G. (1992). Understanding motor events: A neurophysiological study. *Experimental Brain Research*, *91*, 176–180.

Fadiga, L., Fogassi, L., Pavesi, G., & Rizzolatti, G. (1995). Motor facilitation during action observation: A magnetic stimulation study. *Journal of Neurophysiology*, *73*(6), 2608–2611.

Feinberg, T. E. (2001). *Altered egos: How the brain creates the self*. New York: Oxford University Press.

Feinberg, T. E., & Keenan, J. P. (2005). Where in the brain is the self? *Consciousness and Cognition*, *14*, 661–678.

Fenigstein, A., Scheier, M. F., & Buss, A. H. (1975). Public and private self-consciousness: Assessment and theory. *Journal of Consulting and Clinical Psychology*, *36*, 1242–1250.

Flombaum, J. I., & Santos, L. R. (2005). Rhesus monkeys attribute perceptions to others. *Current Biology*, *15*, 447–452.

Frith, C. D., & Frith, U. (1999). Interacting minds: A biological basis. *Science*, *286*, 1692–1695.

Frith, U., & Frith, C. D. (2003). Development and neurophysiology of mentalizing. *Philosophical Transactions of the Royal Society of London, B, Biological Sciences*, *358*, 459–473.

Frith, U., & Happé, F. (1999). Theory of mind and self-consciousness: What is it like to be autistic? *Mind and Language*, *14*(1), 1–22.

Gallagher, H. L., & Frith, C. D. (2003). Functional imaging of "theory of mind." *Trends in Cognitive Sciences*, *7*, 77–83.

Gallese, V. (2001). The "shared manifold" hypothesis: From mirror neurons to empathy. *Journal of Consciousness Studies*, *8*(5–7), 33–50.

Gallese, V. (2003). The roots of empathy: The shared manifold hypothesis and the neural basis of intersubjectivity. *Psychopathology*, *36*, 171–180.

Gallese, V., Fadiga, L., Fogassi, L., & Rizzolatti, G. (1996). Action recognition in the premotor cortex. *Brain*, *119*, 593–609.

Gallese, V., Fogassi, L., Fadiga, L., & Rizzolatti, G. (2002). Action representation and the inferior parietal lobule. In W. Prinz & B. Hommel (Eds.), *Attention and performance. XIX. Common mechanisms in perception and action* (pp. 334–355). New York: Oxford University Press.

Gallese, V., & Goldman, A. (1998). Mirror neurons and the simulation theory of mind-reading. *Trends in Cognitive Sciences*, *2*(12), 493–501.

Gallese, V., Keysers, C., & Rizzolatti, G. (2004). A unifying view of the basis of social cognition. *Trends in Cognitive Sciences*, *8*(9), 396–403.

Gallese, V., & Umiltà, M. A. (2002). From self-modeling to the self model: Agency and the representation of self. *Neuro-Psychoanalysis*, *4*(2), 35–40.

Gallup, G. G., Jr. (1982). Self-awareness and the emergence of mind in primates. *American Journal of Primatology*, *2*, 237–248.

Gallup, G. G., Jr. (1994). Self-recognition: Research strategies and experimental design. In S. T. Parker, R. W. Mitchell, & M. L. Boccia (Eds.), *Self-awareness in animals and humans: Developmental perspectives* (pp. 35–50). New York: Cambridge University Press.

Gallup, G. G., Jr. (1997). On the rise and fall of self-conception in primates. *Annals of the New York Academy of Sciences*, *818*, 72–82.

Gallup, G. G., Jr., Anderson, J. R., & Platek, S. M. (2003). Self-awareness, social intelligence, and schizophrenia. In A. S. David & T. Kircher (Eds.), *The self and schizophrenia: A neuropsychological perspective*. Cambridge: Cambridge University Press.

Gallup, G. G., Jr., Anderson, J. R., & Shillito, D. J. (2002). The mirror test. In M. Bekoff, C. Allen, & G. M. Burghardt (Eds.), *The cognitive animal* (pp. 325–333). Cambridge, MA: MIT Press.

Gallup, G. G., Jr., & Platek, S. M. (2002). Self-face recognition is affected by schizotypal personality traits. *Schizophrenia Research*, *57*, 311–315.

Goldman, A. I., & Sripada, C. S. (2005). Simulationist models of face-based emotion recognition. *Cognition*, *94*, 193–213.

Gómez, J. C. (1998). Assessing theory of mind with nonverbal procedures: Problems with training methods and an alternative "key" procedure. *Behavioral and Brain Sciences*, *21*(1), 119–120.

Gómez, J. C. (2004). *Apes, monkeys, children, and the growth of mind*. Cambridge, MA: Harvard University Press.

Goodall, J. (1990). *Through a window: My thirty years with the chimpanzees of gombe*. Boston: Houghton Mifflin.

Grandin, T. (1995). *Thinking in pictures: And other reports from my life with autism*. New York: Vintage Books.

Gusnard, D. A., Akbudak, E., Shulman, G. L., & Raichle, M. E. (2001). Medial prefrontal cortex and self-referential mental activity: Relation to a default mode of brain function. *Proceedings of the National Academy of Sciences, U.S.A.*, *98*, 4259–4264.

Hamada, Y., & Udono, T. (2002). Longitudinal analysis of length growth in the chimpanzee (*Pan troglodytes*). *American Journal of Physical Anthropology*, *118*, 268–284.

Happé, F. (2003). Theory of mind and the self. In *The self from soul to brain* [Special issue]. *Annals of the New York Academy of Sciences*, *1001*, 134–144.

Hare, B., Addessi, E., Call, J., Tomasello, M., & Visalberghi, E. (2003). Do capuchin monkeys, *Cebus apella*, know what conspecifics do and do not see? *Animal Behaviour*, *65*, 131–142.

Hare, B., Call, J., Agnetta, B., & Tomasello, M. (2000). Chimpanzees know what conspecifics do and do not see. *Animal Behaviour*, *59*, 771–785.

Hare, B., Call, J., & Tomasello, M. (2001). Do chimpanzees know what conspecifics know? *Animal Behaviour*, *61*, 139–151.

Hare, B., & Tomasello, M. (2004). Chimpanzees are more skillful in competitive than in cooperative cognitive tasks. *Animal Behaviour*, *68*, 571–581.

Hari, R., Forss, N., Avikainen, S., Kirveskari, E., Salenius, S., & Rizzolatti, G. (1998). Activation of the human primary motor cortex during action observation: A neuromagnetic study. *Proceedings of the National Academy of Sciences, U.S.A.*, *95*, 15061–15065.

Hulbert, R. T., Happé, F., & Frith, U. (1994). Sampling the form of inner experience in three adults with Asperger's syndrome. *Psychological Medicine*, *24*, 385–395.

Humphrey, N. (1980). Nature's psychologists. In B. Josephson & V. Ramachandran (Eds.), *Consciousness and the physical world* (pp. 57–75). Oxford: Pergamon Press.

Humphrey, N. (1982). Consciousness: A just-so story. *New Scientist*, *19*, 474–477.

Humphrey, N. (1986). *The inner eye*. New York: Oxford University Press.

Iacoboni, M., Molnar-Szakacs, I., Gallese, V., Buccino, G., Mazziotta, J. C., & Rizzolatti, G. (2005). Grasping the intentions of others with one's own mirror neuron system. *Public Library of Science—Biology*, *3*(3), 529–535.

Iacoboni, M., Woods, R. P., Brass, M., Bekkering, H., Mazziotta, J. C., & Rizzolatti, G. (1999). Cortical mechanisms of human imitation. *Science*, *286*, 2526–2528.

Jackson, P. L., & Decety, J. (2004). Motor cognition: A new paradigm to study self-other interactions. *Current Opinion in Neurobiology*, *14*, 29–37.

Jarrold, C., Carruthers, P., Smith, P. K., & Boucher, J. (1994). Pretend play: Is it metarepresentational? *Mind and Language*, *9*, 445–468.

Johnson, A. K., Barnacz, A., Constantino, P., Triano, J., Shackelford, T. K., & Keenan, J. P. (2004). Female deception detection as a function of commitment and self-awareness. *Personality and Individual Differences*, *37*, 1417–1424.

Johnson, A. K., Barnacz, A., Yokkaichi, T., Rubio, J., Racioppi, C., Shackelford, T. K., et al. (2005). Me, myself and lie: The role of self-awareness in deception. *Personality and Individual Differences*, *38*, 1847–1853.

Johnson, S. C., Baxter, L. C., Wilder, L. S., Pipe, J. G., Heiserman, J. E., & Prigatano, G. P. (2002). Neural correlates of self-reflection. *Brain*, *125*, 1808–1814.

Karin-D'Arcy, M. R., & Povinelli, D. J. (2002). Do chimpanzees know what each other see? A closer look. *International Journal of Comparative Psychology*, *15*, 21–54.

Keenan, J. P., Freund, S., Hamilton, R. H., Ganis, G., & Pascual-Leone, A. (2000). Hand response differences in a self-face identification task. *Neuropsychologia*, *38*, 1047–1053.

Keenan, J. P., Gallup, G. G., Jr., & Falk, D. (2003). *The face in the mirror: The search for the origins of consciousness*. New York: HarperCollins.

Keenan, J. P., McCutcheon, B., & Pascual-Leone, A. (2001). Functional magnetic resonance imaging and event related potentials suggest right prefrontal activation for self-related processing. *Brain and Cognition*, *47*, 87–91.

Keenan, J. P., Nelson, A. M., & Pascual-Leone, A. (2001). Self-recognition and the right hemisphere. *Nature*, *409*, 305.

Keenan, J. P., Wheeler, M. A., & Ewers, M. (2003). The neuropsychology of self. In A. S. David & T. Kircher (Eds.), *The self and schizophrenia: A neuropsychological perspective.* Cambridge: Cambridge University Press.

Kelley, W. M., Macrae, C. N., Wyland, C. L., Caglar, S., Inati S., & Heatherton, T. F. (2002). Finding the self: An event-related fMRI study. *Journal of Cognitive Neuroscience, 14,* 785–794.

Kircher, T., Senior, C., Phillips, M. L., Rabe-Hesketh, S., Benson, P., Bullmore, E. T., et al. (2001). Recognizing one's own face. *Cognition, 78,* 1–15.

Kohler, E., Keysers, C., Umiltà, M. A., Fogassi, L., Gallese, V., & Rizzolatti, G. (2002). Hearing sounds, understanding actions: Action representation in mirror neurons. *Science, 297*(5582), 846–848.

Lane, R. D., Reiman, E. M., Ahern, G. L., Schwartz, G. E., & Davidson, R. J. (1997). Neuroanatomical correlates of happiness, sadness, and disgust. *American Journal of Psychiatry, 154,* 926–933.

Larsen, C. S. (2003). Equality for the sexes in human evolution? Early hominid sexual dimorphism and implications for mating systems and social behavior. *Proceedings of the National Academy of Sciences, U.S.A., 100*(16), 9103–9104.

Lawrence, A. D., Calder, A. J., McGowan, S. M., & Grasby, P. M. (2002). Selective disruption of the recognition of facial expressions of anger. *NeuroReport, 13*(6), 881–884.

Leslie, A. M. (1987). Pretense and representation: The origins of "theory of mind." *Psychological Review, 94*(4), 412–426.

Levine, B., Black, S. E., Cabeza, R., Sinden, M., McIntosh, A. R., Toth, J. P., et al. (1998). Episodic memory and the self in a case of isolated retrograde amnesia. *Brain, 121,* 1951–1973.

Lewis, M. (1992). *Shame: The exposed self.* New York: Free Press.

Lewis, M. (1995). Self-conscious emotions. *American Scientist, 83,* 68–78.

Lewis, M. (2000). Self-conscious emotions: Embarrassment, pride, shame and guilt. In M. Lewis & J. Haviland (Eds.), *Handbook of emotions* (2nd ed., pp. 623–636). New York: Guilford Press.

Lewis, M. (2003). The role of the self in shame. *Social Research, 70*(4), 1181–1204.

Lou, H. C., Luber, B., Crupian, M., Keenan, J. P., Nowak, M., Kjaer, T. W., et al. (2004). Parietal cortex and representation of the mental self. *Proceedings of the National Academy of Sciences, U.S.A., 101,* 6827–6832.

Macrae, C. N., Heatherton, T. F., & Kelley, W. M. (2004). A self less ordinary: The medial prefrontal cortex and you. In M. S. Gazzaniga (Ed.), *The Cognitive Neurosciences III* (pp. 1067–1075). Cambridge, MA: MIT Press.

McCabe, K., Houser, D., Ryan, L., Smith, V., & Trouard, T. (2001). A functional imaging study of cooperation in two-person reciprocal exchange. *Proceedings of the National Academy of Sciences, U.S.A., 98*(20), 11832–11835.

Miall, R. C., & Wolpert, D. M. (1996). Forward models for physiological motor control. *Neural Networks, 9,* 1265–1279.

Morris, J. S., Öhman, A., & Dolan, R. J. (1998). Conscious and unconscious emotional learning in the human amygdala. *Nature, 393,* 467–470.

Morrison, I., Lloyd, D., & Roberts, N. (2004). Vicarious responses to pain in anterior cingulated cortex: Is empathy a multisensory issue? *Cognitive, Affective, and Behavioral Neuroscience*, *4*(2), 270–278.

Nishitani, N., & Hari, R. (2000). Temporal dynamics of cortical representation for action. *Proceedings of the National Academy of Sciences, U.S.A.*, *97*, 913–918.

Parr, L. A. (2001). Cognitive and physiological markers of emotional awareness in chimpanzees, *Pan troglodytes*. *Animal Cognition*, *4*, 223–229.

Paukner, A., Anderson, J. R., & Fujita, K. (2004). Reactions of capuchin monkeys (*Cebus apella*) to multiple mirrors. *Behavioural Processes*, *66*, 1–6.

Pelphrey, K. A., Morris, J. P., & McCarthy, G. (2005). Neural basis of eye gaze processing deficits in autism. *Brain*, *128*, 1038–1048.

Perner, J. (1991). *Understanding the representational mind*. Cambridge, MA: MIT Press.

Platek, S. M., Critton, S. R., Myers, T. E., & Gallup, G. G., Jr. (2003). Contagious yawning: The role of self-awareness and mental state attribution. *Brain Research: Cognitive Brain Research*, *17*, 223–227.

Platek, S. M., Keenan, J. P., Gallup, G. G., Jr., & Mohamed, F. B. (2004). Where am I? The neurological correlates of self and other. *Brain Research: Cognitive Brain Research*, *19*, 114–122.

Platek, S. M., Loughead, J. W., Gur, R. C., Busch, S., Ruparel, K., Phend, N., et al. (2006). Neural substrates for functionally discriminating self-face from personally familiar faces. *Human Brain Mapping*, *27*, 91–98.

Platek, S. M., Mohamed, F. B., & Gallup, G. G., Jr. (2005). Contagious yawning and the brain. *Brain Research: Cognitive Brain Research*, *23*, 448–452.

Platek, S. M., Thomson, J. W., & Gallup, G. G., Jr. (2004). Cross-modal self-recognition: The role of visual, auditory, and olfactory primes. *Consciousness and Cognition*, *13*, 197–210.

Povinelli, D. J. (2004). Behind the ape's appearance: Escaping anthropocentrism in the study of other minds. *Daedalus*, *133*(1), 29–41.

Povinelli, D. J., & Cant, J. G. H. (1995). Arboreal clambering and the evolution of self-conception. *Quarterly Review of Biology*, *70*, 393–421.

Preston, S. D., & de Waal, F. B. M. (2002). The communication of emotions and the possibility of empathy in animals. In S. G. Post & L. G. Underwood (Eds.), *Altruism and altruistic love: Science, philosophy, and religion in dialogue* (pp. 284–308). Oxford: Oxford University Press.

Ramachandran, V. S. (1998). *Phantoms in the brain*. New York: William Morrow.

Ramachandran, V. S., & Hirstein, W. (1998). The perception of phantom limbs. The D. O. Hebb Lecture. *Brain*, *121*, 1603–1630.

Reno, P. L., Meindl, R. S., McCollum, M. A., & Lovejoy, C. O. (2003). Sexual dimorphism in *Australopithecus afarensis* was similar to that of modern humans. *Proceedings of the National Academy of Sciences, U.S.A.*, *100*(16), 9404–9409.

Reno, P. L., Meindl, R. S., McCollum, M. A., & Lovejoy, C. O. (2005). The case is unchanged and remains robust: *Australopithecus afarensis* exhibits only moderate skeletal dimorphism. A reply to Plavcan et al. (2005). *Journal of Human Evolution*, *49*, 279–288.

Rizzolatti, G., & Craighero, L. (2004). The mirror-neuron system. *Annual Review of Neuroscience*, *27*, 169–192.

Rizzolatti, G., Fadiga, L., Gallese, V., & Fogassi, L. (1996). Premotor cortex and the recognition of motor actions. *Brain Research: Cognitive Brain Research*, *3*, 131–141.

Roser, M. E., & Gazzaniga, M. S. (2004). Automatic brains: Interpretive minds. *Current Directions in Psychological Science*, *13*, 56–59.

Roser, M. E., & Gazzaniga, M. S. (in press). The interpreter in human psychology. In T. M. Preuss & J. H. Kaas (Eds.), *The evolution of primate nervous systems*. New York: Elsevier.

Sacks, O. (1985). *The man who mistook his wife for a hat*. London: Duckworth.

Singer, T., Seymour, B., O'Doherty, J., Kaube, H., Dolan, R. J., & Frith, C. D. (2004). Empathy for pain involves the affective but not sensory components of pain. *Science*, *303*, 1157–1162.

Sprengelmeyer, R., Young, A. W., Schroeder, U., Grossenbacher, P. G., Federlein, J., Buttner, T., et al. (1999). Knowing no fear. *Proceedings of the Royal Society, B, Biological Sciences*, *266*, 2451–2456.

Suarez, S. D., & Gallup, G. G., Jr. (1986). Social responding to mirrors in Rhesus macaques: Effects of changing mirror location. *American Journal of Primatology*, *11*, 239–244.

Suddendorf, T., & Busby, J. (2003). Mental time travel in animals? *Trends in Cognitive Sciences*, *7*(9), 391–396.

Suddendorf, T., & Whiten, A. (2001). Mental evolution and development: Evidence for secondary representation in children, great apes, and other animals. *Psychological Bulletin*, *127*(5), 629–650.

Suddendorf, T., & Corballis, M. C. (1997). Mental time travel and the evolution of the human mind. *Genetic, Social, and General Psychology Monographs*, *123*(2), 133–168.

Sugiura, M., Kawashima, R., Nakamura, K., Okada, K., Kato, T., Nakamura, A., et al. (2000). Passive and active recognition of one's own face. *NeuroImage*, *11*, 36–48.

Sullivan, M. W., Bennett, D. S., & Lewis, M. (2003). Darwin's view. Self-evaluative emotions as context-specific emotions. *Annals of the New York Academy of Sciences*, *1000*, 304–308.

Tomasello, M., & Call, J. (1997). *Primate cognition*. Oxford: Oxford University Press.

Tomasello, M., Call, J., & Hare, B. (2003). Chimpanzees understand psychological states: The question is which ones and to what extent. *Trends in Cognitive Sciences*, *7*(4), 153–156.

Tomasello, M., Carpenter, M., Call, J., Behne, T., & Moll, H. (in press). Understanding and sharing intentions: The origins of cultural cognition. *Behavioral and Brain Sciences*.

Uddin, L. Q., Kaplan, J. T., Molnar-Szakacs, I., Zaidel, E., & Iacoboni, M. (2005). Self-face recognition activates a frontoparietal "mirror" network in the right hemisphere: An event-related fMRI study. *NeuroImage*, *25*, 926–935.

Umiltà, M. A., Kohler, E., Gallese, V., Fogassi, L., Fadiga, L., Keysers, C., et al. (2001). I know what you are doing: A neurophysiological study. *Neuron*, *31*, 155–165.

Verbeek, P., & de Waal, F. B. M. (1997). Postconflict behavior of captive brown capuchins in the presence and absence of attractive food. *International Journal of Primatology*, *18*(5), 703–725.

Villalobos, M. E., Mizuno, A., Dahl, B. C., Kemmotsu, N., & Müller, R.-A. (2005). Reduced functional connectivity between V1 and inferior frontal cortex associated with visuomotor performance in autism. *NeuroImage*, *25*, 916–925.

Vogeley, K., Bussfeld, P., Newen, A., Herrmann, S., Happé, F., Falkai, P., et al. (2001). Mind reading: Neural mechanisms of theory of mind and self-perspective. *NeuroImage*, *14*, 170–181.

Vogeley, K., & Fink, G. R. (2003). Neural correlates of the first-person-perspective. *Trends in Cognitive Sciences*, *7*(1), 38–42.

Völlm, B. A., Taylor, A. N. W., Richardson, P., Corcoran, R., Stirling, J., McKie, S., et al. (2006). Neuronal correlates of theory of mind and empathy: A functional magnetic resonance imaging study in a nonverbal task. *NeuroImage*, *29*, 90–98.

Vonk, J., & Povinelli, D. J. (in press). Similarity and difference in the conceptual systems of primates: The unobservability hypothesis. In E. Wasserman & T. Zentall (Eds.), *Comparative cognition: Explorations of animal intelligence*. Oxford: Oxford University Press.

Whiten, A., & Byrne, R. W. (1988). Tactical deception in primates. *Behavioral and Brain Sciences*, *11*, 233–273.

Wicker, B., Keysers, C., Plailly, J., Royet, J.-P., Gallese, V., & Rizzolatti, G. (2003). Both of us disgusted in *my* insula: The common neural basis of seeing and feeling disgust. *Neuron*, *40*, 655–664.

Williams, J. H. J., Whiten, A., & Singh, T. (2004). A systematic review of action imitation in autistic spectrum disorder. *Journal of Autism and Developmental Disorders*, *34*(3), 285–299.

Williams, J. H. G., Whiten, A., Suddendorf, T., & Perrett, D. I. (2001). Imitation, mirror neurons and autism. *Neuroscience and Biobehavioral Reviews*, *25*, 287–295.

17 The Assortative Mating Theory of Autism

Simon Baron-Cohen

There are two major ways to predict changing events. If the event is agentive, one can adopt the "intentional stance" (or "empathize"). If the event is nonagentive, one can "systemize." In this chapter I outline a new theory, which holds that the systemizing mechanism has variable settings and that people with autism spectrum conditions are hypersystemizers, who therefore can process only highly systemizable (law-governed) information. In keeping with the focus of this book on evolutionary perspectives, I explore the evidence for the assortative mating theory, according to which autism is the result of both parents being high systemizers.

Systemizing Nonagentive Change

A universal feature in the environment that the brain has to react to is change. There are at least two types of structured change, agentive change and nonagentive change. Regarding the former, if change is perceived to be *self-generated* or *self-propelled* (i.e., there is no apparent external cause), the brain interprets it as agentive, that is, the individual is functioning as an agent with a goal. Goal detection (or intentionality detection, ID) is a fundamental aspect of how the human brain interprets and predicts the behavior of other animals (Baron-Cohen, 1994; Heider & Simmel, 1944; Perrett et al., 1985). Structured nonagentive change, by contrast, is any change that is not self-propelled and where there is a precipitating event (interpreted as a possible cause of the change) or a pattern to the change. Some patterns are cyclical (the pattern repeats every fixed number of units), but there are many other types of pattern.

Structured nonagentive change occurs by degrees. Some change occurs with total (100%) regularity or pattern (e.g., the sun always rises in the east and sets in the west). Other change occurs with a lower frequency or regularity, but there is still a pattern to be discerned. The perception of structured nonagentive change matters because the change might be injurious or have a negative impact (e.g., planting crops in February leads to them withering) or a positive impact (e.g., planting in March leads to the crops thriving). Being able to anticipate change thus allows the organism to avoid negative consequences or benefit from positive change.

Systemizing is the most powerful way to predict change. Systemizing involves law detection via observation of *input-operation-output* relationships (Baron-Cohen, 2002). Systemizing prompts a search for structure (patterns, rules, regularities, periodicity) in data. The goal of systemizing is to test whether the changing data are part of a system. Systems may be mechanical (e.g., machines), natural (e.g., a leaf), abstract (e.g., mathematics), collectible (e.g., a collection), motoric (e.g., a tennis stroke), or even social (e.g., the rules of etiquette). Thus, an engineer, a lawyer, a mathematician, a film editor, a librarian, an astronomer, a meteorologist, a chemist, a musician, a grammarian, a company CEO, and a zoologist all systemize; they are all concerned with formulating laws governing change—laws of physics, laws of nature, mathematical laws, social laws, and so on.

Systemizing allows the brain to predict that event x will occur with probability P—that is, to identify laws driving the system. Some systems are 100% lawful (e.g., an electrical light switch or a mathematical formula). During systemizing, the brain represents the information as input and output separately, so that the pattern emerges (table 17.1). Systems that are 100% lawful have zero (or minimal) variance and can therefore be predicted and controlled 100%. A computer might be an example of a 90% lawful system: the variance is wider, because the operating system may work differently depending on what other software is installed or which version of the software is running, and so on. The weather may be a system with only moderate lawfulness.

A key feature of systemizing is that single observations are recorded in a standardized manner. A meteorologist makes measurements at fixed times and fixed places, measuring rainfall (in a cup), temperature (with a thermometer), pressure (with a barometer), wind speed (with an anemometer), and so on. An astronomer records the position of a planet at fixed times and fixed places, tracking its movement. Such systematic data collection (phase 1 of systemizing) can then lead to the observation

Table 17.1
Two Examples of 100% Lawful Systems

A. Electricity Switch	
Input = Switch position	Output = Light Operation = Switch change
Up Down	On Off

B. Mathematical Rule	
Input = Number	Output = Number Operation = Add 2
2 3 4	4 5 6

of the pattern of law (phase 2 of systemizing). Systemizing thus has the power to reveal the structure or laws of nature.

Systemizing Agentive Change

Some aspects of agentive behavior are highly lawful (e.g., cats typically use their right paw to swipe at a moving object). Some human behavior is also sufficiently *scripted* to be moderately lawful (e.g., ballroom dancing). Human behavior that has been recorded on film is of course highly lawful, since each time the film is replayed, the actors do and say the same thing. But outside of these special cases, if there are laws governing human behavior, they are complex, and the variance is maximal. Maximal variance means that when change occurs, it could occur in a virtually infinite number of ways. Thus, a person's hands, eyes, mouth, posture, and facial expression might change in one of hundreds if not thousands of possible combinations. Nor is there a one-to-one mapping between facial expression and the underlying mental state that might be causing such changes in the face (Baron-Cohen, Golan, & Wheelwright, 2004). Situations do not predict the subtlety of emotions, since in the same situation different people react differently. Finally, humans as moving, changing objects also require the agent they are interacting with to respond. They talk, and their words appear as novel, unique combinations on each occasion, unlike scripted behavior. The right response to their words isn't to reply with a script. Agentive change in the social world is too fast, and the laws—if they exist—are thus too complex to systemize. Skinner (1976) claimed that human behavior could be systemized if one had a complete record of all the historical antecedents (A) and all the consequences (C) for any piece of behavior (B), such that A → B ← C. The real social world, of course, is not a Skinner Box.

Systemizing only works when one can measure or count one thing at a time, ignoring or holding everything else constant. Systemizing is enormously powerful as a way of predicting and controlling events in the nonagentive world and has led to the technological achievements of the modern world. It has this power because nonagentive changes are simple changes to predict: the systems are at least moderately lawful, with narrow variance.

Because ordinary social behavior defies a systematic approach, the second-by-second changes in agentive behavior are more parsimoniously interpreted in terms of the agent's goals (Baron-Cohen, 1994; Heider & Simmel, 1944; Perrett et al., 1985). It appears that humans have specialized, inherited "hardware" for dealing with the complex social world. The "empathizing system" comprises basic instruments—analogous to barometers, thermometers, and anemometers—that come compiled to help the normal infant make sense of the social world and react to it, without having to learn it all from scratch. Empathizing is explained in more detail elsewhere (Baron-Cohen, 1995, 2003, 2005; Baron-Cohen & Goodhart, 1994). Such basic modules or neurocognitive mechanisms give the normal infant a foothold in making sense of and responding to the social world. The neural circuitry of empathizing has been extensively investigated (Baron-Cohen et al., 1999; Frith & Frith, 1999; Happé et al., 1996); key brain areas involved in empathizing include the amygdala, the orbito- and medial frontal cortex, and the superior temporal sulcus. Experience allows us to learn the subtleties of empathy, but such hard-wired, innate mechanisms bootstrap the brain to make rapid sense of social change.

The hypersystemizing theory posits that we all have a systemizing mechanism (SM), which is set at different levels in different individuals. The SM is like a volume control or a dimmer switch. Genes and other biological factors (possibly fetal testosterone) turn this mechanism up or down (Knickmeyer, Baron-Cohen, Raggatt, & Taylor, 2004). In some people the SM is set high, so that they systemize *any* changing input, analyzing it for possible structure. A high systemizer searches all data for patterns and regularities. In other people the SM is set at a medium level, so that they systemize some but not all of the time. In yet other people the SM is set so low that they would hardly notice if regularity or structure was in the input or not.

Systemizing in the General Population (Levels 1–4)

Evidence suggests that within the general population, there are four degrees of systemizing. Level 1 corresponds to having little or no interest or drive to systemize, and consequently persons at this level of SM can cope with total change. This might be expressed as a talent at socializing, joined to a vagueness over details, and the ability to cope with change easily. Most people, however, have some interest in systems, and there are sex differences observable in the level of interest. More females in the general population have the SM turned up to Level 2, and more males have it turned up to Level 3. Those with an SM at Level 2 might show typical female interests (e.g., emotions; Baron-Cohen & Wheelwright, 2003), and those with an SM at Level 3 might show typical male interests (e.g., in mechanics; Baron-Cohen, 2003). These differences can be quite subtle, but, for example, on a test of map reading or mental rotation, males might score higher than females because of the higher-level SM (Kimura, 1999). Some evidence comes from the Systemizing Quotient, on which males score higher than females (Baron-Cohen, Richler, Bisarya, Gurunathan, & Wheelwright, 2003). Another piece of evidence comes from the Physical Prediction Questionnaire, an instrument administered to select applicants for engineering careers. The task involves predicting which direction levers will move when an internal mechanism (consisting of cogwheels and pulleys) is activated. Men score significantly higher on this test than women do (Lawson, Baron-Cohen, & Wheelwright, 2004).

Level 4 denotes a higher than average level of systemization. There is some evidence that above-average systemizers have more autistic traits. Thus, scientists, who by definition are good systemizers, score higher than nonscientists on the Autism Spectrum Quotient (AQ). Mathematicians, who by definition focus on abstract systems, have the highest AQ score of all scientists (Baron-Cohen, Wheelwright, Skinner, Martin, & Clubley, 2001). Another group of people who are above-average systemizers are parents of children with autism spectrum conditions (Baron-Cohen & Hammer, 1997; Happé, Briskman, & Frith, 2001). The genetic implications of this are discussed shortly, as these parents have been described as having the "broader phenotype" of autism (Bolton, 1996). One would expect a person at Level 4 to be talented at understanding systems with moderate variance (the stock market, running a company, the law, engineering).

Systemizing in the Autistic Spectrum (Levels 5–8)

The autistic spectrum comprise at least four subgroups: Asperger syndrome (AS) (Asperger, 1944; Frith, 1991), and high-, medium-, and low-functioning autism (Kanner, 1943). All share the phenotype of social difficulties and obsessional interests (American Psychiatric Association, 1994). An individual with AS has normal or above-average IQ and no language delay. In the three autism subgroups (high, medium, and low functioning), there is invariably some degree of language delay, and the level of functioning is indexed by overall IQ.[1]

Evidence suggests that people on the autistic spectrum have their SM set at levels above those in the general population—anywhere from Level 5 to Level 8. Level 5 can be seen as corresponding to AS: the person can easily systemize totally lawful systems (those that are 100% lawful, such as train timetables or historical chronologies) or highly lawful systems (e.g., computers) (Hermelin, 2002). They might also show an interest in systems like the weather, where the variance is quite high, so that the system is only moderately lawful (perhaps 60% lawful). The clinical literature is replete with anecdotal examples (e.g., one man with AS collected information of the type shown in table 17.2 or figure 17.1), but there is also experimental evidence for superior systemizing in AS: (1) People with AS have a higher than average Systemizing Quotient score (Baron-Cohen et al., 2003). (2) People with AS perform at a normal or high level on tests of intuitive physics (Baron-Cohen, Wheelwright, Skinner et al., 2001; Jolliffe & Baron-Cohen, 1997; Lawson et al., 2004; Shah & Frith, 1983). (3) People with AS can achieve extremely high levels in systemizing domains such as mathematics, physics, or computer science (Baron-Cohen, Wheelwright, Stone, & Rutherford, 1999). (4) People with AS have an "exact mind" when it comes to art (Myers, Baron-Cohen, & Wheelwright, 2004) and show superior attention to

1 High-functioning autism can be thought of as within 1 SD of population mean IQ (i.e., an IQ of 85 or above); medium-functioning autism can be thought of as between 1 and 3 SD below the population mean (i.e., an IQ of 55–84). Low-functioning autism can be thought of below this (i.e., an IQ of 54 or below).

Table 17.2
An Example of Systemizing Hydrangea Coloration

Hydrangea Name	Acidic Soil	Neutral Soil	Alkaline Soil
Annabelle	White	White	White
Ayesha	Blue	Purple	Pink
Alpengluhen	Purple	Red	Red
Altona	Blue	Purple	Red
All Summer Beauty	Blue	Purple	Pink
Ami Pasquier	Purple	Red	Red
Amethyst	Blue	Purple	Pink
Bodensee	Blue	Purple	Pink
Blauer Prinz	Blue	Purple	Purple
Bouquet Rose	Blue	Purple	Pink
Breslenburg	Blue	Purple	Pink
Deutschland	Purple	Red	Red
Domotoi	Blue	Purple	Pink
Dooley	Blue	Purple	Pink
Enziandom	Blue	Purple	Red

Source: http://www.hydrangeasplus.com.

detail (O'Riordan, Plaisted, Driver, & Baron-Cohen, 2001; Plaisted, O'Riordan, & Baron-Cohen, 1998a).

There is some evidence that in people with high-functioning autism, the SM is set at Level 6, in those with medium-functioning autism it is at Level 7, and in low-functioning autism it is at the maximum setting, Level 8. Thus, the high-functioning individuals who try to mentalize are thought to do this by "hacking" (i.e., systemizing) the solution (Happé, 1996), and on a picture-sequencing task they perform above average on sequences that contain temporal or physical-causal (i.e., systemizable) information (Baron-Cohen, Leslie, & Frith, 1986). Medium-functioning individuals, in contrast to their difficulty on the false belief task (an empathizing task), perform normally or above average on two equivalent systemizing tasks, the false photograph task (Leslie & Taiss, 1992) and the false drawings task (Charman & Baron-Cohen, 1992). In the low-functioning group, their obsessions cluster in the domain of systems (Baron-Cohen & Wheelwright, 1999), and, given a set of colored counters, they show their hypersystemizing as extreme pattern imposition (Frith, 1970). Table 17.3 lists 16 behaviors that would be expected if an individual had an SM turned up to the maximum setting of Level 8.

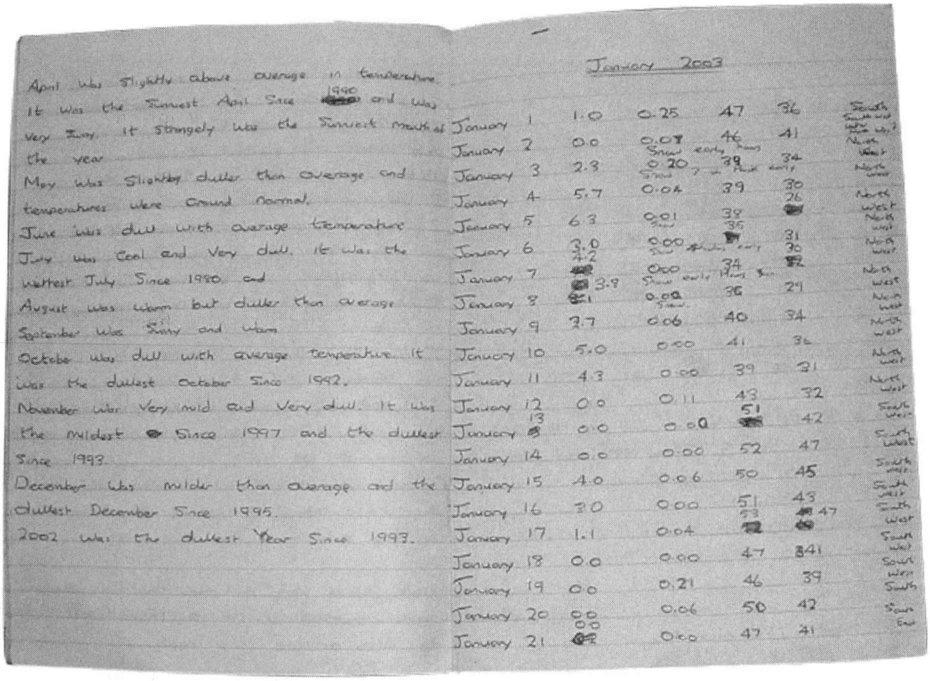

Figure 17.1
An example of systemizing the weather, from the notebook of Kevin Phillips, a man with Asperger syndrome. Reproduced with Mr. Phillips's kind permission.

The hypersystemizing theory thus has the power to explain not only what unites individuals across the autistic spectrum but why the particular constellation of symptoms is seen in this syndrome. It also explains why some people with autism may have more or less language, or a higher or lower IQ, or differing degrees of mindblindness (Baron-Cohen, 1995). This is because, according to the theory, as the SM dial is turned down from the maximum level of 8, at each point on the dial the individual at that point should be able to tolerate a greater amount of change or variance in the system. Thus, if the SM is set at Level 7, the person should be able to deal with systems that are less than 100% lawful but still highly (e.g., at least 90%) lawful. The child could achieve a slightly higher IQ (since there is a little more possibility for learning about systems that are less than 100% lawful), and the child would have a little more ability to generalize than someone with classic

Table 17.3
Systemizing Mechanism at Level 8: Classic, Low-Functioning Autism

What does it mean for one's SM to be turned up to Level 8? The person by definition systemizes everything. Since in the social world the information is too complex to be systemized, such individuals focus on systems that are totally lawful (that is, with zero [or minimal] variance). Key behaviors that follow from extreme systemizing include the following:

- *Highly repetitive behavior* (e.g., producing a sequence of actions, sounds, or set phrases, or bouncing on a trampoline)
- *Self-stimulation* (e.g., a sequence of repetitive bodyrocking, finger-flapping in a highly stereotyped manner, spinning oneself round and round)
- *Repetitive events* (e.g., spinning objects round and round, watching the cycles of the washing machine; replaying the same video 1,000 times; spinning the wheels of a toy car)
- *Preoccupation with fixed patterns or structure* (e.g., lining things up in a strict sequence, electrical light switches being in either an on or off position throughout the house; running water from the taps/faucet)
- *Prolonged fascination with systemizable change* (e.g., sand falling through one's fingers, light reflecting off a glass surface, playing the same video over and over again, preference for simple, predictable material such as *Thomas the Tank Engine* movies)
- *Tantrums at change:* As a means to return to predictable, systemizable input with minimal variance
- *Need for sameness:* The child attempts to impose lack of change onto the world, to turn the world into a totally controlled or predictable environment (a Skinner Box), to make it systemizable
- *Social withdrawal:* Since the social world is unsystemizable
- *Narrow interests:* In just one or two systems (types of windows, catalogues of information)
- *Mindblindness:* Since the social world is largely unsystemizable
- *Immersion in detail:* Since a high-systemizing mechanism needs to record each data point (e.g., noticing small changes)
- *Reduced ability to generalize:* Since high systemizing means a reluctance to formulate a law until there has been massive and sufficient data collection (this could also reduce IQ and breadth of knowledge)
- *Severe language delay:* Since other people's spoken language varies every time it is heard, so it is hard to systemize
- *Islets of ability:* Since the high systemizer will channel attention into the minute detail of one lawful system (the script of a video, or the video player itself, spelling of words, prime numbers), going round and round in this system to obtain evidence of its total lawfulness

autism.[2] The higher the SM level, the less generalization is possible, because systemizing involves identifying laws that might apply only to the current system under observation. Systemizing a *Thomas the Tank Engine* video (a favorite of many children with autism) may not lead to a rule about *all* such videos but just a rule that applies to *this* particular one with *this* unique sequence of crackles and hisses.[3]

At Level 7, some language delay is to be expected, but it might be only moderate, since someone whose SM is set at Level 7 can tolerate a little variance in the way language is spoken and still see meaningful patterns. The child's mindblindness would be less than total. If the SM is set at Level 6, the theory predicts that such an individual would be able to deal with systems that are slightly less (e.g., at least 80%) lawful. This would therefore be expressed as only mild language delay, mild obsessions, mild delay in theory of mind, and stilted social behavior, such as attempts at systemizing social behavior (e.g., asking for affirmation of the rule, "You mustn't shout in church, must you?") (Baron-Cohen, 1992).

Autism as a Result of Assortative Mating of Two High Systemizers

It is well established that autism arises for genetic reasons (Bailey et al., 1995; Folstein & Rutter, 1988; Gillberg, 1991). The evidence for systemizing being part of the genetic mechanism for autism includes the following: fathers and grandfathers of children with autism are twice as likely to work as engineers (chosen as a clear example of a systemizing occupation) than are men in the general population (Baron-Cohen, Wheelwright, Stott, Bolton, & Goodyear, 1997). The implication is that these fathers and grandfathers (both maternal and paternal) have their SM set higher than average (Level 4). Consistent with this observation,

2 I am indebted to Nigel Goldenfeld for suggesting this connection between hypersystemizing and IQ.

3 The "reduced generalization" theory of autism (Plaisted et al., 1998) is thus seen as a consequence of hypersystemizing rather than as an alternative theory. Reduced generalization has been noted in autism for many decades (Rimland, 1964) but is not discussed in any functional or evolutionary context. In contrast, systemizing (an evolved function of the human brain) presumes that one does not generalize from one system to another until one has enough information that the rules of system A are identical to those of system B. Good generalization may be a feature of average or poor systemizers, while "reduced" generalization can be seen as a feature of hypersystemizing.

students in the natural sciences—engineering, mathematics, physics, all of which require developed systemizing in relation to mechanical or abstract systems—have a higher number of relatives with autism than do students in the humanities (Baron-Cohen et al., 1998). If systemizing talent is genetic, such genes appear to cosegregate with genes for autism.

The evidence that autism could be the genetic result of having *two* systemizers as parents (assortative mating) includes the following: (1) Both mothers *and* fathers of children with AS have been found to be strong in systemizing on the Embedded Figures test (Baron-Cohen & Hammer, 1997). This study suggests that both parents may be contributing their systemizing genotypes. (2) Both mothers and fathers of children with autism or AS have elevated rates of systemizing occupations among their fathers (Baron-Cohen et al., 1997). (3) Mothers of children with autism show hypermasculinized patterns of brain activity during a systemizing task (Baron-Cohen et al., 2006). (4) The probability of having a brain of Type S (Level 3) in the male population is 0.44, and the probability of having a brain of Type S in the female population is 0.14 (Goldenfeld, Baron-Cohen, Wheelwright, Ashwin, & Chakrabarti, in press). If autism arises from assortative mating of two strong systemizers, then the probability of autism in the population should be $(0.44 \times 0.14) = 0.062$. This is remarkably close to the actual rate of autism spectrum conditions in the general population (Baird et al., 2000; Fombonne, 2001). It is unlikely that the liability genes for autism in males in the general population are common polymorphisms but that these are relatively rare in females in the general population. Rather, it may be that in males the liability genes interact with some other (endocrine?) factor to increase risk, or that in females there is some protective factor that decreases risk.

Hypersystemizing versus Weak Central Coherence versus Executive Dysfunction Theories

The hypersystemizing theory predicts that when presented with information or tasks that can be systemized, and especially when presented with information that derives from a highly lawful system, people with autism spectrum conditions will perform at an intact or even superior level, always relative to a mental-age-matched control group. Such an account differs from the two dominant theories of the nonsocial features of autism, the weak central coherence theory (Frith, 1989) and the executive dysfunction theory (Russell, 1997).

Regarding the former, people with autism perform well on the Embedded Figures test and on the Block Design subtest (Shah & Frith, 1983, 1993), and these results have been interpreted as signs of weak central coherence. But given that both of these are lawful systems, the same data can be taken as evidence of hypersystemizing. People with autism have been shown to have deficits in contextual processing (Jolliffe & Baron-Cohen, 1999), but such material is harder to systemize. Regarding the latter, people with autism show perseveration on the Wisconsin Card Sorting test (Rumsey & Hamberger, 1988), taken as a sign of an executive dysfunction. But their perseveration on this task suggests that people with autism spectrum conditions are focused on establishing a rule (a key aspect of systemizing), and as good systemizers they would not be expected to stop testing the rule but instead to keep on testing the rule, ignoring the experimenter's request to shift to a new, arbitrary rule. What appears as perseveration may therefore be a sign of hypersystemizing. Equally, people with autism may make more moves on the Tower of London test (or its equivalent) (Hughes, Russell, & Robbins, 1994), but if they are more focused on systemizing the task (identifying any lawful regularities), issues such as solving the task in the minimum number of moves may be irrelevant to them. We should be careful not to attribute a deficit to people with autism spectrum conditions when they may simply be approaching the task from a different standpoint from the experimenter's.

Conclusion

According to the hypersystemizing theory, the core of autism is both a social deficit (since the social world is the ultimate unsystemizable domain) and what Kanner (1943) astutely observed and aptly named "the need for sameness." Autism is the result of a normative systemizing mechanism—the adaptive function of which is to serve as a law detector and a change-predicting mechanism—being set too high. This theory explains why people with autism prefer either no change or systems that change in highly lawful or predictable ways (i.e., systems with simple change, such as mathematics, physics, repetition, objects that spin, routine, music, machines, collections) and why they become disabled when faced with systems characterized by complex change (such as social behavior, conversation, people's emotions, or fiction). Because they cannot systemize complex change, they become "change resistant" (Gomot et al., 2005).

While autism spectrum conditions are disabling in the social world, their strong systemizing can lead to talent in areas that are systemizable. For many people with autism spectrum conditions, the hypersystemizing never moves beyond phase 1: the massive collection of facts and observations (lists of dates and the rainfall on each of these, lists of trains and their departure times, lists of records and their release dates, watching the spin cycle of a washing machine) or highly repetitive behavior (spinning a plate or the wheels of a toy car). But for those who go beyond phase 1 to identify a law or a pattern in the data (phase 2 of systemizing), this can constitute original insight. In this sense, it is likely that the genes for increased systemizing have made remarkable contributions to human history (Fitzgerald, 2000, 2002; James, 2003).

Acknowledgments I am grateful for the support of the Medical Research Council during this work. Matthew Belmonte, Marie Gomot, Johnny Lawson, Jac Billington, Nigel Goldenfeld, and Sally Wheelwright all contributed in important ways to the discussion of these ideas. Portions of this chapter appeared in *Neuropsychopharmacology and Biological Psychiatry* (2005).

References

American Psychiatric Association (1994). *Diagnostic and statistical manual of mental disorders* (4th ed.). Washington, DC: American Psychiatric Association.

Asperger, H. (1944). Die "Autistischen Psychopathen" im Kindesalter. *Archiv für Psychiatrie und Nervenkrankheiten*, *117*, 76–136.

Bailey, A., Le Couteur, A., Gottesman, I., Bolton, P., Simonoff, E., Yuzda, E., et al. (1995). Autism as a strongly genetic disorder: Evidence from a British twin study. *Psychological Medicine*, *25*, 63–77.

Baird, G., Cox, A., Charman, T., Baron-Cohen, S., Wheelwright, S., Swettenham, J., et al. (2000). A screening instrument for autism at 18 months of age: A six-year follow-up study. *Journal of the American Academy of Child and Adolescent Psychiatry*, *39*, 694–702.

Baron-Cohen, S. (1992). The girl who liked to shout in church. In R. Campbell (Ed.), *Mental lives: Case studies in cognition.* Oxford: Basil Blackwell.

Baron-Cohen, S. (1994). How to build a baby that can read minds: Cognitive mechanisms in mindreading. *Cahiers de Psychologie Cognitive/Current Psychology of Cognition*, *13*, 513–552.

Baron-Cohen, S. (1995). *Mindblindness: An essay on autism and theory of mind.* Cambridge, MA: MIT Press/Bradford Books.

Baron-Cohen, S. (2002). The extreme male brain theory of autism. *Trends in Cognitive Sciences*, *6*, 248–254.

Baron-Cohen, S. (2003). *The essential difference: Men, women and the extreme male brain.* London: Penguin.

Baron-Cohen, S. (2005). The empathizing system: A revision of the 1994 model of the mindreading system. In B. Ellis & D. Bjorklund (Eds.), *Origins of the social mind.* Surrey, UK: Guilford Publications.

Baron-Cohen, S., Bolton, P., Wheelwright, S., Short, L., Mead, G., Smith, A., et al. (1998). Does autism occurs more often in families of physicists, engineers, and mathematicians? *Autism, 2,* 296–301.

Baron-Cohen, S., Golan, O., Wheelwright, S., & Hill, J. J. (2004). *Mindreading: The interactive guide to emotions.* London: Jessica Kingsley.

Baron-Cohen, S., & Goodhart, F. (1994). The mindreading system: New directions for research. *Current Psychology of Cognition, 13,* 724–750.

Baron-Cohen, S., Ring, H., Chitnis, X., Wheelwright, S., Gregory, L., Willams, S., Brammer, M. J., & Bullmore, E. T. (2006). Mothers of children with autism: An fMRI study of the "Eyes" and Embedded Figures test and the "extreme male brain" theory. Unpublished manuscript, University of Cambridge.

Baron-Cohen, S., & Hammer, J. (1997). Parents of children with Asperger syndrome: What is the cognitive phenotype? *Journal of Cognitive Neuroscience, 9,* 548–554.

Baron-Cohen, S., Leslie, A. M., & Frith, U. (1986). Mechanical, behavioral and intentional understanding of picture stories in autistic children. *British Journal of Developmental Psychology, 4,* 113–125.

Baron-Cohen, S., Richler, J., Bisarya, D., Gurunathan, N., & Wheelwright, S. (2003). The Systemising Quotient (SQ): An investigation of adults with Asperger syndrome or high functioning autism and normal sex differences. In *Autism: Mind and brain* [Special issue]. *Philosophical Transactions of the Royal Society, Series B, 358,* 361–374.

Baron-Cohen, S., Ring, H., Wheelwright, S., Bullmore, E. T., Brammer, M. J., Simmons, A., et al. (1999). Social intelligence in the normal and autistic brain: An fMRI study. *European Journal of Neuroscience, 11,* 1891–1898.

Baron-Cohen, S., & Wheelwright, S. (1999). Obsessions in children with autism or Asperger syndrome: A content analysis in terms of core domains of cognition. *British Journal of Psychiatry, 175,* 484–490.

Baron-Cohen, S., & Wheelwright, S. (2003). The Friendship Questionnaire (FQ): An investigation of adults with Asperger syndrome or high functioning autism, and normal sex differences. *Journal of Autism and Developmental Disorders, 33,* 509–517.

Baron-Cohen, S., Wheelwright, S., Scahill, V., Lawson, J., & Spong, A. (2001). Are intuitive physics and intuitive psychology independent? *Journal of Developmental and Learning Disorders, 5,* 47–78.

Baron-Cohen, S., Wheelwright, S., Skinner, R., Martin, J., & Clubley, E. (2001). The Autism Spectrum Quotient (AQ): Evidence from Asperger syndrome/high functioning autism, males and females, scientists and mathematicians. *Journal of Autism and Developmental Disorders, 31,* 5–17.

Baron-Cohen, S., Wheelwright, S., Stone, V., & Rutherford, M. (1999). A mathematician, a physicist, and a computer scientist with Asperger syndrome:

Performance on folk psychology and folk physics test. *Neurocase*, *5*, 475–483.

Baron-Cohen, S., Wheelwright, S., Stott, C., Bolton, P., & Goodyear, I. (1997). Is there a link between engineering and autism? *Autism: An International Journal of Research and Practice*, *1*, 153–163.

Bolton, P. (1996). Genetic advances and their implications for child psychiatry. *Child Psychology and Psychiatry Review*, *1*, 82–93.

Charman, T., & Baron-Cohen, S. (1992). Understanding beliefs and drawings: A further test of the metarepresentation theory of autism. *Journal of Child Psychology and Psychiatry*, *33*, 1105–1112.

Connellan, J., Baron-Cohen, S., Wheelwright, S., Ba'tki, A., & Ahluwalia, J. (2001). Sex differences in human neonatal social perception. *Infant Behavior and Development*, *23*, 113–118.

Fitzgerald, M. (2000). Did Ludwig Wittgenstein have Asperger's syndrome? *European Child and Adolescent Psychiatry*, *9*, 61–65.

Fitzgerald, M. (2002). Did Isaac Newton have Asperger's syndrome or disorder? *European Journal of Adolescent Psychiatry.*

Folstein, S., & Rutter, M. (1988). Autism: Familial aggregation and genetic implications. *Journal of Autism and Developmental Disorders*, *18*, 3–30.

Fombonne, E. (2001). Is there an epidemic of autism? *Pediatrics*, *107*, 411–412.

Frith, C., & Frith, U. (1999). Interacting minds: A biological basis. *Science*, *286*, 1692–1695.

Frith, U. (1970). Studies in pattern detection in normal and autistic children. II. Reproduction and production of color sequences. *Journal of Experimental Child Psychology*, *10*(1), 120–135.

Frith, U. (1989). *Autism: Explaining the enigma.* Oxford: Basil Blackwell.

Frith, U. (1991). *Autism and Asperger's syndrome.* Cambridge: Cambridge University Press.

Gillberg, C. (1991). Clinical and neurobiological aspects of Asperger syndrome in six family studies. In U. Frith (Ed.), *Autism and Asperger syndrome.* Cambridge: Cambridge University Press.

Goldenfeld, N., Baron-Cohen, S., Wheelwright, S., Ashwin, C., & Chakrabarti, B. (in press). Empathizing and systemizing in males and females, and in autism spectrum conditions. In T. Farrow (Ed.), *Empathy and mental illness.* Cambridge: Cambridge University Press.

Gomot, M., Bernard, F. A., Davis, M. H., Belmonte, M. K., Ashwin, C., Bullmore, E. T., et al. (2005). Change detection in children with autism: An auditory event-related fMRI study. *NeuroImage*, *29*, 475–495.

Happé, F. (1996). *Autism.* London: University College London Press.

Happé, F., Briskman, J., & Frith, U. (2001). Exploring the cognitive phenotype of autism: Weak "central coherence" in parents and siblings of children with autism. I. Experimental tests. *Journal of Child Psychology and Psychiatry*, *42*, 299–308.

Happé, F., Ehlers, S., Fletcher, P., Frith, U., Johansson, M., Gillberg, C., et al. (1996). Theory of mind in the brain: Evidence from a PET scan study of Asperger syndrome. *Neuroreport*, *8*, 197–201.

Heider, F., & Simmel, M. (1944). An experimental study of apparent behavior. *American Journal of Psychology, 57*, 243–259.

Hermelin, B. (2002). *Bright splinters of the mind: A personal story of research with autistic savants.* London: Jessica Kingsley.

Hughes, C., Russell, J., & Robbins, T. (1994). Evidence for executive dysfunction in autism. *Neuropsychologia, 32*, 477–492.

James, I. (2003). Singular scientists. *Journal of the Royal Society of Medicine, 96*, 36–39.

Jolliffe, T., & Baron-Cohen, S. (1997). Are people with autism or Asperger's syndrome faster than normal on the Embedded Figures task? *Journal of Child Psychology and Psychiatry, 38*, 527–534.

Jolliffe, T., & Baron-Cohen, S. (1999). Linguistic processing in high-functioning adults with autism or Asperger syndrome: Is local coherence impaired? *Cognition, 71*, 149–185.

Kanner, L. (1943). Autistic disturbance of affective contact. *Nervous Child, 2*, 217–250.

Kimura, D. (1999). *Sex and cognition.* Cambridge, MA: MIT Press.

Knickmeyer, R., Baron-Cohen, S., Raggatt, P., & Taylor, K. (2004). Foetal testosterone, social cognition, and restricted interests in children. *Journal of Child Psychology and Psychiatry, 46*, 198–210.

Lawson, J., Baron-Cohen, S., & Wheelwright, S. (2004). Empathising and systemising in adults with and without Asperger syndrome. *Journal of Autism and Developmental Disorders, 34*, 301–310.

Leslie, A. M., & Thaiss, L. (1992). Domain specificity in conceptual development: Evidence from autism. *Cognition, 43*, 225–251.

Myers, P., Baron-Cohen, S., & Wheelwright, S. (2004). *An exact mind.* London: Jessica Kingsley.

O'Riordan, M., Plaisted, K., Driver, J., & Baron-Cohen, S. (2001). Superior visual search in autism. *Journal of Experimental Psychology: Human Perception and Performance, 27*, 719–730.

Perrett, D., Smith, P., Potter, D., Mistlin, A., Head, A., Milner, A., et al. (1985). Visual cells in the temporal cortex sensitive to face view and gaze direction. *Proceedings of the Royal Society of London, B, 223*, 293–317.

Plaisted, K., O'Riordan, M., & Baron-Cohen, S. (1998a). Enhanced discrimination of novel, highly similar stimuli by adults with autism during a perceptual learning task. *Journal of Child Psychology and Psychiatry, 39*, 765–775.

Plaisted, K., O'Riordan, M., & Baron-Cohen, S. (1998b). Enhanced visual search for a conjunctive target in autism: A research note. *Journal of Child Psychology and Psychiatry, 39*, 777–783.

Premack, D. (1990). The infant's theory of self-propelled objects. *Cognition, 36*, 1–16.

Rimland, B. (1964). *Infantile autism: The syndrome and its implications for a neural theory of behavior.* New York: Appleton-Century-Crofts.

Rumsey, J., & Hamberger, S. (1988). Neuropsychological findings in high functioning men with infantile autism, residual state. *Journal of Clinical and Experimental Neuropsychology, 10*, 201–221.

Russell, J. (Ed.). (1997). *Autism as an executive disorder.* Oxford: Oxford University Press.

Scaife, M., & Bruner, J. (1975). The capacity for joint visual attention in the infant. *Nature, 253*, 265–266.

Shah, A., & Frith, U. (1983). An islet of ability in autism: A research note. *Journal of Child Psychology and Psychiatry, 24*, 613–620.

Shah, A., & Frith, U. (1993). Why do autistic individuals show superior performance on the Block Design test? *Journal of Child Psychology and Psychiatry, 34*, 1351–1364.

Skinner, B. F. (1976). *About behaviorism.* New York: Vintage Books.

18 Deception, Evolution, and the Brain

Sean T. Stevens, Kevin Guise, William Christiana, Monisha Kumar, and Julian Paul Keenan

The advantages that deception currently incurs strongly suggest that similar benefits may have existed for our ancestors. The ability to instill false belief in a competitor confers benefits on the deceiver in terms of the acquisition of resources and gaining access to mates. However, such behavior encounters competition in the sense that conspecifics possess the ability to detect instances of deception based on physiological clues. Self-deception may enable one to deceive more effectively in the sense that detection becomes increasingly difficult the more faithfully the deceiver believes her or his own lies. We propose that natural selection favored the evolution of brains capable of deception through the selection of individuals with specific characteristics of memory and other higher-order processes that give rise to self-deception. We also review theories postulating that self-deception is requisite for effective deception.

In a highly social species, deception can be extremely advantageous. Successful deception can provide one with greater access to limited resources, as well as an advantage over conspecifics. Deception can be defined as an instance in which a particular agent, often in an attempt to further her or his own cause, attempts to convince another of information incongruent with the truth (Spence et al., 2001). Deception can take many forms (Ekman, 1985; Hirstein, 2005) and can be carefully constructed in advance or enacted spontaneously. Investigations of the neural correlates of such processes have turned up results that appear to be equally as varied (Ganis, Kosslyn, Stose, Thompson, & Yurgelun-Todd, 2003; Johnson, Barnhardt, & Zhu, 2004; Lee et al., 2002; Spence et al., 2001). For example, Ganis et al. (2003) have suggested that several discrete neural networks may operate to support different forms of deception and that any investigation of the neural correlates of deception must consider different types of deception.

Biologists have long accepted that deception is a near-universal feature of predator–prey relationships. It is now largely accepted that deception is a widespread feature in primate communication (Trivers, 1991), although the extent to which it is intentional remains debated. The sophistication of deception in primates has been found to be positively correlated with neocortical volume (Byrne & Corp, 2004). As language further increases social interaction among humans, one would expect to encounter highly complex forms of deception. The human capacity for self-awareness and the subsequent development of a theory of mind (ToM) provides an individual with the capacity for autobiographical memories, the ability to project one's self into the future or the past, and the ability to model the mental state of others based on prior experience. These abilities allow for highly complex forms of deception between humans (Keenan, Gallup, & Falk, 2003; Trivers, 1991). In addition, because deception is often costly to the deceived, it is believed that selection pressures for heightened detection have correspondingly honed abilities to detect such instances, though it is noted that humans are generally poor at detecting deception (DePaulo, Lindsay, Malone, Muhlenbruck, Charlton, & Cooper, 2003).

Trivers's Evolutionary Theory of Self-Deception

Deception can confer great benefit on the deceiver, though not without cost. Misrepresentation may be used as a means to secure resources as well as opportunities to mate. However, one is also at risk for gaining a reputation that one is not to be trusted, thus decreasing the benefits of deception. Research investigating this relationship has been examined in detail elsewhere (see Whiten & Byrne, 1997), in which traditional cost-benefit analyses have guided such endeavors. Nonetheless, a number of questions remain, primarily concerning the phenomenon of self-deception.

Trivers has proposed an evolutionary theory of self-deception in which self-deception (defined as an active misrepresentation of reality to a conscious mind) has its roots in an "evolutionary arms race" between deception and deception detection. Trivers proposes that there exists a tendency for deception between individuals to generate patterns of self-deception within individuals. In highly social species, the detection of deception involves a careful examination of another's behavior with the intent of making inferences regarding the other's intention or mental state. This close scrutiny increases the stress of the deceiver and may be accompanied by covert displays indicative of one's true intention. Numerous cues to emotion have been established in humans: facial expression, parlance, prosody, voice quality, eye movement, small movements of extremities, and emotional microexpressions. It is possible to detect and use these cues in detecting deception. Therefore, to avoid detection, it is advantageous for an individual to believe his or her own lie, reducing the number of unconscious deception cues that are displayed (Ekman, 1985; Trivers, 1991).

In social environments it may be adaptive for the deceiver to relegate the knowledge of deception to the unconscious. This would better mask any behavioral or physiological indications of attempted deceit, making detection more difficult. Trivers and Ekman both argue that the lies of those who are unaware of deceiving are the most difficult to detect (Ekman, 1985; Trivers, 1991). Therefore, it is proposed that the most effective forms of deception require the deceiver to first deceive herself or himself. In other words, effective deception requires effective self-deception.

Studying Self-Deception

The existence of self-deception has been a topic of debate among philosophers and psychologists for centuries (Gur & Sackheim, 1979). However, based on a number of psychological phenomena, such as confabulation, delusional misidentification syndrome, delusional reduplication syndrome, false memories, and false recognition (Byrne & Kurland, 2001), there is little doubt that the human mind is capable of extreme distortion of current and past experience. As an example, it has been suggested that some larger societies collectively reconstruct their own history (Schacter, 1995). Examination of related phenomena may provide us with the opportunity to study the neural correlates of various forms of self-deception.

Gur and Sackheim (1979) proposed a definition of self-deception, arguing that four criteria are necessary for self-deception to occur: an individual must (1) simultaneously (2) hold two contradictory beliefs (3) one of which the individual is not aware of (4) because of a motivated act. Following this definition, when studying self-deception in humans, Trivers expects to find three features of self-deception: (1) true information and false information are simultaneously stored in the individual; (2) the true information is unconscious, the false information is conscious; and (3) the form of an individual's self-deception can be affected by changing the individual's relationship to others.

How can an individual simultaneously represent two alternative forms of some piece of information, knowing which has a basis in the truth, and still be deceived by herself or himself? Trivers proposes that this ability is attributable to the existence of conscious and unconscious states and processes. The split of consciousness and unconsciousness is proposed to be related to energy efficiency. The ability to selectively focus attention, as conferred by the conscious state, requires a great deal of energy. In contrast, numerous processes operate outside of conscious thought and are less demanding of resources. Conscious monitoring and suppression of possible cues to deceptive communication therefore should be taxing on energy resources, perhaps so much as to negate the benefits reaped upon successful deception. However, if one can represent the factual alternative of some piece of information outside of consciousness yet be consciously aware of only the representation of the false alternative, then the latter may be communicated with confidence by the deceiving individual. Thus, as Ekman (1985) has suggested, those who begin to believe their own lies—and thus are victims of self-deceit—will greatly improve their ability to deceive at lesser cost.

Ramachandran's Evolutionary Theory of Self-Deception

Ramachandran has proposed an alternate theory about the evolution of self-deception, suggesting that the real reason for the evolution of such mechanisms (e.g., confabulation or Freudian defense mechanisms) is to create a coherent belief system for the self. This allows the individual to act in such a manner that stability is imposed on his or her general schema (Ramachandran, 1996; Ramachandran & Blakeslee, 1998). This view takes a priori that each of us has a need for consistency, coherence, and continuity in our behavior and that in response to instances that do not fit our script, we tend to engage in self-deceptive behavior in order to preserve the autobiographical script and make the conflicting information "fit" (Ramachandran, 1996; Ramachandran & Blakeslee, 1998). Ramachandran further argues that, although Trivers's theory probably has some degree of truth to it, its natural conclusion is that self-deception becomes maladaptive and would not provide access to a greater evolutionary advantage. Instead it serves to prevent both the deceiver and the deceived from benefiting.

Ramachandran gives a hypothetical example of two chimpanzees in a zoo, one of which has seen where the zookeeper has placed food (Chimp A) whereas the other has not (Chimp B). In order to gain access to the food, Chimp A will engage in a form of deception and point in the wrong direction, away from the food, in order to deceive Chimp B. Ramachandran argues that, according to Trivers's theory, to prevent Chimp B from detecting the deceptive communication, Chimp A must also engage in self-deception and will now therefore believe the bananas are in the wrong place. This can not, in any way, benefit either chimp (Ramachandran, 1996).

Trivers versus Ramachandran

Recently, Byrne and Kurland (2001) pitted the theories of Trivers and Ramachandran in the context of an evolutionary hawk-dove game to determine which strategy is more stable in evolutionary terms. Self-deception was viewed as a mental state that is acute and transient: once an individual succeeds in deceiving a rival, self-deception ceases and the individual no longer engages in it. This appears to provide a counterargument to Ramachandran's chimpanzee scenario. Now, Chimp A (as before) sees where the zookeeper has placed the food, and deceives Chimp B into thinking that the food is somewhere else through self-deception. However, instead of continuing to deceive himself, Chimp A's self-deception terminates, having accomplished the goal of deceiving a rival. The outcome of the hawk-dove game largely supported Trivers's theory (Byrne & Kurland, 2001).

Investigating the Neural Correlates of Self-Deception

It may be possible to learn a great deal about self-deception and its neural correlates by studying individuals with psychological disorders and comparing their behavior with that of normal subjects. Disorders such as anosognosia, asomatognosia, and the delusional misidentification syndromes and the associated behavior of confabulation are of interest to the study of self-deception, as it is self-deception that perpetuates the confabulation. By investigating the neural correlates of these various psychological phenomena, we may be able determine the neural correlates of self-deception in humans.

Confabulation

Confabulation is defined as lying without the knowledge or intention of doing so. The confabulating individual fully believes her or his own statements even when presented with clear evidence of their falsity. Confabulation differs from forms of deception previously discussed, as there appears to be no intention to alter the thinking of others; it is not an instance of tactical deception. As a result, the statements of confabulating individuals are often unbelievable. Therefore, it is presumed that confabulatory statements are wholeheartedly believed by the individual, as a successful lie needs to be fabricated with regard to what the individual to be deceived may find believable, else effort is wasted. Investigating confabulation in terms of self-deception is often difficult because some instances pose a challenge to Trivers's first criterion for the study of self-deception (see below).

The executive theory of confabulation suggests that self-deceptive confabulatory behavior results from damage to substrates subserving at least two processes: memory and an executive "checker" (Burgess & Shallice, 1996; Kapur & Coughlan, 1980; Stuss, Alexander, Lieberman, & Levine, 1978). According to this theory, first a false memory or deceptive response is produced, then an executive process subsequently fails to check or inhibit the erroneous response. In terms of self-deception, this suggests that for self-deception to occur, first a false memory or deceptive response must be constructed. Subsequently an executive process must facilitate this process either by failing to detect it or by suppressing the unwanted truthful response from consciousness. Convergent data, reviewed later in the chapter, implicate the right hemisphere and the anterior cingulate in the checking of information to be communicated and in subsequent self-deception through the suppression of an unwanted or undesired response.

Asomatognosia

Individuals with asomatognosia often claim that their arm or leg (typically the left) does not belong to them. This condition often occurs in conjunction with left hemiplegia, left hemispatial neglect, or severe sensory loss on the left side of the body, all of which are caused by damage to the right cerebral hemisphere (Feinberg & Keenan, 2005).

Ramachandran and Blakeslee (1998) note that a number of the confabulatory claims made by these patients offer a hint that, subconsciously, the knowledge of paralysis exists. To garner evidence for this theory, Ramachandran duplicated an experiment performed by Bisiach, in which a syringe filled with ice-cold water was discharged into the left external acoustic meatus of a patient, inducing convection currents in the semicircular canals of the inner ear, disrupting vestibular function, and inducing nystagmus. After such treatment, patients who had previously denied their paralysis now admitted that they could not use their left arm. Injecting ice-cold water into the right ear, and thus inducing a different pattern of eye movement, did not produce any such effect regarding confabulation. Ramachandran was able to duplicate this finding in one of his patients, who remained cognizant of the paralysis of her left arm for at least a half-hour after treatment before reverting back to a confabulatory state (Ramachandran & Blakeslee, 1998). The explanation offered by Ramachandran related the induced honesty to the recollection of information during REM sleep; however, Ramachandran admits that he does not place much premium on the theory. These observations confirm Trivers's proposed characteristics of self-deception: both alternative forms of information are stored within the system, with the individual aware only of the false alternative.

Delusional Misidentification Syndromes

The delusional misidentification syndromes (DMS) feature several subdivisions and are characterized by patients holding a belief that the physical or psychological identity (or both) of others and/or the self have changed into something else (Lewis-Lehr, Slaughter, Rupright, & Singh, 2000). These syndromes are categorized as dissociation between recognition and identification (Luaute & Bidault, 1994) and are defined as conditions in which a patient consistently and adamantly misidentifies persons, places, objects, or events.

An individual with Capgras' syndrome claims that people he or she knows have been replaced by impostors. In some cases the delusion is even directed at the self. Such individuals tend to claim that their reflection in the mirror is not themselves (Ellis, Luaute, & Retterstol, 1994). Capgras' syndrome can be distinguished from prosopagnosia (lack of facial recognition) because patients with the latter use other features, such as voice recognition or clothing, to correctly identify the person with whom they are interacting. Patients with Capgras' syndrome, on the other hand, firmly maintain their delusional claims that an individual they know—parent, friend, doctor—has been replaced by an impostor. Declerambault's syndrome often occurs with Capgras' syndrome and typically involves claims that a certain, often famous, person is secretly in love with the patient (O'Dwyer, 1990; Signer & Isbister, 1987). Signer (1994) reviewed 252 cases of Capgras' syndrome and hypothesized that the disorder results from a combination of left temporal and right frontal damage. The possibility of right frontal damage playing a role in DMS or delusional reduplication syndrome (DRS) is also supported by the findings of Feinberg and Keenan (2005), as well as by a number of case studies.

Patients with Fregoli's syndrome claim that several unfamiliar people, despite having a different appearance, are the same person, usually someone they know (Paillere-Martinot, Dao-Castallana, Masure, Pillon, & Martinot, 1994). We can all experience this form of misidentification from time to time. For example, when looking for a friend in a crowd of people we sometimes mistake a stranger for the person we are looking for. However, patients with Fregoli's syndrome experience this form of misidentification at significantly higher rates than the general population. Hirstein (2005) suggests that this syndrome appears to be the opposite of Capgras' syndrome: instead of claiming that impostors have replaced familiar people, patients instead claim that unfamiliar people are someone that they know. In one instance a patient claimed that the hospital's nurses, doctors, and therapists were his sons, daughters-in-law, and co-workers. In another instance a man diagnosed with Fregoli's syndrome mistook both people and places for familiar. He claimed that the hospital was his place of employment, that he dealt with computers, and that he even had an office on the premises. He also misidentified several hospital staff as co-workers from his company (Feinberg & Keenan, 2005).

Another DMS/DRS is Cotard's syndrome, in which patients claim that they or others are "dead" or "empty," that they feel nothing inside, and that they have been catapulted into a parallel reality or hell (Butler, 2000). The illusion of subjective doubles involves a patient claiming that he or she has a *Doppelgänger*. This syndrome differs from Capgras' syndrome because the double is considered a separate entity from the patient. A patient with intermetamorphosis disorder claims that people change into other people, sometimes right before the patient's eyes.

In all the forgoing delusional syndromes, affected individuals are confident that their statements are correct and do not indicate that they have any awareness of the obvious falsity of such claims. In essence, they are confabulating. Ramachandran (1996) has labeled such behavior self-deceptive. Damage to frontal cortical regions has been implicated in both confabulation and DMS. A large body of research suggests that DMS more often results after damage to the right hemisphere than after damage to the left (Feinberg & Keenan, 2005; Fleminger & Burns, 1993; Forstl, Almeida, Owen, Burns, & Howard, 1991; Hakim, Verma, & Greiffenstein, 1988). Other researchers have suggested that DMS results from damage to both frontal and temporal areas (Alexander, Stuss, & Benson, 1979; Signer, 1994).

False Memories and False Recognition

It is now widely accepted that our memories of experiences, thoughts, and facts are not snapshots of the past in which every detail is preserved. They are instead constructed based on the limited information that has been stored, what has been previously recalled, and current conditions (Gonsalves et al., 2004; Schacter, 1995, 2001). A false memory (or reality-monitoring error) can be defined as an event that is only imagined being remembered as if it had happened (Gonsalves et al., 2004). Such memories may arise due to similarities between the imagined and perceived (or "real") events (Gonsalves et al., 2004; Johnson, Hashtoudi, & Lindsay, 1993; Johnson & Raye, 1981). The more vividly imagined the false memory is, the greater its chance of acceptance as truth (Gonsalves et al., 2004; Loftus, 1997).

In 1959, James Deese instructed subjects in an experiment to study a list of words that were strong associates of a nonstudied lure word. Participants were then asked to recall the studied words. Deese reported that participants often produced the nonstudied lure words on the recall test. In other words, a false memory for the strongly associated items had been created. This procedure is now used in various modified forms to induce false memories in subjects in laboratory experiments (Roediger & McDermott, 1995; Schacter, 1995, 2001) and in conjunction with neuroimaging techniques to explore the neural correlates of true and false memories (Gonsalves & Paller, 2000; Gonsalves et al., 2004; Schacter, 1995; Slotnick & Schacter, 2004).

In the 1950s, the practice of cingulectomy, a bilateral ablation of the anterior cingulate (Whitty & Lewin, 1957) was used to treat severe obsessional neurosis. The procedure produced some unexpected results: patients reported difficulty in distinguishing their thoughts from events occurring in the external world. A common complaint among patients was that their lives had become like a waking dream, or that dreams could not be distinguished from reality (Whitty & Lewin, 1957).

Studies have indicated that true and false memories produce different patterns of neuronal activation, suggesting that such memories are differentially represented in the brain (Fabiani, Stadler, & Wessels, 2000). In addition, it has been found that hemispheric differences exist in the false recognition of words not previously studied (Fabiani et al., 2000; Ito, 2001). Slower reaction time was found for false recognition of nonstudied words when the words were presented to the right hemisphere (Fabiani et al., 2000). Ito (2001) found the left hemisphere to be significantly more accurate at detecting targets from a series of presentations of both target and lure words.

Functional magnetic resonance imaging (fMRI) and event-related potential (ERP) studies have extended these findings and suggested that the recollection of true memories and the recognition of false memories, although sharing some common substrates, are associated with a distinct pattern of cortical activation (Gonsalves & Paller, 2000; Gonsalves et al., 2004; Okado & Stark, 2003). In these studies, participants were presented with a word and instructed to visualize a common object in response. Following a certain subset of words, a photograph was presented. Participants were engaged in a recall task and asked whether a word was previously accompanied by a photograph (Gonsalves & Paller, 2000; Gonsalves et al., 2004). fMRI results indicated significantly greater levels of activation in the anterior cingulate, right inferior parietal cortex, and the precuneus in relation to words falsely remembered as being accompanied by a photograph as compared to correct rejections; a larger response associated with correct rejections as compared to false memories in the left inferior frontal gyrus; larger activations in the left inferior frontal gyrus and the left anterior hippocampus associated with remembered photographs as compared to forgotten photographs; and a larger response for later forgotten photographs than for remembered ones in the right inferior parietal cortex and the precuneus (Gonsalves et al., 2004).

Okado and Stark (2003) utilized a similar paradigm, presenting subjects with spoken words and accompanying half of them with pictures and half of them with blank rectangles. Participants were instructed to visualize an image of the word presented and indicate whether it was larger or smaller than a shoebox. Between the study and recall phases, participants underwent a misinformation phase in which a lie test was administered. Participants were presented with recorded words and asked whether they had seen a picture of the object in the test phase. A point system designed to encourage lying was used during this phase. Immediately afterward, the recall task was administered. This task featured words previously presented with and without pictures as well as new words not previously encountered. fMRI results indicated greater activity in the right anterior cingulate for imagined pictures endorsed as seen as compared to pictures correctly endorsed as previously viewed or imagined pictures correctly rejected (Okado & Stark, 2003).

Recently, researchers have demonstrated that true and false memories have distinct unconscious sensory signatures (Slotnick & Schacter, 2004). They suggest that the early visual processing regions (BA 18, BA 19) show greater activity during true recognition than during false recognition. Further, they suggest that this early visual processing activity is a type of implicit memory, implying that the sensory signature differentiating true and false recognition may be inaccessible to conscious awareness (Slotnick & Schacter, 2004).

fMRI results indicate that false recognition of related shapes compared to true recognition of shapes shows greater activation in the anterior cingulate gyrus (BA 24, BA 32), as well as in a number of other regions. Interestingly, the left anterior cingulate gyrus (BA 32), in addition to other regions, also shows greater activation for both true recognition of shapes and false recognition of related shapes when compared to new shapes. Both the left and right anterior cingulate gyrus were associated with false recognition compared to true recognition, but only the left anterior cingulate gyrus was associated with both true recognition and false recognition when compared to new correct rejections. This pattern of activation suggests that the anterior cingulate gyrus may have access to both true and false information in an individual. Finally, areas of the right frontal lobe showed greater activation for false recognition than it did for true recognition, suggesting a right hemisphere role in false recognition (Slotnick & Schacter, 2004).

Over a variety of recall and recognition tasks, different patterns of brain activation occur during accurate and inaccurate memories. Across a number of tasks, regions of the left hemisphere are primarily active during accurate memories, whereas regions of the right hemisphere are primarily active during false memories and false recognition. These results suggest a dominant role of the right hemisphere in the experience of false memories and false recognition and a role of the left hemisphere in accurate recall and recognition. A role of the anterior cingulate consistent with its hypothesized role in the executive theory of confabulation (see previously) is also supported.

Case Study: The case of B.G., a man who had sustained right frontal lobe infarction, involved an extremely high rate of occurrence of false recognition across a number of different remember/know recognition tests for words presented visually or aurally, environmental sounds, pseudowords, and pictures (Schacter, Curran, Galluccio, Milberg, & Bates, 1996). Further tests extended this extremely high false recognition rate to include yes/no recognition tasks (Curran, Schacter, Norman, & Galluccio, 1997). These increased false alarm rates have been suggested to be the result of overreliance on general similarity. As B.G. sustained considerably greater damage to the right hemisphere, this case strengthens the above neuroimaging results, providing a causal link between the right frontal lobe and experiences of false recall and false recognition.

False Memory as a Form of Self-Deception

Might it be possible that false memory constitutes an instance of self-deception? Neuroimaging data demonstrating similar neural correlates for false memory and self-deception suggest this may be so. For example, the anterior cingulate and precuneus have been implicated in both deception and false memory by means of fMRI (Ganis et al., 2003; Gonsalves & Paller, 2000; Gonsalves et al., 2004; Johnston et al., 2004; Kozel, Padgett, & George, 2004; Langleben et al., 2002). Data further implicating the anterior cingulate in deception demonstrate that it is also involved in conflict monitoring (Botvinick, Nystrom, Fissel, Carter, & Cohen, 1999), the suppression of an unwanted thought (Wyland, Kelley, Macrae, Gordon, & Heatherton, 2003), and the conscious self-regulation (inhibition) of sexual arousal (Beauregard, Levesque, & Bourgouin, 2001). All of these activities may be thought of as forms of self-deception, because participants made conscious decisions to suppress unwanted thoughts and inhibit sexual arousal.

The Seven Sins of Memory

Human memory does not typically encode and store all details of an experience but rather the general essence of the experience and a small subset of details specifically attended to. Some may see this as a flaw in the system, but others have suggested it may be evolutionarily adaptive

(Schacter, 2001). In relation to the study of self-deception, these so-called flaws of memory may facilitate the adaptive nature of self-deception. Schacter has proposed the existence of seven "sins of memory," which he further divides into two groups: the sins of omission (transience, absent-mindedness, and blocking) and the sins of commission (misattribution, suggestibility, bias, and persistence).

Sins of Omission

Transience is defined as the weakening or loss of memory over time, or simply forgetting. Shortly after some experience, memory for that instance is initially recalled with great detail. However, the details begin to fade over time, and we may remember the gist of the experience but not the minute details. In one particular fMRI study investigating the neural correlates of transience, subjects were to make a decision on the concreteness or abstractness of a previously presented word at a later time during the experiment. Greater activations were found in the left parahippocampal gyrus and left frontal lobe for decisions made on correctly remembered words compared to forgotten words (Wagner et al., 1998). Slotnick and Schacter (2004) also observed that left frontal activation was greater during true versus false recall and recognition. Right frontal areas were more active during false versus true recall and recognition.

Absent-mindedness entails a breakdown between attention and memory. Forgetting where one has placed some personal possession is an example of absent-mindedness and occurs when attention is not focused on the task at hand. This differs from transience in that a memory has not been encoded prior to the retrieval attempt.

A failure in an attempt to retrieve information from memory constitutes Schacter's third sin of memory, blocking. One of the most common manifestations is the tip-of-the-tongue phenomenon, in which an individual blocks on a specific word he feels he knows, but cannot retrieve. Blocking may also include forms of repression or suppression. A recent fMRI study found significantly greater levels of activation in the anterior cingulate cortex, dorsolateral prefrontal cortex, ventrolateral prefrontal cortex, premotor areas (BA 6, BA 6/9), the intraparietal sulcus, and the right putamen during suppression of unwanted memories compared to retrieval. The activation of these areas was accompanied by reduced activation of the hippocampus (Ochsner et al., 2004).

Other fMRI studies have found greater activations in the anterior cingulate in connection with the suppression of a particular thought as compared to thinking freely. The researchers suggest that the anterior cingulate may function as a vigilance monitor for intrusions by unwanted thoughts (Wyland et al., 2003). The right anterior cingulate has also been implicated in the conscious inhibition of sexual arousal. This was found in a study conducted by Beauregard et al. (2001) in which participants were asked to respond in a typical manner to excerpts of an erotic film or to inhibit their sexual arousal. These results are consistent with the hypothesis that the anterior cingulate plays a role in conflict monitoring (Botvinick et al., 1999; Kerns et al., 2004).

Further support for the hypothesis of blocking as a form of self-deception comes from data indicating that individuals who employ a repressive coping style report low levels of distress while simultaneously exhibiting signs of high physiological reactivity during an otherwise stressful situation (Asendorpf & Scherer, 1983; Gudjonsson, 1981; Weinberger, Schwartz, & Davidson, 1979).

Sins of Commission

The assignment of a memory to an incorrect source—mistaking dreams or fantasy for reality or even confusing the actual source of a memory (e.g., learned in class versus watched on television outside of class)—has been defined by Schacter as misattribution. We have already considered some examples of misattribution: DMS/DRS, false memories (Gonsalves & Paller, 2000; Gonsalves et al., 2004; Loftus, 1997), and the behavior of patients who have undergone cingulectomy (Whitty & Lewin, 1957). Eyewitness testimony is another area where misattribution may occur frequently (Schacter, 2001). Converging evidence implicates the anterior cingulate in misattribution. For example, a cingulectomy produced a waking dream state in which patients confused their dreams with reality (Whitty & Lewin, 1957), supporting a role for the anterior cingulate in misattribution. Damage to the anterior cingulate may produce a form of disinhibition, allowing for more false memories to be expressed consciously.

The right cerebral hemisphere is thought to be implicated in disorders such as Capgras' syndrome and Fregoli's syndrome. Though rare, it appears that each disorder only results from damage to the right cerebral hemisphere (Feinberg & Keenan, 2005; Signer, 1994). Thus it is suggested that the right hemisphere is involved in the maintenance of the concept of self, both in relation to the outside world and internally. Furthermore, it is suggested that its damage may produce extreme disorders of the self in which self-deception continues until it is no longer advantageous.

Suggestibility, the fifth sin of memory, concerns how memory may be corrupted by suggestions or comments during an attempt to recall information. Schacter points out that, like misattribution, this characteristic of memory has profound ramifications for the legal system.

Memory's sixth sin, bias, concerns itself with how our current thoughts, knowledge, and beliefs influence our recollection of the past. Like misattribution, the sin of bias is extremely relevant to the study of self-deception. Schacter suggests five types of biases that influence our memory of past events: (1) consistency bias, (2) change bias, (3) hindsight bias, (4) egocentric bias, and (5) stereotypical bias.

One study demonstrating consistency bias involved a comparison of undergraduate students' evaluations of a significant other over an interval of 2 months. The results indicated that students whose evaluations become more negative over the 2-month interval reported their initial evaluation to be more negative than what they felt they remembered (Schacter, 2001). As both Ramachandran and Schacter have pointed out, there are numerous instances in everyday life in which people use consistency biases to impose a sense of stability or logic on the perceived world in a self-deceptive manner.

Change bias refers to one's tendency to believe that one has improved to a greater degree than one may have when improvement is expected. For example, Schacter (2001) cites a study in which, after completion of a program designed to improve college entrance examination scores, students tended to rate their abilities prior to the program lower than their own initial self-ratings. Students who were on a waiting list for the duration of the program and who thus did not participate showed no such effect.

It is often said that "hindsight is 20/20." Schacter notes that hindsight bias is a frequent occurrence in everyday society and affects everyone from political pundits to sports radio callers, jurors, and students. We all display a tendency to report that we knew all along something was going to happen the way it did. This is because the past is reconstructed to fit the present, again consistent with Ramachandran's view of self-deception as imposing stability on one's script.

Hindsight bias has been demonstrated in the laboratory. A common paradigm is to ask subjects to rate the likelihood of alternative endings of a story to which they already know the ending, while pretending to be naive (Carli, 1999; Schacter, 2001). It has been found that once an individual knows the outcome of a scenario, that scenario is rated as being the most likely outcome to said situation. Schacter believes the precedence of hindsight bias to be troubling in light of the possible social ramifications, the high degree of fallibility of eyewitness testimony being one important example.

Self-deception, hindsight bias, and transience may be advantageous characteristics of the human mind, instilling in a person a false sense of past victory, an exaggerated view of the self, and an inflated ego. Such overconfidence may in turn aid in the ability to deceive others and reap the associated benefits. If the effects of these combined characteristics are successful in this sense, then we should expect to see a tendency toward such behaviors over evolutionary time.

The tendency to hold the fidelity of one's own memory in higher regard than another's is an example of what Schacter referred to as egocentric bias. Schacter points out that this reflects the important role that the self plays in organizing our thoughts and our internal mental life. Many memory aids involve relating material to the self at the time of encoding so that future recall is facilitated. The self, however, is not an objective or neutral observer. This type of bias is another example of how we inflate our own importance and abilities, another instance of self-deception.

We make use of stereotypes to quickly organize concepts into mental sets in order to effectively store some form of knowledge (Wood, Romero, Knutson, & Grafman, 2005). These are generalizations based on past experience that we use to categorize current novel experiences. According to Wood et al. (2005), the ability to quickly organize concepts into larger sets may have been favored by natural selection, as this would allow an individual to act quickly in some situation in which survival was threatened. Social psychologists have long been interested in how stereotypical biases can have undesirable effects, the most glaring examples being racism and sexism. Such biases also affect our memory, as Schacter (2001) has pointed out. Recall of a particular trait is typically better when the trait conforms well to a stereotype held by the perceiver. Such stereotype bias influences our current perception of situations, as well as our recollections of the past.

Schacter's final sin of memory is persistence. Persistence involves the repeated recollection of information or events that one would prefer to forget. Obsessive-compulsive disorder demonstrates this well. Patients with this disorder cannot rid themselves of a particular thought (the obsession) without partaking in a particular behavior (the compulsion) (Schacter, 2001).

As the sins of memory establish, our memory is not perfect; in fact, it is far from it. It is vulnerable to a number of biases, as well as to source-monitoring and reality-monitoring errors. These flaws may be adaptive, or may have adaptive consequences, by facilitating self-deception and thus enhancing the ability to effectively deceive. Without the features of a number of Schacter's sins—transience, blocking, misattribution, suggestibility, and bias—self-deception may not have developed into a sufficiently robust psychological force.

Neural Correlates of Self-Deception

Although some behavior may not have a single isolable neural correlate (e.g., deception or self-deception), a large body of evidence implicates the anterior cingulate as a key structure involved in self-deception. The anterior cingulate has been implicated in such processes as conflict monitoring (Botvinick et al., 1999; Kerns et al., 2004), inhibition, suppression of thought (Ochsner et al., 2004; Wyland et al., 2003), and conscious self-regulation of arousal (Beauregard et al., 2001). Whitty and Lewin's (1957) study of patients who had undergone a cingulectomy found that removal of the anterior cingulate produced a waking dream state in which patients were unable to discern reality from fantasy, which implicates the anterior cingulate in reality monitoring.

Given the involvement of the anterior cingulate in conflict monitoring, it is likely that this structure is integral to the selection for a deceptive response. The activity in the anterior cingulate during deception (Johnson et al., 2004) and "recollection" of false memories (Gonsalves et al., 2004; Slotnick & Schacter, 2004) lends support to this. Other neuroimaging findings extend these observations even further, suggesting a role of the anterior cingulate in the suppression of unwanted thoughts (Wyland et al., 2003), the suppression of unwanted memories (Ochsner et al., 2004), the conscious self-regulation of emotions (Beauregard et al., 2001), and the inhibition of positive attitudes (Wood et al., 2005).

Conclusion

Self-deception involves simultaneously holding both true and false information, and the unconscious suppression of truth. This allows both sets of information to exist without conscious awareness of the conflict. This ultimately serves to improve one's ability to deceive without detection and thus to reap the associated benefits. When appropriate, the anterior cingulate may be responsible for the suppression of factual information in favor of false information. Further connecting this structure in self-deception, a large body of research implicates the anterior cingulate in conflict monitoring (Botvinick et al., 1999; Kerns et al., 2004), suppression of thought (Beauregard et al., 2001; Wyland et al., 2003), and false responses (Nunez, Casey, Egner, Hare, & Hirsch, 2005), all of which are processes contributing to deceptive behavior.

Once a deceptive response or communication has been made, has the anterior cingulate failed in checking for a false memory, as implied in the executive theory of confabulation? Perhaps it has done its job well, by suppressing factual information while checking for a deceptive response when one is appropriate. Maybe confabulation is an example of the anterior cingulate overperforming. Further research is needed to better limn the details of the role of the anterior cingulate in self-deception and confabulation and the relationship between both phenomena.

Deception, deception detection, ToM, and self-awareness have all been suggested to be lateralized to the right hemisphere (Keenan et al., 2003; Malcolm & Keenan, 2003, 2005). Converging evidence from the study of self-deception further indicates that the right hemisphere plays a large role in supporting one's concept of self and the accurate representation of reality. Evidence from numerous experiments and case studies indicates that right hemisphere damage, specifically right frontal damage, results in disorders of the self that may be considered extreme cases of self-deception. DMS/DRS and their associated confabulatory statements are clear examples of self-deceptive behavior and typically arise only after right frontal damage.

Ramachandran has proposed that the left hemisphere is primarily responsible for imposing consistency on one's script, while the right hemisphere is the "anomaly detector." The left hemisphere therefore is the source of self-deceptive behaviors, with intentions of imposing stability on the script. This viewpoint is based on studies of anosognosic patients who have suffered damage to the right cerebral hemisphere. These patients will vehemently deny the paralysis of their left arm following a stroke, often giving reasons such as "I have arthritis in that shoulder" or "I'm not very ambidextrous" (Ramachandran, 1996). It was concluded that damage to the right hemisphere significantly weights information from the left hemisphere, resulting in extremes of self-deception. With damage to the right hemisphere, the patients could no longer maintain an accurate representation of reality and thus engaged in excessive confabulatory statements and committed an increased number of reality-monitoring errors.

Hirstein (2005) has hypothesized that these patients and their confabulations provide ample ground to suggest a possible ToM deficit—a process also suggested to be lateralized to the right hemisphere (Keenan et al., 2003). In primates, successful deception is most effective when the deceiver can effectively "put himself in the other person's shoes." By being able to see the situation from the rival's perspective, a deceiver is better able to deceive. DMS/DRS patients make absolutely no attempt to produce coherence in their confabulatory statements, supporting Hirstein's conjecture that a ToM deficit is present (Hirstein, 2005). If one's sense of self or one's representation of reality is altered, then an accurate model of the mind of another will be difficult to construct (see also Gallup, 1982).

The sins of blocking, misattribution, bias, and suggestibility all play a role in self-deception and the deception of others. Such characteristics of memory allow us to inflate our self-esteem, instilling within us optimism and confidence. Though there are clearly instances in which self-deception is adaptive, there are also instances in which self-deception proves detrimental to the existence of an organism. Trivers addresses the possible maladaptive nature of self-deception, indicating the potential for it to occur in various social situations. This arises out of the cost of self-deception—an impaired ability to comprehend or deal with reality (Trivers, 1991). For example, Trivers and Newton (2002) theorize that the crash of Flight 90, a Tampa Bay–bound Air Florida 737, was doomed by self-deception. They point to the transcript of the conversation between the pilot and copilot during the flight's final 30 minutes. On this particular flight the copilot was acting as the primary pilot as part of his training. The pilot appropriately attempted to instill confidence in his copilot, playing down many of his qualms about taking off in heavy snow. At some moments he even showed unnecessary bravado and a willingness to disregard policy for a risky ground maneuver. Only when the plane was descending perilously and fatally did he finally admit the reality of the situation to himself (Trivers & Newton, 2002). This situation is clearly an example of a maladaptive aspect of self-deception, as the pilot's distortion of reality contributed a great deal to the fatal crash (Trivers & Newton, 2002).

When viewed in terms of Dawkins's (1976) selfish gene theory, it is easy to see how the human mind may have evolved to produce effective deception via deception of the self. Dawkins proposed a concept of natural selection such that the individual is a temporary vehicle for the primary replicator (DNA). Such vehicles are constructed from genetically encoded "recipes." An effective recipe will result in vehicles that act in the appropriate manner so that they and their progeny will leave more decendants than other vehicles competing for the same resources. Thus, it logically follows that all animal behavior is intrinsically selfish. Because deception confers advantage on the deceiver, aiding in access to resources and opportunities to mate, natural selection should favor the construction of brains capable of effectively deceiving. However, the ability to deceive is an emergent property of several processes; there is no deception center of the brain. So it seems that the ability to deceive is dependent on several characteristics of memory, self-awareness, and ToM processes. Therefore, natural selection should favor the presence of such traits or processes giving rise to deceptive behavior, and so on and so forth, recursively, with natural selection favoring the most well-defined functions that ultimately give rise to deceptive behavior. In short, the end result is the ability to deceive, one's self and thus the other, effectively, in proper balance with communication of the truth, and a myriad of other traits, in order to achieve maximal propagation of one's genes. Thus, humans, as we know them, lie.

References

Alexander, M. P., Stuss, D. T., & Benson, D. F. (1979). Capgras' syndrome a reduplicative phenomenon. *Neurology*, *29*, 334–339.

Asendorpf, J. A., & Scherer, K. R. (1983). The discrepant repressor: Differentiation between low anxiety, high anxiety, and repression of anxiety by autonomic facial-verbal patterns of behavior. *Journal of Personality and Social Psychology*, *45*, 1334–1346.

Beauregard, M., Levesque, J., & Bourgouin, P. (2001). Neural correlates of conscious self-regulation of emotion. *Journal of Neuroscience*, *21*, 165–170.

Botvinick, M., Nystrom, L. E., Fissell, K., Carter, C. S., & Cohen, J. D. (1999). Conflict monitoring versus selection-for-action in the anterior cingulate cortex. *Nature*, *402*, 179–181.

Burgess, P. W., & Shallice, T. (1996). Confabulation and the control of recollection. *Memory*, *4*, 359–411.

Butler, P. V. (2000). Diurnal variation in Cotard's syndrome (copresent with Capgras' delusion) following traumatic brain injury. *Australian and New Zealand Journal of Psychiatry*, *34*, 684–687.

Byrne, C. C., & Kurland, J. A. (2001). Self-deception in an evolutionary game. *Journal of Theoretical Biology*, *212*, 457–480.

Byrne, R. W., & Corp, N. (2004). Neocortex size predicts deception rate in primates. *Proceedings of the Royal Society of London*, *271*, 1693–1699.

Carli, L. L. (1999). Cognitive reconstruction, hindsight, and reactions to victims and perpetrators. *Personality and Social Psychology Bulletin*, *25*, 966–979.

Curran, T., Schacter, D. L., Norman, K. A., & Galluccio, L. (1997). False recognition after a right frontal lobe infarction: Memory for general and specific information. *Neuropsychologia*, *35*(7), 1035–1049.

Dawkins, R. (1989). *The selfish gene* (2nd ed.). New York: Oxford University Press.

Deese, J. (1959). On the prediction of occurrence of particular verbal intrusions in immediate recall. *Journal of Experimental Psychology*, *58*, 17–22.

DePaulo, B. M., Lindsay, J. J., Malone, B., Muhlenbruck, L., Charlton, K., & Cooper, H. (2003). Cues to deception. *Psychological Bulletin*, *126*, 74–118.

Ekman, P. (1985). *Telling lies: Clues to deceit in the marketplace, politics, and marriage* (2nd ed.). New York: Norton.

Ellis, H. D., Luaute, J. P., & Retterstol, N. (1994). Delusional misidentification syndromes. *Psychopathology*, *27*, 117–120.

Fabiani, M., Stadler, M. A., & Wessels, P. M. (2000). True but not false memories produce a sensory signature in human lateralized brain potentials. *Journal of Cognitive Neuroscience*, *12*(6), 941–949.

Feinberg, T. E., & Keenan, J. P. (2005). Where in the brain is the self? *Consciousness and Cognition*, *14*, 461–463.

Fleminger, S., & Burns, A. (1993). The delusional misidentification syndromes in patients with and without evidence of organic cerebral disorder: A structured review of case reports. *Biological Psychiatry*, *33*, 22–32.

Forstl, H., Almeida, O. P., Owen, A., Burns, A., & Howard, R. (1991). Psychiatric, neurological and medical aspects of misidentification syndromes: A review of 260 cases. *Psychological Medicine*, *21*, 905–950.

Gallup, G. G. (1982). Self-awareness and the emergence of mind in primates. *American Journal of Primatology*, *2*, 237–248.

Ganis, G., Kosslyn, S. M., Stose, S., Thompson, W. L., & Yurgelun-Todd, D. A. (2003). Neural correlates of different types of deception: An fMRI investigation. *Cerebral Cortex*, *13*(8), 830–836.

Gonsalves, B., & Paller, K. A. (2000). Neural events that underlie remembering something that never happened. *Nature Neuroscience*, *3*(12), 1316–1321.

Gonsalves, B., Reber, P. J., Gitelman, D. R., Parrish, T. B., Marsel-Mesulam, M., & Paller, K. A. (2004). Neural evidence that vivid imagining can lead to false remembering. *Psychological Science*, *15*(10), 655–660.

Gudjonsson, G. H. (1981). Self-reported emotional disturbance and its relation to electrodermal reactivity, defensiveness and trait anxiety. *Personality and Individual Differences*, *2*, 47–52.

Gur, C. R., & Sackheim, H. A. (1979). Self-deception: A concept in search of a phenomenon. *Journal of Personality and Social Psychology*, *37*, 147–169.

Hakim, H., Verma, N. P., & Greiffenstein, M. F. (1988). Pathogenesis of reduplicative paramnesia. *Journal of Neurology, Neurosurgery, and Psychiatry, 51*, 839–841.

Hirstein, W. (2005). *Brain fiction: Self-deception and the riddle of confabulation.* Cambridge, MA: MIT Press.

Ito, Y. (2001). Hemispheric asymmetry in the induction of false memories. *Laterality, 6*(4), 337–346.

Johnson, M. K., Hashtoudi, S., & Lindsay, D. S. (1993). Source monitoring. *Psychological Bulletin, 114*, 3–28.

Johnson, M. K., & Raye, C. L. (1981). Reality monitoring. *Psychological Review, 88*, 67–85.

Johnston, R., Barnhardt, J., & Zhu, J. (2004). The contribution of executive processes to deceptive responding. *Neuropsychologia, 42*, 878–901.

Kapur, N., & Coughlan, A. K. (1980). Confabulation and frontal lobe dysfunction. *Journal of Neurology, Neurosurgery, and Psychiatry, 43*, 461–463.

Keenan, J. P., Gallup, G., Jr., & Falk, D. (2003). *The face in the mirror: The search for the origins of consciousness.* New York: HarperCollins.

Kerns, J. G., Cohen, J. D., MacDonald, A. W., III, Cho, R. Y., Stenger, V. A., & Carter, C. S. (2004). Anterior cingulate conflict monitoring and adjustments in control. *Science, 303*(5660), 1023–1026.

Kozel, F. A., Padgett, T. M., & George, M. S. (2004). A replication study of the neural correlates of deception. *Behavioral Neuroscience, 118*(4), 852–856.

Langleben, D. D., Schroeder, L., Maldjian, J. A., Gur, R. C., McDonald, S., Ragland, J. D., et al. (2002). Brain activity during simulated deception: An event-related functional magnetic resonance study. *NeuroImage, 15*, 727–732.

Lee, T. M. C., Liu, H-L., Tan, L-H., Chan, C. C. H., Mahankali, S., Feng, C.-M., et al. (2002). Lie detection by functional magnetic resonance imaging. *Human Brain Mapping, 15*, 157–164.

Lewis-Lehr, M. W., Slaughter, J. R., Rupright, J., & Singh, A. (2000). The man who called himself "hockey stick": A case report including misidentification delusions. *Brain Injury, 14*(5), 473–478.

Loftus, E. F. (1997). Creating false memories. In A. R. Damasio (Ed.), *The Scientific American book of the brain.* Guilford, CT: Lyons Press.

Luaute, J. P., & Bidault, E. (1994). Capgras' syndrome: Agnosia of identification and delusion of reduplication. *Psychopathology, 27*(3–5), 186–193.

Malcolm, S., & Keenan, J. P. (2003). My right I: Deception detection and hemispheric differences in self-awareness. *Social Behavior & Personality: An International Journal, 31*(8), 767–772.

Malcolm, S., & Keenan, J. P. (2005). Hemispheric asymmetry and deception detection. *Laterality, 10*(2), 103–110.

Nunez, J. M., Casey, B. J., Egner, T., Hare, T., & Hirsch, J. (2005). Intentional responding shares neural substrates with response conflict and cognitive control. *NeuroImage*, 25, 267–277.

Ochsner, K. N., Kuhl, B., Cooper, J., Robertson, E., Gabrieli, S. W., Glover, G. H., et al. (2004). Neural systems underlying the suppression of unwanted memories. *Science, 303*(5655), 232–235.

O'Dwyer, J. M. (1990). Coexistence of Capgras' and de Clerambault's syndromes. *British Journal of Psychiatry, 156*, 575–577.

Okado, Y., & Stark, C. (2003). Neural processing associated with true and false memory retrieval. *Cognitive, Affective & Behavioral Neuroscience, 3*(4), 323–334.

Paillere-Martinot, M. L., Dao-Castallana, M. H., Masure, M. C., Pillon, B., & Martinot, J. L. (1994). Delusional misidentification: A clinical, neuropsychological, and brain imaging case study. *Psychopathology, 27*, 200–210.

Platek, S. M., Thomson, J. W., Irani, F., Shelky, E., & Keenan, J. P. (2006). Unpublished manuscript.

Ramachandran, V. S. (1996). The evolutionary biology of self-deception, laughter, dreaming and depression: Some clues from anosognosia. *Medical Hypotheses, 47*, 347–362.

Ramachandran, V. S., & Blakeslee, S. (1998). *Phantoms in the brain: Probing the mysteries of the human mind.* New York: HarperCollins.

Roediger, H. L., III, & McDermott, K. B. (1995). Creating false memories: Remembering words not presented in lists. *Journal of Experimental Psychology: Learning, Memory, and Cognition, 21*, 803–814.

Schacter, D. L. (1995). Memory distortion: History and current status. In D. L. Schacter (Ed.), *Memory distortion: How minds, brains, and societies reconstruct the past.* Cambridge, MA: Harvard University Press.

Schacter, D. L. (2001). *The seven sins of memory.* Boston: Houghton Mifflin.

Schacter, D. L., Curran, T., Galluccio, L., Milberg, W. P., & Bates, J. F. (1996). False recognition and the right frontal lobe: A case study. *Neuropsychologia, 34*(8), 793–808.

Schacter, D. L., Verfaellie, M., & Pradere, D. (1996). The neuropsychology of memory illusions: False recall and recognition in amnesic patients. *Journal of Memory and Language, 35*, 319–334.

Signer, S. F. (1994). Localization and lateralization in the delusion of substitution. *Psychopathology, 27*, 168–176.

Signer, S. F., & Isbister, S. R. (1987). Capgras' syndrome, de Clerambault's syndrome and *folie a deux*. *British Journal of Psychiatry, 151*, 402–404.

Slotnick, S. D., & Schacter, D. L. (2004). A sensory signature that distinguishes true from false memories. *Nature Neuroscience, 7*(6), 664–672.

Spence, S. A., Farrow, T. F. D., Herford, A. E., Wilkinson, I. D., Zheng, Y., & Woodruff, P. W. R. (2001). Behavioural and functional anatomical correlates of deception in humans. *Neuroreport, 12*(13), 2849–2853.

Stuss, D. T., Alexander, M. P., Lieberman, A., & Levine, H. (1978). An extraordinary form of confabulation. *Neurology, 28*, 1166–1172.

Trivers, R. L. (1991). Deceit and self-deception: The relationship between communication and consciousness. In M. Robinson & L. Tiger (Eds.), *Man and beast revisited* (pp. 175–191). Washington, DC: Smithsonian Institution Press.

Trivers, R. L., & Newton, H. P. (2002). The crash of Flight 90: Doomed by self deception? In *Natural selection and social theory: Selected papers of Robert Trivers* (pp. 262–271). New York: Oxford University Press.

Wagner, A. D., Schacter, D. L., Rotte, M., Koutstaal, W., Maril, A., Dale, A. M., et al. (1998). Building memories: Remembering and forgetting of verbal experiences as predicted by brain activity. *Science*, *281*, 1188–1191.

Weinberger, D. A., Schwartz, G. E., & Davidson, R. J. (1979). Low-anxious, high-anxious, and repressive coping styles: Psychometric patterns and behavioral responses to stress. *Journal of Abnormal Psychology*, *88*, 369–380.

Whiten, A., & Byrne, R. W. (1997). *Machiavellian intelligence II: Extensions and evaluations*. Cambridge: Cambridge University Press.

Whitty C. W. M., & Lewin, W. (1957). Vivid day-dreaming: An unusual form of confusion following anterior cingulectomy. *Brain*, *80*, 72–76.

Wood, J. N., Romero, S. G., Knutson, K. M., & Grafman, J. (2005). Representation of attitudinal knowledge: Role of prefrontal cortex, amygdala and parahippocampal gyrus. *Neuropsychologia*, *43*, 249–259.

Wyland, C. L., Kelley, W. M., Macrae, C. N., Gordon, H. L., & Heatherton, T. F. (2003). Neural correlates of thought suppression. *Neuropsychologia*, *41*, 1863–1867.

19 On the Evolution of Human Motivation: The Role of Social Prosthetic Systems

Stephen M. Kosslyn

This chapter develops an analysis and set of speculations about social evolution that provide a new way to regard some aspects of human motivation. The theory proposed here hinges on the concept of *social prosthetic systems*, which are human relationships that extend one's emotional or cognitive capacities. In such systems, other people serve as prosthetic devices, filling in for lacks in an individual's cognitive or emotional abilities. Social prosthetic systems can be short-term (created to accomplish a specific task) or long-term (part of enduring relationships). According to this theory, humans are motivated to behave in ways that create, extend, and support social prosthetic systems, in part because the "self" becomes distributed over other people, who function as long-term social prosthetic systems. Moreover, some human behavior may be directed toward establishing conditions that will induce others to serve as one's social prosthetic systems. The implications and predictions of this theory are explored.

An interesting consensus emerged from a symposium at Harvard's Kennedy School of Government in 2002, still in the direct shadow of the attacks of September 11, 2001. The experts agreed that education per se would not reduce the threat of terrorism in the world, that it was rooted instead in something far deeper—a need to be valued and respected. I won't go through the details of their reasoning here (which included the fact that most of the terrorists who participated in the September 11 attack were well educated), but the experts' basic message struck a chord. When it comes to life-organizing decisions, our motivations are not simple. Religion, for example, motivates us to behave in certain ways, and it's no accident that religion plays such a central role in most of the world's current conflicts. In fact, some people place priority on actions that they perceive as important to their group as a whole, even if—as too often demonstrated by the suicide bombers in the Middle East—that action actually ends their lives.

As many have noted, at first glance such behaviors might seem flagrantly to disregard the most basic goal of evolution by natural selection: self-preservation. If each of us is a bundle of "selfish genes," as Dawkins maintains, Maslow's famous theory of a hierarchy of motivations should hold: People should focus first on survival, minding essential biological functions, and only thereafter be swayed by more abstract motivations. But we need only read the newspapers to realize that this just isn't so.

In the same vein, the problem of altruism has been a perennial burr in the side of evolutionary psychology. If the point is to propagate one's own genes, why would one ever hold others' interests above one's own? One possible account centers on the idea that animals behave in such a way as to propagate the genes they carry—even if these genes are harbored by another individual (Hamilton, 1964). This idea can explain why one would sacrifice oneself for two brothers or eight cousins, because of the number of one's genes they share. But how can we understand why someone would sacrifice himself for total strangers? What sort of motivation would overcome even the urge to survive?

In this chapter I explore one possible way to address this vexing problem. I will assume that natural selection is not focused solely on the individual, but also on the group. We might ask, How useful would language be if only one person had it? About as useful as the Internet if only one computer existed in the world. But the answer is farther reaching: language has both productive and receptive aspects, and both had to be tailored in expectation of the other. We speak in ways that can be properly registered by others, and others speak in ways that we can understand. Language, be it serious discussion or aimless jabbering, is a group activity. And thus language had to evolve in the context of the group. But this observation only scratches the surface. I argue that in a very deep sense, we have been shaped by evolution not only so that we work well in groups, but also so that our personal identity depends on our relationships with others.

The World in the Brain

Evolution by natural selection has given birds hard beaks and aerodynamically designed wings, fishes fins, and ivy the ability to orient toward the sun. Evolution by natural selection has given humans hands and an upright posture—and more powerfully, it has made humans the most adaptable creature on the planet. We owe our extraordinary adaptability to our brain. It is no exaggeration to say that the brain is the body's most flexible organ. This flexibility in part is the flip side of a limitation: we simply don't have enough genes to program the brain fully in advance. So, as in most other animals, both the function and structure of our brains are not entirely determined by our genes but instead depend in part on characteristics of the environment in which we find ourselves.

A concrete example is useful here. We determine the distance of objects partly by using information from our two eyes. If we look at an object that is within 10 feet or so of where we are sitting, then cover first one eye and then the other, the image appears to shift slightly as we switch the viewing eye. Because our eyes do not occupy exactly the same place on our face, each has a different view. A process called *stereovision* makes use of these differences to compute how far away something is (the difference becomes smaller for objects that are farther away; see, e.g., Pinker, 1997). The interesting point is that the precise difference between the views seen by each eye depends on exactly how far apart the eyes are—but there is no way the genes can "know" that in advance. Depending on the mother's diet before birth and the child's diet after birth, the head will grow at different rates; thus, the precise distance between the eyes depends partly on environmental factors, which cannot be anticipated in advance by the genes. Thus, it is not really a question of having to make do with a limited number of genes, for even with an unlimited number of genes, it is impossible for the genes to program exactly how the brain should carry out stereovision.

How does stereovision develop? The solution hit on by the genes is exceedingly clever and remarkably simple: the genes simply overpopulate the brain with connections, some that will work properly in one scenario, others that will work properly in another, and so on. Over time, the outputs from some neural circuits provide estimates of distance that successfully guide reaching, eye movements, and the like, and the outputs from other circuits turn out not to be useful. In the case of stereovision, the circuits that provide useful information about the environment survive and the others are eliminated (pruned away; for a recent overview, see Huttenlocher, 2002). Just as the muscles of the legs were configured by natural selection on the basis of the "assumption" that gravity exists (and thus we don't need as much strength to put our legs down as to lift them up), the brain evolved with the "assumption" that it would exist in a world of three-dimensional objects. In other words, the genes have been selected to produce a brain that in turn can be shaped by the environment in specific ways.

In fact, our heads are stuffed with extra neural connections up to about age 8. The environment (including its indirect effects, for example, on bone growth) literally sculpts our brains (cf. Comery, Stamoudis, Irwin, & Greenough, 1996; Kleim et al., 1998; Merzenich et al., 1983). In my view, this sculpting occurs not simply because we move around and interact with physical objects (which tunes our visual and motor systems), it also occurs because we learn language—and all the concepts and attitudes embedded in it—from others and are bathed in a full cultural array of practices and procedures, beliefs and bylaws. Such early experience leads us to carry the world within us; the boundaries between "the world"—especially the social world—and "us" are fuzzier than they might appear.

The implications of the idea that the environment sculpts the developing brain are far-reaching. For example, the Austrian philosopher Ludwig Wittgenstein (1999) asserted, "If a lion could speak, we would not understand him." What he meant was that lions have a different "form of life" than humans: they run on four legs, try to bring down antelope, must navigate through tall grass, and so on, whereas we are bipedal, manipulate objects with our hands, live largely in a carpentered world, and so on. The concepts needed to negotiate the two environments shaped the brains of the species, lions and humans, and because the environments are so different, the concepts would be, too. And thus the lion's words would rely on different concepts than we have, which are rooted in the very structure of the brain. I wonder if this might figure into an explanation of why at least some "wild children," such as the Wild Boy of Aveyron and his contemporary counterparts, could never be taught to communicate well with other people (Newton, 2002). These children were never socialized, growing up either solely in the company of lower animals or virtually isolated from all forms of social contact.

The point I want to make here is that the boundary between the brain and world is blurred. Although genes hard-wire the brain so that it can develop in specific ways (to have stereovision, use language, and so on), the genes build a brain that can develop in numerous ways. The brain is "initialized" by the environment in which it finds itself, allowing it to conceptualize and function within that environment.

But what does this have to do with why we are motivated in some ways but not others? To see the link, we first need to look at the brain-environment relation from the other direction, not the world in the brain but rather the brain in the world.

The Brain in the World

My key notion is that the brain uses the world and other people as extensions of itself, and we are motivated to behave in ways that help us use the world and other people in this way. I want to argue that "you" are not confined to what is in your head but are in part represented in things around you, including other people. If so, then it makes sense why natural selection operates in part at the level of the group—so does the individual! Let's see where this takes us.

Shaping the Environment

I will begin with an anecdote that, while not conclusive, is at least evocative. Traffic in the Boston area rivals that of Rome; as a pedestrian you take your life in your hands by jaywalking. This is nothing new, of course, and long ago city fathers decided that pedestrians needed an "environmental support" to increase their chances of successfully negotiating traffic when crossing streets. Thus arose (I surmise) the invention of the crosswalk. Crosswalks are a feature of the environment that are designed to help humans accomplish a task. However, some crosswalks are better than others at fulfilling this role. Specifically, I recently noticed that crosswalks in Cambridge, Massachusetts, were gradually being repainted as "zebra stripes" (horizontal stripes about 18 inches wide and 6 feet long, with each stripe alternating with an equal amount of unpainted pavement). The first time I had seen such crosswalks was on the cover of the Beatles' *Abbey Road* album shortly after it was released. At that time, crosswalks in the United States consisted of two long lines that traversed the street, indicating the boundaries of the pedestrian corridor. But now, those European-style crosswalks are commonplace in Cambridge. In fact, they are displacing the old-style ones, and for good reason: the zebra-striped ones are objectively better at doing what crosswalks are supposed to do. From the point of view of the driver, they reflect more light and indicate better the zone where pedestrians may be crossing. From the point of view of the pedestrian, they delimit better the walking corridor.

But more than that, the zebra-striped crosswalks are also better for another, deeper reason: they engage brain systems that help one to walk automatically. This is demonstrated by studies of people who have Parkinson's disease. Such patients often shuffle because they lack a key neurotransmitter, dopamine, in the frontal lobes, which is used during voluntary movement; however, if they are asked to walk over a set of boards placed crosswise at regular intervals in front of them (like railroad ties), they walk smoothly. The visible stimuli engage another part of the brain, the cerebellum, which plays a key role in automatic movements. The stripes on the crosswalk seem to do the same. In fact, if you look at that *Abbey Road* album cover, you'll notice that the Beatles are walking with wide strides, which seem to be calibrated to the width of the stripes. The crosswalk not only delimits the pedestrian corridor well, it may also engage automatic movements to help walkers cross more swiftly. It is clear that, given what crosswalks are designed to do, for the human brain the zebra-stripe version is better than the two-line version.

This example illustrates a broader point. For many tasks, humans do better when the environment is properly engineered. Nobody would argue with this general assertion, but the question immediately arises: What does it mean for the environment to be properly engineered? An important aspect of human nature is that we are limited. Humans can only run so fast, jump so far, and can't sprout wings—if we could run faster, jump 50 feet in a single leap, or fly, we probably wouldn't need crosswalks to help us get across even busy boulevards. We engineer the environment to make up for our limitations. And the most important of such limitations are in our brains. Our brains can process information only within limited bounds. For example, if speech is sped up faster and faster (in such a way that the pitch is held constant—no chipmunk voices here), the brain areas that process it work progressively harder and harder. But at some point the speech becomes incomprehensible, and at just that point the various brain areas stop working so hard (Poldrack et al., 2001). Our brains are limited not only in their ability to process information but also in the amount of information that can be held in mind at the same time. For example, if you are asked to multiply 3,976 by 5,222, you will probably find this difficult to do in your head. This observation leads to the central part of my argument, namely, that we use the world as an extension of ourselves, to make up for our limitations.

Social Prosthetic Systems

There are many obvious and dramatic illustrations of ways that the environment can make up for a deficit. For example, if you were missing a leg, you would receive a substitute—the modern equivalent of the proverbial wooden leg. This leg is a prosthesis, it fills in for a missing part of your anatomy. But don't think that only the unfortunate rely on prosthetic devices. All of us do, much (perhaps even most) of the time. For example, consider how you would actually go about multiplying those two large numbers, 3,976 by 5,222. Because we humans have a limited working memory capacity (Baddeley, 1986; Smith & Jonides, 1997) and thus cannot hold such numbers in mind and work on them effectively, we rely on external aids, such as pencil and paper or a calculator. These socially created aspects of the world are prosthetic devices. They compensate for a deficit, just as a prosthetic leg would do if a leg had been amputated.

But notepads and calculators are not the most powerful or common prosthetic devices we use. More interesting, in my view, is that we use other people as extensions of ourselves. Specifically, we rely on other people to extend our cognitive and emotional capacities. Others help us formulate alternatives, evaluate options, and make decisions; others also help us interpret and control our emotions. Evolution has allowed our brains to be configured during development so that we are "plug compatible" with other humans, so that others can help us extend ourselves.

When someone devotes time and energy to helping you, you are literally using part of their brain. Your self extends beyond your own head and into those of others who work with you. Such situations constitute what I call *social prosthetic systems* (SPS). An SPS is a socially created system that extends one's emotional or cognitive capacities. The idea of an SPS must be distinguished from the idea of a team, where members work together to achieve a common goal, or of "distributed cognition," where people take on different aspects of a cognitive task in order to improve how well the group as a whole performs (e.g., Wegner, 1995; see also Hollingshead, 1998). My idea is different, in two ways. First, in an SPS, one person has a goal and draws other people in as prosthetic devices to help him or her accomplish that goal. The other people function as extensions of the person, and in fact may or may not desire to accomplish that task for its own sake (we'll get to their motivations shortly). Second, I conceive of a goal in the broadest possible way, so broadly that whenever two or more people interact they are, in my view, attempting to achieve at least one goal; we always have agendas, even if they are as mundane as simply relaxing or wanting emotional support. An SPS can be set up to help you relax, but few of us are lucky enough to have a team devoted to achieving such a goal.

We can distinguish between two types of SPSs. On the one hand, a short-term SPS is set up in a specific context with the aim of accomplishing a specific task. For example, if you are moving a heavy box, a friend might lend you a hand. Or if you are stuck while writing a computer program, a person at the Help Desk might point out your error. These sorts of SPSs are transient: once the task is accomplished, the SPS ceases to exist. On the other hand, a long-term SPS comes to be used not just for one specific task but for a class of tasks. Such SPSs rely on establishing enduring relationships. For example, your spouse can help you handle interpersonal problems at work and help you plan your next vacation together. Such long-term SPSs are like tools in a toolbox. Once you have them, they remain available to be used in the appropriate circumstances.

Two Sources of Motivation

What does all of this have to do with human motivation? I want to argue that this perspective provides insight into two sources of motivation. First, we are motivated to create, extend, and support specific SPSs in part because the "self" becomes distributed over other people, who function as our long-term SPS. Second, we are motivated to create conditions that will induce others to serve as our SPS. The next sections address each type of motivation in turn.

The Self in Social Context

I propose that humans are motivated to establish, extend, and deepen long-term SPSs. This makes sense from a strictly functionalist perspective because one will be better able to cope with future events if (1) existing SPSs are further strengthened, making another person more willing or able to help one, or (2) more SPSs are established, with a greater range of skills and abilities for you to draw on. The toolbox analogy is useful again: the better the tools, and the more of them and more varied they are, the better positioned you are for the future.

However, there is a deeper reason why we find it so gratifying to establish relationships with others where we can count on their loyalty and assistance in the future. One result of setting up deeper, more varied long-term SPSs is that your "self" is more firmly entrenched outside your body. When another person assumes the role of a long-term SPS, he or she has "gotten to know you," and has learned how to behave in ways that help you. What this comes down to is that his or her brain has become configured to operate as an extension of yours! All learning involves changes in the brain, and this particular sort of learning involves changing a brain so that it operates well in conjunction with another brain.

According to this view, then, to the degree that you become imbedded in a network of SPSs, your self is not confined to the neural tissue nestled between your own ears; rather, the self extends into other people's brains. This idea is consistent with the fact that people have remarkably few stable traits, such as being very honest or having a strong desire to be neat and orderly; instead, much of the way we behave depends on the context in which we find ourselves (cf. Mischel, 1999). These effects of the environment run deep. In fact, different environments upregulate and downregulate genes in the brain that produce neurotransmitters and neuromodulators, thereby greasing the mental wheels for some kinds of processing but not others. Who we are depends on where we are and what we are doing. This makes sense because in different contexts, we draw on different prostheses—and we are the sum of what we carry with us and what we use externally.

Inducing Others to Serve

An SPS helps one only insofar as other people are willing to serve as SPSs. Why would others be willing to lend you not just an ear, but part of their brains? First, others may serve as an SPS because they share the same immediate goal, in which case they also benefit from accomplishing an immediate task; in addition, they may garner an immediate reward, such as being paid (as occurs for psychotherapists delivering psychotherapy). Second, others may serve as a long-term SPS because they are investing in the future, expecting reciprocal behavior from you. One reason others may become your SPS is because they expect that you later will serve as their SPS.

This leads to a final idea. Humans are motivated to improve skills and abilities because this prepares us to cope better with future challenges. There are obvious evolutionary advantages for having such inclinations. But working to improve oneself leads to a secondary gain: by developing skills and abilities, we also make ourselves more valuable to others, which in turn leads them to be more willing to serve as our SPS. If others perceive that we have something to offer, they will be willing to invest their time and effort for us in hopes that in the future, we will be a valuable SPS for them.

Implications of the Theory

The SPS theory has several interesting implications. First, the theory leads us to infer that diversity is not a luxury but rather is essential in many walks of life. An analogy is the many different tools a carpenter has in his toolbox. It is impossible to know in advance what challenges the environment will produce and what abilities will need to be marshaled; and if you need abilities you don't possess, you will need to draw on others as prosthetic devices. Variety is more than the spice of life, it's the essence of life. Hypothetically, I can imagine a stripe of eugenicists who would be inclined to exterminate dark-skinned, hunched, hairy little men, but if another ice age descended, those people might well be the very ones best suited to survive. Just as the genes were smart enough to populate the brain with the potential to cope with many eventualities, we should be smart enough to preserve a wide variety of people in our societies, to cope with many eventualities. We cannot know in advance what sorts of SPS we will need to help us negotiate future challenges.

Second, the theory explains why many people are motivated by a desire "to make a difference." To the extent that your views change how others think, feel, and behave, they can function as an SPS for you. And to the extent that others are configured to be receptive to being your SPS, you have created an environment in which you can recruit more ways to distribute yourself—and the more distributed you are, the more likely you are to survive. This notion might also explain why fundamentalists of all stripes proselytize so vigorously, and why many of us try so hard to convince others of our most deeply held views.

To return to the issue I broached at the outset, why did the terrorists willingly kill themselves? According to this theory, each of them was a member of a set of SPSs that are embedded in a particular culture, and thus their identity was distributed over many other people. And because the self is distributed over those who serve as prostheses, to the extent that they felt tightly embedded in this group—which they perceived to be their culture at large—they would perceive that essential components of their selves would continue to exist even when their bodies died. Thus, they sacrificed their bodies to preserve their "greater selves." The same reasoning would explain why a soldier would willingly throw himself on a grenade to save the lives of his comrades-in-arms. Objectively, these behaviors may be misguided, because no matter how distributed one's self becomes, there is still more of it embedded in one's own brain. Nevertheless, if one perceives that one's identity is distributed over the group, one will not perceive one's death as obliterating the self—and to some extent, this perception is not erroneous.

As another example, why would someone help an old lady cross the street, even if she never mentioned it and nobody else observed this generous act? If one feels virtuous, this feeling alone may increase one's sense of self-worth. And the greater one's perceived value, the more easily one can induce others to serve as SPSs. The same line of reasoning allows me to explain why some people are motivated to perform an act of vengeance (such as the man who kills his wife's murderer): if one views this act as enforcing the social order, one will feel that one has increased one's value to other members of the society. However, precisely the same motivation could lead others *not* to perform an act of vengeance. It all depends on how one perceives the consequences of one's act.

Such reasoning can be used to explain why people are motivated to perform any behavior that increases their competence (I don't mean competence in the linguistic sense, which is distinguished from performance, but rather in the sense of increasing one's skills and abilities). For example, consider an amateur artist who takes great pride in her paintings. She might be highly motivated to learn to paint the sky well, even if this act would be considered rather minor in the greater scheme of things. Why is she so motivated? To the extent that we develop competence, we make ourselves valuable as SPSs to others, and that in turn means that others will be more willing to allow themselves to be used as our SPSs. Although you and I might not consider learning to paint a valuable skill, someone who wants a portrait of their children or a painting of a favorite vista might. Moreover, because it is impossible to know which particular competences will be valued most in the future, we are motivated by our own perceptions of value.

This perspective leads me to reconsider Hamilton's (1964) notion of inclusive fitness. Behaving altruistically to those with shared genes may account for the behavior of lower animals, and may have been the origins of some forms of human behavior. But after humans developed their current range of cognitive and emotional competences, selection may have focused on something else: the ability to interact with others so that they function as SPSs. For humans, the number of shared genes may be less important than the likelihood that the others can function as your SPSs. Specifically, those who share our temperaments (which apparently are in part under genetic control; Kagan, Snidman, Arcus, & Reznick, 1994) and who live in close proximity are good candidates to be long-term SPSs. As a thought experiment, suppose you had an identical twin who was separated from you at birth, whisked off to Afghanistan and raised as an especially fanatical member of the Taliban. At age 35 you meet your sibling for the first time and are forced to choose between the life of your spouse, who shares none of your genes, and your identical twin. Which will it be? My strong intuition is that in that situation I would choose my SPS over the stranger who happens to share all of my genes. In fact, I would make this decision in a flash, with no conflict or anguish. One might object that my spouse is valuable because she has the potential to have children, and thus perpetuate my genes. To counter this possibility, imagine that your spouse is sterile or your children are grown. Would that change the equation? I would make the same prediction for homosexual couples who have established a long-term relationship, even though there is no issue of a confound with genetics.

Conclusion

This chapter has developed an argument with the following logic: (1) evolution shaped our brains to "assume" that they would exist in particular kinds of environments; (2) the environment, particularly the social environment, sculpts the child's brain so it can work well in that environment; (3) as adults, we use the environment as a crutch, to help us accomplish tasks; (4) a critical way in which we use the environment is by using other people to extend our skills and abilities, creating SPSs; (5) we can distinguish between two kinds of SPSs: short-term and long-term; (6) humans are motivated to behave in ways that create, extend, and support long-term social prosthetic systems in part because the self becomes distributed over other people who function as long-term SPSs; and (7) finally, people are motivated to increase their skills and abilities in part because it makes them more valuable as potential SPSs, which in turn provides motivation for others to serve as their SPSs.

I suggest that motivation can be divided into two very general categories. The first concerns behaviors that either allow one to gain sensory rewards or to avoid unpleasant states; we do things because they feel good or they have consequences that feel good, or because they help us avoid things that feel bad or have consequences that feel bad. I include all of the basic biological drives in this category (such as eating and sex), as well as activities driven by aesthetics (such as listening to music). The second class concerns behaviors that develop competences. I divide this class into two parts, those behaviors that enhance skills and abilities that can be directly applied in the world (such as learning carpentry or money management) and those behaviors that develop competences that create, extend, and support SPSs or that will increase one's probability of recruiting SPSs. This last category of motivations may be uniquely human.

ACKNOWLEDGMENTS The ideas in this chapter have evolved over the course of many years and reflect the influences of many people. I'm sure that some of Anne Harrington's notions have shaped my views (she likes to talk about "culture under the skin," for example), but she would not necessarily endorse what I've done with them. And I would never have spent so much time thinking about such matters but for Jeffrey Epstein's persistent and insightful prodding. Moshe Bar, Philip Clayton, Richard Hackman, Elizabeth Knoll, Sam Moulton, and Jennifer Shephard offered helpful comments on an earlier draft, as did Justin Kosslyn and Robin Rosenberg (and without question, many of the ideas behind these ideas were developed in collaborating on our jointly authored psychology textbooks). Finally, I wish to thank Philip Clayton and Mark Richardson for taking this project seriously and inviting me to participate in the Science and the Spiritual Quest program, which gave me an excuse to put this down on paper.

References

Baddeley, A. (1986). *Working memory*. Oxford: Clarendon Press.

Comery, T. A., Stamoudis, C. X., Irwin, S. A., & Greenough, W. T. (1996) Increased density of multiple-head dendritic spines on medium-sized neurons of the striatum in rats reared in a complex environment. *Neurobiology of Learning and Memory, 66*, 93–96.

Hamilton, W. (1964). The evolution of altruistic behavior. *American Naturalist, 97*, 354–356.

Hollingshead, A. B. (1998). Retrieval processes in transactive memory systems. *Journal of Personality & Social Psychology, 74*, 659–671.

Huttenlocher, P. (2002). *Neural plasticity*. Cambridge, MA: Harvard University Press.

Kagan, J., Snidman, N., Arcus, D., & Reznick, J. S. (1994). *Galen's prophecy: Temperament in human nature*. New York: Basic Books.

Kleim, J. A., Swain, R. A., Armstrong, K. E., Napper, R. M. A., Jones, T. A., & Greenough, W. T. (1998). Selective synaptic plasticity within the cerebellar cortex following complex motorskill learning. *Neurobiology of Learning and Memory, 69*(3), 274–289.

Merzenich, M. M., Kaas, J. H., Wall, J., Sur, M., Nelson, R. J., & Felleman, D. (1983). Progression of changes following median nerve section in the cortical representation of the hand in areas 3b and 1 in adult owl and squirrel monkeys. *Neuroscience, 10*, 639–665.

Mischel, W. (1999). *An introduction to personality* (6th ed.). New York: Harcourt Brace Jovanich.

Newton, M. (2002). *Savage girls and wild boys: A history of feral children.* London: Faber.

Pinker, S. (1997). *How the mind works.* New York: Norton.

Poldrack, R. A., Temple, E., Protopapas, A., Nagarajan, S., Tallal, P., Merzenich, M., et al. (2001). Relations between the neural bases of dynamic auditory processing and phonological processing: Evidence from fMRI. *Journal of Cognitive Neuroscience, 13*, 687–697.

Smith, E. E., & Jonides, J. (1997). Working memory: A view from neuroimaging. *Cognitive Psychology, 33*, 5–42.

Smith, E. E., & Medin, D. L. (1983). *Concepts and categories.* Cambridge, MA: Harvard University Press.

Wegner, D. M. (1995). A computer network model of human transactive memory. *Social Cognition, 13*, 319–339.

Wittgenstein, L. (1999). *Philosophical investigations* (trans. G. E. M. Anscombe). New York: Prentice Hall.

VI Theoretical, Ethical, and Future Implications for Evolutionary Cognitive Neuroscience

In the Northeast of the United States, professional hockey is popular. Whereas one of the editors is an ardent supporter of the New Jersey Devils, another is a supporter of the rival Philadelphia Flyers. While we have yet to come to blows over our affiliations, we have exchanged a number of barbed emails and comments to each other. As two educated individuals, we can easily admit that our affiliations are superficial and somewhat artificial. However, such bonds to our respective sports teams were formed easily, and they play a large role in our lives.

As of this writing, the world is in political turmoil. The bonds and affiliations we have to our nations, religions, and causes are deepening and becoming more prevalent on a global scale. At the beginning of the twenty-first century we are facing challenges that, at first glance, might appear unique. However, these challenges are not exclusive to our generation or even to our species. Terrorism, for example, is now a global contest, with its reach expanding seemingly every day. Yet this may be of little surprise to those who study evolutionary theory. Humans have clearly been a violent species for many generations, as evidenced by the fossils left by both late *Homo* and Neanderthal (which may or may not be our descendant; if not, it is probably because *Homo* played an active role in their demise). The aggression of the chimpanzee (*Pan troglodytes*) has been well documented. Aggression and violence appear early in life, and there are direct neural and hormonal components that can be experimentally manipulated to increase or decrease their prevalence. Violent crimes are documented in almost every society from modern to primitive. Further, while the genes for aggression are not yet fully identified (in part because of the broadness of the term), they are certainly there, and breeding aggressive animals is a clear demonstration of the power of genetics in influencing aggression.

On top of this propensity for violence and hostility, there is clearly a genetic and cognitive link to affiliation, with a relation to perpetuating one's own genes. Social neuroscience has provided evidence of distinct brain processes associated with affiliation. Social psychologists have described for years how easy it is to create social bonds, even when the members of the "groups" are selected at random. Taken together, aggression based on even weak affiliations is not surprising. Those individuals that bonded together and defended their bonded group left more decedents. One need not travel back millennia to see such a link, as many of us are the product of the social bonding and aggression of our European ancestors. Earlier chapters emphasized more directly how social affiliation and aggression may have evolved—for example, how a change in social structure and dietary demands could have directly influenced our brain and cognitive development. At a very basic level, there is a relationship between and among calories, social structure, brain size, cognitive ability, and reproductive success. Although these variables are complex—and the complexity includes significant interactions within each variable—it is clear that they play a role in present-day behavior and cognition.

One would not be surprised to discover that our cognitions, rooted in brains that have been developed as such, might lead an individual to act aggressively in defense of his or her affiliated group. Yet rather than providing a negative (and simplistic) overview to cognition and behavior, we believe that understanding and hopefully correcting our behavior comes from understanding its origins at both a proximate and an ultimate level. For example, in the case of religious fanaticism or the equally alarming reaction to it, evolutionary cognitive neuroscientists can provide clear antecedents for its existence and its prevalence. Instead of responding to perpetrators as evil-doers, we become interested in the role of time (i.e., evolved cognitive structures), genes, neurological structures and functions, and the environment in creating cognitive schemas that would lead to such events.

The development of our research programs clearly touches on ethical grounds. People were shocked (pun intended) when Jose Delgado demonstrated that deep brain stimulation could influence cognition and behavior. By stimulating certain brain regions, he was able to influence or instigate behavior, including aggression. His work was deemed unethical, as it was thought to be "mind control." However, as of this writing, deep brain stimulation is being used as a potential treatment for Parkinson's disease and major depression, based in part on his advances of the technical means. Thus, while sometimes touching on sensitive issues, our work will better society if we do what science directs us to do, which is to understand the truths of nature.

What are the ethical implications of evolutionary studies? What are the implications of such research? At an ultimate level, this field expands our knowledge of modern cognition. We can understand present behavior by examining its roots and origins across multiple levels. By having such an understanding, we hope to clarify how our cognitions came to be, how they exist, and how they may be in the future. The future of our field likely includes cybernetics, human genetic manipulations, cross-species neural implants, and a host of other applications that we are just seeing developed. These research areas will affect the lives of an incalculable number of people in the future, and they have the potential to create a better world. It is somewhat ironic that although research in evolutionary neuroscience scrutinizes past events to build hypotheses, we see no field contributing more significantly to the future. For example, disentangling the 2% difference between us and chimpanzees may provide some of the greatest contributions in the advancement of neuropsychopharmacology research.

Ethically, it is important that we approach these issues with a sense of responsibility. Equally important, we should not back down from research areas that may provide benefit in the future, however remote such a possibility may be. We insist that there be no political motivation or thought of individual gain. We implore that the research continue to further our knowledge and address important questions. In the United States as in other countries, we face challenges in performing such research. Whereas the conflict between religion and evolution is longstanding, brain research has also been flooded recently with neuroethical challenges, as exemplified by the debate over stem cell research. These issues are not going away. If research on affiliation tells us anything, it is that bonds, however abstract, can be strong and easily integrated into a person's cognitive self-identification. Therefore, we hope that our readers and contributors continue to further this field by performing the studies that have made evolutionary neuroscience so richly diverse and important.

20 Philosophical, Ethical, and Social Considerations of Evolutionary Cognitive Neuroscience

Michael B. Kimberly and Paul Root Wolpe

Evolutionary cognitive neuroscience (ECN) has important implications for our understanding of the human mind and for how we conceive of ourselves and our place in the world. As research has further explicated the role evolution plays in the development of the human brain and thus the content of our thoughts, it has begun to influence the way we think about things like society, morality, responsibility, race, gender, and even science itself. The earlier chapters in this volume addressed many of these issues from a scientific and descriptive perspective. Previous chapters introduced the methodology of ECN and evidence for genetic determinants of brain physiology and cognition; explored the evolutionary interplay among the brain, culture, and the sexes; and considered the neural basis for consciousness and thought. In this chapter we consider the implications of ECN for philosophy, ethics, and society as a whole. We hope to initiate a broad-based discussion that will demonstrate the profound connection of ECN to humanity.

In the first section of this chapter we explore some of the philosophical questions embedded in ECN. Does research in ECN have implications for what we consider moral and immoral? Does it encourage a brand of genetic behavioral determinism that renders free will an illusion and moral responsibility merely a product of inevitable evolutionary forces? In addressing these philosophical questions, we do not intend any normative observations. That is, in the first section we do not attempt to delineate the kinds of questions that ECN should or should not explore, or the ends to which ECN research should and should not be put. Rather, we explore what ECN means for the philosophy of morality and moral accountability.

In the second section we consider studies relating to controversial knowledge. Research in ECN might establish, for instance, categorical differences corresponding to divisions of race, ethnicity, or gender. We explore what such differences might mean for society and science. With this potential for impact in mind, we consider whether the scientific project is inherently valuable and implies an imperative to pursue knowledge for its own sake, or if the scientific project is instead a means to an end and ought instead be directed by other social values.

Drawing on the previous two sections, we conclude in the third section with an ethical consideration of ECN and its interplay with society, media, and the justice system. We document and discuss examples of misrepresentation of research results and data by the national media, followed by a consideration of the ways in which ECN, together with advances in neurotechnology, may find public and private application. We then discuss ECN researchers' obligation to ensure that their science is accurately and fairly represented. Finally, we discuss the potential for misuse of ECN research as diagnostic evidence in criminal proceedings. We consider whether research may function as grounds for illegitimately determining innocence or guilt, and conclude by addressing the ethical obligation of researchers and legal practitioners alike to ensure that ECN is not used inappropriately in the justice system.

Evolutionary Cognitive Neuroscience and Morality

Because it brings together three contentious areas of debate—genetics, the nature of human cognition, and biological and genetic determinism—ECN raises two primary issues of interest to philosophical inquiry. The first issue assumes that because our conceptions of the good are tied to human nature, morality is defined, at least in part, by who we are; and if who we are is determined by our genes and what they tell us to do, our conception of the good must be genetically determined as well. By elucidating in exactly which ways our genes influence and perhaps even determine our behavior, thoughts, and feelings, ECN may redefine morality. If some humans are genetically determined to be racist or violent or aggressive, for instance, then perhaps some naturalists will argue that it is *right* that these people are this way, or at least that it is not wrong.

The second issue is the flip side of the first. Rather than defining right and wrong, genetic determination of behavior may make ascribing praise and blame for right and wrong incoherent. Instead of redefining morality, ECN may render it meaningless. Moral accountability requires the free will to make choices and act on those choices. If ECN establishes that certain, perhaps all, behaviors are genetically determined, then free will is illusory, and we have no grounds on which to hold criminals and other moral transgressors accountable for their actions. As experimental exploration into ECN progresses, so too should our consideration of what ECN means for morality.

Evolution and Directionality

Several theories of morality rely on human nature as a centerpiece for understanding how moral agents ought to act. Among these theories, natural law ethics is not only the best-defined and most often discussed, it also spans both religious and secular reasoning. Simply stated, natural law theory proposes that humans have a certain, perfect state (*telos*) to which they are naturally inclined, and that the pursuit of this state and action in accordance with it defines morality. John Finnis (1980), a contemporary natural lawyer, has argued that human reason serves as a tool for determining conformity with this ideal human nature. Echoing Thomas Aquinas, the father of natural law ethics, Finnis contends that were human nature different, the moral duties of humans would necessarily be different as well. Natural law theory requires that human nature supervene on human morality in the same way that the arrangement of pixels on a monitor supervenes on the image that monitor projects: a change in the former property necessarily implies a change in the latter. Human nature, according to natural law and other such theories of morality, therefore has a reductionistic causal relationship to human duties.

On the grounds of such a natural law theory of ethics, we may reasonably express concern about the implications of ECN for morality and the good. If ECN successfully establishes a genetic foundation for certain behavioral dispositions—dispositions toward which humans are naturally inclined—then a natural law theory may have to accommodate those dispositions in its formulation of morality. For instance, recent studies conducted in the United States and the United Kingdom have shown neurobiological correlates for aggressive behavior (e.g., Raine, Buchsbaum, & LaCasse, 1997), for lack of empathy (e.g., Moll, de Oliveira-Souza, & Eslinger, 2003), and even for psychopathy (e.g., Blair, 1995; Raine, Lencz, Bihrle, LaCasse, & Colletti, 2000). If these studies withstand the scrutiny of their dissenters, and if some humans are in fact physiologically and genetically predetermined to be racist or violent or aggressive, then might those who posit moral value in human nature have to incorporate racism, violence, and aggression into their formulations of the good?

The problem demonstrated by this troubling proposition is known as the *naturalistic fallacy*, or the illegitimate derivation of *ought* from *is*. The naturalistic fallacy states simply that prescription (*ought*) and proscription (*ought not*) cannot coherently be derived from description (*is*). Because our genetic composition is the product of the valueless processes of natural and sexual selection (a description), it cannot arbitrarily serve as the foundation for moral behavior (pre- and proscription).

For the sake of argument, let us assume that humans, like every other organism that exhibits ethological patterns, are subject to innate behavior. Despite our learning how to act and respond by imitating our parents or siblings, and later moderating our behavior by using our rationality to deduce the way we think we ought to act, some of our behavior is indeed genetically wired. Although clear distinctions between nature and nurture cannot easily be drawn, examples sometime arise. For instance, adolescents need not be taught to desire sexual intercourse, nor how to go about doing it. Babies need not be taught to cry when they are hungry, and parents need not be taught to be protective of their children. These innate behaviors are determined by our genes, which have in turn been determined by evolution. Yet herein lies the fallacy inherent in claiming that genetics determines the good: evolution is not a goal-oriented process. Its only concern is reproductive fitness. Edward O. Wilson, one of the earliest writers on sociobiology, notes that "no species, ours included, poses a purpose beyond the imperatives created by its genetic history. Species may have vast potential for guidance and mental progress, but they lack any imminent purpose or guidance from agents beyond their immediate environment or even an evolutionary goal toward which their molecular architecture automatically steers them" (Wilson, 1982, p. 2).

That is to say, evolution does not guide humans, or any other species, in any particular direction. The only factor that evolution values in its progressive effect is reproductive fitness, or those characteristics, behavioral or physical, that endow their bearers with a better ability to produce and/or protect offspring. Those behaviors encoded in genes that lead to a greater proliferation of progeny lead, *eo ipso*, to a greater proliferation of those behaviors themselves throughout the species population. Were those behaviors to become reproductively disadvantageous, however, say, by the introduction of some new environmental variable, then natural selection would remove them from the population.

These genetically determined behaviors thus have no inherent value, and they endure only as instrumental means of survival. We leave it to the other contributors of this volume to argue whether evolution truly does play a hand in human's cognition and behavior—whether ECN is real as a description of how the genetic design of our brains influences or even determines how we make moral choices. But no matter the outcome of these discussions, our genes determine *ought* no more and no less than *ought* determines our genes.

Richard Dawkins (1996b) has reframed this distinction by arguing that all that science, including ECN, can do is answer *how*, not *why*. Humans are driven to seek purpose in the world and often confound science's satisfying answers to *how* questions with their insatiable desire for *why* answers. But, Dawkins argues, the mere fact that a question can be asked does not make it sensible or appropriate. Asking *why* of the direction and outcome of the evolutionary process is like asking for the temperature of emotions or the color of choices. The why question posed of a boulder makes no sense, and neither does it posed of evolution or genetics. Elsewhere, Dawkins (1996a, p. 5) concludes that "Natural selection, the blind, unconscious, automatic process which Darwin discovered, and which we now know is the explanation for the existence and apparent purposeful form of all life, has no purpose in mind. It has no mind at all and no mind's eye. It does not plan for the future. It has no vision, no foresight, no sight at all. If it can be said to play the role of watchmaker in nature, it is a *blind* watchmaker." That is to say, natural and sexual selection does not steer our species toward any particular virtuous form. If this conclusion is true, then *ought* cannot be derived from a description of humankind's genetic behavioral predispositions. ECN therefore poses no new problems for morality, no matter how powerful a tool it turns out to be in furthering our understanding ourselves as evolutionary products, and serves only as further evidence of the incoherence of the naturalistic fallacy.

Genetic Determinism and Moral Responsibility

Whether humans enjoy freedom of will to make real choices or instead exist within a determined framework of cause and effect is one of the most important and complex questions of philosophy. ECN may offer new scientific input into the free will debate by demonstrating that certain behaviors are determined by genetic influence over the brain. Successfully demonstrating this influence may threaten the reality of free will by showing that moral agents are not actually free from the behavioral constraints imposed by their genes and, in turn, physiology. If ECN helps show that free will is limited—or even altogether false—then it will have important implications for holding people morally and legally accountable for what they do.

Given the complicated nature of the debate, the implications of ECN for free will and moral accountability are therefore more subtle and harder to evaluate than they are for a natural derivation of morality itself. Beginning with Hobbes and Hume several centuries ago, focus on the dilemma of determinism has concerned causal determinism. In its simplest terms, causal determinism argues that every effect in the universe has a sufficient cause that necessitates the effect. For instance, for a billiard ball to move across a pool table (an effect), it must have a cause. This cause—say, a player hitting the ball with her cue—moreover necessitates the effect. That is, given the cause, the ball *had* to move across the table. Causal determinism holds that, like the billiard ball, a person's will to do something is necessarily caused by other things. A person's choice to eat a bagel, for instance, is caused, *inter alia*, by her hunger, the fact that a bagel is available to her, and the fact that she likes to eat bagels. But according to deterministic theory, because it is the very nature of a cause that it generates a necessary effect, the choice to act according to those causes is not really a choice at all. Realizing that those other causes are themselves effects requiring of further upstream causation, the analysis extends ad infinitum to the original state of the universe at the beginning of time, the conditions of which were sufficient to produce the current universe, including all humans and their decisions in it. This universe represents the only mediate state of affairs that could have arisen and had to arise from those first causes at the beginning of time. Thus, that nature of cause and effect indicates that determinism is true and that humans have no freedom of will, insofar as they *must* make the choices that they do.

Causal determinism can neither be altogether confirmed nor altogether refuted by ECN. Instead, if successful at demonstrating genetic determinism, ECN may fit in as a component of causal determinism by showing that one of the many causes generating effects of the will (of choices and behavior) is the particular genetic composition of the agent, a genetic composition caused upstream by natural and sexual selection. For ECN to seriously threaten the notion of moral accountability, however, it would have to prove a surprisingly rigorous version of genetic (and downstream brain-physiological) determinism. To explore this claim, let us consider the nature of moral decision making as it bears on moral accountability, and how genetics and brain physiology may influence it.

First, let us consider that for an agent to be a moral agent coherently subject to praise and blame, she must have access to moral reasoning and enjoy the free will to employ that reasoning. An agent lacking either of these capacities cannot be considered a moral agent, and it would therefore be incoherent to subject her to moral assessment. For instance, an elephant cannot employ the powers of reason to make a moral conclusion. Thus, when it kills a bystander at the zoo even though the bystander has done nothing wrong, we cannot hold the elephant accountable for failing to meet a supposed moral obligation not to murder humans. And regarding free will, when an epileptic harms an innocent bystander in a fit of involuntary convulsions, we cannot hold her accountable for causing that harm because her convulsions do not represent actions resulting from a choice. Only when a rational agent's action expresses a choice can we logically hold that agent open to the force of moral judgment.

For the conclusions of ECN to bear on the coherence of moral accountability, then, they must describe limitations on either our ability to reason about obligations (Position A) or, as previously discussed, our ability to exercise free will in the making of the choices that motivate the bodily movements we make (Position B) (Figure 20.1). The paramount obstacle to overcome for the sake of proving either version of

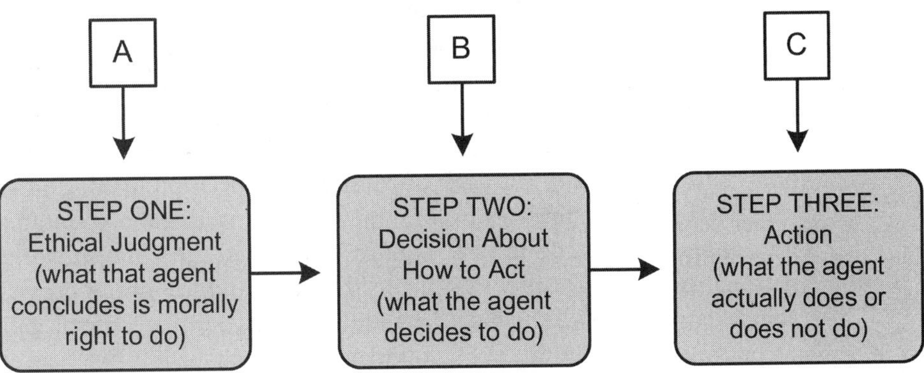

Figure 20.1
Free will implies three steps from a judgment to an action. Physical coercion limits free will at Position C, while genetic determinism may limit it at Position A or Position B.

causal determinism is the distinction between the exertion of influence at either of these points and the setting of hard-and-fast limits.

Significant research in ECN bears on this question. For example, work done by Greene, Sommerville, Nystrom, Darley, and Cohen (2001) shows that emotions are highly influential in certain kinds of moral judgment making, and that the activation of these emotions correlates with increased brain activity in the medial frontal gyrus, the posterior cingulate gyrus, and the angular gyrus bilaterally. These findings have implications for establishing the extent to which brain physiology and genetics promote or demote emotions or logical reasoning in the formation of judgments; arguably, deficits in the correlated areas of the brain would limit moral emotions and their influence over moral reasoning.

Other research bears on the extent to which genes and physiology incline certain people to violence and aggression. A study by Raine et al. (1997) demonstrated reduced glucose metabolism in the bilateral prefrontal cortex, posterior parietal cortex, and corpus callosum in the brains of murderers pleading "not guilty by reason of insanity." The authors reasoned that the amygdala, prefrontal cortex, and hippocampus make up part of the limbic system governing the expression of emotion, and that the thalamus relays information from the subcortical limbic structures to the prefrontal cortex. Therefore, abnormalities of brain function in murderers may indicate deficits in making appropriately emotionally informed decisions, thereby establishing a biological predisposition to violence and aggression.

Further work by Raine et al. (2000) demonstrated the extent to which brain physiology and genetics promote psychopathy (also known as antisocial personality disorder). They found a statistically significantly smaller volume of prefrontal gray matter among people with psychopathy, and conjectured that these structural deficits underlie psychopaths' poor fear conditioning and lack of conscience in decision making.

Relating these findings back to their impact on moral accountability, the *Merck Manual* (Beers & Berkow, 2004) states that the antisocial personality exhibits "callous disregard the rights and feelings of others" and that persons with the disorder "exploit others for materialistic gain or personal gratification." Moreover, psychopaths "have a well-developed capacity for glibly rationalizing their behavior or for blaming it on others," indicating that, like the other physiologically founded conditions of disinhibition, antisocial personality disorder functions at Position A, influencing the ways in which agents reason, and not how they translate their reasoning about right or wrong into choices.

In another example, Moll et al. (2002) demonstrated that patients with dysfunction of the orbitofrontal cortex are often able to make moral judgments (since their higher-order cognitive abilities are intact) but fail to effectively translate their judgments into moral behavior. The same authors showed elsewhere (2003) that certain other types of psychopaths can tell right from wrong, but they suffer a dissociation of knowing good from acting good. Each of these examples indicates that brain physiology also acts at Position B, influencing the ways in which agents are able to translate their deliberations over right and wrong into choices about how to act given their judgments. The open question is whether such factors influence normal decision making in the nonpathological populations, that is, whether differential activation of these areas might correlate with differences in moral judgments and decisions about how to act in the general population.

The functional question in exploring whether the brain physiologies of the violent, aggressive, and uninhibited personalities described in these studies is whether they create insurmountable limits or merely influence behavior. The very thing that makes each of us humans unique, after all, is the aggregation of influences—biological and environmental—on our behavior. There are those inclined to a general weakness of will who know what is right to do but cannot bring themselves to do it, and there are those who are highly principled and supererogatorily sacrifice personal gain for a sense of what is right. Certainly the influences operating on these two kinds of people (and the range of people between them) are very different. And certainly those differences guide each in different directions. But speaking of those influences as though there were no choice but to act according to them is, by definition, mistaken. The person with a weak will is still perfectly capable of doing what is right, just as the principled person is perfectly capable of doing what is wrong. No mandate, genetic or otherwise, precludes either from choosing and doing anything within the physical limits of their situations.

Indeed, the very task at the heart of morality is to overcome countervailing influences. Were morality easy, were every influence over our behavior to incline us toward right actions and proper concern for others, it would hardly be a topic worth consideration; everyone would be effortlessly moral. But precisely because this is not the case, precisely because we are all influenced by our desires and inclinations to act *despite* our moral judgments, and moreover have the *choice* of whether or not to do so, considering the conditions and requirements of morality is a worthwhile project.

Consider, for instance, a teenager with $40 of allowance money who wants a trendy pair of shoes that costs $60. She has no way of raising the necessary funds for the purchase in time for a big party over the coming weekend, and so she contemplates shoplifting the shoes. She knows that stealing from the store owner is wrong, but she wants desperately to fit in with the cool crowd at the party and believes that wearing the $60 shoes will help to that end. Not stealing the shoes requires overcoming powerful influences urging her to act counter to what she believes to be the moral choice. Nonetheless, it is no less a blameworthy choice when she takes the shoes that she is subject to such strong inclinations and preferences to look good and be accepted socially. She is perfectly capable of leaving the shoes and walking out of the store empty-handed—nothing forces her hand to pick them up and walk out. Instead, she *chooses*, under the range of influences on her behavior, to take them. Her will to resist fails to overcome those influences, a failure for which she is fully responsible. Indeed, were we to excuse her from blame because of the power of those influences, we would miss the entire point of moral accountability.

The influences of brain physiology are not as cut-and-dried. Clearly, some conditions do represent true mental incapacity. In fact, there is likely a point along the continuum of all of the deficits cited in this chapter at which true incapacity does limit the range of choices open to individuals suffering from major brain defects. We know, for instance, that anencephalics will never make any decisions at all, and that people with severe brain damage, like the infamous Phineas Gage, have little control over their emotions and asocial behavior. Autistic and developmentally disabled adults and others with mental and neurological illnesses clearly suffer from a physiologically identifiable impulse control deficit. The brain, like other organs, functions well or less well across its range of capacities.

There is also a range of morphological variation in the brain that falls within the normal range of mental capacity. These variations include different lobe sizes and proportions, different levels of neurotransmitters, and so on. People with low-volume prefrontal gray matter, for instance, may have a more difficult time mediating their behavior with conscience. According to present research, they will have a harder time considering the perspectives of others and using their moral emotions to appropriately inform their decisions. But in the range of people falling outside the boundaries of true mental defect, each of our idiosyncrasies of the brain is nothing more than influence. The man who mugs a student for his cellular telephone may have a smaller prefrontal cortex than the woman who stops to help the student after the mugger has fled. But does this somehow determine their actions, or somehow set hard-and-fast limits on the choices they could have made? Just as with the teenager who stole the shoes, the answer is no. The mugger was perfectly capable of making the right choice not to mug the student, and the bystander was just as capable of walking past without helping the student after the attack. The physiology of both their brains in no way limited the range of choices open to them but rather influenced which option each chose, just as physiology influences every choice they and we all make every day.

Controversial Knowledge and the Research Imperative

Evolutionary cognitive neuroscience also has important implications regarding the pursuit of controversial knowledge—that is, scientific facts that may generate or reinforce social controversy and particular social attitudes and prejudices. For the past half-century, a great deal of scientific effort has attempted to discover cognitive differences among races, for instance, and to show that one race is more or less intelligent than others. Arthur Jensen has been a vocal figure in the measurement of intelligence, for instance, beginning with a famous article in the *Harvard Educational Review* in 1969. The paper suggested that intelligence quotient, or IQ, is genetically determined, and moreover that blacks as a population have an average IQ 15 points lower than the white population. This study and the many like it that followed, conducted in a racially charged time, made some of the same fallacious assumptions discussed earlier in this chapter: that evolution is a unilinearly directional process that points all organisms toward utopian standard of cognitive function, for instance (Gould, 1996).

But even regardless of the quality and validity of their findings, such studies raise important questions about the pursuit of knowledge and what some have called the research imperative (e.g., Callahan, 2003). In such a time as 1969, might there have been an ethical obligation to withhold research results that were certain to feed the racial unrest choking the nation? More generally, is the scientific project an inherently valuable endeavor in its own right, one that warrants the freedom to investigate questions unchecked by the potentially biased restraints of social and political values? Or is the scientific project instead only valuable as means to its various ends, such as better medicine and more efficient technology, and therefore ought it be directed by the quality, worth, and value of those ends? How we characterize the research imperative for ECN as driven by inherent or instrumental value, and consequently how those who practice ECN formulate their obligations to society, will depend on a wide range of factors.

There is little doubt that the pursuit of new knowledge, information, and theory in ECN has inherent value. Much of the research presented in the earlier chapters of this book is, for lack of a better word, *exciting*. Though there are major steps yet to be made, our understanding of the brain has become impressively deep. Moreover, research in ECN employs a range of remarkable new technologies inexorably connected with research itself, such as functional magnetic resonance imaging (fMRI). It would trivialize such impressive imaging technology to consider it nothing more than a tool for research and medicine. To see a computer model of one's brain—the three-dimensional lattice of the blood vessels infusing every cubic millimeter, the unique and maze-like compilation of folds and grooves composing the cortex—evokes more than respect for a tool. It evokes awe and stands as an inherently valuable testament to the ever-receding limits of human achievement. In this way, it seems to better society just to have the ability to produce such images and to use them to expand our knowledge of brain physiology, chemistry, and genetics. Coming to understand ourselves, our biology, and our minds is a goal in and of itself.

And yet such technology and the results of research that come of it have the ability to improve or harm the human condition in a more important way than through the inherent value of knowledge by itself. Research and technology are ultimately conceived to put knowledge to use. For instance, imaging technology previously limited to radiology clinics might be used for new national security measures (Feder, 2001) and for exciting advances in medical and psychological therapies. Biomedical engineering teams across the country are working to develop new, practical technologies for brain imaging in airports, in emergency rooms, and even on the battlefield (Kimberly, Wolpe, Platek, Bunce, & Caplan, 2004). Technology such as functional near-infrared optical brain imaging promises to be as portable as a personal digital assistant and as inexpensive as a personal computer. Such developments have enormous ramifications for new kinds of research, as well as new kinds of day-to-day applications for the findings of that research. Employers may be better able to assign employees to tasks for which they are particularly well-suited, and schools may be better able to identify children predisposed to depression or attention deficit/hyperactivity disorder before either becomes a problem. These and the countless other potential applications for ECN research represent uses that can help people improve their lives, and clearly represent the instrumental value of the field.

But ECN may also be used for more dubious ends. Significant current research in ECN describes neuroscientific differences among races (e.g., Phelps et al., 2000) and between the sexes (e.g., Gaab, Keenan, & Schlaug, 2003). Although the nature of the research itself may raise no flags, the ends to which it might be put could raise concern. Racism and sexism are both alive and well in the twenty-first century, especially with respect to equality-of-opportunity measures like education and income (Blank, Dabady, & Citro, 2004), and research findings that could be used to defend or perpetuate systematic discrimination give reason to pause. Although it would be both impossible and unreasonable for researchers in any field to predict and consider the malefic ends to which their findings may be put to use, it will nonetheless be important that, to the extent they are components of and considerations within ECN research, distinctions of race and sex do not influence the direction of the research itself, and that those researchers who participate in ECN sensitive to race and gender are neither silently nor openly guided by the demand characteristics of prejudice. Moreover, in cases where such intent is clear, and certainly despite any inherent value of the acquisition of knowledge, a recognition of the value of the ends of research may imply an imperative to refrain from such research.

Even independent of the possible applications of ECN research, research on the brain and the way we think has an impact on our daily lives in another profound way. Ideas, particularly ideas bolstered (legitimately or illegitimately) by empirical evidence, are not purely theoretical in their content. Even the most academic ideas are inextricably linked with day-to-day reality—they describe it, they affect it, and they are part of it. The ideas that researchers in ECN are inclined to explore are particularly prone to exert influence on society. After all, what is society if not the vast and intricate mesh of interconnected social, political, and legal relationships? As ECN research recasts the definition of those relationships in terms of genetics and fMRI images, the relationships themselves may be affected as we look at them differently and impose on them new judgments and feelings (Winner, 1988). It may sometimes be tempting to retreat to an intellectual isolation from the reality reflected in ECN, remaining ostensibly loyal to pure science and above the fray of politics and social reality, but such isolation is fundamentally impossible. It is an illusion. Although the physics that described the first atomic bomb is pure and elegant science, treating that science or the scientists as though they were isolated from the bomb would be disingenuous—or perhaps an example of denial (Lawler, 1978). It is unlikely that ECN will ever lead to discoveries that facilitate such mass destruction as an atomic bomb, but ECN nonetheless has profound social, political, psychological, and legal implications that cannot be divorced from the science itself. That is to say, conceiving the research imperative in ECN as motivated entirely by means-based value isolated from the consequences of the research would entail an untenable denial of the connection between the theory of ECN and social reality.

Evolutionary Cognitive Neuroscience and Responsibility to Society

Even the most abstract theories within ECN are inextricably connected to real and tangible ends, insofar as they influence our thinking about social, political, psychological, and legal realities. But should these potential consequences influence the work of researchers in ECN? Does their impact affect the kinds of questions we ought and ought not to ask, or the ways in which we ought and ought not to go about answering and representing them? In this final section we consider how research in ECN might be used (and misused) by social institutions such as the national media and the criminal justice system, and explore possible guidelines for ECN researchers to consider in the representation and use of their work in social settings.

Cognitive Neuroscience and Technology in the Media

The national and international media today often overhype stories, from the criminal trial du jour to the latest breakthrough in science. In seeking greater market share, the print and broadcast media often resort to fear and anxiety to sell their messages (Altheide, 2002). Science reporting is not immune to the tendency to sacrifice accuracy to sensationalism, especially since science is uniquely positioned in society to evoke feelings of both great hope (e.g., new medical treatments) and visceral fear (e.g., human cloning). Science news is also riddled with abstract ideas like statistical relationships and microscopic neurons and molecular neurotransmitters, which can be difficult to reduce accurately to the digestible sound bites that dominate media today.

Examples of such inaccuracies regarding neuroscience abound. For instance, one *USA Today* article (Vergano, 2000) claimed that "Brain deficits predispose individuals towards violence," although the brain imaging research on which the article was based (Raine et al., 2000) failed to establish causation and even warned that causation cannot be drawn from the correlation of antisocial personality disorder with a small prefrontal cortex. A similar article in *The Boston Globe* on the same subject (Foreman, 2002) claimed that damage to the prefrontal cortex causes people to be impulsively violent, although the studies cited again only established correlation.

In another example (Smith, 2004), an article from a prominent Australian newspaper claimed that neuroscience research has established a neural basis for identifying "racist attitudes, lying, people's responses to movies and erotica, and even why they prefer Coke to Pepsi," and raised the possibility of "spying on the cortex." A similar article in *Science Daily* (Connelly, 2000) titled "Brain Imaging Technology Can Reveal What a Person Is Thinking About" quoted a neuroscientist as claiming, "What we've shown is that we can actually tell, on a moment-by-moment basis, what an individual is thinking about." Both of these articles reflect concern that the tentative conclusions of ECN will be taken as hard facts. Of course, fMRI cannot reveal what people are thinking about, or even if they are thinking rationally or emotionally, or why they prefer one cola over another. It cannot blindly flag a racist or even yet serve as a useful lie detector. At its base, all fMRI shows is which areas of the brain are using oxygen to metabolize ATP. Although data from fMRI studies establish statistically significant correlations of activity in certain areas of the cortex with particular states of cognition, how the areas of metabolic activity correlate with the aggregated cognitive state of lying or truth-telling will remain a matter for interpretation. Moreover, those interpretations are based on average population-level data inherently insensitive to the cognitive and neurological uniqueness of each individual, such as the uniquenesses of brain size, proportion, shape, and activity addressed earlier. That is, although population-level fMRI data may significantly correlate, say, anxiety with activity in a certain location on the cortex, this correlation does not represent an absolute rule for analysis of individuals—aberrations will and do arise. Nonetheless, laypeople, including juries and the TV-watching, magazine-reading public at large, are likely neither to hear this explanation nor to understand or apply it on those rare occasions when they do (Dumit,

1999). When they read that neuroscientists can read people's minds or that brain properties cause certain behaviors, they are likely to take these claims at face value.

As ECN progresses further in its sophistication, researchers have a profoundly important ethical and professional responsibility to report their results accurately, precisely, and with due moderation. Explanations of the origins or the nature of our minds can fuel or refute popular misconceptions of the nature of human cognition, affect, and even intangibles like free will or religious impulse. As they report exciting and encouraging progress in their field, ECN researchers should be careful and explicit in explaining the limits of their findings, and should be circumspect in speculating about the implications of their theories for controversial social and psychological policy.

Cognitive Neuroscience and Technology in the Criminal Justice System

There are few domains in which practitioners are as eager for theories and technologies of the mind as in criminal justice. Brain-based theories of criminal behavior have a long pedigree, tracing back in the modern day at least to Cesare Lombroso, "the father of modern criminology." Lombroso postulated that criminal behavior was in large part due to the birth of "atavists," evolutionarily arrested individuals whose primitive brains gave them a predisposition to violence, sexual immorality, and greed (Lombroso, 1876). In the century and a half after Lombroso, genetic and neuroscientific theories of the criminal mind flourished. When translated into penal policy, the results were often lamentable; a large percentage of the more than 60,000 psychosurgical procedures performed in the United States between 1936 and 1956 were imposed on "moral degenerates" and "hereditary defectives" (Feldman & Goodrich, 2001). The desire for diagnostic and corrective neurotechnologies is no less intense today.

The advent of neuroimaging has reinvigorated the quest for morphological or functional correlates to criminal behavior (e.g., Bassarath, 2001; George et al., 2004; Kiehl et al., 2004). At present, the pursuit of technologies such as neuroimaging-based lie detection has outpaced the desire for justifiable theoretical underpinnings for the technologies. However, the pressure to develop these technologies is great. The development of neuroimaging techniques for lie detection has largely been funded by the Department of Homeland Security and the Department of Defense, eager to use the technology for security screening. As technologies and theories tend to develop more or less in tandem in science, there will undoubtedly be an attempt to understand neurotechnological findings through the development and modification of theories of the mind.

The danger here is in the use of ECN and other theories of the mind to justify changes in jurisprudence that may not be in the best interest of a fair system of criminal justice. For instance, the use of neuroscientific evidence in criminal proceedings by defense lawyers is on the rise (Kulynych, 1997), but the ethical appropriateness of its use is dubious, for two reasons. In the first place, neuroscientific evidence stands on uncertain scientific theory and data and is vulnerable to manipulation and distortion (Reeves et al., 2003). Because brain imaging, for instance, requires standardization against the individual's baseline neural activity as well as against population averages, the generation of the image itself implies a range of assumptions and compromises at all levels of analysis (Reeves et al., 2003). As previously mentioned, although activity or inactivity at certain points on the cortex correlate with behaviors at the population level, variation within a population makes individual data difficult to interpret. Relative to "average" images of brain activity, a defendant's brain scan may indeed *look* abnormal, even though it may really fall within the range of normalcy that underlies that average. Despite this critical caveat, evidence indicates that juries find neuroscientific evidence highly persuasive (Dumit, 1999; Kulynych, 1997; Reeves et al., 2003) and are unlikely to understand the interpretive nuances that make its introduction suspect.

Moreover, there are two components to neuroscientific evidence, the technical interpretation of data and the inferential assessment of the effect of a neurophysiological or cognitive abnormality on the person's behavior (Kulynych, 1997). The validity of the second component falls squarely within the purview of ECN and is not a matter of scientific fact but rather an ongoing debate. The justice system is ultimately predicated on the truth of free will, and the inclusion of neuroscience evidence in psychiatric evaluations constitutes an attempt to illustrate brain-physiological limits on free will. As discussed earlier in the chapter, however, nothing short of hard-and-fast limits on the range of choices open to an agent will serve this function. *Influence* over behavior does not mitigate responsibility; rather, failing to overcome influence is the very basis *for* culpability. That is, an agent's failure to overcome influence necessarily implies responsibility, whether that influence is physiological, social, emotional, financial, or whatever else. It is therefore important that neuroscientists involve themselves in debate over the proper use of neurotechnology and neuroscientific theories in the criminal justice system, and that if they are called on to participate in evaluations for criminal proceedings, they temper their conclusions with a recognition of the limitations of the science.

On the other hand, thoughtful use of ECN can illuminate aspects of human behavior, including, perhaps, antisocial and criminal behavior. The goal of ethical introspection is neither to inhibit the development of ECN theory nor even to restrict its application in criminal justice where appropriate. The danger lies in the premature or unjustified importation of neuroscientific theories into the justice system to support preexisting notions of culpability or punishment. The potential for cognitive neuroscience to do harm through misapplication is as great as its potential to contribute to a humane and scientifically defensible jurisprudence.

Conclusion

The scientific effort to understand the interplay among genes, the brain, and behavior is laudable. Although we may never untangle the entire mystery of neurological evolution or understand the complexity of human behavior, the attempt to connect the two is a fascinating and inspiring scientific enterprise. Touching the most basic components of human life—cognition, affect, and behavior—the theory and understanding generated by ECN will have a longlasting influence over many aspects of our lives. In truth, however, we will never know the full effect of ECN on the texture of human life. Certainly its impact extends beyond philosophy, the media, and the justice system as we have narrowly construed it here. The broader and ultimately more important impact of ECN is unknowable, and to the extent that it is known, it is not within our purview or that of ECN scientists to assess. Such judgments about the philosophical, ethical, and social implications of ECN belong to society as a whole. We hope we have helped to begin an informed conversation to that end.

References

Altheide, D. L. (2002). *Creating fear: News and the construction of crisis.* New York: Aldine de Gruyter.

Bassarath, L. (2001). Neuroimaging studies of antisocial behaviour. *Canadian Journal of Psychiatry—Revue Canadienne de Psychiatrie, 46*(8), 728–732.

Beers, M., & Berkow, R. (Eds.). (2004). Personality disorders. In *The Merck manual of diagnosis and therapy* (17th ed., sect. 15, chap. 191). New York: Wiley.

Blair, R. J. R. (1995). A cognitive developmental approach to morality: Investigating the psychopath. *Cognition, 57*, 1–29.

Blank, R. M., Dabady, M., & Citro, C. F. (2004). *Measuring racial discrimination.* Washington, DC: National Academies Press.

Callahan, D. (2003). *What price better health? Hazards of the research imperative.* Los Angeles: University of California Press.

Connelly, K. (2000, November 11). Brain imaging technology can reveal what a person is thinking about. *Science Daily.* Available: http://www.sciencedaily.com/releases/2000/11/001110073236.htm

Dawkins, R. (1996a). *The blind watchmaker: Why the evidence of evolution reveals a universe without design.* New York: Norton.

Dawkins, R. (1996b). *River out of Eden: A Darwinian view of life* (reprint ed.). (Science Masters Series). New York: HarperCollins.

Dawkins, R. (1998). *Unweaving the rainbow: Science, delusion, and the appetite for wonder.* New York: Houghton Mifflin.

Dumit, J. (1999). Objective brains, prejudicial images. *Science in Context, 12*(1), 173–201.

Feder, B. J. (2001, October 9). Truth and justice, by the blip of a brain wave. *The New York Times.*

Feldman, R. P., & Goodrich, J. T. (2001). Psychosurgery: A historical overview. *Neurosurgery, 48*, 647–657.

Finnis, J. (1980). The illicit inference from facts to norms. In *Natural law and natural rights* (pp. 33–36). New York: Oxford University Press.

Foreman, J. (2002, March 26). Brain scans draw a dark image of the violent mind. *The Boston Globe.*

Gaab, N., Keenan, J. P., & Schlaug, G. (2003). The effects of gender on the neural substrates of pitch memory. *Journal of Cognitive Neuroscience, 15*(6), 810–820.

George, D. T., Rawlings, R. R., Williams, W. A., Phillips, M. J., Fong, G., Kerich, M., et al. (2004). A select group of perpetrators of domestic violence: Evidence of decreased metabolism in the right hypothalamus and reduced relationships between cortical/subcortical brain structures in position emission tomography. *Psychiatry Research, 130*(1), 11–25.

Greene, J. D., Sommerville, R. B., Nystrom, L. E., Darley, J. M., & Cohen, J. D. (2001). An fMRI investigation of emotional engagement in moral judgment. *Science, 293*, 2105–2108.

Gould, S. J. (1996). *The mismeasure of man.* New York: Norton.

Haddock, V. (2004, May 16). Gay panic defense in Araujo case. *San Francisco Chronicle*, p. E1.

Jensen, A. (1969). How much can we boost IQ and scholastic achievement? *Harvard Educational Review, 39*, 1–123.

Kiehl K. A., Smith, A. M., Mendrek, A., Forster B. B., Hare, R. D., & Liddle, P. F. (2004). Temporal lobe abnormalities in semantic processing by criminal psychopaths as revealed by functional magnetic resonance imaging. *Psychiatry Research, 130*(3), 297–312.

Kimberly, M. B., Wolpe, P. R., Platek, S., Bunce, S. C., & Caplan, A. (2004, October 16). Ethical consideration of detecting deception using functional NIR optical brain imaging [abstract]. Abstracts of the Biomedical Engineering Society Conference, Philadelphia.

Kulynych, J. (1997). Psychiatric neuroimaging research: A high-tech crystal ball? *Stanford Law Review, 49*, 1249–1270.

Lawler, J. (1978). *IQ, heritability and racism*. New York: International Publishers.

Lombroso, C. (1876). *L'uomo delinquente*. Milan: Hoepli.

Moll, J., de Oliveira-Souza, R., & Eslinger, P. J. (2003). Morals and the human brain: A working model. *Neuroreport, 14*, 299–305.

Moll, J., de Oliveira-Souza, R., Eslinger, P. J., Bramati, I. E., Mourão-Miranda, J., Andreiuolo, P. A., & Pessoa, L. (2002). The neural correlates of moral sensitivity: A functional magnetic resonance imaging investigation of basic and moral emotions. *Journal of Neuroscience, 22*, 2730–2736.

Phelps, E. A., O'Connor, K. J., Cunningham, W. A., Funayama, E. S., Gatenby, J. C., Gore, J. C., & Banaji, M. R. (2000). Performance on indirect measures of race evaluation predicts amygdala activity. *Journal of Cognitive Neuroscience, 12*(5), 729–738.

Raine, A., Buchsbaum, M. S., & LaCasse, L. (1997). Brain abnormalities in murderers indicated by positron emission tomography. *Biological Psychiatry, 42*, 495–508.

Raine, A., Lencz, T., Bihrle, S., LaCasse, L., & Colletti, P. (2000). Reduced prefrontal gray matter volume and reduced autonomic activity in antisocial personality disorder. *Archives of General Psychiatry, 57*, 119–127.

Reeves, D., Mills, M. J., Billick, S. B., & Brodie, J. D. (2003). Limitations of brain imaging in forensic psychiatry. *Journal of the American Academy of Psychiatry and the Law, 31*, 89–96.

Smith, D. (2004, April 10). Science's mind games. *The Age*.

Vergano, D. (2000, June 6). Brain abnormalities may compel senseless violence: When people from "good" homes go bad, defects are found. *USA Today*.

Wilson, E. O. (1982). *On human nature*. New York: Bantam.

Winner, L. (1988). *The whale and the reactor: A search for limits in an age of high technology*. Chicago: University of Chicago Press.

21 Evolutionary Cognitive Neuroscience

Julian Paul Keenan, Austen L. Krill, Steven M. Platek, and Todd K. Shackelford

Paying homage to the brain is an exercise in futility and understatement, and as such, it is perhaps best approached with humor by the likes of the American comedian Woody Allen who once rhetorically noted, "My brain? That's my second favorite organ." The complexity of the brain in the modern human is certain; there is no need to point out the brain's sophistication to those who even casually study it. At the risk of invoking a second attempt at humor, I can recall one student at the end of an arduous semester of physiological psychology claiming that, "The only thing that made sense in this class was learning on the first day that that the frontal lobes are in the front of the brain."

Each of the three fields invoked by the term *evolutionary cognitive neuroscience* (ECN) is sufficiently complex on its own. When these three fields are joined together in a single systems approach, it quickly becomes obvious that the complexity has expanded exponentially rather than arithmetically. The clear reward of a systems approach is the synergy resulting from melding different subdisciplines and the more integrated overview of what we wish to examine, namely, the organization, function, maintenance, and development of neural systems. The danger of such an approach lies in its very generality, and the consequent danger of losing sight of the contributions of the subdisciplines when studied in isolation. In effect, researchers risk becoming generalists in everything, specialists in nothing.

The tension between a systems approach and studying specifically defined subfields in isolation is reflected daily in discussions among researchers in ECN. We examine the brain and its related systems with the understanding that the brain is derived in part from modularized networks and regions that sometimes provide specific functions, and that these modules do not function in isolation, because they are influenced by both internal and external factors.[1] Similarly, evolution, cognition, and neuroscience are modules of academic study that, like the components of the nervous system, are also part of a larger system. By integrating these fields, we are aware that a meta-process is occurring. The evolution of this new field mirrors the system we study.

The benefit of understanding science from an ECN perspective is that we can consider simultaneously a significant number of variables within any given model, as would occur with a widening of the lens in any combination of fields. The cost is that one must exert significantly more energy creating, understanding, and elucidating those models. By allowing for a system that focuses on both proximate and ultimate mechanisms, we derive substantial benefits but incur significant costs as well. As with the evolved human nervous system, we assume that the benefits outweigh the costs.

The work of Canli highlights the benefits of the ECN (Canli, Omura, Haas, Fallgatter, Constable, & Lesch, 2005). The serotonin transporter gene (5-HTT, *SLC6A4*) is known to be involved in a number of psychological processes, including affective regulation (the short form of the allele is associated with depression, for example). It is also known that the amygdala modulates affective response, including regulation of prefrontal functioning. Using fMRI, Canli and colleagues found that the carriers of the short-form allele of the 5-HTT transporter gene showed dysfunctional activation in the amygdala during stimuli presentation. In particular, when negative words were presented, there was an increase in activation. However, such an increase in activity was not the whole story. Rather, what was apparently occurring was that carriers of the 5-HTT short-form allele actually demonstrated a blunted activation to neutral words. This lessening of response to neutral words made the activation during the negative word trials seem greater (Canli et al., 2005).

This study demonstrates a number of principles that appear indicative of the future of ECN. First, these studies are inherently complex. Even an advanced student of fMRI would be hard-pressed to discover in this work that the increased activation in the amygdala may be due

---

1 One author (J.P.K.) has previously applied the analogy of the mobile to the brain. The actions of brain modules are dependent on each other, as well as on the function of the entire mobile. Further, the mobile is influenced by its environment, and if we measure any given position of the mobile, it is clearly state-dependent. Further, disruption in one region may influence distal modules, even if the influence is subtle. Although incomplete, this analogy provides a visual image for understanding the brain as a system.

to a blunted baseline response; only careful manipulation and analysis on the part of the researchers made this finding possible.[2] Second, there is inherent complexity in the interpretation of the research. In this case, examining group differences (based on 5-HTT allele variant) added a level of sophistication, because actual genotyping was performed. Canli's group has since added yet another layer to this line of research that involves an analysis of the components at the behavioral level. Such an increase in complexity is expected to lead to a more complete understanding of the system being examined. We can now speculate on the nature of affective responses from both a genetic and a neural perspective. As an example, we now have the tools to understand and appreciate that the evolution and expression of an affective dysfunction (e.g., depression) may be due in part to a blunted amygdala response that is not based on negative events but instead might be due to tonic reactivity.

Studies such as these point to a promising future for ECN. By combining the technologies, methodologies, and the theories of each field, researchers can achieve a deeper understanding of the question at hand. For example, the "Hobbits" of Flores (*Homo floresiensis*) have proved to be somewhat of a mystery. Briefly, the discovery of an apparently isolated group of *Homo* with brains similar in size to the common chimpanzee (*Pan troglodytes*; in terms of overall size, ca. 400 cm^3) came as a surprise, as did the fact that they may have used and made tools. They existed relatively recently (about 15,000 years ago) and have as yet only been found in a single region, an island in Indonesia. The implications of this finding at first appeared tremendous, as researchers had to rethink some common ideas about brain and body scaling in humans, its relation to behavior, and the cyclical role played by cortical expansion, brain size, and body size. However, some rejected the hypothesis that *H. floresiensis* represented a new species that was closely tied to our recent ancestor (*H. erectus*), and it has been alternatively suggested that *H. floresiensis* was a pathological human microcephalic. Employing advanced scanning technology, Falk et al. (2005) found it statistically unlikely that the brain of *H. floresiensis* was a miniaturized version of *H. erectus* or

2 The detection of such subtleties is inherent to the notion of expertise. Familiarity with a technique does provide significant advantages in being a critical consumer. For example, in fMRI, significant activations between conditions may be due to an increase in the firing of inhibitory neurons. Thus, although a researcher may conclude that a region has significantly more activation, she often avoids stating that such activation is due to the firings of excitatory neurons.

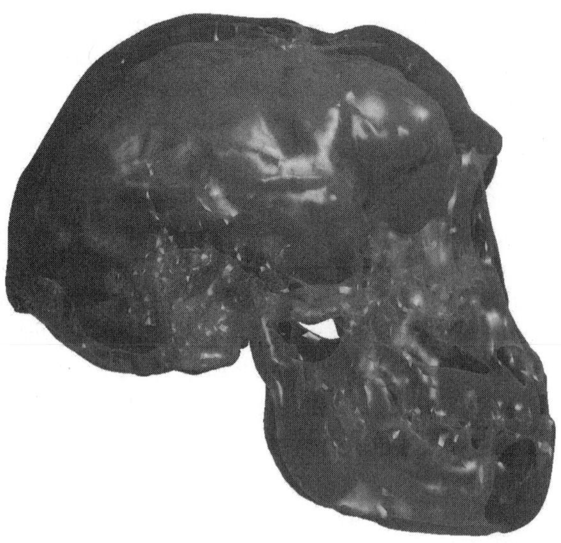

Figure 21.1
The virtual brain of *Homo floresiensis*. Because the brains of *Homo floresiensis* are not preserved, a recreated brain was developed derived from skull fossils. Image by Kirk Smith, Mallinckrodt Institute of Radiology; courtesy of Dean Falk. Used with permission.

sapiens. Instead, it appeared that this species was related to *erectus* but was composed of specific biological traits that did not appear to suggest a "scaled-down" *erectus*. The analysis revealed a set of derived features and evidence that *H. floresiensis* was likely not microcephalic (Figure 21.1). The ability to create "virtual brains" using advanced technology allows us to advance our knowledge in ways previously thought to be impossible.

We envision similar discoveries in the future, and as each discovery occurs, we will probably have to rethink our ideas about ECN. Further, we suspect that each discovery will spur debate among colleagues, and that such debate will foster new discoveries. Although this is the hallmark of many fields, we look forward to ECN expanding as the disciplines within the field gain in popularity and importance.[3]

3 The political climate in the United States is not conducive to pursuing some of the ideas presented in this field. In particular, the evolution question has become, almost inconceivably, an educational issue in our schools. Although this book is not the appropriate forum for discussing the matter, the future of this

John Huglings Jackson (1835–1911) is one of the most interesting figures in the history of neuroscience. Like the nervous system he studied, his theories were complex and organized in a manner that make them difficult to condense. That he never produced a final treatise of his work is another point of comparison with neural systems, as there never was (or will be) a final product. More than a century ago, Jackson speculated about the evolution of the nervous system in terms of function and structure. He thought that the brain functioned as a sensorimotor assemblage, and that even at the highest levels of consciousness, there was sensorimotor integration. The brain was a structure that was divided into different regions with the "lowest, middle, and highest" each indicating different evolutionary levels (1932, p. 41). Current views, particularly on scaling, often mirror his postulations that different networks reflect differences in evolution and that there is communication between regions that establishes higher-order cognitive abilities. It is this view that leads one down a usually fruitful but sometimes fatal path, as one is tempted to assume that there are general trends in brain complexity that lead to behavioral flexibilities *without exception*. Although our general principles are typically reliable, they are, in fact, incomplete, as Jackson seemingly assumed. As Lori Marino once asserted, "There is more than one way to evolve an intelligent species." This is an important idea to remember. As modularized as we may think the brain is, the evidence from the hemispherectomy literature leads us to understand that plasticity and adaptiveness are equally strong players, and one must consider these influences as well when considering cognition and evolution. Further, we must understand at some level the global function of the brain (Gray, Konig, Engel, & Singer, 1989) and come to terms with a significant number of issues, such as the problem of consciousness (Samsonovich & Ascoli, 2005), if we are to describe in any completeness the evolution of cognitive neural systems. Because general principles in ECN are difficult to establish and because those that exist have exceptions, there is always room for interpretation. That is why we assume a cautionary approach. Although allometric scaling applies well in primates, the correlation coefficients are never perfect, and we must be sensitive to error variance.

---

and many other fields depends on the education of students in the fundamentals of science and scientific theory. As we integrate these principles into classroom teaching and study, we must be sensitive to the role that the external environment plays not only in teaching evolutionary principles but in applying them as well.

This book has demonstrated that, as in all disciplines, the simplest questions may lead to complex answers. Genetic markings may help to provide an answer in some cases, new technology in others. Some of these questions are best addressed using animal models, and others can only be answered using a shovel, a pick axe, and a brush. By bringing together a truly diverse set of techniques, we can elaborate and expand on such "simple" questions: "Are chimpanzees right-handed?" or "Does adding more neurons expand the number of neuronal connections geometrically?" The future of ECN lies in addressing questions such as these, but more important, in addressing the questions that are generated by this research As teachers and researchers, we are excited by these possibilities: our students and their students will assuredly answer questions that we never thought could even be addressed, much less resolved.[4]

From stem cells to neural implants, ECN will provide practical and clinical applications. The ideas generated by this field do not live only in academic tomes but instead can have impacts beyond scholarly debate. Further, the techniques invented and refined by researchers in ECN will continue to have a significant impact in medical, educational, and clinical arenas. Obvious examples include the development of neuroimaging techniques and psychoneuroimmunology advances. However, there are more subtle applications, such as taking an ultimate perspective in our approach to studying human behavior. For example, we can determine how the brain may have evolved in creating and maintaining a disorder such as autism (Baron-Cohen, Lutchmaya, & Knickmeyer, 2004) and how even in nonautistic populations, such brain differences may be involved in maintaining current selection pressures (Kanazawa & Vandermassen, 2005).

The future of ECN is both promising and exciting. We hope that this book will inspire and encourage researchers in the field. We also hope that tomorrow's students will continue to pass on this field to their students, perhaps with modifications that are derived from competitive forces, such that ECN continues to evolve.

References

Baron-Cohen, S., Lutchmaya, S., & Knickmeyer, R. (2004). *Prenatal testosterone in mind: Amniotic fluid studies*. Cambridge, MA: MIT Press.

4 The editors of this volume are all at institutions that value teaching and research equally, and as such, we are particularly sensitive to issues involving classroom teaching.

Canli, T., Omura, K., Haas, B. W., Fallgatter, A., Constable, R. T., & Lesch, K. P. (2005). Beyond affect: A role for genetic variation of the serotonin transporter in neural activation during a cognitive attention task. *Proceedings of the National Academy of Sciences, U.S.A., 102*, 12224–12229.

Falk, D., Hildebolt, C., Smith, K., Morwood, M. J., Sutikna, T., Brown, P., Jatmiko, Saptomo, E. W., et al. (2005). The brain of LB1, *Homo floresiensis. Science*, *308*, 242–245.

Gray, C. M., Konig, P., Engel, A. K., & Singer, W. (1989). Oscillatory responses in cat visual cortex exhibit inter-columnar synchronization which reflects global stimulus properties. *Nature*, *338*, 334–337.

Jackson, J. H. (1932). *Selected writings of John Hughlings Jackson* (edited by J. Taylor). London: Hodder and Stoughton.

Kanazawa, S., & Vandermassen, G. (2005). Engineers have more sons, nurses have more daughters: An evolutionary psychological extension of Baron-Cohen's extreme male brain theory of autism. *Journal of Theoretical Biology*, *233*, 589–599.

Samsonovich, A. V., & Ascoli, G. A. (2005). The conscious self: Ontology, epistemology and the mirror quest. *Cortex*, *41*, 621–636.

Index

Note: Page numbers followed by f indicate figures; those followed by t indicate tables; those followed by n indicate notes.